Nanosystems

With this book, Drexler has established the field of molecular nanotechnology. The detailed analyses show quantum chemists and synthetic chemists how to build upon their knowledge of bonds and molecules to develop the manufacturing systems of nanotechnology, and show physicists and engineers how to scale down their concepts of macroscopic systems to the level of molecules.

—William A. Goddard III
Professor of Chemistry and Applied Physics
Director, Materials and
    Molecular Simulation Center
California Institute of Technology

Devices enormously smaller than before will remodel engineering, chemistry, medicine, and computer technology. How can we understand machines that are so small? *Nanosystems* covers it all: power and strength, friction and wear, thermal noise and quantum uncertainty. This is *the* book for starting the next century of engineering.

—Marvin Minsky
Professor of Electrical Engineering
    and Computer Science
Toshiba Professor of Media Arts and Sciences
Massachusetts Institute of Technology

This work provides the scientific and technological foundation for the emerging field of molecular systems engineering...It is essential for anyone contemplating research in this area...a milestone in the development of the technologies that will underpin the final industrial revolution.

—John Walker
Cofounder of Autodesk, Inc.

What the computer revolution did for manipulating data, the nanotechnology revolution will do for manipulating matter, juggling atoms like bits...This multidisciplinary synthesis opens the door to the new field of molecular manufacturing.

—Ralph Merkle
Member of the Research Staff
Computational Nanotechnology Project
Xerox Palo Alto Research Center

Nanosystems

Molecular Machinery, Manufacturing, and Computation

K. Eric Drexler

Research Fellow
Institute for Molecular Manufacturing
Palo Alto, California

A WILEY-INTERSCIENCE PUBLICATION
John Wiley & Sons, Inc.
NEW YORK • CHICHESTER • BRISBANE • TORONTO • SINGAPORE

Excerpt from "There's Plenty of Room at the Bottom"
by Richard P. Feynman © 1960 by *Engineering & Science*,
California Institute of Technology; used with permission.

*In recognition of the importance of preserving what has
been written, it is a policy of John Wiley & Sons, Inc.
to have books of enduring value published in the United
States printed on acid-free paper, and we exert our best
efforts to that end.*

Library of Congress Cataloging-in-Publication Data

Drexler, K. Eric.
   Nanosystems : molecular machinery, manufacturing,
and computation / K. Eric Drexler.
     p.   cm.
   "A Wiley-Interscience publication."
   Includes bibliographical references and index.
   ISBN 0-471-57547-X. -- ISBN 0-471-57518-6 (pbk.)
   1. Nanotechnology.   I. Title.
T174.7.D74   1992
620.4--dc20   92-30870
                                        CIP

Printed in the United States of America

10 9 8 7 6 5 4 3 2 1

To experimentalists, engineers, and software builders:
they do the hard parts

Contents

in the transverse-continuum approximation. Mechanisms
of energy dissipation. Sleeve bearings in molecular detail.
Less symmetrical sleeve bearings.

Part III   Implementation Strategies

Preface

Manufactured products are made from atoms, and their properties depend on how those atoms are arranged. This volume summarizes 15 years of research in *molecular manufacturing,* the use of nanoscale mechanical systems to guide the placement of reactive molecules, building complex structures with atom-by-atom control. This degree of control is a natural goal for technology: Microtechnology strives to build smaller devices; materials science strives to make more useful solids; chemistry strives to synthesize more complex molecules; manufacturing strives to make better products. Each of these fields requires precise, molecular control of complex structures to reach its natural limit, a goal that has been termed *molecular nanotechnology.*

It has become clear that this degree of control can be achieved. The present volume assembles the conceptual and analytical tools needed to understand molecular machinery and manufacturing, presents an analysis of their core capabilities, and explores how present laboratory techniques can be extended, stage by stage, to implement molecular manufacturing systems. It says little about applications other than computation (describing 10^9-instruction-per-second submicron scale CPUs executing $\sim 10^{16}$ instructions per second per watt) and manufacturing (describing desktop devices able to produce precisely structured, kilogram-scale products from simple chemical feedstocks). Surveys of broader implications appear elsewhere (Drexler, 1986a; 1989c; Drexler et al., 1991).

The intended readership

Molecular manufacturing is linked to many areas of science and technology. In writing this volume, I have been guided by an imaginary committee of readers with differing demands.

One is a reader with a general science background, interested in the basic principles, capabilities, and nature of molecular nanotechnology, but not in the mathematical derivations. Accordingly, I have attempted to summarize the chief results in descriptions, diagrams, and example calculations, and have included comparisons of this field to others that are more familiar. Such a reader can skip many sections without becoming lost.

Another is a student considering a career in the field. This reader demands an introduction to the foundations of molecular nanotechnology presented in terms of the basic physics, calculus, and chemistry taught to students in other fields. Accordingly, I have grounded most derivations in basic principles, developing intermediate results as needed.

The rest of the committee includes a physicist, a chemist, a molecular biologist, a materials scientist, a mechanical engineer, and a computer scientist. Each has deep professional knowledge of a particular field. Each demands answers to special questions that presuppose specialized knowledge. Each knows the exceptions that hide behind most generalizations, and the approximations that hide behind most textbook formulas. Accordingly, the discussion sometimes dives into a topic that readers outside the relevant discipline may find opaque. Skipping past these topics will seldom impair comprehension of what follows.

Each of these specialists also represents a community of researchers able to advance the development of molecular nanotechnology. Accordingly, many of the discussions implicitly or explicitly highlight open problems, inviting work in theoretical analysis, in computer-aided design and modeling, and in laboratory experimentation. I hope that this volume will be seen both as a guide and as an invitation to a promising new field.

The nature of the subject

Our ability to model molecular machines—of specific kinds, designed in part for ease of modeling—has far outrun our ability to make them. Design calculations and computational experiments enable the theoretical study of these devices, independent of the technologies needed to implement them. Work in this field is thus (for now) a branch of theoretical applied science (Appendix A).

Molecular manufacturing applies the principles of mechanical engineering to chemistry (or should one say the principles of chemistry to mechanical engineering?) and uses results drawn from materials science, computer science, and elsewhere. But interdisciplinary studies can foster misunderstandings. From every disciplinary perspective, a superficial glance suggests that something is wrong—applying chemical principles leads to odd-looking machines, applying mechanical principles leads to odd-looking chemistry, and so forth. The following chapters offer a deeper view of how these principles interact.

Criticism of criticism

Research in molecular nanotechnology requires a design perspective because it aims to describe workable systems. It is easy to describe unworkable systems, and criticisms of a critic's own bad design have on occasion been presented as if they were criticisms of molecular nanotechnology as a whole. Some examples: assuming the use of flexible molecules, then warning that they will have no stable shape; assuming the manipulation of unbound reactive atoms, then warning that they will react and bond to the manipulator; assuming the use of materials with unstable surfaces, then warning that the surfaces will change; assuming that reactive gases permeate nanosystems, then warning that reactions will occur; assuming that nanomachines must "see," then warning that light waves are too long and x-rays too energetic; assuming that nanomachines swim from point to point, then warning that Brownian motion makes such navigation impossible; assuming that nanomachines dissipate enormous power in a small volume, then warning of overheating; and so on, and so forth. These observations constitute not criticisms, but rediscoveries of elementary engineering constraints.

Use of tenses

In ordinary discourse, "will be" suggests a prediction, while "would be" suggests a conditional prediction. Using these future-tense expressions is inappropriate when discussing the time-independent possibilities inherent in physical law.

In speaking of spacecraft trajectories to Pluto, for example, to say that they "will be" is to predict the future of spaceflight; to say that they "would be" is to remind readers of the uncertainties of budgets and life. Both phrases distract from the analysis of celestial mechanics and engineering trade-offs. The present tense is more serviceable: One can say that as-yet unrealized spacecraft trajectories to Pluto "are of two kinds, direct and gravity assisted," and then analyze their properties without distraction. Similarly, one can say that as-yet unrealized nanomachines of diamondoid structure "are typically stiffer and more stable than folded proteins." Much of the discussion in this volume is cast in this timeless present tense; this is not intended to imply that devices like those discussed in Parts I and II presently exist.

Citations and apologies

It is much easier to grasp and apply the main results of a field than it is to provide a balanced guide to the recent work, omitting no useful citations. I am sure that my discussions of chemistry and protein engineering, for example, omit papers fully as valuable as the best included. I apologize to authors I have slighted.

Less forgivable are those instances (which I cannot yet identify) in which I may have rederived some result that should be attributed to an earlier author, perhaps well known in some specialty. In interdisciplinary research, one cannot spend a professional lifetime immersed in a single literature, and such failures of attribution become likely—mathematics often yields results more easily than does a library. Any such lapses brought to my attention will be corrected in future editions; their most likely locations are Chapters 5, 6, and 7.

Aside from these lapses, material presented without citation falls into two categories that I trust are distinct. First, well-known principles and results from established fields—physics, statistical mechanics, chemistry—are used without citing Newton, Boltzmann, Pauling, or their kin. Second, the designs, concepts, and analytical results that are both specific to nanomechanical systems and not attributed to someone else are to the best of my knowledge original contributions, many presented for the first time in this volume.

It also seems necessary to apologize for doing theoretical work in a world where experimental gains are often so hard-won. If this theoretician's description of possibilities seems to make light of experimental difficulties, I can only plead that it would soon become tedious to say, at every turn, that laboratory work is difficult, and that the hard work is yet to be done.

Acknowledgments

The research behind this volume began in 1977, stimulated by the growing literature on biological molecular machines. Basic results appeared in a refereed paper (Drexler, 1981). The present work began as notes for a course taught at Stanford in 1988 at the invitation of Nils Nilsson; early versions of some chapters did service as a doctoral thesis at MIT in 1991. During this long gestation, many people have contributed through discussion, criticism, and detailed review.

I thank the participants of the monthly series of nanotechnology seminars (some centered on draft chapters of this volume) organized by Ralph Merkle at the Xerox Palo Alto Research Center for wide-ranging discussion and criticism. These have included Lakshmikantan Balasubramaniam, David Biegelsen, Ross Bringans, David Fork, Babur Hadimioglu, Stig Hagstrom, Conyers Herring, Tad Hogg, Warren Jackson, Noble Johnson, Martin Lim, Jim Mikkelsen, John Northrup, K. V. Ravi, Paolo Santos, Mathias Schnabel, Bob Street, Lars-Erik Swartz, Eugen Tarnower, Dean Taylor, Rob Tow, and Chris Van der Walle.

For reviews, suggestions, and specific pieces of help, I thank Jeff Bottaro, Randall Burns, Jamie Dinkelacker, Greg Fahy, Jonathan Goodman, Josh Hall, Robin Hanson, Norm Hardy, Ted Kaehler, Markus Krummenacker, Arel Lucas, Tim May, John McCarthy, Mark Samuel Miller, Chip Morningstar, Russell Parker, Marc Stiegler, Eric Dean Tribble, John Walker, and Leonard Zubkoff.

For discussion and suggestions that helped in preparing a 1989 draft paper that became Section 15.3, I thank Joe Bonaventura, Jeff Bottaro, William DeGrado, Bruce Erickson, Barbara Imperiali, Jim Lewis, Danute Nitecki, Chris Peterson, Fredric Richards, Jane Richardson, and Kevin Ulmer. For similar help in developing ideas in Section 15.4, I thank Tom Albrecht, John Foster, Paul Hansma, Jan Hoh, Ted Kaehler, Ralph Merkle, Klaus Mosbach, and Craig Prater. For sponsoring the initial 1981 publication, I thank Arthur Kantrowitz.

For remarkable efforts while this work was on its way to fulfilling the thesis requirement of an interdepartmental doctoral program hosted by the Media Arts and Sciences Section at MIT, I thank the committee's chair, Marvin Minsky (Department of Electrical Engineering and Computer Science; Media Arts and Sciences Section), as well as committee members Alexander Rich (Department of Biology), Gerald Sussman (Department of Electrical Engineering and Computer Science), Rick Danheiser (Department of Chemistry), and Steven Kim (Department of Mechanical Engineering), with special thanks to Steve Benton and Nicholas Negroponte (Media Arts and Sciences Section) for making the program possible in an environment hospitable to new research directions.

Ralph Merkle has helped greatly by providing steady encouragement and extensive opportunities for discussion during the writing of this volume, by reviewing it (and helping to obtain other reviews), and by collaborating on several of the design studies described. Special thanks also go to Jeffrey Soreff, whose checking of mathematical results and physical reasoning has been just *barely* incomplete enough for him to escape blame for the remaining errors: with his other help, this places his contribution in a class by itself. Barry Silverstein, John Walker, and the Institute for Molecular Manufacturing each provided essential support for a major portion of this work. Diane Cerra and Bob Ipsen of John Wiley & Sons made publication a pleasure. Chris Peterson, my spouse and partner, provided essential support of kinds too numerous to list.

At the other pole of involvement, the most general thanks go to members of the hundred or more audiences at universities and industrial laboratories in the U.S., Europe, and Japan who have listened to presentations of these ideas and aided in their development by intelligent questioning. I hope that this volume provides many of the answers they sought.

Palo Alto, July 1992

K. Eric Drexler

1

Introduction and Overview

1.1. Why molecular manufacturing?

The following devices and capabilities appear to be both physically possible and practically realizable:

- Programmable positioning of reactive molecules with ~0.1 nm precision
- Mechanosynthesis at $>10^6$ operations/device·second
- Mechanosynthetic assembly of 1 kg objects in $<10^4$ s
- Nanomechanical systems operating at ~10^9 Hz
- Logic gates that occupy ~10^{-26} m^3 (~10^{-8} μ^3)
- Logic gates that switch in ~0.1 ns and dissipate $<10^{-21}$ J
- Computers that perform 10^{16} instructions per second per watt
- Cooling of cubic-centimeter, ~10^5 W systems at 300 K
- Compact 10^{15} MIPS parallel computing systems
- Mechanochemical power conversion at $>10^9$ W/m^3
- Electromechanical power conversion at $>10^{15}$ W/m^3
- Macroscopic components with tensile strengths $>5 \times 10^{10}$ GPa
- Production systems that can double capital stocks in $<10^4$ s

Of these capabilities, several are qualitatively novel and others improve on present engineering practice by one or more orders of magnitude. Each is an aspect or a consequence of molecular manufacturing.

1.2. What is molecular manufacturing?

This volume describes the fundamental principles of molecular machinery and applies them to nanomechanical devices and systems, including molecular manufacturing systems and computers. At present, however, these are unfamiliar topics. New fields often need new terms to describe their characteristic features, and so it may be excusable to begin with a few definitions: *Molecular manufacturing* is the construction of objects to complex, atomic specifications using sequences of chemical reactions directed by nonbiological molecular machinery. *Molecular nanotechnology* comprises molecular manufacturing together with its techniques, its products, and their design and analysis; it describes the field as a whole. *Mechanosynthesis*—mechanically guided chemical synthesis—is fundamental to molecular manufacturing: it guides chemical reactions on an atomic

1

scale by means other than the local °steric* and electronic properties of the °reagents; it is thus distinct from (for example) enzymatic processes and present techniques for organic synthesis.†

At the time of this writing, positional chemical synthesis is at the threshold of realization: precise placement of atoms and molecules has been demonstrated (for example, see Eigler and Schweizer, 1990), but flexible, extensible techniques remain in the domain of design and theoretical study (Part III), as does the longer-term goal of molecular manufacturing (Chapter 14). Accordingly, the implementation of molecular nanotechnologies like those analyzed in Part II awaits the development of new tools. This volume is addressed both to those concerned with identifying promising directions for current research, and to those concerned with understanding and preparing for future technologies.

The following chapters form three parts: Part I describes the chief physical principles and phenomena of importance in molecular machinery and manufacturing. Part II applies the results of Part I to the design and analysis of components and systems (yielding the conclusions summarized in Section 1.1). Part III then describes implementation pathways leading from our current technology base to systems like those described in Part II.

The rest of the present section attempts to clarify the nature of the topic by discussing an example of a nanomechanical device and by presenting a chemical perspective on molecular manufacturing. Sections 1.3 to 1.5 present a set of comparisons between this and other fields (mechanical engineering, microtechnology, chemistry, and molecular biology), a discussion of overall approach (including objectives, level, scope, and assumptions), and an overview of the later chapters and how they fit together. Table 1.1 lists some of the known problems and constraints that are addressed elsewhere in this volume.

1.2.1.   Example: a nanomechanical bearing

As discussed in Section 1.3, the mechanical branch of molecular nanotechnology forms a field related to, yet distinct from, mechanical engineering, microtechnology, chemistry, and molecular biology. An example may serve as a better introduction than would an attempt at a written definition.

Figure 1.1 shows several views of one design for a nanomechanical bearing like those discussed in greater depth in Chapter 10. (Figure 1.2 describes some conventions used in illustrations like Figure 1.1) In a functional context, many of the bonds shown as hydrogen terminated would instead link to other moving parts or to a structural matrix. Several characteristics are worthy of note:

- The components are °polycyclic, more nearly resembling the fused-ring structures of diamond than the open-chain structures of biomolecules such as °proteins.

---

* Words appearing in the Glossary are sometimes prefixed with a small circle, e.g., °steric.
† Considering the words in isolation, the terms *molecular nanotechnology* and *molecular manufacturing* could instead be interpreted to include much of chemistry, and *mechanosynthesis* could be interpreted to include substantial portions of enzymology and molecular biology. These established fields, however, are already named; the above terms will serve best if reserved for the fields they have been coined to describe, or for borderline cases that emerge as these fields are developed.

Table 1.1. Some known issues, problems, and constraints.

Thermal excitation
Thermal and quantal positional uncertainty
Quantum-mechanical tunneling
Bond energies, strengths, and stiffnesses
Feasible chemical transformations
Electric field effects
Contact electrification
Ionizing radiation damage
Photochemical damage
Thermomechanical damage
Stray reactive molecules
Device operational reliabilities
Device operational lifetimes
Energy dissipation mechanisms
Inaccuracies in molecular mechanics models
Limited scope of molecular mechanics models
Limited scale of accurate quantal calculations
Inaccuracy of semiempirical models
Providing ample safety margins for modeling errors

- Accordingly, each component is relatively stiff, lacking the numerous opportunities for internal rotation about bonds that make °conformational analysis difficult in typical biomolecules.
- Repulsive, nonbonded interactions strongly resist both rotations of the shaft away from its axial alignment with the ring, and displacement along that axis or perpendicular to it.
- Rotation of the shaft about its axis within the ring encounters negligible energy barriers, indicating a nearly complete absence of static friction.
- The combination of °stiffness in five degrees of freedom with facile rotation in the sixth makes the structure act as a good bearing, in the conventional mechanical engineering sense of the term.
- The absence of significant static friction in a system that places bumpy surfaces in firm contact with no intervening lubricant is surprising by in conventional mechanical engineering standards.
- No solvent is illustrated, and there is no reason to think that the bearing structure shown would in fact be soluble.
- Neither of the components of the bearing is a plausible target for synthesis using reagents diffusing in solution; barring unprecedented chemical cleverness, their construction requires mechanosynthetic control.

How typical are these characteristics? Stiff, polycyclic structures are ubiquitous in the designs presented in Part II. Many components are designed to combine stiff constraints in some degrees of freedom with nearly free motion in others, thereby fulfilling roles familiar in mechanical engineering; nonetheless,

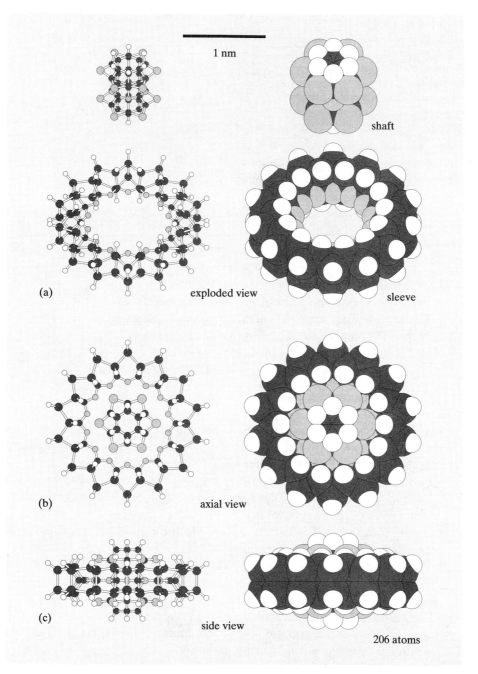

1 nm

shaft

(a)   exploded view   sleeve

(b)   axial view

(c)   side view

206 atoms

Figure 1.1. End views and exploded views of a sample overlap-repulsion bearing design (shown in both ball-and-stick and space-filling representations). Geometries represent energy minima determined by the MM2/CSC molecular mechanics software. Note the six-fold symmetry of the shaft structure and the fourteen-fold symmetry of the surrounding ring; this relatively prime combination results in low energy barriers to rotation of the shaft within the ring. Bearing structures are discussed further in Chapter 10. (MM2/CSC denotes the Chem3D Plus implementation of the MM2 molecular mechanics force field. The MM2 model is discussed in Section 3.3.2; Chem3D Plus is a product of Cambridge Scientific Computing, Cambridge, Massachusetts.)

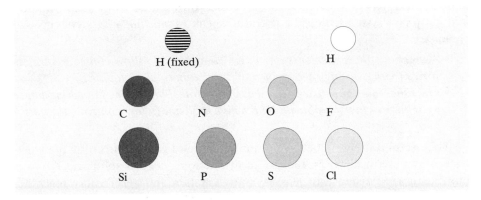

Figure 1.2. Conventions for atom representation using shading and relative sizes. The "H (fixed)" atom represents a hydrogen atom held at fixed spatial coordinates; these are used to model some mechanical constraints applied by a larger structure (e.g., in stiffness calculations). All radii are set equal to the values for 0.1 nN compressive contacts given by Eq. (3.20).

a detailed understanding of how those roles are fulfilled requires analyses based on uniquely molecular phenomena. The operating environment assumed for the nanomechanical and mechanosynthetic systems discussed in Part II is high vacuum, rather than a solvent. Finally, the designs in Part II (unlike those described in Part III) are consistently of a scale and complexity that precludes synthesis using present techniques.

The bearing shown in Figure 1.1 suggests the nature of other systems described in Part II. For example, the combination of a bearing and shaft suggests the possibility of extended systems of power-driven machinery. The outer surface of the bearing suggests the possibility of a molecular-scale gear. The controlled rotary motion of the shaft within the ring, together with the concept of extended systems of machinery, suggests the possibility of controlled molecular transport and positioning, which is necessary for advanced mechanosynthesis.

1.2.2. A chemical perspective on molecular manufacturing

Chemistry today (and chemical synthesis in particular) focuses chiefly on the behavior of molecules diffusing and colliding in solution. Reaction rates in solution-phase chemistry are determined by multiple influences, including concentration-dependent collision frequencies, and steric and electronic effects local to the reacting molecules.

Although based on the same principles of physics, mechanosynthesis performed by molecular machinery in vacuum differs greatly from conventional chemistry. Concepts developed to describe diffusing molecules in a gas or liquid (or immobile molecules in a solid) often must be modified in describing systems characterized by nondiffusive mobility. The concept of "concentration," for example, in the familiar sense of "number of molecules of a particular type per unit of macroscopic volume" becomes inapplicable to calculations of reaction rates. Local steric and electronic effects remain important, but the decisive influence on reaction rates becomes mechanical positioning aided by applied force.

a. Machine-phase systems.   To emphasize differences from solid-, liquid-, and gas-phase systems, it can be useful to speak of *machine-phase* systems and chemistry:

- *A machine-phase system is one in which all atoms follow controlled trajectories (within a range determined in part by thermal excitation).*
- *Machine-phase chemistry describes the chemical behavior of machine-phase systems, in which all potentially reactive moieties follow controlled trajectories.*

The useful distinction between liquid phase and gas is blurred by the existence of supercritical fluids; the useful distinction between solid and liquid is blurred by the existence of glasses, liquid crystals, and gels. Where machine-phase chemistry is concerned, the definitional ambiguities are chiefly associated with the words *all* and *controlled*. In a conventional chemical reaction or an enzymatic active site, a moderate number of atoms in a small region can be said to follow somewhat-controlled trajectories, but these examples fall outside the intended bounds of the definition. In a good example of a machine-phase system, large numbers of atoms follow paths that seldom deviate from a nominal trajectory by more than an atomic diameter while executing complex motions in an extended region from which freely-diffusing molecules are rigorously excluded. Machine-phase conditions can be termed °*eutactic* ("well arranged," from the Greek *eus,* "good," and *taktikos,* "of order or arrangement"). Eutactic conditions are quite unlike those of solution-phase chemistry.*

Mechanosynthesis of the sort discussed in Chapters 8 and 13 is a machine-phase process (Part III discusses mechanosynthesis in a solvent environment). Eutactic mechanosynthesis offers novel chemical capabilities, such as position-based discrimination among chemically equivalent sites, strong suppression of side reactions, and new sources of °activation energy.

Chemistry in the machine phase shares characteristics of gas-, solution-, and solid-phase chemistry, and yet displays unique characteristics; these similarities and differences are discussed further in Sections 6.4.2 and 8.3. Since experience shows that habits of thought developed in the study of liquid- and gas-phase systems can yield misleading conclusions if hastily applied to machine-phase systems, frequent recourse to fundamental principles is necessary. The Index of this volume includes an entry (Chemistry: contrasts between machine and solution phase) that cites discussions of this issue. Table 1.2 provides a compact summary.

1.2.3. Exposition vs. implementation sequence

The implementation sequence for molecular manufacturing might proceed as follows: The ability to make complex molecular objects in solution is extended to objects of greater size and complexity. These molecular objects are used as components in molecular machines capable of directing the mechanosynthesis of yet larger and more complex machines. Through a series of steps, solution-

---

* D. Cram has introduced the concept of an *inner phase* to describe the interior of container molecules in which (for example) single, isolated molecules of cyclobutadiene are stable at room temperature because the container walls block intermolecular collisions (Cram et al., 1991). Inner-phase systems prevent collisions; machine-phase systems control them.

Table 1.2. Contrasts between solution-phase and mechanosynthetic chemistry (a more detailed comparison appears in Table 8.1).

Characteristic	Solution-phase chemistry	Mechanosynthetic chemistry
Reagent transport	Diffusion	Mechanical conveyance
Reaction site selectivity	Structural influences	Direct positional control
Reaction environment	Solution	Mechanisms in vacuum
Control of reaction environment	Relatively little	Relatively great
Intermolecular reactions	Ubiquitous opportunities	Strictly controlled opportunities
Unwanted reactions	Inter- and intramolecular	Chiefly intramolecular
Sensitivity to energy differences	10 maJ is large	10 maJ is (often) small
Typical product size	10 to 100 atoms	$>10^{10}$ atoms

based mechanosynthetic methods are replaced by methods that require an inert environment, then polymeric building materials are replaced by diamondoid materials. Further increases in scale and capability yield advanced molecular manufacturing systems under computer control. (Each of these steps is discussed in Chapter 16.)

The expository sequence of this volume is quite different. It begins by describing fundamental principles of broad applicability (Part I), then applies them to the design and analysis of advanced systems (Part II). Finally, having described principles and objectives, it turns to implementation pathways (Part III). Thus, the means considered are guided by the objectives pursued.

1.3. Comparisons

Molecular nanotechnology resembles and overlaps with other fields, yet differs substantially. A discussion of resemblances can illustrate the applicability of existing knowledge and emerging developments; a discussion of differences can warn of false analogies and consequent misunderstandings.

Table 1.3 compares several existing production processes—conventional fabrication, microfabrication, solution-phase chemistry, and biochemistry—with molecular manufacturing. The following sections provide a more detailed comparison of molecular manufacturing and molecular nanotechnology with these other processes and their products. Appendix B focuses on areas of these fields having special relevance molecular manufacturing; the present section makes broad comparisons to the mainstream.

1.3.1. Conventional fabrication and mechanical engineering

a. Similarities: components, systems, controlled motion, manufacturing. Many mechanical engineering concepts apply directly to nanomechanical systems. As shown in Chapters 3 to 6, methods based on classical mechanics suffice for much of the required analysis. As shown in Chapters 9 to 11, beams, shafts, bearings, gears, motors, and the like can all be constructed on a nanometer scale to serve familiar mechanical functions. As a consequence, macromechanical

Table 1.3. Typical characteristics of conventional machining, micromachining, solution-phase chemistry, biochemistry, and molecular manufacturing.[a]

Characteristic	Conventional fabrication	Micro-fabrication	Solution chemistry	Bio-chemistry	Molecular manufacturing
Molecular precision?	no	no	yes	yes	yes
Positional control?	yes	yes	no	partial	yes
Typical feature scale	1 mm	$1\,\mu$	0.3 nm	0.3 nm	0.3 nm
Typical product scale	1 m	10 mm	1 nm	10 nm	100 nm +
Typical defect rate[b]	10^{-4}	10^{-7}	10^{-2}	10^{-11}	10^{-15}
Typical cycle times	1 s	100 s	1000 s	10^{-3} s	10^{-6} s
Products described by	materials and shapes	materials and shapes	atoms and bonds	monomer sequences	atoms and bonds

[a] Product scale, defect rates, and cycle times vary widely from process to process within most of these families; feature scale varies widely within the first two.
[b] The defect rate listed for biochemistry corresponds to high-reliability DNA replication processes that include kinetic proofreading (Watson et al., 1987); most biochemical defect rates are higher. All rates are on a per-component basis.

engineering and nanomechanical engineering share many design issues and analytical techniques.

Both molecular and conventional manufacturing systems use machines to perform planned patterns of motion: they shape, move, and join components to build complex three-dimensional structures. Systems of both kinds can manufacture machines, including machines used for manufacturing (Chapter 14).

b. Differences: scale, molecular phenomena.   Despite these similarities, nanomechanical engineering forms a distinct field. The familiar model of objects as made of homogeneous materials, though still useful (Chapter 9), often must be replaced by models that treat objects as sets of bonded atoms (Chapters 2 and 3). Thermally excited vibrations are of major importance, and quantum effects are sometimes significant (Chapters 5, 6, and 7). Further, nanomachines suffer from molecular damage mechanisms (Chapter 6); molecular phenomena permit (and demand) novel bearings (Chapter 10); scaling laws favor electrostatic over electromagnetic motors (Chapters 2 and 11); and the basic operations of manufacturing on a molecular scale are chemical transformations (Chapters 8 and 13). These transformations typically move the system between discrete states, leading to structures that are either exactly right or clearly wrong; this resembles digital logic more closely than it does metalworking.

1.3.2. Microfabrication and microtechnology

a. Similarities: small scale, electronic quantum effects.   Microtechnology has enabled the fabrication of micron-scale mechanical devices. These share basic scaling properties with nanomechanical devices—and so, for example, electrostatic motors are preferred over electromagnetic motors in both micro- and nanotechnology (Section 2.4.3). Further, quantum electronic devices of kinds now being explored with microfabrication technologies may become targets for molecular manufacturing.

b. Differences: fabrication technology, scale, molecular phenomena. Microfabrication relies on technologies (photolithographic pattern definition, etching, deposition, diffusion) essentially unrelated to those of molecular manufacturing. In a sense these two fields are moving in opposite directions: microfabrication attempts to make bulk-material structures *smaller* despite fabrication irregularities; molecular manufacturing will emerge from attempts to make molecular structures *larger* without losing the atomic precision characteristic of stereospecific chemical synthesis. Making structures consisting of a few dozen precisely-arranged atoms seems unachievable using microfabrication, but is routine in chemical synthesis. The gears, bearings, and motors described in Chapters 10 and 11 differ in volume from their closest microfabricated counterparts by a factor of $\sim 10^9$, and rely on molecular structures and phenomena in their operation.*

1.3.3. Solution-phase chemistry

a. Similarities: molecular structure, processes, fabrication. Chemical principles describe the basic steps of molecular manufacturing, since each consists of a chemical transformation. Chemical knowledge can help in evaluating the stability of products, and chemical research has produced the most useful models of the mechanical behavior of molecular objects. Organic chemistry is particularly relevant owing to the superiority of carbon-based structures for most mechanical applications. Fundamental chemical concepts such as °bonding, °strain, °reaction rates, °transition states, °orbital symmetry, and °steric hindrance are all applicable; familiar chemical entities such as °alkanes, °alkenes, °aromatic rings, °radicals, and °carbenes are all useful.† Solution-based organic synthesis can make precisely structured molecular objects; it has even been used to make molecular gears (Mislow, 1989), although of a sort having no obvious utility for nanomechanical engineering.

---

* The term *nanotechnology*, first widely used to refer to what is here termed *molecular nanotechnology*, has since been applied to many small-scale technologies, including conventional microfabrication techniques working in the submicron size range. Accordingly, discussions of the history, status, and prospects of so-called nanotechnology often confuse essentially dissimilar concepts, as if the term *ornithology* were used to describe the study of flying things, thereby stirring together birds, bats, spacecraft, balloons, and bombers into a single conceptual muddle.

† Discussions phrased in terms of controlling and building with "individual atoms" (Drexler, 1986a) have fostered a perception that molecular manufacturing would employ individual, *unbonded*, and hence highly reactive carbon atoms. This rightly strikes chemists as implausible. Indeed, unbound atoms would be difficult to produce and control; more conventional reactants seemed appropriate from the start. The same volume speaks of using "reactive molecules" as tools "to bond atoms together...a few at a time," and the first paper on the subject (Drexler, 1981) speaks of positioning "reactants" and "reactive groups." Controlling the motions and reactions of individual molecules, of course, implies controlling the motions and destinations of their individual constituent atoms.

b. Differences: machine-phase systems, mechanosynthesis. The chief differences between the present subject and conventional chemistry stem from the properties of machine-phase systems and of mechanosynthesis. These have been summarized in Section 1.1.2 and are discussed at length in Chapter 8.

1.3.4. Biochemistry and molecular biology

a. Similarities: molecular machines, molecular systems. Molecular biology, like molecular nanotechnology, embraces the study of molecular machines and molecular machine systems. Ribosomes—like mechanisms in flexible molecular manufacturing systems—can be viewed as numerically controlled machine tools following a series of instructions to produce a complex product. Molecular biology and biochemistry stimulated the train of thought that led to the concept of molecular manufacturing (Drexler, 1981), and their techniques offer paths to the development of molecular manufacturing systems (Section 15.2).

b. Differences: materials, machine phase, general mechanosynthesis. Biology is a product of evolution rather than design, and molecular biologists study systems that differ greatly from the eutactic systems described here. Unlike molecular manufacturing systems, the molecular machines found in cells can synthesize only relatively small molecules and a stereotyped set of polymers; they cannot synthesize a broad class of °diamondoid structures. Larger biological structures typically acquire their shapes through the action of weak forces (°hydrogen bonds, °salt bridges, °van der Waals attraction, °hydrophobic forces). As a consequence of stronger bonding, the strength and °modulus of diamondoid components can exceed those of biological structures by orders of magnitude. The bearings, gears, motors, and computers discussed in Part II are accordingly quite different from the bacterial flagellar motor, the actin-myosin system, systems of neurons, and so forth. Biological and nanomechanical systems are organized in fundamentally different ways. For example, cells rely on diffusion in a liquid phase—although they contain molecular machines, they are not machine-phase systems.*

1.4. The approach in this volume

1.4.1. Disciplinary range, level, and presentation

As Section 1.2 indicates, the study of molecular nanotechnology spans multiple disciplines. This circumstance has hampered both evaluation of the existing concepts and research aimed at extending and superseding them. One purpose

---

* It is sometimes suggested that artificial molecular machine systems cannot improve on biological systems because the latter have been shaped by billions of years of evolution. In specific engineering parameters, however, the products of evolution have already been surpassed: graphite composites are stronger than bone, copper is more conductive than axonal cytoplasm, and so forth. Biological systems do, however, excel in their capacity to evolve, and it can be shown that several of their characteristic differences from eutactic nanomechanical systems (including the use of diffusive transport rather than mechanical conveyance) are important to this capacity (Drexler, 1989a).

of the present volume is to assemble a large portion of the necessary core knowledge in a form that requires no specialized knowledge of the component disciplines. An effort has been made (and a glossary provided) to make key chemical concepts accessible to nonchemists, solid-state physics concepts accessible to nonphysicists, and so forth, assuming only a basic background in both chemistry and physics (and a willingness to skip past the occasional obscure observation aimed at a reader in a different discipline). The intended contribution of this work consists not in extending the frontiers of existing fields, but in combining their basic results to lay the foundations of a new field.

To facilitate understanding, several mathematical results in Part I are derived from fundamental principles. Many of these results appear in existing textbooks; others (so far as is known) are novel, being motivated by new questions. The exposition of these mathematical models includes an unusually large number of graphs that illustrate equations in the text; these are provided to facilitate design, which is a synthetic as well as an analytic process. In the analysis of a given system, a calculation based on an equation with a single set of parameter values frequently suffices. In synthesis, however, the designer usually wishes to understand how system properties will vary as controllable parameters are changed; for this, a graph can be more useful than a bare equation.

Different fields have applied different energy units to molecular-scale phenomena, including the kilocalorie per gram-mole of items ($\approx 6.95 \times 10^{-21}$ J per single item) and the kilojoule per gram-mole of items ($\approx 1.66 \times 10^{-21}$ J per single item) of chemistry, and the electron volt ($- eV \approx 160 \times 10^{-21}$ J) of physics. The standard SI unit of energy, of course, is the joule itself. To avoid allusions to hypothetical moles of identical systems or to electrons not involved in the problem, and (more important) to enable mechanical calculations involving force, work, kinetic energy, and so forth to proceed without frequent unit conversions, this volume adheres to the joule as the unit of energy. The attojoule ($= aJ = 10^{-18}$ J) and milli-attojoule ($= maJ = 10^{-21}$ J) are convenient fractional units.

1.4.2. Levels of abstraction and approximation

In an ideal world, chemists would be able to predict the behavior of molecules by applying quantum electrodynamics (QED) to suitably defined assemblages of nuclei and electrons, and engineers would be able to predict automotive performance in the same manner. In this regard, the world is far from ideal. Although no experiment has yet shown an imperfection in QED as a description of the properties of ordinary matter (setting nuclear and gravitational interactions aside), it is computationally intractable as a description even of small molecules. In the real world, chemists and engineers describe systems using a hierarchy of levels of abstraction and approximation (Table 1.4). It is worth surveying this hierarchy because it provides a framework for practical analysis.

As discussed in Chapter 3, the most rigorous models ordinarily used by chemists apply *ab initio* molecular orbital methods; these approximate the Schrödinger equation, which approximates the Dirac equation, which approximates QED, which approximates the unknown exact, universal laws of physics. In describing the mechanical properties of large molecular structures, however, chemists abandon molecular orbital methods in favor of molecular mechanics

Table 1.4. Levels of abstraction and approximation
in molecular systems engineering.

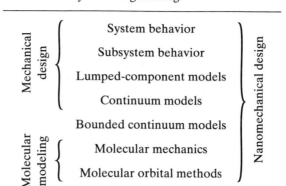

methods of more limited applicability but lower computational cost; these too
are discussed in Chapter 3.

At the upper levels of the hierarchy, engineers set objectives in terms of system behavior and use these objectives to determine requirements for subsystem
behavior (this can proceed through several layers of subsystems). Systems are
commonly analyzed in terms of subsystem capabilities, which are analyzed in
terms of lumped-component models, which in turn are analyzed in terms of
continuum models. For example, a modern computer is described by its subsystems—processor, memory, disk, bus, cooling, power supply, and so forth. A
processor (give or take some intermediate levels) is described as an interconnected network of discrete transistors and other lumped components. Transistors are described by continuum models that consider gate geometries, dopant
distributions, electron transport, and so forth. Individual atoms and electrons
are neglected in describing transistors, and one never describes a computer by
describing electron transport within transistors.

Nanomechanical systems are subject to a similar analysis, describing system-
level objectives served by subsystem capabilities implemented using lumped
components. Continuum models, however, become problematic on the nanometer scale. Chapter 9 develops *bounded continuum models* that take sufficient
account of surface effects to enable the analysis of a broad range of nanomechanical designs in less than atomic detail. To design the smallest devices, however, detailed molecular mechanics models are necessary, and to provide a first-
principles analysis of a mechanochemical process, there is no substitute for
molecular orbital methods.

1.4.3. Scope and assumptions

The present volume adheres to design constraints that may not limit future engineering practice. Each constraint excludes possibilities that are presently difficult to analyze, but that may prove both feasible and desirable. The following

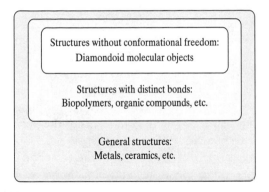

Figure 1.3. Diamondoid structures are a subset of covalently bonded structures, which are a subset of the broad range of solid structures.

assumptions and limitations are thus °conservative, resulting in what are likely to be underestimates of future capabilities:

a.   A narrow range of structures.   From the broad range of materials (metals, ionic crystals, molecular crystals, polymers, etc.), the present work selects the class of diamondoid °covalent solids as its focus (Section 9.3.1f). These structures form a small subset of those that are possible (Figure 1.3), and contain atoms drawn chiefly from the shaded region of Figure 1.4. Diamond itself is the strongest and stiffest structure presently known at ordinary pressures (Kelly, 1973), making it and similar materials attractive on engineering grounds. Since many components can be regarded as polycyclic organic molecules, much of the vast base of knowledge developed by organic chemists is immediately applicable. Small components are subject to large surface effects, but typical organic molecules are, in effect, *all* surface; accordingly, surface effects are an integral part of molecular models.

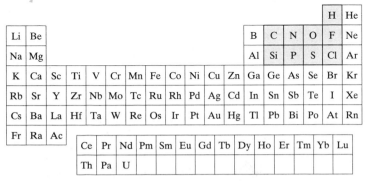

Figure 1.4.   Periodic table of the atoms, with cells shaded to indicate those of greatest importance to nanomechanical design: hydrogen (H), carbon (C), nitrogen (N), oxygen (O), fluorine (F), silicon (Si), phosphorus (P), sulfur (S), and chlorine (Cl). Other elements have applications, but few of the structures discussed in the following chapters contain them. [H is more often placed above lithium (Li) than above F, but hydrocarbons resemble the stable fluorocarbons far more than they do the reactive lithiocarbons. Like F, H is only one electron short of a closed-shell configuration.]

b. No nanoelectronic devices. On a macroscale, mechanical systems are quite distinct from electronic systems: they involve the motion of materials, rather than of electrons and electromagnetic fields. On a nanoscale, mechanical motions are identified with the displacements of nuclei, but electronic activity can cause such motions. Nevertheless, many systems (e.g., the bearing in Figure 1.1) can be described by molecular mechanics models that take no explicit account of electronic degrees of freedom, instead subsuming them into a potential energy function (Chapter 3) defined in terms of the positions of nuclei. This volume focuses on mechanical systems of this sort.

Some systems are strongly electronic in character, relying on changes of electronic state to change other electronic states, with the associated nuclear motions being of small amplitude. Molecular nanotechnologies will surely include nanometer-scale electronic devices exploiting quantum phenomena to achieve (for example) switching and computation. Research relevant to this class of devices is already in progress.

Although nanoelectronic devices are likely to be important products of molecular manufacturing, they fall beyond the scope of the present work. There are several reasons for this. First, the analytical methods required for quantum electronics differ from those required for molecular machinery; including them would have made this volume larger and later. Second, these analytical methods require approximations that frequently render the results dubious, reducing their value as a premise for further reasoning (the existence in 1992 of at least three competing classes of theories to explain high-temperature superconductivity, observed and studied in cuprates since 1987, suggests the magnitude of the difficulties). Finally, while nanomechanical devices can build nanoelectronic devices, the latter cannot return the favor; thus, nanomechanical devices are in a technological sense more fundamental. Accordingly, the nanocomputers discussed in Chapter 12 are based on mechanical devices, although electronic devices will surely permit greater speed.

Chapter 11 briefly discusses nanoscale electromechanical systems. These use conductors, insulators, and tunneling junctions as components of motors, actuators, and switches. Quantum phenomena are important in nanoscale electromechanical systems, but the gross results (switching, interconversion of electrical and mechanical energy) do not rely on subtle quantum effects. Finally, despite their likely utility, machine-phase electrochemical processes are mentioned only in passing.

c. Machine-phase chemistry. Because it can guide reactions (and can avoid most competing reactions) by tightly constraining molecular motions, machine-phase chemistry can be simpler, in some respects, than is solution-phase chemistry. Mechanosynthesis and other operations can be conducted by systems of molecular machines immersed in a solution environment, and there are sometimes advantages to doing so. These less controlled, more complex chemical systems fall beyond the scope of the present work, save for the discussion of implementation strategies in Part III.

d. Room temperature processes. Reduced temperatures decrease thermally excited displacements (Chapter 5), thermal damage rates (Chapter 6), and phonon-mediated drag (Chapter 7). The opportunities and problems presented by low-temperature systems are nonetheless not explored in the present work. Operation at high temperatures is desirable in some circumstances, and can facilitate both desired and undesired chemical reactions, but discussions of the associated technological opportunities and problems are likewise omitted.

e. No photochemistry. Photochemical damage mechanisms are discussed in Section 6.5. Design of molecular machines for photochemical damage resistance is an important challenge, but for simplicity the following will instead assume that devices operate in optically shielded environments. Deliberate use of photochemistry is mentioned only in passing.

f. The single-point failure assumption. The design of small components that can tolerate atomic-scale damage and defects is worthy of attention, but the following will instead assume that any such flaw causes component-level (and usually subsystem-level) failure, unless the components are relatively large. The use of redundant systems to achieve damage tolerance is proposed and analyzed only at higher levels of system organization. Diverse damage mechanisms are reviewed and modeled in Chapter 6.

1.4.4. Objectives and nonobjectives

Because this volume is a work of theoretical applied science (Appendix A), some objectives appropriate to pure and applied experimental studies, or to pure theoretical studies, are inappropriate here. The following paragraphs outline both familiar objectives that are neglected and less-familiar objectives that are pursued. Appendix A describes these issues in more depth.

a. Not describing new principles and natural phenomena. An enormous literature describes new natural phenomena, but this volume describes the implications of known phenomena for new technologies.

b. Seldom formulating exact physical models. In analyzing functional systems, estimates should be accurate or conservative, but need not be exact. The goal is to gain a quantitative understanding of the relationship between structure and function, not to formulate exact physical models for their own sakes.

c. Seldom describing immediate objectives. Most of the scientific literature discusses either past achievements or objectives achievable using existing techniques. Few publications examine objectives that require substantial preparatory development (space science and high-energy physics produce many of the exceptions). Constructing the physical systems discussed in Chapters 8–14 and 16, however, is well beyond current capabilities. In this volume, only Chapter 15 describes objectives suited to the constraints of present laboratory technique. Our ability to model has outstripped our ability to make, and these studies exploit that fact to analyze ambitious objectives.

d. Not portraying specific future developments. This volume describes systems that can deliver performance orders of magnitude beyond that possible with current technology. Nonetheless, as research in this area expands, better designs will in most instances supersede these systems prior to their realization.

Accordingly, they should be regarded not as portrayals of specific future developments, but merely as examples of what can be done.

e.   Seldom seeking an optimal design in the conventional sense.   In mature fields of technology, competitive pressures encourage a search for designs that are nearly optimal. In the exploratory phase of design, however, the more modest goal of workability suffices. A design can depart from optimality either (1) by being overdesigned and inefficient, or (2) by being underdesigned and unreliable. Here, (1) is acceptable, (2) is not.

f.   Seldom specifying complete detail in complex systems.   Given an established technology base, a designer can often describe a system at a high level of abstraction (Section 1.4.2) with confidence that compatible sets of components can be specified before placing the unit into production. This is easier if the product can be overdesigned and inefficient, because margins of safety in the initial design can then compensate for uncertainties in component performance. The approach pursued in this volume amounts to the analytical (as distinct from physical) development of a technology base with capabilities known within some error margins. The later chapters describe systems at various levels of abstraction, likewise using margins of safety to compensate for uncertainties in component performance.

g.   Favoring false-negative over false-positive errors in analysis.   In modeling and analyzing proposed designs, one would ideally distinguish workable from unworkable designs with perfect accuracy. But since models are never exact and accurate, errors of some sort are inevitable. These errors can be of two kinds: *False-positive* evaluations wrongly *accept* an unworkable design; *false-negative* evaluations wrongly *reject* a workable design. In exploring a new domain of technology, conclusions regarding the feasibility of systems rest on conclusions regarding the feasibility of subsystems, forming a hierarchical structure of analysis that can have several layers. A substantial rate of false-positive assessments at a subsystem level would make false-positive assessments at the system level quite likely: designs that rely on unworkable subsystems will not work, and may be beyond repair. False-*negative* assessments, in contrast, are relatively benign. Correcting a false-negative assessment of a rejected subsystem cannot invalidate the analysis of a system whose design omits that subsystem; indeed, correcting this error merely expands the list of workable designs. It is desirable to minimize errors of both kinds, but where uncertainties remain, it is better to bias analytical models and criteria toward safe, false-negative conclusions. This strategy guides the following chapters.

h.   Describing technological systems of novel kinds and capabilities.   The objective of this volume, then, is to provide an analytical framework adequate for designing nanomechanical systems (Chapters 3 to 10), and to begin to exploit that framework by designing systems capable of processing both information (mechanical nanocomputers) and matter (molecular manufacturing systems). A further objective is to show how these systems can be implemented, starting from our present technology base (Chapters 15 and 16). Achieving these objectives can define fruitful goals for experimentation, simulation, and software development, and can provide a better basis for understanding human capabilities and choices in the coming years.

1.5. Overview of following chapters

In a work of this length in a new field, the relationship among topics can easily become lost in a mass of detail. This section provides a chapter-by-chapter overview, describing relationships and indicating areas where a detailed analysis merely shows that a simpler analysis is sufficient.

1.5.1. Overview of Part I

Chapter 2 summarizes classical scaling laws for mechanical, electrical, and thermal systems, describing the magnitudes they imply for various physical parameters, and describing where and how these laws break down (requiring the substitution of molecular and quantum mechanical models). These relationships are provided for perspective and as an aid in making preliminary estimates of physical quantities. They are seldom used in later chapters, where most calculations proceed directly from physical principles, rather than from a scaling analysis.

Chapter 3 provides an overview of molecular and intersurface potential energy functions, describing in some detail the molecular mechanics models that later chapters apply to the description of molecular machines. The concepts developed in this chapter are fundamental to much of what follows, since the potential energy function of a molecular system provides the basis for an essentially complete description of its mechanical properties. Its chief conclusion is that the limitations and inaccuracies of molecular mechanics models, although a serious handicap in solution-phase chemistry, are compatible with the use of these models in designing certain classes of molecular machinery.

Chapter 4 provides an overview of various models of molecular dynamics and describes the basis for the choice of models made in later chapters. Its chief conclusion is that classical mechanics and classical statistical mechanics, with occasional quantum mechanical corrections, provide an adequate basis for analyzing most molecular mechanical systems operating at ordinary temperatures.

Chapter 5 examines several classes of mechanical structures and derives relationships that describe the positional uncertainties resulting from the combined effects of quantum mechanics and thermal excitation; later chapters use these results to calculate error rates. This is the most heavily mathematical chapter in the book, but it chiefly concludes that classical statistical mechanics adequately describes the positional uncertainties of nanometer-scale mechanical systems at ordinary temperatures. Only Eq. (5.4) is directly applied in later chapters.

Chapter 6 examines various processes that cause transitions among potential wells in a nanomechanical system, including transitions that cause errors and structural damage. It describes standard theoretical models used to calculate chemical reaction rates based on potential energy functions; these are applied later in analyzing molecular manufacturing processes. Regarding damaging transitions, the chapter concludes that systems at room temperature can be designed to limit rates of damage caused by thermal, mechanical, and photochemical effects to low enough levels that the overall rate of damage is chiefly determined by the background level of ionizing radiation. Damage caused by radiation imposes major constraints on the design of large systems.

Chapter 7 examines various processes that degrade mechanical energy into heat, causing frictional losses; among these are acoustic radiation, phonon scattering, shear-reflection drag, phonon viscosity, thermoelastic damping, nonisothermal compression of mobile components, and transitions among potential wells that vary with time. It develops a set of analytical models applied in later chapters to estimate the power dissipated by various nanomechanical devices.

Chapter 8 compares and contrasts the established capabilities of solution chemistry to those expected from mechanochemical systems, considering speed, efficiency, versatility, and reliability; it also examines an illustrative set of mechanochemical processes in detail. It concludes that a large set of mechanochemical reactions can be made extremely reliable, with error rates $<10^{-15}$. (Although not strictly necessary, this degree of reliability substantially simplifies molecular manufacturing.) It also concludes that mechanochemical processes can be used to construct a wide range of diamondoid structures; this motivates the consideration of diamondoid components in Part II.

- Overall, Part I expends substantial effort in describing physical effects that are of negligible importance in nanomechanical design. In established fields of engineering, training and experience focus attention on the important physical effects; most trivial effects are automatically ignored. In surveying a new field, however, insignificant effects must often be examined before they can be recognized as insignificant.

1.5.2. Overview of Part II

Chapter 9 discusses nanoscale structural components and relates molecular mechanics models to descriptions based on bulk material properties. It develops the concept of a *bounded continuum model* that omits atomic detail while treating surfaces in a way that takes account of interatomic potentials. It concludes that diamondoid structures are desirable nanoscale components, that bounded continuum models can provide useful preliminary descriptions of component properties, and that a wide range of shapes can be constructed on a nanometer scale, despite the discrete nature of atoms and bonds.

Chapter 10 uses molecular mechanics models and analytical models developed in Part I to describe the mechanical properties and performance characteristics of active devices such as gears, bearings, and drive mechanisms. It concludes that structures on a several-nanometer scale can serve as mechanical devices of most classes familiar on the macroscopic scale, and that these nanomechanical devices can in many instances move with negligible static friction. Models from Chapter 7 are applied to describe energy dissipation from dynamic friction. The conditions for low-friction motion derived in this chapter are assumed as background in Chapters 11 to 14.

Chapter 11 describes various subsystems of intermediate size and complexity. These include devices capable of measuring and distinguishing molecular features, stiff drive mechanisms, fluid handling systems (including walls, seals, and vacuum pumps), cooling systems, and electromechanical devices such as motors and actuators. These subsystems and their capabilities are applied or assumed as background in Chapters 12 to 14.

Chapter 12 describes nanomechanical computer systems. It starts with mechanical digital logic systems, including logic gates, signal transmission channels, registers, and their integration into finite-state machines (with an analysis of thermally induced error rates). It then discusses carry chains, random access memory, tape-based storage, power supplies, clock distribution, information input and output to macroscopic systems, and overall system performance (volume, speed, and power dissipation). Its chief conclusion is that a 1000 MIPS computer can occupy less than one cubic micron and consume less than 0.1 microwatt of power.

Chapter 13 describes systems for converting impure feedstock solutions into diamondoid molecular objects, using molecular concentration and purification systems, followed by special-purpose mechanochemical systems (molecular mills) capable of performing repetitive operations efficiently, and by general-purpose mechanochemical systems (i.e., molecular manipulators) capable of performing a complex series of operations under programmable control. A key conclusion is that, after an initial purification and ordering process, molecular assembly can be sufficiently reliable that cycles of inspection and rework to correct failures can be avoided. The conclusions of this chapter are directly exploited in the next.

Chapter 14 describes molecular manufacturing systems that use purification systems, mills, and manipulators to transform an impure feedstock solution into any one of a large set of macroscopic (kilogram-scale) products within a few thousand seconds. It focuses on systems integration and overall performance, and discusses a range of issues including design complexity and product cost.

- Part II describes diamondoid nanomechanical systems of increasing complexity, ending with a description of molecular manufacturing systems that can build complex diamondoid nanomechanical systems. This supports the nontrivial proposition that molecular manufacturing systems are feasible, given molecular manufacturing systems with which to build them. Part III discusses the implementation of molecular manufacturing systems in the absence of preexisting systems of the same kind. It begins from our present technology base.

1.5.3. Overview of Part III

Chapter 15 describes current capabilities for the design and fabrication of multinanometer scale molecular objects (by chemical and biochemical means) and discusses approaches for combining and extending these capabilities to enable the construction of larger and more complex systems. The major approaches considered are (1) protein engineering, (2) the engineering of protein-like molecules designed to fold more predictably and stably, and (3) the development of mechanosynthetic systems that make use of atomic force microscope (AFM) technologies. The first two approaches would construct large molecular systems by °Brownian assembly (self assembly) of smaller units; the AFM approach would construct them directly from small monomers. This chapter concludes that means can be developed for constructing molecular mechanical systems containing 10^5 monomers, operating in a solution environment.

Chapter 16 describes a development pathway leading from small molecular mechanical systems operating in solution, through larger and better-isolated systems, to ~100 nm scale mechanisms able make complex diamondoid structures by means of mechanosynthesis. It proposes the use of externally generated pressure fluctuations to provide power and control to these intermediate-technology devices, and the use of optically probed, environmentally modulated fluorescent molecules to enable prompt sensing of the results of attempted mechanosynthetic operations. This chapter concludes that accessible (though not necessarily easy) development paths lead from present capabilities to advanced molecular manufacturing.

  • Part II begins and ends with long-term technologies, but Part III begins
    with current technologies. Readers more interested in short-term develop-
    ments and practical foundations may wish to start with Part III. It is rela-
    tively independent of the rest, although it occasionally cites results from
    previous chapters.

1.5.4. Overview of Appendices

Appendix A discusses methodologies appropriate to the study of technological possibilities, comparing and contrasting them to the methodologies appropriate to the study of natural phenomena (i.e., standard scientific problems) and to the design of products for immediate implementation (i.e., standard engineering problems).

Appendix B discusses related work by other researchers. It returns to the fields of mechanical engineering, microtechnology, chemistry, and molecular biology (adding protein engineering and proximal probe technology), focusing on the work in these fields that has advanced furthest toward engineering complex molecular systems.

1.5.5. Open problems

Many chapters end with a discussion of open problems. These range from developing specific examples and analyses to developing major software systems and laboratory research programs. The discussions are not exhaustive, but are intended to highlight useful directions for research.

Part I

Physical Principles

2

Classical Magnitudes and Scaling Laws

2.1. Overview

Most physical magnitudes characterizing nanoscale systems differ enormously from those familiar in macroscale systems. Some of these magnitudes can, however, be estimated by applying scaling laws to the values for macroscale systems. Although later chapters seldom use this approach, it can provide orientation, preliminary estimates, and a means for testing whether answers derived by more sophisticated methods are in fact reasonable.

The first of the following sections considers the role of engineering approximations in more detail (Section 2.2); the rest present scaling relationships based on classical continuum models and discuss how those relationships break down as a consequence of atomic-scale structure, mean-free-path effects, and quantum mechanical effects. Section 2.3 discusses mechanical systems, where many scaling laws are quite accurate on the nanoscale. Section 2.4 discusses electromagnetic systems, where many scaling laws fail dramatically on the nanoscale. Section 2.5 discusses thermal systems, where scaling laws have variable accuracy. Finally, Section 2.6 briefly describes how later chapters go beyond these simple models.

2.2. Approximation and classical continuum models

When used with caution, classical continuum models of nanoscale systems can be of substantial value in design and analysis. They represent the simplest level in a hierarchy of approximations of increasing accuracy, complexity, and difficulty.

Experience teaches the value of approximation in design. A typical design process starts with the generation and preliminary evaluation of many options, then selects a few options for further elaboration and evaluation, and finally settles on a detailed specification and analysis of a single preferred design. The first steps entail little commitment to a particular approach. The ease of exploring and comparing many qualitatively distinct approaches is at a premium, and drastic approximations often suffice to screen out the worst options. Even the final design and analysis does not require an exact calculation of physical behavior: approximations and compensating safety margins suffice. Accordingly, a design process can use different approximations at different stages, moving toward greater analytical accuracy and cost.

Approximation is inescapable because the most accurate physical models are computationally intractable. In macromechanical design, engineers employ approximations based on classical mechanics, neglecting quantum mechanics, the thermal excitation of mechanical motions, and the molecular structure of matter. Since macromechanical engineering blends into nanomechanical engineering with no clear line of demarcation, the approximations of macromechanical engineering offer a point of departure for exploring the nanomechanical realm. In some circumstances, these approximations (with a few adaptations) provide an adequate basis for the design and analysis of nanoscale systems. In a broader range of circumstances, they provide an adequate basis for exploring design options and for conducting a preliminary analysis. In a yet broader range of circumstances, they provide a crude description to which one can compare more sophisticated approximations.

2.3. Scaling of classical mechanical systems

Nanomechanical systems are fundamental to molecular manufacturing and are useful in many of its products and processes. The widespread use in chemistry of molecular mechanics approximations together with the classical equations of motion (Sections 3.3, 4.2.3a) indicates the utility of describing nanoscale mechanical systems in terms of classical mechanics. This section describes scaling laws and magnitudes with the added approximation of continuous media.

2.3.1. Basic assumptions

The following discussion considers mechanical systems, neglecting fields and currents. Like later sections, it examines how different physical magnitudes depend on the size of a system (defined by a length parameter L) if all shape parameters and material properties (e.g., strengths, moduli, densities, coefficients of friction) are held constant.

A description of scaling laws must begin with choices that determine the scaling of dynamical variables. A natural choice is that of constant stress. This implies scale-independent °elastic deformation, and hence scale-independent shape; since it results in scale-independent speeds, it also implies constancy of the space-time shapes describing the trajectories of moving parts. Some exemplar calculations are provided, based on material properties like those of diamond (Table 9.1): density $\rho = 3.5 \times 10^3$ kg/m^3; °Young's modulus $E = 10^{12}$ N/m^2; and a low working stress (~0.2 times tensile strength) $\sigma = 10^{10}$ N/m^2. This choice of materials often yields large parameter values (for speeds, accelerations, etc.) relative to those characteristic of more familiar engineering materials.

2.3.2. Magnitudes and scaling

Given constancy of stress and material strength, both the strength of a structure and the force it exerts scale with its cross-sectional area

$$total\ strength \propto force \propto area \propto L^2 \qquad (2.1)$$

Nanoscale devices accordingly exert only small forces: a stress of 10^{10} N/m^2 equals 10^{-8} N/nm^2, or 10 nN/nm^2. Stiffness in °shear, like stretching stiffness, depends on both area and length

$$shear\ stiffness \propto stretching\ stiffness \propto \frac{area}{length} \propto L \qquad (2.2)$$

and varies less rapidly with scale; a cubic nanometer block of $E = 10^{12}$ N/m^2 has a stretching stiffness of 1000 N/m. The bending stiffness of a rod scales in the same way

$$bending\ stiffness \propto \frac{radius^4}{length^3} \propto L \qquad (2.3)$$

Given the above scaling relationships, the magnitude of the deformation under load

$$deformation \propto \frac{force}{stiffness} \propto L \qquad (2.4)$$

is proportional to scale, and hence the shape of deformed structures is scale invariant.

The assumption of constant density makes mass scale with volume,

$$mass \propto volume \propto L^3 \qquad (2.5)$$

and the mass of a cubic nanometer block of density $\rho = 3.5 \times 10^3$ kg/m^3 equals 3.5×10^{-24} kg.

The above expressions yield the scaling relationship

$$acceleration \propto \frac{force}{mass} \propto L^{-1} \qquad (2.6)$$

A cubic-nanometer mass subject to a net force equaling the above working stress applied to a square nanometer experiences an acceleration of $\sim 3 \times 10^{15}$ m/s^2. Accelerations in nanomechanisms commonly are large by macroscopic standards, but aside from special cases (such as transient acceleration during impact and steady acceleration in a small flywheel) they rarely approach the value just calculated. (Terrestrial gravitational accelerations and stresses usually have negligible effects on nanomechanisms.)

Modulus and density determine the acoustic speed, a scale-independent parameter [along a slim rod, the speed is $(E/\rho)^{1/2}$; in bulk material, somewhat higher]. The vibrational frequencies of a mechanical system are proportional to the acoustic transit time

$$frequency \propto \frac{acoustic\ speed}{length} \propto L^{-1} \qquad (2.7)$$

The acoustic speed in diamond is $\sim 1.75 \times 10^4$ m/s. Some vibrational modes are more conveniently described in terms of lumped parameters of stiffness and mass,

$$frequency \propto \sqrt{\frac{stiffness}{mass}} \propto L^{-1} \qquad (2.8)$$

but the scaling relationship is the same. The stiffness and mass associated with a cubic nanometer block yield a vibrational frequency characteristic of a stiff, nanometer-scale object: $[(1000 \text{ N/m})/(3.5 \times 10^{-24} \text{ kg})]^{1/2} \approx 1.7 \times 10^{13}$ rad/s.

Characteristic times are inversely proportional to characteristic frequencies

$$time \propto frequency^{-1} \propto L \qquad (2.9)$$

The speed of mechanical motions is constrained by strength and density. Its scaling can be derived from the above expressions

$$speed \propto acceleration \cdot time = constant \qquad (2.10)$$

A characteristic speed (only seldom exceeded in practical mechanisms) is that at which a flywheel in the form of a slim hoop is subject to the chosen working stress as a result of its mass and centripetal acceleration. This occurs when $v = (\sigma/\rho)^{1/2} \approx 1.7 \times 10^3$ m/s (with the assumed σ and ρ). Most mechanical motions considered in this volume, however, have speeds between 0.001 and 10 m/s.

The frequencies characteristic of mechanical motions scale with transit times

$$frequency \propto \frac{speed}{length} \propto L^{-1} \qquad (2.11)$$

These frequencies scale in the same manner as vibrational frequencies, hence the assumption of constant stress leaves frequency ratios as scale invariants. At the above characteristic speed, crossing a 1 nm distance takes $\sim 6 \times 10^{-13}$ s; the large speed makes this shorter than the motion times anticipated in typical nanomechanisms. A modest 1 m/s speed, however, still yields a transit time of only 1 ns, indicating that nanomechanisms can operate at frequencies typical of modern micron-scale electronic devices.

The above expressions yield relationships for the scaling of mechanical power

$$power \propto force \cdot speed \propto L^2 \qquad (2.12)$$

and mechanical power density

$$power\ density \propto \frac{power}{volume} \propto L^{-1} \qquad (2.13)$$

A 10 nN force and a 1 nm^3 volume yield a power of 17 μW and a power density of 1.7×10^{22} W/m^3 (at a speed of 1.7×10^3 m/s) or 10 nW and 10^{19} W/m^3 (at a speed of 1 m/s). The combination of strong materials and small devices promises mechanical systems of extraordinarily high power density, even at low speeds (an example of a mechanical power density is the power transmitted by a gear divided by its volume).

Most mechanical systems use bearings to support moving parts. Macromechanical systems frequently use liquid lubricants, but (as noted by Feynman, 1961), this poses problems on a small scale. The above scaling law ordinarily holds speeds and stresses constant, but reducing the thickness of the lubricant layer increases shear rates and hence viscous shear stresses:

$$viscous\ stress\ at\ constant\ speed \propto shear\ rate \propto \frac{speed}{thickness} \propto L^{-1} \qquad (2.14)$$

In Newtonian fluids, shear stress is proportional to shear rate. Molecular simulations indicate that liquids can remain nearly Newtonian at shear rates in excess of 100 m/s across a 1 nm layer (e.g., in the calculations of Ashurst and Hoover, 1975), but they depart from bulk viscosity (or even from liquid behavior) when film thicknesses are less than 10 molecular diameters (Israelachvili, 1992; Schoen et al., 1989), owing to interface-induced alterations in liquid structure. Feynman suggested the use of low-viscosity lubricants (such as kerosene) for micromechanisms (Feynman, 1961); from the perspective of a typical nanomechanism, however, kerosene is better regarded as a collection of bulky molecular objects than as a liquid. If one nonetheless applies the classical approximation to a 1 nm film of low-viscosity fluid ($\eta = 10^{-3}$ N·s/m^2), the viscous shear stress at a speed of 1.7×10^3 m/s is 1.7×10^9 N/m^2; the shear stress at a speed of 1 m/s, 10^6 N/m^2, is still large, dissipating energy at a rate of 1 MW/m^2.

The problems of liquid lubrication motivate consideration of dry bearings (as suggested by Feynman, 1961). Assuming a constant coefficient of friction,

$$\textit{frictional force} \propto \textit{force} \propto L^2 \qquad (2.15)$$

and both stresses and speeds are once again scale-independent. The frictional power,

$$\textit{frictional power} \propto \textit{force} \cdot \textit{speed} \propto L^2 \qquad (2.16)$$

is proportional to the total power, implying scale-independent mechanical efficiencies. In light of engineering experience, however, the use of dry bearings would seem to present problems (as it has in silicon micromachine research). Without lubrication, efficiencies may be low, and static friction often causes jamming and vibration.

A yet more serious problem for unlubricated systems would seem to be wear. Assuming constant interfacial stresses and speeds (as implied by the above scaling relationships), the anticipated surface erosion rate is independent of scale. Assuming that wear life is determined by the time required to produce a certain fractional change in shape,

$$\textit{wear life} \propto \frac{\textit{thickness}}{\textit{erosion rate}} \propto L \qquad (2.17)$$

and a centimeter-scale part having a ten-year lifetime would be expected to have a 30 s lifetime if scaled to nanometer dimensions.

Design and analysis have shown, however, that dry bearings with atomically precise surfaces need not suffer these problems. As shown in Chapters 6, 7, and 10, dynamic friction can be low, and both static friction and wear can be made negligible. The scaling laws applicable to such bearings are compatible with the constant-stress, constant-speed expressions derived previously.

2.3.3. Major corrections

The above scaling relationships treat matter as a continuum with bulk values of strength, modulus, and so forth. They readily yield results for the behavior of iron bars scaled to a length of 10^{-12} m, although such results are meaningless because a single atom of iron is over 10^{-10} m in diameter. They also neglect the influence of surfaces on mechanical properties (Section 9.4), and give (at best) crude estimates regarding small components, in which some dimensions may be only one or a few atomic diameters.

Aside from the molecular structure of matter, major corrections to the results suggested by these scaling laws include uncertainties in position and velocity resulting from statistical and quantum mechanics (examined in detail in Chapter 5). Thermal excitation superimposes random velocities on those intended by the designer. These random velocities depend on scale, such that

$$thermal\ speed \propto \sqrt{\frac{thermal\ energy}{mass}} \propto L^{-3/2} \qquad (2.18)$$

where the thermal energy measures the characteristic energy in a single degree of freedom, not in the object as a whole. For $\rho = 3.5 \times 10^3$ kg/m^3, the mean thermal speed of a cubic nanometer object at 300 K is ~55 m/s. Random thermal velocities (commonly occurring in vibrational modes) often exceed the velocities imposed by planned operations, and cannot be ignored in analyzing nanomechanical systems.

Quantum mechanical uncertainties in position and momentum are parallel to statistical mechanical uncertainties in their effects on nanomechanical systems. The importance of quantum mechanical effects in vibrating systems depends on the ratio of the characteristic quantum energy ($\hbar\omega$, the quantum of vibrational energy in a °harmonic oscillator of angular frequency ω) and the characteristic thermal energy (kT, the mean energy of a thermally excited harmonic oscillator at a temperature T, if $kT \gg \hbar\omega$). The ratio $\hbar\omega/kT$ varies directly with the frequency of vibration, that is, as L^{-1}. An object of cubic nanometer size with $\omega = 1.7 \times 10^{13}$ rad/s has $\hbar\omega/kT_{300} \approx 0.4$ ($T_{300} = 300$ K; $kT_{300} \approx 4.14$ maJ). The associated quantum mechanical effects (e.g., on positional uncertainty) are smaller than the classical thermal effects, but still significant (see Figure 5.2).

2.4. Scaling of classical electromagnetic systems

2.4.1. Basic assumptions

In considering the scaling of electromagnetic systems, it is convenient to assume that electrostatic field strengths (and hence electrostatic stresses) are independent of scale. With this assumption, the above constant-stress, constant-speed scaling laws for mechanical systems continue to hold for electromechanical systems, so long as magnetic forces are neglected. The onset of strong field-emission currents from conductors limits the electrostatic field strength permissible at the negative electrode of a nanoscale system; values of ~10^9 V/m can readily be tolerated (Section 11.6.2).

2.4.2. Major corrections

Chapter 11 describes several nanometer scale electromechanical systems, requiring consideration of the electrical conductivity of fine wires and of insulating layers thin enough to make tunneling a significant mechanism of electron transport. These phenomena are sometimes (within an expanding range of conditions) understood well enough to permit design calculations.

Corrections to classical continuum models are more important in electromagnetic systems than in mechanical systems: quantum effects, for example, become dominant and at small scales can render classical continuum models useless even as crude approximations. Electromagnetic systems on a nanometer scale commonly have extremely high frequencies, yielding large values of $\hbar\omega/kT_{300}$. Molecules undergoing electronic transitions typically absorb and emit light in the visible to ultraviolet range, rather than the infrared range characteristic of thermal excitation at room temperature. The mass of an electron is less than 10^{-3} that of the lightest atom, hence for comparable confining energy barriers, electron °wave functions are more diffuse and permit longer-range tunneling. At high frequencies, the inertial effects of electron mass become significant, but these are neglected in the usual macroscopic expressions for electrical circuits. Accordingly, many of the following classical continuum scaling relationships fail in nanoscale systems. The assumption of scale-independent electrostatic field strengths itself fails in the opposite direction, when scaling up from the nanoscale to the macroscale: the resulting large voltages introduce additional modes of electrical breakdown. In small structures, the discrete size of the electronic charge unit, $\sim 1.6 \times 10^{-19}$ C, disrupts the smooth scaling of classical electrostatic relationships (Section 11.7.2c).

2.4.3. Magnitudes and scaling: steady-state systems

Given a scale-invariant electrostatic field strength,

$$voltage \propto electrostatic\ field \cdot length \propto L \qquad (2.19)$$

At a field strength of 10^9 V/m, a one nanometer distance yields a 1 V potential difference. A scale-invariant field strength implies a force proportional to area,

$$electrostatic\ force \propto area \cdot (electrostatic\ field)^2 \propto L^2 \qquad (2.20)$$

and a 1 V/nm field between two charged surfaces yields an electrostatic force of ~ 0.0044 nN/nm^2.

Assuming a constant resistivity,

$$resistance \propto \frac{length}{area} \propto L^{-1} \qquad (2.21)$$

and a cubic nanometer block with the resistivity of copper would have a resistance of $\sim 17\ \Omega$. This yields an expression for the scaling of currents,

$$ohmic\ current \propto \frac{voltage}{resistance} \propto L^2 \qquad (2.22)$$

which leaves current density constant. In present microelectronics work, current densities in aluminum interconnections are limited to $<10^{10}$ A/m^2 or less by electromigration, which redistributes metal atoms and eventually interrupts circuit continuity (Mead and Conway, 1980). This current density equals 10 nA/nm^2 (as discussed in Section 11.6.1b, however, present electromigration limits are unlikely to apply to well-designed eutactic conductors).

For field emission into free space, current density depends on surface properties and the electrostatic field intensity, hence

$$field\ emission\ current \propto area \propto L^2 \qquad (2.23)$$

and field emission currents scale with ohmic currents. Where surfaces are close enough together for tunneling to occur from conductor to conductor, rather than from conductor to free space, this scaling relationship breaks down.

With constant field strength, electrostatic energy scales with volume:

$$electrostatic\ energy \propto volume \cdot (electrostatic\ field)^2 \propto L^3 \qquad (2.24)$$

A field with a strength of 10^9 V/m has an energy density of ~4.4 maJ per cubic nanometer ($\approx kT_{300}$).

Scaling of capacitance follows from the above,

$$capacitance \propto \frac{electrostatic\ energy}{(voltage)^2} \propto L \qquad (2.25)$$

and is independent of assumptions regarding field strength. The calculated capacitance per square nanometer of a vacuum capacitor with parallel plates separated by 1 nm is ~9×10^{-21} F; note, however, that electron tunneling causes substantial conduction through an insulating layer this thin.

In electromechanical systems dominated by electrostatic forces,

$$electrostatic\ power \propto electrostatic\ force \cdot speed \propto L^2 \qquad (2.26)$$

and

$$electrostatic\ power\ density \propto \frac{electrostatic\ power}{volume} \propto L^{-1} \qquad (2.27)$$

These scaling laws are identical to those for mechanical power and power density. Like them, they suggest high power densities for small devices (see Section 11.7).

The power density caused by resistive losses scales differently, given the above current density:

$$resistive\ power\ density \propto (current\ density)^2 = constant \qquad (2.28)$$

The current density needed to power an electrostatic motor, however, scales differently from that derived from a constant-field scaling analysis. In an electrostatic motor, surfaces are charged and discharged with a certain frequency, hence

$$motor\ current\ density \propto \frac{charge}{area}\ frequency \propto field \cdot frequency \propto L^{-1} \quad (2.29)$$

and the resistive power losses climb sharply with decreasing scale:

$$motor\ resistive\ power\ density \propto (motor\ current\ density)^2 = L^{-2} \qquad (2.30)$$

Accordingly, the efficiency of electrostatic motors decreases with decreasing scale. The absence of long conducting paths (like those in electromagnets) makes resistive losses smaller to begin with, however, and a detailed examination (Section 11.7) shows that efficiencies remain high in absolute terms for motors of submicron scale. The above relationships show that electromechanical systems cannot be scaled in the simple manner suggested for purely mechanical systems, even in the classical continuum approximation.

Electromagnets are far less attractive for nanoscale systems, since

$$magnetic\ field \propto \frac{current}{distance} \propto L \qquad (2.31)$$

At a distance of 1 nm from a conductor carrying a current of 10 nA, the field strength is 2×10^{-6} T. The corresponding forces,

$$magnetic\ force \propto area \cdot (magnetic\ field)^2 \propto L^4 \qquad (2.32)$$

are minute in nanoscale systems: two parallel, 1 nm long segments of conductor, separated by 1 nm and carrying 10 nA, interact with a force of 2×10^{-23} N. This is 14 orders of magnitude smaller than the strength of a typical covalent bond and 11 orders of magnitude smaller than the characteristic electrostatic force just calculated. Magnetic forces between nanoscale current elements are usually negligible. Magnetic fields generated by magnetic materials, in contrast, are independent of scale: forces, energies, and so forth follow the scaling laws described for constant-field electrostatic systems. Nanoscale current elements interacting with fixed magnetic fields can produce more significant (though still small) forces: a 1 nm long segment of conductor carrying a 10 nA current experiences a force of up to 10^{-17} N when immersed in a 1 T field.

The magnetic field energy of a nanoscale current element is small:

$$magnetic\ energy \propto volume \cdot (magnetic\ field)^2 \propto L^5 \qquad (2.33)$$

The scaling of inductance can be derived from the above, but is independent of assumptions regarding the scaling of currents and magnetic field strengths:

$$inductance \propto \frac{magnetic\ energy}{(current)^2} \propto L \qquad (2.34)$$

The inductance per nanometer of length for a fictitious solenoid with a 1 nm^2 cross sectional area and one turn per nanometer of length would be $\sim 10^{-15}$ h.

2.4.4. Magnitudes and scaling: time-varying systems

In systems with time-varying currents and fields, skin-depth effects increase resistance at high frequencies; these effects complicate scaling relationships and are ignored here. The following simplified relationships are included chiefly to illustrate trends and magnitudes that *preclude* the scaling of classical AC circuits into the nanometer size regime.

For *LR* circuits,

$$inductive\ time\ constant \propto \frac{inductance}{resistance} = L^2 \qquad (2.35)$$

Combining the characteristic 17 Ω resistance and 10^{-15} h inductance calculated above yields an *LR* time constant of $\sim 6 \times 10^{-17}$ s. This time constant is nonphysical: it is, for example, short compared to the electron °relaxation time in a typical metal at room temperature ($\sim 10^{-14}$ s). In reality, current decays more slowly because of electron inertia (which has effects broadly similar to those of inductance) and because of the related effect of finite electron relaxation time.

With the approximation of scale-independent resistivity,

$$capacitative\ time\ constant \propto resistance \cdot capacitance = constant \qquad (2.36)$$

This implies that the time required for a capacitor to discharge through a resistor in a pure RC circuit is independent of scale; with the scale dependence of the LR time constant, however, a structure with fixed proportions can change from a nearly pure RC circuit (if built on a small scale) to a nearly pure LR circuit (if built on a large scale). The nanometer-scale RC time constant indicated by this expression is $(17\ \Omega) \times (9 \times 10^{-21}\ F) \approx 1.5 \times 10^{-19}$ s, but this result is nonphysical because it neglects the effects of electron inertia and relaxation time.

The LC product defines an oscillation frequency

$$oscillation\ frequency \propto \sqrt{\frac{1}{inductance \cdot capacitance}} \propto L^{-1} \qquad (2.37)$$

The characteristic inductance and capacitance calculated above would yield an LC circuit with an angular frequency of $\sim 3 \times 10^{17}$ rad/s. Alternatively, in structures such as waveguides,

$$oscillation\ frequency \propto \frac{wave\ speed}{length} \propto L^{-1} \qquad (2.38)$$

To propagate in a hypothetical waveguide 1 nm in diameter, an electromagnetic wave would require a frequency of $\sim 9 \times 10^{17}$ rad/s or more. Even the lower of the two frequencies just mentioned corresponds to quanta with an energy of ~ 30 aJ, that is, to photons in the x-ray range with energies of ~ 200 eV. These frequencies and energies are inconsistent with physical circuits and waveguides (metals are transparent to x-rays, electrons are stripped from molecules at energies well below 200 eV, etc.). Quantum effects and electron inertia make Eq. (2.38) inapplicable in the nanometer range.

Scale also affects the quality of an oscillator:

$$Q \propto oscillation\ frequency \frac{inductance}{resistance} \propto L \qquad (2.39)$$

Since Q is a measure of the damping time relative to the oscillation time, small AC circuits will be heavily damped unless nonclassical effects intervene.

Where the following chapters consider electromagnetic systems at all, they describe systems with currents and fields that are slowly varying by the relevant standards. High-frequency quantum electronic devices, though undoubtedly of great importance, are not discussed here.

2.5.   Scaling of classical thermal systems

2.5.1.   Basic assumptions

The classical continuum model assumes that volumetric heat capacities and thermal conductivities are independent of scale. Since heat flows in nanomechanical systems are typically a side effect of other physical processes, no independent assumptions are made regarding their scaling laws.

2.5.2. Major corrections

Classical, diffusive models for heat flow in solids can break down in several ways. On sufficiently small scales (which can be macroscopic for crystals at low temperatures) thermal energy is transferred ballistically by °phonons for which the mean free path would, in the absence of bounding surfaces, exceed the dimensions of the structure in question. In the ballistic transport regime, interfacial properties analogous to optical reflectivity and emissivity become significant. Radiative heat flow is altered when the separation of surfaces becomes small compared to the characteristic wavelength of blackbody radiation, owing to coupling of nonradiative electromagnetic modes in the surfaces. In gases, separation of surfaces by less than a mean free path again modifies conductivity. The following assumes classical thermal diffusion, which can be a good approximation for liquids and for solids of low thermal conductivity, even on scales approaching the nanometer range.

2.5.3. Magnitudes and scaling

With a scale-independent volumetric heat capacity,

$$heat\ capacity \propto volume \propto L^3 \qquad (2.40)$$

A cubic nanometer volume of a material with a (typical) volumetric heat capacity of 10^6 J/m^3·K has a heat capacity of 1 maJ/K.

Thermal conductance scales like electrical conductance, with

$$thermal\ conductance \propto \frac{area}{length} \propto L \qquad (2.41)$$

and a cubic nanometer of material with a (fairly typical) thermal conductivity of 10 W/m·K has a thermal conductance of 10^{-8} W/K.

Characteristic times for thermal equilibration follow from these relationships, yielding

$$thermal\ time\ constant \propto \frac{heat\ capacity}{thermal\ conductance} \propto L^2 \qquad (2.42)$$

For a cubic nanometer block separated from a heat sink by a thermal path with a conductance of 10^{-8} W/K, the calculated thermal time constant is ~10^{-13} s, which is comparable to the acoustic transit time. (In an insulator, a calculated thermal time constant approaching the acoustic transit time signals a breakdown of the diffusive model for transport of thermal energy and the need for a model accounting for ballistic transport; in the fully ballistic regime, time constants scale in proportion to L, and thermal energy moves at the speed of sound.)

The scaling relationship for frictional power dissipation, Eq. (2.16), implies a scaling law for the temperature elevation of a device in thermal contact with its environment,

$$temperature\ elevation \propto \frac{frictional\ power}{thermal\ conductance} \propto L \qquad (2.43)$$

This indicates that nanomechanical systems are more nearly isothermal than analogous systems of macroscopic size.

Table 2.1.   Summary of classical continuum scaling laws.

Physical quantity	Scaling exponent	Typical magnitude	Scaling accuracy
Area	2	$10^{-18}\,\mathrm{m}^2$	Definitional
Force, strength	2	$10^{-8}\,\mathrm{N/m}^2$	Good
Stiffness	1	$1000\,\mathrm{N/m}$	Good
Deformation	1	$10^{-11}\,\mathrm{m}$	Good
Mass	3	$10^{-24}\,\mathrm{kg}$	Good
Acceleration	-1	$10^{15}\,\mathrm{m/s}^2$	Good
Vibrational frequency	-1	$10^{13}\,\mathrm{rad/s}$	Good
Stress-limited speed	0	$10^3\,\mathrm{m/s}$	Good
Motion time	-1	10^{-12} to $10^{-9}\,\mathrm{s}$	Good
Power	2	$10^{-8}\,\mathrm{W}$	Good
Power density	-1	$10^{19}\,\mathrm{W/m}^3$	Good
Viscous stress	-1	$10^6\,\mathrm{N/m}^2$	Moderate to poor
Frictional force	2	(see Ch. 10)	Moderate to inapplicable
Wear life	1	(see Ch. 6, 10)	Moderate to inapplicable
Thermal speed	$-3/2$	$100\,\mathrm{m/s}$	Good
Voltage	1	$1\,\mathrm{V}$	Good at small scales
Electrostatic force	2	$10^{-12}\,\mathrm{N}$	Good at small scales
Resistance	-1	$10\,\Omega$	Moderate to poor
Current	2	$10^{-8}\,\mathrm{A}$	Moderate to poor
Electrostatic energy	3	$10^{-21}\,\mathrm{J}$	Good at small scales
Capacitance	1	$10^{-20}\,\mathrm{F}$	Good
Magnetic field	1	$10^{-6}\,\mathrm{T}$	Good
Magnetic force	4	$10^{-23}\,\mathrm{N}$	Good
Inductance	1	$10^{-15}\,\mathrm{h}$	Good
Inductive time constant	2	$<10^{-16}\,\mathrm{s}$	Bad[a]
Capacitive time constant	0	—	Moderate to poor[b]
Elect. oscill. frequency	1	$>10^{18}\,\mathrm{rad/s}$	Bad[a]
Oscillator Q	1	—	Moderate to poor[b]
Heat capacity	3	$10^{-21}\,\mathrm{J/K}$	Good
Thermal conductance	1	$10^{-8}\,\mathrm{W/K}$	Good to poor
Thermal time constant	2	$10^{-13}\,\mathrm{s}$	Good to poor

[a] Values included only to illustrate the failure of the scaling law.
[b] Values omitted; realistic geometries would require several arbitrary parameters.

2.6.   Beyond classical continuum models

This chapter has described the scaling laws implied by classical continuum models for mechanical, electromagnetic, and thermal systems, together with the magnitudes they suggest for the physical parameters of nanometer scale systems. It has also considered limits to the validity of these models, imposed by statistical mechanics, quantum mechanics, the molecular structure of matter, and so forth. Different classical models fail at different length scales, with the most dramatic failures appearing in AC electrical circuits.

The following chapters go beyond classical continuum models. Chapters 3 and 4 examine models of molecular structure, dynamics, and statistical mechanics from a nanomechanical systems perspective. Chapters 5 and 6 examine the combined effects of quantum and statistical mechanics on nanomechanical systems, first analyzing positional uncertainty in systems subject to a restoring force, and then analyzing the rates of transitions, errors, and damage in systems that can settle in alternative states. Chapter 7 examines mechanisms for energy dissipation. These chapters provide a foundation for analyzing specific nanomechanical systems. Later chapters examine not only nanomechanical systems, but a few specific electrical and fluid systems; where analysis of the latter must go beyond classical continuum approximations, the needed principles are discussed in context.

2.7. Conclusions

The accuracy of classical continuum models and scaling laws to nanoscale systems depends on the physical phenomena considered. It is low for electromagnetic systems with small calculated time constants, reasonably good for thermal systems and slowly varying electromagnetic systems, and often excellent for purely mechanical systems, provided that the component dimensions substantially exceed atomic dimensions. Scaling principles indicate that mechanical components can operate at high frequencies, accelerations, and power densities. The adverse scaling of wear lifetimes suggests that bearings are a special concern. Later chapters support these expectations regarding frequency, acceleration, and power density; Chapter 10 describes suitable bearings. Table 2.1 summarizes many of the relationships discussed in this chapter.

Potential Energy Surfaces

3.1. Overview

The concept of a molecular potential energy surface (PES) is fundamental to practical models of molecular structure and dynamics. The PES describes the potential energy in terms of the molecular geometry, which in turn is defined by the positions of the atomic nuclei. In the classical approximation, molecular motions are determined by forces (and hence accelerations) corresponding to gradients of the PES, and equilibrium molecular structures correspond to minima of the PES. The term *potential energy surface* stems from a visualization in which a potential energy function in N dimensions (that is, in *configuration space*) is described as a surface in $N + 1$ dimensions, with energy corresponding to height. (When $N > 2$, the visualization is necessarily somewhat nonvisual.)

The significant features of a PES are its °potential wells and the passes (termed °*cols*) between them. A point representing the state of a stiff, °stable structure resides in a well with steep walls (i.e., with a large second derivative of the energy, the basis of stiffness, for all displacements) and no low, accessible cols leading to alternative wells (i.e., no low-energy routes to other structures, the basis of stability). A point representing the initial state of a chemically reactive structure, in contrast, resides in a well linked to another well by a col of accessible height. A point representing a mobile nanomechanical component commonly moves in a well with a long, level floor. Molecular systems in which all transitions occur between distinct potential wells resemble transistor systems in which all transitions occur between distinct logic states: in both instances, the application of correct *design* principles can yield reliable behavior even though other systems subject to the same *physical* principles behave erratically.

Physicists and chemists describe molecular behavior using a hierarchy of approximations of varying accuracy; the PES concept itself is one such approximation. The following sections move from extremely accurate but impractical theories to less accurate but more useful approximations. Section 3.2 discusses both exact quantum mechanical theories and chemically useful approximations to them. Section 3.3 discusses molecular mechanics methods at some length: for nanomechanical engineering, approaches like those described in Sections 3.3 to 3.5 often provide the most useful approximations to the PES, from the standpoint both of accuracy and of computational feasibility. (A discussion of the accuracy of these methods relative to different requirements appears in Section 4.4.) Section 3.4 discusses specialized potential energy functions describing chemical reactions, and Section 3.5 discusses models of the interaction energy of large objects, neglecting atomic detail. Most of the topics introduced in this chapter are discussed at length in the literature; suggestions for further reading are appended.

3.2. Quantum theory and approximations

The analysis of nanomechanical systems requires models describing the behavior of molecular scale systems. The appropriate models more closely resemble those used in chemistry and materials science than (for example) those used in particle physics. For perspective, however, it may be useful to view the hierarchy of approximations from the heights of modern physical theory before moving into the domain of practical calculation. Note that the following overview of quantum mechanics *is* offered chiefly for perspective; although some analytical models in this volume describe quantum mechanical effects, none makes direct use of wave equations or of the mathematical apparatus of quantum mechanics itself.

3.2.1. Overview of quantum mechanics

a. Relativistic theories.   In the 1940s, Feynman, Schwinger, and Tomonaga each independently developed formulations of the theory of quantum electrodynamics (QED), which describes electromagnetic fields and electrons in a unified way. Where the mathematics of QED can be manipulated to yield precise predictions, its predictions (e.g., of the magnetic properties of free electrons and the spectrum of the hydrogen atom) have been confirmed to the last measurable detail. It correctly predicts that the electron g-factor is ~2.002319304 rather than 2 as expected from previous theory, and it predicts the Lamb shift in the hydrogen spectrum (which changes an energy level by less than one part in 10^6) to an accuracy of many decimal places. There is every reason to think that the theory would be equally precise in other areas, if it could be applied. In practice, owing to the difficulty of calculations, chemists and material scientists do not use QED. Its great contributions have been in high-energy physics and in its use as a prototype for other physical theories, such as quantum chromodynamics.

Earlier, in 1931, Dirac had developed a fully relativistic quantum mechanics which predicted electron spin (with a g-factor of 2) and provided a correct explanation for the splitting of certain spectral lines in hydrogen, reflecting shifts in energy levels by about one part in 10^4. Relativistic effects are large in the inner electron shells of heavy elements, but these shells are so tightly bound to the nucleus that they are chemically inert; most relativistic effects on °valence electrons (especially those of light elements) are chemically negligible.* Since

---

* Relativistic effects in heavy atoms cause strong spin-orbit coupling, which can increase the rate of electronic transitions such as the flipping of unpaired electron spins in free radicals, with consequent effects on the rates of chemical reactions (Section 8.4.4b). For typical molecules of interest, however, spin-orbit coupling does not significantly affect structure or dynamics before or after the transition.

Dirac's theory is difficult to apply and describes effects that can often be neglected, it is not used in standard quantum mechanical calculations in chemistry and materials science.

b. Schrödinger's theory. Earlier still, in 1926, Schrödinger had developed a formulation of nonrelativistic quantum mechanics that remains the basis for calculations in quantum chemistry. Schrödinger described matter in terms of a wave equation (here shown with a scalar potential),

$$-\frac{\hbar^2}{2}\sum_j \frac{1}{m_j}\frac{\partial^2}{\partial r_j^2}\psi(\mathbf{r},t)+\mathcal{V}(\mathbf{r},t)\,\psi(\mathbf{r},t)=i\hbar\frac{\partial}{\partial t}\psi(\mathbf{r},t) \tag{3.1}$$

in which a system of N particles with masses $m_0, m_1, m_2,..., m_{N-1}$ is described (neglecting spins) by the coordinate vector $\mathbf{r}$ with components $r_0, r_1, r_2,..., r_{3(N-1)}$ equaling $x_0, y_0, z_0, x_1, y_1, z_1, x_2, y_2, z_2,..., z_{(N-1)}$, and a potential energy function $\mathcal{V}(\mathbf{r},t)$. This is a partial differential equation in a $3N$-dimensional configuration space, and physically valid solutions, $\psi(\mathbf{r},t)$, are subject to a set of boundary, continuity, normalization, and particle-exchange symmetry conditions. Molecular structure calculations seek solutions describing bound, time-invariant systems, but this problem cannot be solved exactly even for a molecule as simple as H_2. Nonetheless, because no simpler theory provides an acceptable description of the quantum mechanical behavior of molecular matter, the Schrödinger equation has become the basis of a host of approximation schemes. These approximations represent steps backward in fundamental physical theory, but steps forward in understanding real physical systems.

c. Schrödinger's theory applied to molecules. In order to understand the general nature of these approximation schemes, it is necessary to say a little more about Schrödinger's theory as applied to molecules.* In calculating the structure of isolated systems, the potential energy function $\mathcal{V}(\mathbf{r},t)$ is time independent, and is simply the total Coulomb energy for the interaction of each pair of charged particles

$$\mathcal{V}(\mathbf{r})=\sum_{i<j}\frac{q_i q_j}{4\pi\varepsilon_0 d_{ij}} \tag{3.2}$$

where d_{ij} is the distance between particles i and j, q_i and q_j are their charges, and ε_0 is the permittivity of free space. The potential energy function is thus based on a picture of particles with precise positions in ordinary space. The °wave function determines the °probability density function

$$f_{\mathbf{r},t}(\mathbf{r},t)=\psi^*(\mathbf{r},t)\,\psi(\mathbf{r},t) \tag{3.3}$$

---

* The basic nature of the equation suggests why calculations are difficult. Using samples at 10 points to approximate the integral of a one-dimensional function is often easy, though not always very accurate. But consider a function in a 126-dimensional space: sampling a $10\times10\times10\times10\times\ldots$ grid would require an impossible 10^{126} function evaluations. Solving a differential equation with boundary conditions is generally more difficult than integration, and finding a wave function for the electrons in benzene, for example, presents a problem of this sort in a 126-dimensional space.

This expression is defined over the $3N$ dimensional configuration space, hence it yields not only the particle density (and hence the charge density) at each point in 3-dimensional space, but also the probability density associated with each spatial configuration of particles. Owing to particle-exchange symmetry conditions and electrostatic repulsion, electron positions are not independent, but correlated. This precludes solving the $3N$ dimensional problem as a set of N coupled problems in three dimensions. (In the steady-state solutions of greatest chemical interest, this correlation of particle positions and motions is described by a wave function that is time independent, save for a time-varying phase factor.)

3.2.2. The Born–Oppenheimer PES

In chemistry and nanomechanical engineering, the full wave function gives more information than is necessary; indeed, the wave function *per se* is of little interest. Approximations and partial solutions can accordingly be of great value. Because each particle adds three dimensions and disproportionate computational cost, it is useful to partition problems into simpler subproblems when the resulting inaccuracies are not too severe.

Most computational techniques exploit the Born–Oppenheimer approximation, which treats the motion of electrons and nuclei separately. Even the lightest nucleus has ~1836 times the inertia of an electron. The characteristic speeds and frequencies of nuclear motion are accordingly much lower than those of electronic motion. The Born–Oppenheimer approximation treats nuclei as motionless and computes the wave function and energy for a system of electrons in the presence of a fixed nuclear configuration. In this approximation, each nuclear configuration corresponds to a single electronic ground-state energy. This defines the ground-state Born–Oppenheimer potential energy function $\mathcal{V}(\mathbf{r}_n)$, where the vector $\mathbf{r}_n$ specifies only nuclear coordinates and $\mathcal{V}(\mathbf{r}_n)$ is quite unlike the simple Coulomb potential. In analyzing nuclear motions using this potential function, electronic motions are implicitly assumed to adjust without a time lag. The Born–Oppenheimer approximation breaks down when nuclear motions are fast enough (e.g., in high-energy collisions), when excited states occur at very low energies (this is rare in stable molecules), and when small changes in nuclear coordinates cause large changes in the electron wave function (for example in nearly degenerate states where the Jahn–Teller effect becomes important, as occurs in some symmetrical transition-metal complexes). Under ordinary conditions in which nuclear kinetic energies are less than electronic kinetic energies (and nuclear speeds are accordingly much smaller), the Born–Oppenheimer approximation usually gives an excellent account of molecular behavior.

In most nanomechanical systems, as in most chemical reactions, electron wave functions change smoothly with changes in molecular geometry. Under these conditions, there are no abrupt changes in electron distribution and energy, that is, no electronic transitions. In the absence of electronic transitions, and within the Born–Oppenheimer approximation, molecular dynamics depends only on the Born–Oppenheimer potential. If one knows this potential, or has an adequate approximation to it, then one can analyze molecular dynamics without reference to electronic behavior. (When electronic transitions occur, they place the system on another Born–Oppenheimer potential surface.) The

Born–Oppenheimer potential defines the potential energy surface, and can be used in both classical and quantum models of dynamics.

3.2.3. Molecular orbital methods

Practical calculations require further approximations. The most popular approaches result in a family of techniques known as molecular °orbital methods; these make different approximations, yielding computations of widely varying accuracy and cost.

a.   The independent-electron approximation.   Molecular orbital methods begin with the *independent electron approximation,* in which each electron is treated as moving in the electrostatic potential that would result from the time-average distribution of the other electrons; this neglects the electrostatic *electron correlation* effects mentioned in Section 3.2.1c. Together with the Born–Oppenheimer approximation, this approximates the problem of solving a single Schrödinger equation in $3N$-dimensional configuration space as that of solving N coupled Schrödinger equations in three-dimensional space, where N is the number of electrons. The coupling is treated by the *self-consistent field* method: conceptually, each newly calculated electron wave function results in a new distribution of charge density, which changes the potential experienced by the other electrons, and so demands a new calculation; the resulting iterative process converges toward a set of single-electron wave functions, each consistent with the electrostatic potential of the rest (in practice, more efficient iterative procedures are employed). Each single-electron wave function corresponds to a molecular orbital having a particular energy and charge distribution. In this process, the overall many-electron wave function is described by a determinant based on these orbitals; this imposes particle-exchange symmetry conditions.

b.   Approximate wave functions.   In a typical quantum chemistry program, single-electron wave functions are approximated as weighted sums of simple (e.g., "Gaussian type") basis functions, and the programs vary the weighting of each basis function to form a wave function of minimum energy. Increasing the number of basis functions provides increased flexibility in shaping the wave function, enabling the construction of more accurate molecular orbitals; the number of basis functions must equal or (far better) exceed the number of electrons in the molecule. Computations on moderately large molecules are prohibitively expensive and time-consuming, however, because the cost of computing the energy of a single molecular geometry by these methods varies roughly as the fourth power of the number of basis functions.

c.   Electron correlation and "levels of theory."   Molecular orbital computations can be made more accurate by taking electron correlation into account using *configuration interaction* (CI) and related perturbation theory methods.* In CI methods, a "configuration" refers not to a spatial configuration of electrons (in the sense of a configuration space) but to a pattern of orbital occupancy: that

---

* Calculations using different basis sets or corrections for electron correlation are described as being at different "levels of theory." They might better be described as different levels of approximation, with the more accurate levels converging on the Born–Oppenheimer approximation to the Schrödinger equation.

is, to a wave function formed from a particular set of single-electron wave functions. Judiciously mixing multiple wave functions, some representing excited states and each calculated using the independent-electron approximation, is equivalent to permitting correlated electron motions and hence reduces the errors of the independent-electron approximation. The set of possible configurations is combinatorially large; practical calculations often require selecting the most important configurations from a set of millions. The practical importance of electron correlation varies with the system and the phenomenon considered; it is the source of attractive forces between neutral, nonpolar molecules and is particularly important in calculating the energy of bonds far from °equilibrium geometries (i.e., during the transition state of a chemical reaction). Configuration interaction calculations often converge slowly and expensively.

d. Semiempirical methods. The techniques just described (termed *ab initio* methods) make approximations in determining wave functions, but they use no parameters other than fundamental physical constants. Semiempirical molecular orbital methods (such as MNDO, AM1, and PM3) make further approximations in computing wave functions, but they compensate by introducing parameters for different atom types to fit results to experimental data. These methods neglect certain integrals that arise in more complete calculations and treat the compact, high-energy orbitals of inner-shell electrons as fixed. Semiempirical methods have a far lower computational cost, by a factor of 50 to 500 even for a molecule as small as propane (Clark, 1985), but vary in their accuracy in a complex manner (e.g., having special problems with hydrogen bonds, or with boron). There is considerable room for cleverness in semiempirical methods, which continue to improve (Stewart, 1990).

e. Results and accuracy. Molecular orbital methods can be used to calculate parameters such as charge distribution, polarizability, and ionization energies, but their chief application in the present context is to evaluate the energy as a function of molecular configuration: that is, the Born–Oppenheimer potential. This information suffices to determine equilibrium molecular geometries, corresponding to local or global energy minima, but finding them requires energy evaluations (often augmented by direct calculations of energy gradients) at many points in configuration space. More complex structures require evaluation at more points, hence computational cost scales with size even more adversely than the costs of single-point calculations would suggest.

The challenge of applying molecular orbital methods to chemical reactions is suggested by the great difference between the total molecular binding energy calculated by these methods (the energy required to separate a molecule into isolated electrons and nuclei) and the energy differences of chemical interest. The total binding energy for a small molecule is typically on the order of 10^{-15} J, and the energy of a single chemical bond is on the order of 10^{-18} J, but the energy required to rearrange bonds in a chemical reaction is typically $<10^{-19}$ J, and the success of a solution-phase chemical process can be critically dependent on energy differences of $<10^{-20}$ J. Thus, chemists often require that errors be less than 10^{-4} or 10^{-5} of the total energy, although larger errors in *total* energy can often be tolerated so long as energy *differentials* between similar structures are computed accurately enough.

This accuracy can be achieved using molecular orbital methods (for small structures), and such calculations can be used to design chemically-active groups for use in reactive devices in molecular manufacturing systems. Nonetheless, limitations associated with the expense, accuracy, and scalability of molecular orbital methods motivate the continuing popularity of empirically based models of molecules and chemical reactions.

3.3. Molecular mechanics

An alternative to performing quantum mechanical calculations of electronic structure is to approximate the Born–Oppenheimer potential directly in terms of the molecular geometry. Molecular mechanics methods (among others) take this approach, which is well suited to most problems of nanomechanical design and modeling.

3.3.1. The molecular mechanics approach

a.   Overview.   Organic chemists traditionally represent molecular structures with ball-and-stick models in which each ball represents an atom and each stick represents a bond. In quantum mechanical calculations, there is no *a priori* concept of a "chemical bond," and the *behavior* of bonding emerges afresh in each calculation. Molecular mechanics calculations, in contrast, begin with the concept of bonds, then use them as a basis for modeling the molecular potential energy surface. Molecular mechanics thus builds directly on a familiar and useful visualization of molecular structure.

Molecular mechanics programs use an approximate PES to calculate the properties of equilibrium molecular structures (that is, of local minima on the surface). This PES is developed by choosing functional forms and parameters that yield structures with properties that closely approximate the experimentally measured geometries, energies, and vibrational frequencies of molecules in the laboratory. Structural geometries are determined experimentally by (for example) x-ray crystallography, microwave spectroscopy, and gas-phase electron diffraction; heats of formation by calorimetry; and vibrational frequencies by infrared spectroscopy. *Molecular mechanics thus builds directly on a large body of experimental results regarding the mechanical properties of molecules.*

Burkert and Allinger (1982) provide data on the typical accuracy of calculations performed on hydrocarbons using the MM2 model: estimated bond lengths typically match experimental values to within a few times 10^{-13} m ($\sim 0.1\%$), estimated bond angles typically match within about 0.01 rad ($\approx 0.6°$, or 0.5%), and energies within a few times 10^{-21} J. These values are comparable in accuracy to the experimental data itself. Accuracies are lower among nonhydrocarbon structures, but are often good by nanomechanical standards (Section 4.4.3).

Burkert and Allinger compare the computational costs of molecular mechanics calculations to those of *ab initio* quantum mechanical calculations: for a small molecule (propylamine, with 13 atoms) cost favors molecular mechanics by a factor of 10^3. For larger molecules the difference becomes more pronounced: computational costs for molecular mechanics methods increase at between the second and third power of the number of atoms, rather than the higher powers characteristic of molecular orbital methods. As of 1992, molecular mechanics calculations on 10^5 atom systems have become routine, and systems of 10^2 to 10^3 atoms are readily modeled using personal computers.

b. **Limitations, content, and applications.** Molecular mechanics systems have, however, been successfully applied to only a narrow range of molecular structures in configurations not too far from equilibrium. They use energy functions and parameters tailored to specific, local arrangements of atoms. Fortunately for nanomechanical engineering efforts, the most advanced molecular mechanics methods have been developed to model a class of structures that includes those most suitable for use as nanomechanical components—that is, structures built largely of carbon atoms (often augmented with one or more of the elements H, N, O, F, Si, P, S, and Cl) joined by strong, directional, covalent bonds. Structures of this class are the focus of organic chemistry; a subset of these structures comprises most of the molecular machinery of living systems.

Within this set, limitations remain. Aside from the small inaccuracies found in all structures, standard molecular mechanics programs cannot realistically describe certain structures. For example, they can model many stable structures, even when °strained, but cannot model chemical transformations or systems on the verge of such transformations. Computational results must be examined with an eye for such invalidating conditions.

Molecular mechanics programs differ in their intended applications, with some designed for speed and others for accuracy, with some intended for broad classes of organic structures and others specialized for biomolecules. In general, systems intended for large biomolecules also place a premium on speed and give poor descriptions of structures with large strain energies; popular examples are AMBER (Weiner et al., 1984; Weiner et al., 1986) and CHARMM (Brooks et al., 1983). A widely used molecular mechanics model intended for a broad range of organic structures (including structures with large strain energies) is MM2 (Allinger, 1977), developed by Norman Allinger and his coworkers. Although in the process of being superseded by the similar, but more complex and accurate MM3 (Allinger et al., 1990; Allinger et al., 1989; Lii and Allinger, 1989a; Lii and Allinger, 1989b; Lii and Allinger, 1991), MM2 gives good results for a broad range of structures and is a standard model in the chemical literature. After evaluation and with some caveats, the MM2 model is used as an engineering approximation in much of the present work. Computations performed by MM2-based software are used to characterize the minimum-energy configurations of structures, their stiffness, bearing properties, and the like; MM2 parameters are also used as a basis for several analytical models.

Molecular mechanics programs treat the total potential energy as a sum of terms accounting chiefly for bond stretching, bending, and torsion, and for van der Waals, overlap, and electrostatic interactions among nonbonded atoms. To develop a physical intuition for the mechanical behavior of molecular systems, one needs a feel for the nature and magnitude of these forces. Although physical intuition cannot substitute for a more precise analysis, it can be of great value in shaping designs and choosing what to analyze. The relationships and parameters that follow provide a good basis for physical understanding and also enable pocket-calculator estimates of the stiffness, force, and energy associated with molecular deformations. In the exploratory phase of design, one must test and

Table 3.1.   Some atom types used in the MM2 program (Allinger, 1986) along with (rationalized) MM2 nonbonded interaction parameters. Note that the lone pair electrons associated with nitrogen and oxygen atoms are modeled by treating each pair as a pseudoatom. The full set of atom types described by MM2 is roughly twice as large as this list.

MM2 code	Symbol	Type	ε_{vdw} (maJ)	r_{vdw} (10^{-10} m)	Mass (amu)	Mass (10^{-27} kg)
1	C	sp^3	0.357	1.90	12.000	19.925
2	C	sp^2 alkene	0.357	1.94	12.000	19.925
3	C	sp^2 carbonyl	0.357	1.94	12.000	19.925
4	C	sp acetylene	0.357	1.94	12.000	19.925
5	H	hydrocarbon	0.382[a]	1.50[a]	1.008	1.674
6	O	C—O—[H,C]	0.406	1.74	15.999	26.565
7	O	carbonyl	0.536	1.74	15.999	26.565
8	N	sp^3	0.447	1.82	14.003	23.251
11	F	fluoride	0.634	1.65	18.998	31.545
12	Cl	chloride	1.950	2.03	34.969	58.064
13	Br	bromide	2.599	2.18	78.918	131.038
14	I	iodide	3.444	2.32	126.900	210.709
15	S	sulfide	1.641	2.11	31.972	53.087
19	Si	silane	1.137	2.25	27.977	46.454
20	LP	lone pair	0.130	1.20	0.000	0.000
21	H	alcohol	0.292	1.20	1.008	1.674
22	C	cyclopropane	0.357	1.90	12.000	19.925
25	P	phosphine	1.365	2.18	30.994	51.464

[a] See Section 3.3.2e for correction factors.

modify tentative concepts. This requires quick, accessible estimates of energies, forces, and stiffnesses, and of how they vary. For this, experience shows the value of graphs such as Figures 3.3 to 3.9.

3.3.2.   The MM2 model

In molecular mechanics, bonds are defined by the types of the atoms they join, and these atom types can depend on both atomic number and bonding environment: for example, carbon atoms that participate only in single bonds are classified as a different type from those that participate in double or triple bonds. Table 3.1 lists some of the atom types used by the MM2 program and their nonbonded-interaction parameters (Section 3.3.2e).

*a.   **Bond stretching.***   Bonds resist stretching and compression, tending toward an equilibrium length. The MM2 expression for the potential energy of bond stretching $\mathcal{V}_s$ includes a cubic term to account for anharmonicity

$$\mathcal{V}_s = \tfrac{1}{2} k_s (r - r_0)^2 \left[1 - k_{cubic} (r - r_0) \right] \qquad (3.4)$$

In this expression, r_0 is the equilibrium bond length, r is the actual bond length, k_s is the stretching stiffness, and k_{cubic} is the magnitude of a cubic correction, for which the MM2 model uses an invariant value of 2×10^{10} m^{-1}. (For ease of use,

Table 3.2. MM2 bond-stretching parameters (Allinger, 1986) for some common bond types; the full set is roughly six times larger.

Bond type (MM2 codes)	k_s (N/m)	r_0 $(10^{-10}$ m)	Bond type
1—5	460	1.113	C—H
1—1	440	1.523	C—C
2—2	960	1.337	C=C
4—4	1560	1.212	C≡C
1—6	536	1.402	C—O
1—8	510	1.438	C—N
3—7	1080	1.208	C=O
1—11	510	1.392	C—F
1—12	323	1.795	C—Cl
1—13	230	1.949	C—Br
1—14	220	2.149	C—I
8—20	610	0.600	N—LP
8—8	560	1.381	N—N
6—20	460	0.600	O—LP
6—21	460	0.942	O—H
6—6	781	1.470	O—O
1—19	297	1.880	C—Si
1—25	291	1.856	C—P
1—15	321.3	1.815	C—S
19—19	185	2.332	Si—Si
22—22 (cyclopropane)	440	1.501	C—C

all MM2 expressions and parameters have been converted into SI units.) Table 3.2 lists some MM2 bond-stretching parameters.

b. ***Bond angle-bending.*** Two bonds to a shared atom define a bond angle (four bonds to a shared atom define six such angles). In the structures for which MM2 provides a good approximation, bond angles can be described as having preferred values with displacements countered by restoring forces; this describes amines poorly,* and can describe trivalent but not pentavalent phosphorus (Burkert and Allinger, 1982). In its description of the potential energy $\mathcal{V}_\theta$ of

---

* Modeling of °lone pairs as pseudoatoms with bond-bending interactions incorrectly prohibits the inversion of °amines, a process in which the lone pair smoothly redistributes to the other face of the group. To picture a lone pair, first imagine a methane molecule, tetrahedral CH_4. Now imagine that one of the H nuclei (a proton) is somehow moved to the carbon nucleus: the result is ammonia, NH_3. What happens to the two electrons that previously formed a CH bond? They remain in the same region, but having no proton to attract them, they spread to occupy more space. This bulge of electron density is the nitrogen lone pair, modeled in MM2 by burying an atomlike object in the outer regions of the N atom. Oxygen is treated as having two lone-pair pseudoatoms, but in fluorine the three lone pairs blend into a distribution treated as symmetric; neon has perfect spherical symmetry.

Table 3.3. MM2 bond angle-bending parameters (Allinger, 1986) for some common bond types; the full set is roughly ten times larger.

Angle type (MM2 codes)	k_θ (aJ/rad^2)	θ_0 (deg)	θ_0 (rad)	Angle type
1—1—1	0.450	109.47	1.911	C—C—C
1—1—5	0.360	109.39	1.909	C—C—H
5—1—5	0.320	109.40	1.909	H—C—H
1—1—11	0.650	109.50	1.911	C—C—F
11—1—11	1.070	107.10	1.869	F—C—F
1—2—1	0.450	117.20	2.046	C—C(sp^2)—C
2—1—2	0.450	109.47	1.911	C(sp^2)—C—C(sp^2)
1—2—2	0.550	121.40	2.119	C—C=C
2—2—5	0.360	120.00	2.094	C=C—H
2—2—2	0.430	120.00	2.094	C=C—C(sp^2)
1—4—4	0.200[a]	180.00	3.142	C—C≡C
1—3—7	0.460	122.50	2.138	C—C=O
1—6—1	0.770	106.80	1.864	C—O—C
1—8—1	0.630	107.70	1.880	C—N—C
1—1—6	0.700	107.50	1.876	C—C—O
1—1—8	0.570	109.47	1.911	C—C—N
1—6—20	0.350	105.16	1.835	C—O—LP
1—8—20	0.500	109.20	1.906	C—N—LP
20—6—20	0.240	131.00	2.286	LP—O—LP
19—19—19	0.350	111.30	1.943	Si—Si—Si
19—1—19	0.400	115.50	2.016	Si—C—Si
1—19—1	0.480	110.80	1.934	C—Si—C
12—1—12	1.080	111.70	1.950	Cl—C—Cl
1—1—15	0.550	109.00	1.902	C—C—S
1—15—1	0.720	96.30	1.902	C—S—C

[a] Allinger and Pathiaseril (1987).

bond angle-bending, MM2 includes a sextic term

$$\mathcal{V}_\theta = \tfrac{1}{2} k_\theta (\theta - \theta_0)^2 \left[1 + k_{\text{sextic}} (\theta - \theta_0)^4 \right] \tag{3.5}$$

In this expression, θ_0 is the equilibrium bond angle, θ is the current bond angle, k_θ is the angular spring constant, and k_{sextic} is the magnitude of a sextic bending constant (taken to be 0.754 rad^{-4} in all cases). Table 3.3 lists some MM2 bond angle-bending parameters.

To compare bond angle-bending to bond stretching, one needs a measure of stiffness in terms of spatial (not angular) displacement. To first order, bond angle-bending can be described by the contribution of the angular stiffness k_θ to an ordinary stiffness $k_{s\perp}$ characterizing displacements of an atom perpendicular to a chosen bond and in the plane of the corresponding angle:

$$k_{s\perp} = k_\theta / r_0^2 \tag{3.6}$$

Table 3.4. Some MM2 parameters for out-of-plane bending.

Angle type (MM2 code ranges)	k_θ (aJ/rad^2)
2–(1...9)	0.05
3–(1...9)	0.80
9–(1...4)	0.05
9–(6...9)	0.05

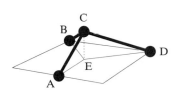

Figure 3.1. Out-of-plane bending geometry in the MM2 model. In calculating angle-bending energies centered on sp^2 atoms (such as C in the above diagram), each angle (such as ACB) is described in terms of an in-plane component (the angle AEB) and an out-of-plane component (the angle CDE); the line CE is perpendicular to the ABD plane.

Bond angle-bending stiffnesses are substantially less than bond-stretching stiffnesses. For example, in the bending of a C—C—C bond angle, $k_{s\perp} = 19.4$ N/m, or about 1/20 the stretching stiffness of a C—C bond (all carbons $^\circ sp^3$). In an sp^3 tetrahedral geometry, bending of one bond with respect to the other three is characterized by a stiffness 1.5 times greater than this.

Systems containing sp^2 atoms have a preferred plane and are subject to a restoring force when bonds are bent away from this plane. Table 3.4 lists some force constants for out-of-plane bending. Figure 3.1 illustrates the definition of the in-plane and out-of-plane bending angles; the three atoms surrounding an sp^2 carbon (e.g., in formaldehyde) define three angles of each kind.

c. ***Bond torsion.*** Rotation about bonds is not *free rotation* [in the technical sense used by chemists (Mislow, 1989)], but can often permit a freely turning motion in a mechanical engineer's sense. The MM2 potential describes the variation in energy associated with rotation about a bond as a sum of four-atom torsional potentials, in which the torsion angle is defined as shown in Figure 3.2. The six hydrogen atoms in ethane define nine such angles. Each four-atom torsional potential takes the form

$$\mathcal{V}_\omega = \tfrac{1}{2}\left[V_1(1+\cos\omega)+V_2(1-\cos2\omega)+V_3(1+\cos3\omega)\right] \tag{3.7}$$

Some parameters are listed in Table 3.5.

As with bond angle-bending, torsional restoring forces can be described (to first order, for configurations at the bottom of a potential well) by a stiffness associated with small displacements tangent to the torsional motion. The maximum stiffness contribution from a single torsional term for a set of four sp^3 carbons (associated with the torsional displacement of one terminal carbon atom) is about 0.36 N/m. Ordinary torsional stiffness terms (and related forces and energies) are thus about 1/50 the magnitude of bond angle-bending stiffness terms, or about 1/1000 the magnitude of bond-stretching stiffnesses. In nanomechanical structures designed for rigidity, torsional contributions to stiffness are

Table 3.5. MM2 parameters for bond torsion (the full set is roughly 80 times larger).

Torsion type (MM2 codes)	V_1 (maJ)	V_2 (maJ)	V_3 (maJ)	Torsion type
1—1—1—1	1.39	1.88	0.65	C—C—C—C
1—1—1—5	0.00	0.00	1.85	C—C—C—H
5—1—1—5	0.00	0.00	1.65	H—C—C—H
1—1—1—11	0.00	−0.60	6.46	C—C—C—F
1—2—2—1	−0.69	69.47	0.00	C—C=C—C
2—2—2—2	−6.46	55.58	0.00	$C(sp^2)$—C=C—$C(sp^2)$
5—2—2—5	0.00	104.21	0.00	H—C=C—H

Figure 3.2. Torsion geometry.

seldom crucial (any structure in which they *were* a chief source of rigidity would not be very rigid). Double bonds have substantial torsional stiffness, and are an exception to this generalization. The MM2 model includes many torsional energy parameters; MM2/CSC adds additional torsional parameters, and the present work has supplemented this set as needed.

d. Electrostatic interactions. Structures containing atoms of substantially different °electronegativity (i.e., most molecules other than hydrocarbons) have significant bond dipole moments, and ionic structures have charge. MM2 describes electrostatic interactions in molecules by summing dipole-dipole interactions (and so forth) in accord with classical electrostatics, assuming a uniform value of the dielectric constant that can be adjusted to reflect the properties of different surroundings. Since the surrounding structures are inhomogeneous on the same scale as the interacting entities, the assumption of a uniform dielectric constant is a relatively crude approximation.

An example of a strongly polarized bond is C—F, with a dipole moment of $\sim 4.7 \times 10^{-30}$ C·m. If this dipole is treated as resulting from atom-centered fractional charges, the corresponding charge on the F atom is about $-0.2e$. Figures 3.6 and 3.7 include a comparison of the magnitude of electrostatic and nonbonded interactions for C—F, C—Cl, and C—Br groups, calculated using atom-centered fractional charges.

e. Nonbonded interactions. Corresponding to each atom type (Table 3.1) is a set of parameters describing the attractive and repulsive forces experienced by pairs of uncharged, nonbonded atoms. Physicists usually apply the term "van der Waals force" to the attractive component alone, that is, to the London dispersion force (along with various polar interactions between molecules). This volume follows a common usage in computational chemistry, treating polar interactions separately but including the °overlap repulsion (a.k.a. exchange-force, hard-core, Born, or steric repulsion) as part of a single nonbonded potential. The MM2 model describes these overall interactions with a *Buckingham* (or *exp–6*) potential

$$\mathcal{V}_{vdw} = \varepsilon_{vdw}\left[2.48\times10^5\exp\left(-12.5\frac{r}{r_{vdw0}}\right)-1.924\left(\frac{r}{r_{vdw0}}\right)^{-6}\right] \qquad (3.8)$$

where r is the distance between the atoms, the parameter ε_{vdw} for the interaction between atoms 1 and 2 equals $(\varepsilon_{vdw1} + \varepsilon_{vdw2})/2$, and r_{vdw0} equals $r_{vdw1} + r_{vdw2}$. This function has a minimum of $-\varepsilon_{vdw}$ at $r = r_{vdw0}$. The forces between atoms bonded to a common atom (i.e., 1–3 interactions) are included not in the nonbonded interaction energy, but in the bond angle-bending energy; all other pairs are included in the sum. (The above expression has been recast to make ε_{vdw} equal the binding energy at $r = r_{vdw0}$.)

Equilibrium separations between atoms in different molecules (and in larger structures) are usually smaller than the pairwise equilibrium separations defined by the nonbonded interaction parameters: The short-range nature of the exponential component allows only nearest neighbors to make a significant contribution to the repulsive side of the balance, but the r^{-6} forces have a longer range, allowing many atoms to make a significant contribution to the attractive side (Section 3.5). Equilibrium separations shrink accordingly.

The absence of nucleocentric spherical symmetry in bonded atoms leads to certain complications, however. Calculation of nonbonded interactions for sp^3 nitrogen and oxygen atoms in the MM2 model includes the effects of associated lone-pair pseudoatoms. When hydrogen atoms form bonds, an unusually large fraction of their electron density shifts toward the other atom; in calculating nonbonded interactions, MM2 models this by moving the effective position of the hydrogen atom inward to 0.915 of the full bond length. Describing pairwise interactions using the above mean and sum expressions for ε_{vdw} and r_{vdw0} is a rough approximation; for the nonbonded interaction between hydrogen and sp^3 carbon, r_{vdw0} is corrected to 0.982 of the summed value, and ε_{vdw} is corrected to 1.011 of the mean value.

f.   Complications and conjugated systems.   The MM2 model includes a number of small corrections and special cases. For example, when bond angles are reduced, equilibrium bond lengths increase. This is described in the MM2 potential by a stretch-bend interaction term; for bonds A and B forming an angle θ,

$$\mathcal{V}_{s\theta} = k_{s\theta}(\theta-\theta_0)[(r_A-r_{A0})+(r_B-r_{B0})] \qquad (3.9)$$

Some parameters are listed in Table 3.6. A reduction in bond angle by 0.1 rad yields a modest 0.00014 nm change in the equilibrium lengths of the two associated bonds. A larger (but static) correction to equilibrium bond length is applied in the presence of adjacent bonded atoms of differing electronegativity. The extreme case is F, which shortens an adjacent C—C bond by 0.0022 nm, or about 1.4%. The MM2 model includes special-case parameters for three- and four-membered rings, and for °hydrogen bond interactions.

Table 3.6.   Stretch-bend parameters.

Angle type[a]	$k_{s\theta}$ (10^{-9} N/rad)
X—First—Y	1.20
X—Second—Y	2.50
X—First—H	0.90
X—Second—H	−4.00

[a] X, Y = first or second row atoms; First = first row atom; Second = second row atom; H = hydrogen.

A major complication arises in °conjugated double-bond systems, in which the treatment of bonds as entities with properties depending only on their near neighbors breaks down. Where single and double bonds alternate along a chain or ring, delocalization of °pi electrons can greatly affect the potential energy function; benzene is a classic example. MM2 describes conjugated systems by performing a minimal quantum mechanical analysis of the participating pi electrons; it uses the results to estimate the magnitude of fractional bonding between pairs of atoms, and then adjusts the force field parameters (equilibrium bond lengths, stiffnesses, torsional energies, etc.) accordingly. (Further complications in the MM2 model are omitted from this discussion.)

g.   Notes on MM2 in light of MM3.   Known shortcomings of MM2 are discussed in a set of papers describing a preliminary version of the MM3 force field (Allinger et al., 1989; Lii and Allinger, 1989a; Lii and Allinger, 1989b). Aside from problems noted in Sections 3.3.1b and 3.3.2b, the greatest shortcoming of MM2 is its inaccurate prediction of molecular vibrational frequencies; accuracy in this was sacrificed to achieve greater accuracy in describing the energy and geometry of equilibrium configurations. Since frequencies depend on molecular stiffnesses, this failure is of significance to the design of molecular machines.

MM3 succeeds in fitting vibrational frequencies while improving accuracy in other areas by virtue of a more complex functional form (e.g., a stretch-torsion interaction, a cubic bending term, a quartic stretching term) and additional parameters. MM3 predicts greater stiffness than MM2, by factors of ~1.5 or more for angle bending (Table 3.7). Since MM3 is a better model, it is clear that MM2 stiffness values are substantially lower than those of real molecules. Greater stiffness is usually an asset in molecular machinery, hence this defect in MM2 usually provides a conservative bias: it tends to generate false-negative rather than false-positive assessments of device feasibility. An exception to this rule is in structures where angle-bending strain relieves bond stretching: the low angle-bending stiffness of the MM2 model may then result in a false-positive assessment of the stability of a stretched bond.

From a nanomechanical perspective, the other major modification in MM3 (again indicating a shortcoming in MM2) is its treatment of nonbonded interactions. MM3 uses an exp–6 potential, but with 12.0 in place of 12.5 in the exponent [see Eq. (3.8)], and a smaller weighting on the exponential term. The atomic parameters developed thus far have larger atomic radii (by a factor of

Table 3.7. Ratios of MM3 and MM2 stiffness parameters.

Bond-stretching stiffness parameters:	
C—C	1.02
C—H	1.03
Bond angle-bending stiffness parameters:	
C—C—C	1.49
C—C—H	1.64
H—C—H	1.72

~1.07) and smaller energy scale factors (by ~0.55). The net result is a softer interaction, with lower energies, forces, and stiffnesses in the deep repulsive regime; typical differences are on the order of tens of percent. This difference has mixed effects on nanomechanical systems.

The softer interactions decrease the calculated stiffness of bearings and other devices using nonbonded contacts, but the stiffness of these interactions is sensitive to load and hence is subject to design control. Softer, longer-range interactions both decrease phonon transmission and make surfaces smoother, thereby decreasing several drag mechanisms that occur in bearings (Section 10.4.6). So long as a substantial margin of safety is provided in the design of stiff, nonbonded interfaces, this shortcoming of MM2 should only rarely result in false-positive assessments of the feasibility of a nanomechanical system.

3.3.3. Energy, force, and stiffness under large loads

Nanomechanical engineering and conventional chemistry place different demands on potential energy functions and emphasize different properties. Some of these differences are discussed in more detail in Section 4.4.3c, which compares the accuracies demanded by solution-phase and °machine-phase chemistry. Other differences result from the nanomechanical emphasis on force as a controllable parameter and on stiffness as a determinant of positioning errors; in conventional chemistry, stiffness is of interest chiefly as a determinant of vibrational frequencies in spectroscopy, and force is rarely mentioned. Further, many nanomechanical systems apply large forces to bonds and to nonbonded interfaces. Although strained organic molecules can experience large bonded and nonbonded forces, potential functions developed for chemistry must be examined for suitability before applying them to problems of nanomechanical design involving large loads.

a. ***Bonds under large tensile loads.*** For describing bonds under large tensile loads, the MM2 bond-stretching function is clearly inadequate: the cubic term eventually results in a repulsive force of unbounded magnitude. Higher-order terms, unimportant in most molecules, were omitted to reduce computational expense. Where stresses and separations are larger, the potential energy of stretching in covalent bonds commonly is modeled using the Morse function

$$\mathcal{V}_{morse} = D_e \left(\{1 - \exp[-\beta(r - r_0)]\}^2 - 1 \right) \tag{3.10}$$

Table 3.8. Bond dissociation energies, stiffnesses, lengths, and Morse β parameters. Values of r_0 and k_s from Table 3.2.

Bond	D_0 (aJ)	D_e (aJ)	k_s (N/m)	β (10^{10} m^{-1})	r_0 (10^{-10} m)	Compound	ref.
C—H	0.642	0.671	460	1.851	1.113	H—C(CH$_3$)$_3$	a
C—C	0.545	0.556	440	1.989	1.523	CH$_3$—C(CH$_3$)$_3$	b
C=C	1.190	1.207	960	1.994	1.337	H$_2$C=CH$_2$	a
C≡C	1.594	1.614	1560	2.198	1.212	HC≡CH	a
C—N	0.498	0.509	510	2.238	1.438	C$_2$H$_5$—N(CH$_3$)$_2$	b
C—O	0.564	0.575	536	2.159	1.402	C$_2$H$_5$—OCH$_3$	b
C=O	1.327	1.343	1080	2.005	1.208		c
C—F	0.876	0.887	510	1.695	1.392	F—C$_2$H$_5$	a
N—H	0.631	0.664	610	2.143	1.020	H—N(CH$_3$)$_2$	b
N—N	0.405	0.417	560	2.592	1.381	H$_2$N—N(CH$_3$)$_2$	a
O—H	0.725	0.753	460	1.747	0.942	H—OC(CH$_3$)$_3$	b
O—O	0.259	0.272	781	3.789	1.470	(CH$_3$)$_3$CO—OC(CH$_3$)$_3$	a
C—Si	0.616	0.624	297	1.543	1.880	CH$_3$—Si(CH$_3$)$_3$	a
C—P	0.439	0.446	291	1.806	1.856		c
C—S	0.532	0.539	321	1.726	1.815	CH$_3$—SCH$_3$	a
C—Cl	0.583	0.591	323	1.653	1.795	Cl—CH$_3$	b
C—Br	0.482	0.488	230	1.536	1.949	Br—CH$_3$	a
C—I	0.393	0.398	220	1.662	2.149	I—CH$_3$	b
Si—Si	0.555	0.559	185	1.286	2.332	(CH$_3$)$_3$Si—Si(CH$_3$)$_3$	a

[a] Kerr, 1990
[b] McMillen and Golden, 1982
[c] Huheey, 1978

with the corresponding forces and stiffnesses

$$F_{\text{morse}} = -\frac{\partial}{\partial r}\mathcal{V}_{\text{morse}} = 2\beta D_e\left\{\exp[-2\beta(r-r_0)] - \exp[-\beta(r-r_0)]\right\} \quad (3.11)$$

$$k_{s,\text{morse}} = \frac{\partial^2}{\partial r^2}\mathcal{V}_{\text{morse}} = 2\beta^2 D_e\left\{2\exp[-2\beta(r-r_0)] - \exp[-\beta(r-r_0)]\right\} \quad (3.12)$$

In Eqs. (3.10) to (3.12),

$$\beta = \sqrt{k_s/2D_e} \quad (3.13)$$

where the energy D_e represents the depth of the potential well. The value of D_e cannot be directly measured, but is the sum of the energy required to break the bond at zero K, shown as D_0, and the zero point vibrational energy associated with the bond. It can be approximated as

$$D_e \approx D_0 + \frac{\hbar}{2}\sqrt{\frac{k_s}{\mu}} = D_0 + \frac{\hbar\omega}{2}; \quad \mu = \frac{m_1 m_2}{m_1 + m_2}. \quad (3.14)$$

where m_1 and m_2 are the masses of the bonded atoms. The zero-point energy of a harmonic oscillator of frequency ω is $\hbar\omega/2$, and Eq. (3.14) treats a bonded pair of atoms as if it were a harmonic oscillator having the frequency of an analogous isolated diatomic molecule. In the case of a C—C bond, D_e is about $1.02\,D_0$. Table 3.8 lists values for D_e based on tabulated values for bond dissociation energies and the bond stiffness values of the MM2 potential, using Eq. (3.14) to approximate the zero-point energy after correction to zero K. These values can be substantially modified by the surrounding molecular structure.

Figure 3.3 plots the C—C Morse potential based on the values in Table 3.8, along with the MM2 bonded and nonbonded potentials; Figures 3.4 and 3.5 plot additional Morse potentials for easy reference. Although its form is motivated by quantum mechanical considerations, the Morse function grows increasingly inaccurate far from the equilibrium separation. From a structural perspective,

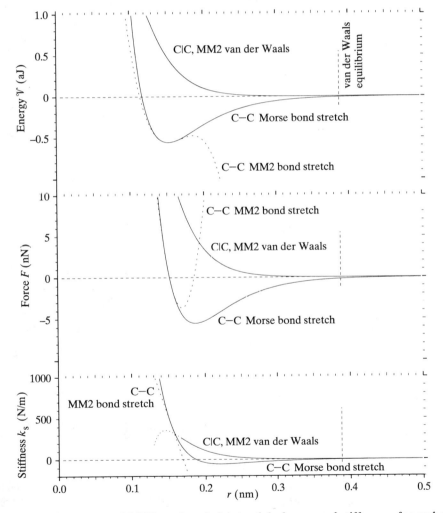

Figure 3.3. Morse and MM2 nonbonded potentials, forces, and stiffnesses for carbon-carbon interactions. The Morse curves are based on parameters for C—C bonds (Table 3.8); a dotted curve shows the MM2 bond-stretching potential for comparison. The curves for C|C contacts are based on parameters from Table 3.1 for sp^2 carbon atoms, which have better-exposed surfaces than do typical sp^3 atoms. Note the breakdown in the MM2 exp–6 model at distances around 0.15 nm; the dotted extensions of solid curves represent clearly nonphysical regions.

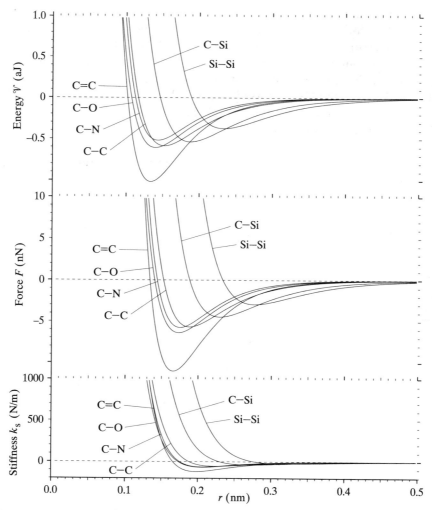

Figure 3.4. Morse potentials, forces, and stiffnesses for a representative sample of bond types useful in structural frameworks. Parameters from Table 3.8.

the shape of the Morse function is of interest chiefly within the separation defined by the inflection point, $r = r_0 + (\ln 2)/\beta$; a Morse bond stretched beyond this length *by a position-independent force* becomes mechanically unstable (for C—C, at ~0.187 nm). Beyond the inflection point, the stiffness becomes negative, reaching a most-negative value of $-0.125 k_s$.

At larger separations, a more accurate potential is the Lippincott function,

$$\mathcal{V}_{\text{lippincott}} = D_e \left[1 - \exp\left(-\frac{k_s r_0 (r - r_0)^2}{2 D_e r} \right) \right], \qquad r \geq r_0 \tag{3.15}$$

which better fits data from vibrational spectroscopy (Eggers et al., 1964) and accurate *ab-initio* molecular orbital calculations (Brown and Truhlar, 1985), although its accuracy is poor for $r < r_0$. In comparison to the Morse function, the Lippincott function commonly predicts a greater magnitude for both the bond breaking force and the most-negative stiffness ($< -0.125 k_s$). Chapter 6 examines

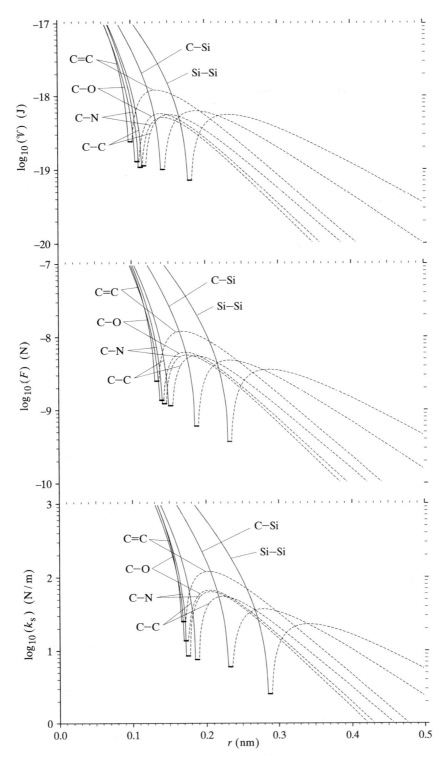

Figure 3.5. The relationships in Figure 3.4 regraphed on a logarithmic scale to facilitate reading of values. Dashed lines represent negative values.

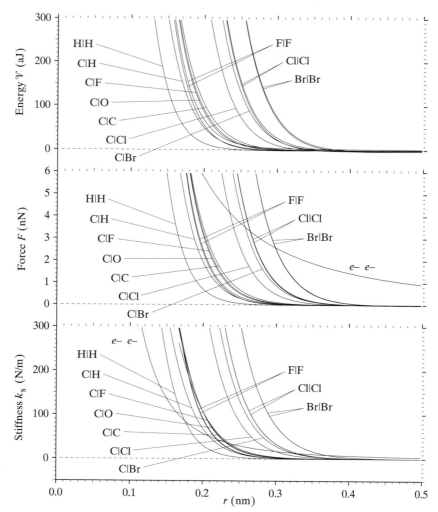

Figure 3.6. MM2 nonbonded potentials, forces, and stiffnesses for a representative sample of pairwise nonbonded interactions (the C|O curves omit the lone-pair contribution). The solid curves for F|F, Cl|Cl, and Br|Br nonbonded interactions are accompanied by dotted curves representing combined nonbonded and electrostatic effects for the collinear dipole systems C—F|F—C, C—Cl|Cl—C, and C—Br|Br—C. Also shown are interactions between two isolated electrons (off scale in the upper graph).

bond strengths using Morse potentials (which underestimate strength); Chapter 8 examines bond-breakage processes using Lippincott potentials (which provide more adverse estimates of stiffness).

b. Nonbonded interactions under large compressive loads. Nonbonded interactions (Figures 3.6 and 3.7) can be divided into multiple components, of which the MM2 model takes account of three: electrostatic forces, attractive van der Waals forces, and overlap repulsion forces. The latter two forces are represented by the r^{-6} and exponential terms of the exp–6 potential, both of which have forms motivated by approximate quantum theories. More thorough treatments of these forces include small attractive r^{-8}, r^{-10} (etc.) terms, three-body interactions, induced dipole effects, different rules of combination for estimating

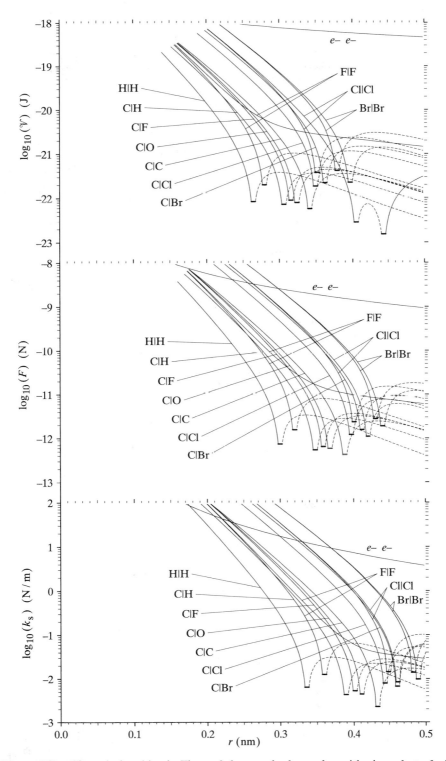

Figure 3.7. The relationships in Figure 3.6 regraphed on a logarithmic scale to facilitate reading of values. Dashed lines represent negative values.

the well depths and equilibrium radii of pairwise interactions, and so forth. Modeling interactions within and between molecules in terms of atom-by-atom pairwise potentials of any kind has no basic theoretical justification and guarantees some inaccuracy. Discussions of alternative models may be found in the literature (Maitland et al., 1981; Rigby et al., 1986). Nonetheless, molecular mechanics potentials including the three components just mentioned have proved accurate enough for a wide range of purposes. The additional interaction terms just mentioned are inconsequential when repulsive forces are dominant, as is common at interfaces within nanomechanical devices.

Several well-known potential forms are unsuitable for analyzing nanomechanical systems that involve large repulsive energies. The Lennard-Jones 6–12 potential, although time honored, has a repulsive interaction that lacks theoretical motivation and is unrealistically steep. The Maitland and Smith potential (Maitland et al., 1981), although excellent in the low-energy range, is again too steep in the deep repulsive regime.

The MM2 expression for nonbonded interactions has two parameters: the equilibrium separation of two atoms and the depth of the well at that point. Well depths commonly are on the order of 1 maJ, but repulsive interaction energies of nanomechanical interest range upward to >100 maJ. This gap raises questions regarding the realism of the MM2 potential at higher energies, as does the behavior of the MM2 potential at distances below $0.323 r_{\text{vdw0}}$, where the exponential repulsion is overwhelmed by a nonphysical extension of the r^{-6} attraction. The general realism of the MM2 exp–6 function at intermediate separations and high energies can be tested by comparing its description of noble gas interactions [using well depths and equilibrium separations drawn from Rigby et al. (1986)] to the interactions measured in collision experiments using neutral particle beams (Amdur and Jordan, 1966). This comparison indicates that the MM2 potential approximates the experimentally measured potential reasonably well (within tens of percent) at radii as small as $0.5 r_{\text{vdw0}}$, and thus at energies >100 maJ. This is an ample range and accuracy for most nanomechanical design work.

As is detailed in Chapter 5, stiff interactions are often necessary in order to minimize quantum and thermal uncertainties in position. Nanomechanical components frequently are constrained not by bonds, but by overlap repulsion. This makes nonbonded forces and stiffnesses important. In the MM2 approximation

$$
\begin{aligned}
F_{\text{vdw}} &= -\frac{\partial}{\partial r} \mathcal{V}_{\text{vdw}} \\
&= \varepsilon_{\text{vdw}} \left[\frac{3.1 \times 10^6}{r_{\text{vdw0}}} \exp\left(-12.5 \frac{r}{r_{\text{vdw0}}} \right) - \frac{11.54}{r} \left(\frac{r}{r_{\text{vdw0}}} \right)^{-6} \right]
\end{aligned}
\tag{3.16}
$$

and

$$
\begin{aligned}
k_{\text{s,vdw}} &= \frac{\partial^2}{\partial r^2} \mathcal{V}_{\text{vdw}} \\
&= \varepsilon_{\text{vdw}} \left[\frac{3.88 \times 10^7}{r_{\text{vdw0}}^2} \exp\left(-12.5 \frac{r}{r_{\text{vdw0}}} \right) - \frac{80.81}{r^2} \left(\frac{r}{r_{\text{vdw0}}} \right)^{-6} \right]
\end{aligned}
\tag{3.17}
$$

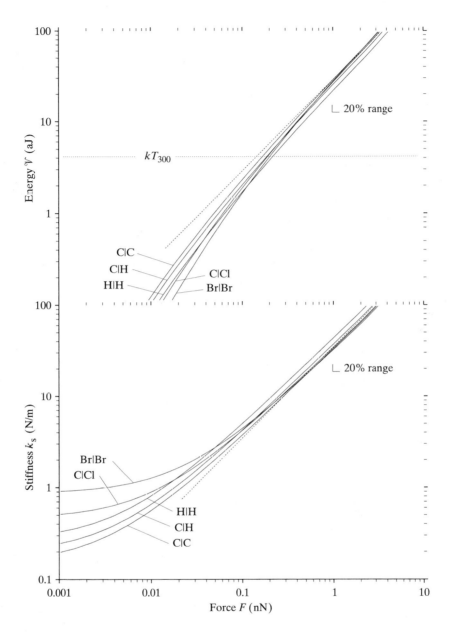

Figure 3.8. MM2 nonbonded potential energies and stiffnesses as a function of the pairwise force for a range of pairwise interactions. The dotted diagonal lines represent Eq. (3.18) (bottom) and Eq. (3.19) (top) for $r_{vdw0} = 0.36$ nm.

Figures 3.6 and 3.7 plot energy, force, and stiffness as functions of distance for various pairwise nonbonded interactions. In the repulsive regime, stiffness increases with decreasing separation and increasing force; the achievable compressive force commonly limits the achievable stiffness. Figure 3.8 plots stiffness and energy as functions of compressive force for a representative set of pairwise interactions. Where the exponential term dominates, stiffness, force, and energy are proportional. For strongly repulsive interactions in the MM2 model

$$k_{s,vdw} \approx \frac{12.5}{r_{vdw0}} F_{vdw} \approx 3.5 \times 10^{10} F_{vdw} \tag{3.18}$$

A similar expression describes the energy in the strongly repulsive regime:

$$\mathcal{V}_{vdw} \approx 0.08 \, r_{vdw0} F_{vdw} \approx 2.9 \times 10^{-11} F_{vdw} \tag{3.19}$$

Under these conditions, the wide range of ε_{vdw} values (a factor of ~10) is of no significance. Only the smaller range of r_{vdw0} values (a factor of ~1.5, omitting the MM2 lone-pair pseudoatom) affects the stiffness and stored energy resulting from a given compressive load. Note that the form of these relationships makes it a matter of indifference whether a compressive load is concentrated on a single atom or spread over many, so long as each single-atom load is large enough for the approximation to apply.

Chemists define various atomic sizes, including covalent, ionic, and van der Waals radii. These have the property that, under the relevant conditions (covalent bonding, ionic contact, and zero-load nonbonded contact) the distance between atoms of any two types can be approximated as the sum of their separate radii. For nanomechanical work, it is convenient to define analogous summable nonbonded radii for atoms under mechanical load. The following

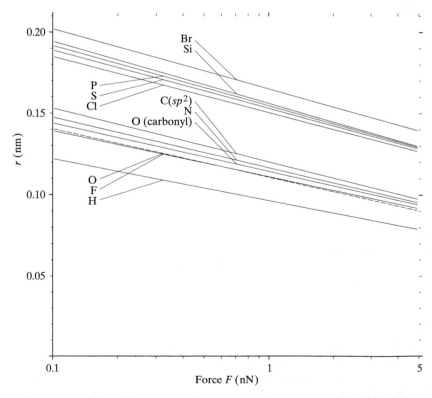

Figure 3.9. Approximate summable nonbonded radii, based on Eq. (3.20): for a given force between two atoms, their separation is approximately the sum of the above radii.

approximate expression defines loaded contact radii as a function of the applied force F

$$r_{\text{loaded}}(F) \approx \frac{r_{\text{vdw}}}{13.6} \ln\left(\frac{2.5 \times 10^6 \varepsilon_{\text{vdw}}}{r_{\text{vdw}} F} \right) \tag{3.20}$$

In the range $F = 0.1$ to 5 nN, this approximation yields separations within 4% of the values implied by the MM2 nonbonded interaction model for all pairs of atoms drawn from the list in Table 3.1 save for those including iodine or the lone-pair pseudoatom. These functions are graphed in Figure 3.9.

3.4. Potentials for chemical reactions

3.4.1. Relationship to other methods

Because molecular-mechanics methods (Section 3.3) are based on the notion of structures with well-defined bonds, they cannot describe transformations that make or break bonds, and cannot predict chemical instabilities (discussed in Chapter 6). Potential energy functions describing chemical reactions have been the subject of separate study (reviewed in Truhlar and Steckler, 1987). Techniques that combine molecular mechanics potentials for describing large structures with reaction potentials [or with quantum mechanical methods applied to small regions (Singh and Kollman, 1986)] are useful in describing nanomechanisms that make and break bonds.

Bond cleavage and formation present computational challenges for molecular orbital methods. Accurate calculations often require extensive use of CI methods to account for electron correlation effects, raising the cost of computations and preventing computations of useful accuracy on complex systems. Even where computations are feasible at many points on the PES, subsequent analytical studies often demand that the surface be described by some function fitted to those points. Accordingly, studies of molecular reaction dynamics usually rely on approximate potential energy surfaces. These are described either by fitting complicated functions to quantum mechanical calculations, or by adjusting a few parameters in a fixed functional form to make calculated reaction rates (and their temperature dependence) fit experimental data. Only the simplest reactions and PES approximations are described here; a growing literature is concerned with elaborating potentials for polyatomic reaction dynamics (Truhlar and Steckler, 1987).

3.4.2. Bond cleavage and radical coupling

The simplest reactions break a single bond and yield two °radicals that undergo no subsequent rearrangement. These reactions have potential energy surfaces that are extensions of bond stretching and bond angle-bending. The Morse function, Eq. (3.10), can serve as an approximate potential for homolytic reactions (i.e., bond cleavage yielding a pair of radicals rather than a pair of ions). For the design of reactive nanomechanisms, the important region of the Morse function lies between the bottom of the potential well ($r = r_0$) and the distance at which the stiffness is most negative [which occurs at $r = r_0 + (\ln 4)/\beta$]. Toward the outer end of this range, however, the Lippincott potential appears to give a more accurate description.

Bond formation by radical coupling (the inverse of homolytic bond cleavage) requires paired, antiparallel electron spins (termed a °*singlet* state, for reasons rooted in spectroscopy). With paired spins, this process follows a Morse potential; systems with parallel, unpaired spins (*triplet* states) experience a potential approximated by the repulsive, anti-Morse function

$$\mathcal{V}_{\text{anti-morse}} = \frac{1}{2} D_e \left(\{1 + \exp[-\beta(r - r_0)]\}^2 - 1 \right) \tag{3.21}$$

and form no bond. Energetic considerations favor spin pairing and bond formation, but the pairing of initially unpaired spins (an electronic transition between potential surfaces termed *intersystem crossing*) can be slow on a molecular time scale.

3.4.3. Abstraction reactions

More complex reactions make and form bonds simultaneously. The most studied class involves the transfer of a single atom from a molecule to a radical, such as the symmetrical hydrogen °abstraction reaction

$$\text{H}_3\text{C} + \text{HCH}_3 \rightarrow \text{H}_3\text{CH} + \text{CH}_3 \tag{3.22}$$

Abstraction reactions often have been modeled using the *London–Eyring–Polanyi–Sato* (LEPS) potential (Sato, 1955), or using the related *extended-LEPS* potential (Kuntz et al., 1966). The latter is based on the expression

$$\mathcal{V}_{\text{LEPS}} = \frac{Q_{\text{AB}}}{1 + S_{\text{AB}}} + \frac{Q_{\text{BC}}}{1 + S_{\text{BC}}} + \frac{Q_{\text{AC}}}{1 + S_{\text{AC}}} \\ - \sqrt{\frac{1}{2}\left[\left(\frac{J_{\text{AB}}}{1 + S_{\text{AB}}} - \frac{J_{\text{BC}}}{1 + S_{\text{BC}}} \right)^2 + \left(\frac{J_{\text{BC}}}{1 + S_{\text{BC}}} - \frac{J_{\text{AC}}}{1 + S_{\text{AC}}} \right)^2 + \left(\frac{J_{\text{AC}}}{1 + S_{\text{AC}}} - \frac{J_{\text{AB}}}{1 + S_{\text{AB}}} \right)^2 \right]} \tag{3.23}$$

a function of the interatomic distances r_{AB}, r_{BC}, and r_{AC} (see following expressions) which reduces to the LEPS potential when the Sato parameters S_{AB}, S_{BC}, and S_{AC} are all equal, and further reduces to the London equation when all three equal zero. In this expression, A, B, and C represent the three atoms directly participating in bond breaking and formation, and the energies Q_{XY} and J_{XY} represent the *coulomb* and *exchange* integrals between atoms X and Y (these quantities arise in quantum mechanical descriptions of bonding). In the LEPS model, bonding interactions between atoms X and Y are described by the Morse function

$$\frac{Q_{\text{XY}} + J_{\text{XY}}}{1 + S_{\text{XY}}} = D_{e\text{XY}} \left(\{1 - \exp[-\beta_{\text{XY}}(r_{\text{XY}} - r_{0\text{XY}})]\}^2 - 1 \right) \tag{3.24}$$

and antibonding interactions are described by the anti-Morse function

$$\frac{Q_{\text{XY}} - J_{\text{XY}}}{1 - S_{\text{XY}}} = \frac{1}{2} D_{e\text{XY}} \left(\{1 + \exp[-\beta_{\text{XY}}(r_{\text{XY}} - r_{0\text{XY}})]\}^2 - 1 \right) \tag{3.25}$$

yielding the expressions

$$Q_{XY} = \frac{1}{2} D_{eXY} \left\{ \frac{1}{2} (S_{XY} + 3) \exp[-2\beta_{XY}(r_{XY} - r_{0XY})] \right. $$
$$\left. - (3S_{XY} + 1) \exp[-\beta_{XY}(r_{XY} - r_{0XY})] \right\} \tag{3.26}$$

and

$$J_{XY} = \frac{1}{2} D_{eXY} \left\{ \frac{1}{2} (3S_{XY} + 1) \exp[-2\beta_{XY}(r_{XY} - r_{0XY})] \right. $$
$$\left. - (S_{XY} + 3) \exp[-\beta_{XY}(r_{XY} - r_{0XY})] \right\} \tag{3.27}$$

The bond properties of isolated molecules define the pairwise Morse parameters. For any given choice of Morse and Sato parameters, the LEPS potential can be expressed as a function of the three interatomic distances, r_{AB}, r_{BC}, and r_{AC}; for a collinear system, it can be expressed as a function of two such distances. In the limit as one atom is removed to infinity while the other pair remains in close proximity, the LEPS potential reduces to the Morse potential for that pair of atoms.

Despite (and because of) their limited flexibility, LEPS and extended LEPS potentials have been a basis for work in molecular reaction dynamics (Levine and Bernstein, 1987) and transition state theory (Bérces and Márta, 1988). They are used in Section 8.5.4a to examine the effects of mechanical forces on abstraction reactions. A wide range of alternative functional forms has been explored in the literature, but improved fits are usually purchased at the expense of substantially greater mathematical complexity (Bérces and Márta, 1988).

3.5. Continuum representations of surfaces

Nanomechanical components often are subject to forces dominated by bonding and overlap repulsion, but van der Waals attractions (London dispersion forces) between nanometer-scale objects can also be substantial. The long-range nature of these forces motivates the use of continuum approximations. At small spacings, surface-surface interactions require a model that separately accounts for the first atomic layer, including overlap forces.

3.5.1. Continuum models of van der Waals attraction

a. The Hamaker constant.   In terms of the MM2 parameters, the long-range interatomic pairwise potential is

$$\mathcal{V}_{vdw} = -\frac{C}{r^6}; \quad C = 1.924 \varepsilon_{vdw} r_{vdw\,0}^6 \tag{3.28}$$

For interactions involving a pair of solid bodies, the continuum model merges the contributions of the individual atoms and describes them in terms of the Hamaker constant, $\mathcal{A}$. In the simplest description, the interacting materials are assumed to be separated by vacuum, and only nonretarded pairwise interactions are considered.* The Hamaker constant can then be expressed as

$$\mathscr{A}_{12} = \pi^2 C n_{v1} n_{v2} \tag{3.29}$$

where n_{v1} and n_{v2} are the number densities of atoms in the interacting bodies. (Where a body contains several atom types, it can be treated as the superposition of several bodies each with a single atom type.) This and the more rigorous Lifshitz theory are described in Israelachvili (1992). Values of the Hamaker constant for condensed media are typically in the range of 30 to 500 maJ.

For nonpolar insulators of different Hamaker constant interacting across vacuum,

$$\mathscr{A}_{12} \approx \sqrt{\mathscr{A}_{11} \mathscr{A}_{22}} \tag{3.30}$$

(Note that the MM2 model uses the arithmetic mean for calculating pairwise dispersion interactions between unlike atoms; the geometric mean has better theoretical justification and has been adopted in MM3.) For interactions between nonpolar insulators 1 and 2 across a medium 3,

$$\mathscr{A}_{132} \approx \left(\sqrt{\mathscr{A}_{11}} - \sqrt{\mathscr{A}_{33}}\right)\left(\sqrt{\mathscr{A}_{22}} - \sqrt{\mathscr{A}_{33}}\right) \tag{3.31}$$

When one or more phases are polar (e.g., water), an estimate of the Hamaker constant can be derived from an approximation to the Lifshitz theory (Israelachvili, 1992), which includes contributions from dipole-dipole interactions:

$$\mathscr{A}_{132} \approx \frac{3\hbar\omega}{8\sqrt{2}} \frac{\left(n_1^2 - n_3^2\right)\left(n_2^2 - n_3^2\right)}{\left(\sqrt{n_1^2 + n_3^2} + \sqrt{n_2^2 + n_3^2}\right)\sqrt{\left(n_1^2 + n_3^2\right)\left(n_2^2 + n_3^2\right)}} \\ + \frac{3}{4}kT\left(\frac{\varepsilon_1 - \varepsilon_3}{\varepsilon_1 + \varepsilon_3}\right)\left(\frac{\varepsilon_2 - \varepsilon_3}{\varepsilon_2 + \varepsilon_3}\right) \tag{3.32}$$

In this expression, n is the optical refractive index, ε is the zero-frequency dielectric constant, and ω is the absorption frequency of the medium (or where frequencies differ, the mean). For many materials, $\hbar\omega \approx 2$ aJ. Table 3.9 lists values of n and ε for several materials, together with the resulting Hamaker constants for interactions across vacuum between two bodies of identical composition.

b. Interactions between objects. Figure 3.10 presents approximate expressions for the attractive potential between objects of various shapes. Note that the potential energy always diverges as the distance $s \to 0$; a minimum separation $s_{min} \approx 0.2$ nm commonly is assumed in the literature. This can be rationalized in terms of a ~0.1 nm gap between a fictitious *Hamaker surface* (which bounds the

---

* The Schrödinger equation in the form given by Eq. (3.1) assumes electro*static* interactions between all particles everywhere, which cannot be quite right: electrons move, and the speed of light is finite, and so the fields must lag. In a theoretician's world in which fields propagate instantaneously, interactions are termed *nonretarded* and often have simpler mathematical descriptions. Owing to retardation, London dispersion forces at large separations fall off as r^{-7} rather than the r^{-6} stated by Eq. (3.28); this is described by the *Casimir–Polder* equation. In solution, retardation effects are important at separations greater than ~5 nm (Israelachvili, 1992).

Table 3.9. Refractive index, zero-frequency dielectric constant, and Hamaker constant [from Eq. (3.32)] for several materials; $\hbar\omega$ taken as 2 aJ.

Material	n	ε	$\mathscr{A}_{11}$ (maJ)
Polytetrafluoroethylene	1.35	2.1	38
Polyethylene	1.52	2.3	76
Diamond	2.40	5.5	340
Fused silica	1.49	3.8	69
Water	1.33	78.5	37
Glycerol	1.47	42.5	66
Metals (Au, Ag, Cu)			300 to 500[a]

[a] Hamaker constants for metals from Israelachvili (1992). Values for n, ε from Gray (1972), Lide (1990).

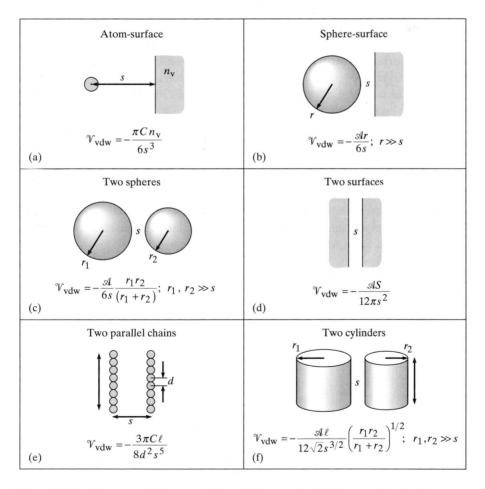

Atom-surface

$$\mathscr{V}_{\text{vdw}} = -\frac{\pi C n_{\text{v}}}{6s^3}$$

(a)

Sphere-surface

$$\mathscr{V}_{\text{vdw}} = -\frac{\mathscr{A}r}{6s}; \quad r \gg s$$

(b)

Two spheres

$$\mathscr{V}_{\text{vdw}} = -\frac{\mathscr{A}}{6s}\frac{r_1 r_2}{(r_1 + r_2)}; \quad r_1, r_2 \gg s$$

(c)

Two surfaces

$$\mathscr{V}_{\text{vdw}} = -\frac{\mathscr{A}S}{12\pi s^2}$$

(d)

Two parallel chains

$$\mathscr{V}_{\text{vdw}} = -\frac{3\pi C \ell}{8d^2 s^5}$$

(e)

Two cylinders

$$\mathscr{V}_{\text{vdw}} = -\frac{\mathscr{A}\ell}{12\sqrt{2}\,s^{3/2}}\left(\frac{r_1 r_2}{r_1 + r_2}\right)^{1/2}; \quad r_1, r_2 \gg s$$

(f)

Figure 3.10. The potential energy, $\mathscr{V}_{\text{vdw}}$ of the van der Waals attraction in the nonretarded continuum approximation; after Israelachvili (1992). In the two-surface expression, S = area. The more complex expressions for geometries with $r \approx s$ appear in Hiemenz (1986).

region of high polarizability in an object) and an approximate *excluded-volume surface* (describing the limits of motion imposed by overlap repulsion). The gap occurs because polarizability chiefly arises from regions of high electron density within bonds and atomic core regions, while the overlap repulsion is substantial even outside these regions. Section 3.5.2 describes flat-surface models that include the effects of overlap repulsion.

Taking $\mathscr{A} = 400$ maJ (a fairly high value) and $s = 0.2$ nm (i.e., contact), the attractive dispersion force between two spheres of $r = 1$ nm is ~0.83 nN. For parallel surfaces at this separation, the force per unit area is ~2.7×10^9 N/m^2 (= 2.7 nN/nm^2), ~1/20 the tensile strength of diamond. Doubling s (for example, by interposing scattered atomic-scale bumps as spacers) reduces this force to ~3.3×10^8 N/m^2.

3.5.2. Transverse-continuum models of surfaces

Chapters 7 and 10 examine the mechanical properties of sliding, unreactive, atomically precise interfaces in which van der Waals attraction and overlap repulsion are the dominant forces.* This requires a model for the interaction energy of two such surfaces at small separations. In the interfaces of greatest practical interest, the structures of the two surfaces are out of register (owing to differences in lattice spacing or to a relative rotation). Neglecting elastic deformations resulting from interfacial forces, this results in a surface-surface potential that can be approximated in terms of the interactions of layers in which the effects of the constituent atoms have been spread uniformly over a plane. In a further approximation, all but the outermost layers can be spread uniformly in the third dimension and their overlap repulsions (which are subject to rapid exponential decay) can be neglected, yielding the standard Hamaker-constant model for interactions involving the interior regions (Figure 3.11).

With this model, the total surface-interaction energy is the sum of the pairwise interactions of the two explicitly treated planes, of each plane and the opposed continuum, and of the two continua:

$$\mathscr{V}_{\text{surf}} = \mathscr{V}_{\text{a1-a2}} + \mathscr{V}_{\text{a1-v2}} + \mathscr{V}_{\text{a2-v1}} + \mathscr{V}_{\text{v1-v2}} \tag{3.33}$$

with obvious generalizations when more layers of atoms are treated explicitly.

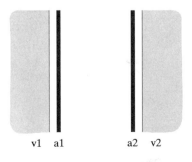

v1   a1   a2   v2

Figure 3.11. Model of two surfaces, showing the outermost atomic layers described as uniform planes and deeper regions approximated as uniform volumes.

---

* This is not universal: metal surfaces (for example) would undergo bonding rather than repulsion; potential functions are discussed in Israelachvili (1992).

The plane-plane interaction energy can be calculated from integrals describing the point-plane interaction energy, using an exp–6 potential. In terms of the densities n_{a1} and n_{a2} of atoms in the two planes, the energy per unit area is

$$\frac{\mathcal{V}_{a\text{-}a}}{S} = n_{a1}n_{a2}\pi\left[\frac{2A}{b^2}(bs+1)\exp(-bs_{a\text{-}a}) - \frac{C_{a\text{-}a}}{2s_{a\text{-}a}^4}\right] \tag{3.34}$$

where C is the same as in Eq. (3.28), and A and b are parameters for the exponential component of the interatomic interaction, when expressed in the form

$$\mathcal{V}_{\text{rep}} = A\exp(-bs) \tag{3.35}$$

In the MM2 model,

$$A = 2.48\times10^5\,\varepsilon_{\text{vdw}}; \quad b = \frac{12.5}{d_{\text{vdw0}}} \tag{3.36}$$

For all other interactions, overlap repulsion is neglected. The interaction between atoms in a plane and in a volume [from Eq. (3.29) and the expression in Figure 3.10a] contributes

$$\frac{\mathcal{V}_{a\text{-}v}}{S} = -n_{a1}n_{v2}\frac{\pi C_{a\text{-}v}}{6s_{a\text{-}v}^3} \tag{3.37}$$

The interaction between the two continuum regions [from Eq. (3.29) and the expression in Figure 3.10d] contributes

$$\frac{\mathcal{V}_{v\text{-}v}}{S} = -n_{v1}n_{v2}\frac{\pi C_{v\text{-}v}}{12s_{v\text{-}v}^2} \tag{3.38}$$

The gap d_g between the plane and the surface of the continuum is chosen so as to count the effects of each atom exactly once, in the long-range attraction limit. Figure 3.12 graphs the energy, force, and stiffness resulting from this model for several sets of parameters.

3.5.3. Molecular models and bounded continuum models

Molecular mechanics methods can be used to calculate the mechanical properties of °diamondoid bulk materials, and to relate bulk properties (such as elastic °modulus) to the properties of nanoscale components of similar structure (Section 9.4). The mechanical behavior of stable diamondoid structures is often well approximated by °linear elastic models; the atomic details of their interiors are then irrelevant to their external properties and can be neglected in much of the design process. Note that such approximations are of little use in standard organic chemistry, where large, stiff, regular structures are rare, and likewise in °protein chemistry, where structures are flexible and highly inhomogeneous.

For irregular surfaces in contact, there often is no substitute for modeling at the level of interatomic potentials. Nonetheless, Eqs. (3.18) and (3.19) provide approximate formulas for the stiffness and energy of repulsive interactions as a function of the applied compressive force. These interactions are relatively insensitive to details of surface structure (within the class of structures of chief

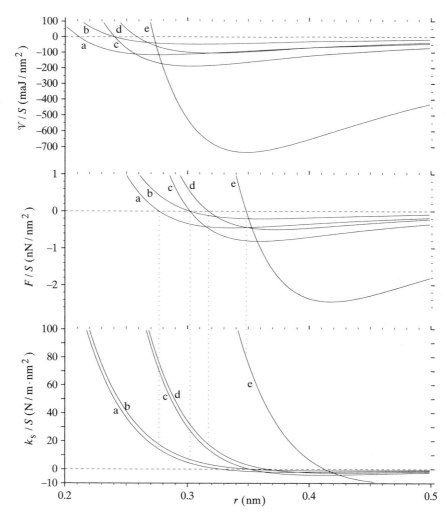

Figure 3.12. Interaction of pairs of surfaces according to Eqs. (3.33)–(3.38), with the following parameters (identical for both surfaces):

Curve	n_a $(10^{19}/m^2)$	Atom type (MM2)	n_v $(10^{29}/m^3)$	Atom type (MM2)	d_g $(10^{-9} m)$
a	0.9	C(1)	1.50	C(1)	0.09
b	0.9	C(1)	0.75	C(1)	0.09
c	1.8	C(1)	1.50	C(1)	0.09
d	1.8	C(1)	0.75	C(1)	0.09
e	1.4	Si(19)	1.50	C(1)	0.13

interest here). In the attractive regime, the comparatively long-range nature of van der Waals forces motivates the use of a continuum approximation, leading to the formulas presented in Figure 3.10. Based on the these observations, Chapter 9 develops bounded continuum models that permit a limited but atomically motivated return to continuum approximations like those used throughout macromechanical engineering.

3.6. Conclusions

The potential energy surface of a ground-state system determines its mechanical properties. Molecular orbital methods familiar in chemistry can provide descriptions of the PES for analyzing mechanochemical processes and small nanomechanical structures, but computational costs grow steeply with system size. For designing nanomechanical components, molecular mechanics models such as MM2 are appropriate, available, and computationally affordable. Molecular mechanics methods and models based on them are used throughout later chapters. Specialized potential energy functions for bond cleavage and abstraction reactions are useful in analyzing some mechanochemical processes, and are applied in Chapter 8. The MM2 potential energy function has here been used to develop a PES for surface-surface interactions that omits atomic detail, providing a basis for continuum models that describe the interaction of adjacent nanoscale components. These models are developed further in Chapters 9 and 10.

3.7. Further reading

This chapter and the next outline topics that are foundational not only to molecular nanotechnology, but to conventional chemistry and chemical physics, hence an extensive literature already exists. This section suggests some points of departure for those wishing to explore these topics in more depth.

Molecular quantum mechanics.   The foundations of molecular quantum mechanics have not changed in decades, and many textbooks teach the subject. Examples include *Atoms and Molecules* (Karplus and Porter, 1970) and *Molecular Quantum Mechanics* (Atkins, 1970). A useful volume for orientation to the subject is *Quanta: a handbook of concepts* (Atkins, 1974), a cross-referenced, alphabetically organized work containing essays, equations, and diagrams on most quantum mechanical phenomena in molecules.

Computational methods in quantum mechanics have advanced greatly over the decades. An introduction to the subject from a user's perspective is *A Handbook of Computational Chemistry* (Clark, 1985); this volume includes concrete examples of the use of popular programs along with less perishable information on techniques and pitfalls in the field. The methods of *ab initio* molecular orbital theory are described in *Ab Initio Molecular Orbital Theory* (Hehre et al., 1986) together with descriptions of numerous computational results.

Potential energy surfaces.   A general introduction to molecular mechanics methods is given by *Molecular Mechanics* (Burkert and Allinger, 1982). *A Handbook of Computational Chemistry* (Clark, 1985) includes descriptions of MM2, although its main focus is on quantum chemistry.

Molecular mechanics methods are evolving rapidly. A common operation is to find a local minimum on the PES; so-called Newton methods have usually required storage space proportional to N^2 (where N is the number of atoms) and a computational time proportional to N^3, but more recent algorithms have better scaling laws: $N^{1.5}$ and N^2 respectively (Ponder and Richards, 1987a). For engineering work, fast, rough approximations are of considerable value in the early stages of design and analysis; a method has been reported that gives results similar to MM2 using simpler, faster code based purely on pairwise, interatomic central forces (Saunders and Jarret, 1986). Quantum chemistry methods are being used to improve molecular mechanics models, for example, in work supported by the Consortium for Research and Development of Potential Energy Functions (Hagler et al., 1989).

Intermolecular forces are complex, though most are weak enough to be neglected in nanomechanical engineering. A good introduction is *The Forces Between Molecules* (Rigby et al., 1986); a more advanced book is *Intermolecular Forces: Their Origin and Determination* (Maitland et al., 1981). An excellent text that emphasizes interactions between objects and forces in liquids is *Intermolecular and Surface Forces* (Israelachvili, 1992). The broad literature on surface science and surface chemistry contains much that is relevant, but often focuses on surfaces that are unstable and reactive; these need not be used in nanomechanical devices.

Potential energy surfaces for atom-transfer (abstraction) reactions are discussed in Levine and Bernstein (1987) and at greater length in Bérces and Márta (1988); both include extensive references to the literature. Truhlar and Steckler (1987) review potential energy surfaces for more complex reacting systems.

4

Molecular Dynamics

4.1. Overview

Molecular dynamics is fundamental to molecular machinery and has received extensive study in physical chemistry and chemical physics. Chapters 5–8 all deal with specific aspects of molecular dynamics; the present chapter provides a brief overview of the topic, examining the applicability of various approaches to nanomechanical problems. Section 4.2 reviews methods used in nonstatistical descriptions of molecular dynamics, considering both quantum mechanical and classical models for the calculation of system trajectories. Section 4.3 reviews statistical descriptions of molecular dynamics (both classical and quantum mechanical), including a discussion of the relationships among °entropy, information, and computation. Section 4.4 returns to the issue of PES approximations, using dynamical principles to examine the differing requirements for accuracy that arise in different applications. As in the previous chapter, suggestions for further reading are appended.

4.2. Nonstatistical mechanics

This section outlines several nonstatistical descriptions of molecular dynamics used in scientific work, commenting on their applicability to nanomechanical engineering problems.

4.2.1. Vibrational motions

Molecular vibrations have been extensively studied in connection with infrared spectroscopy, and observed infrared vibrational frequencies are a major constraint used in determining parameters for molecular mechanics energy functions. Small displacements of a stiff structure from its equilibrium geometry are associated with nearly linear restoring forces. This permits the use of a harmonic approximation, in which the vibrational dynamics can be separated into a set of independent normal modes and the total motion of the system can be treated as a linear superposition of normal-mode displacements. This common approximation is of considerable use in describing nanomechanical systems.

Since both classical and quantum mechanics permit exact solutions for the harmonic oscillator, the time evolution of systems can readily be calculated in the harmonic approximation. In real systems, nonlinear effects permit energy exchange among vibrational modes, causing relaxation to thermal equilibrium. The equilibrium state, in turn, is best described by the methods of statistical mechanics. A nonstatistical description of vibrational motion is of interest chiefly during (or within a few relaxation times of) the excitation of a vibrational mode by a nonthermal energy source.

4.2.2.   Reactions and transition rates

a.   Molecular beam experiments.   Vibrations involve motion within a potential well; reactions involve transitions between potential wells. Molecular reaction dynamics has been extensively studied by means of crossed molecular beams, and a major application (and test) of potential energy functions for chemical reactions has been the calculation of vibrational states and angular distributions resulting from reactive scattering.

With high-quality beams of simple molecules, one can observe quantum mechanical oscillations in the angular distribution of scattered trajectories. These result from interference among alternative collision paths that yield indistinguishable outgoing molecular trajectories and states. A broader distribution of initial energies and angles will obliterate these fine features, however, yielding a smooth distribution of product trajectories.

In practice, calculations of molecular reaction dynamics commonly are based on the *quasiclassical approximation* (Levine and Bernstein, 1987). In this approximation, initial molecular states of vibration and rotation are described classically, but are chosen with initial energies and angular momenta that match quantum constraints. Quantum mechanical uncertainties in position are modeled by calculating many trajectories with randomly chosen vibrational and rotational phase angles. Trajectories are then computed by integrating the classical equations of motion. Finally, quantization of outcomes is modeled by lumping final trajectories into bins in phase space such that each bin corresponds to a permissible quantum state of the product. These calculations cannot yield quantum interference patterns or resonances, but give reasonably accurate descriptions of the coarse dynamical features of a reactive collision, such as the overall distribution of scattering angles and vibrational excitations.

b.   Solution and machine phase vs. molecular beams.   Coarse features from a reaction dynamics perspective are, however, fine features from the perspective of a chemist or a nanomechanical engineer. Chemical reactions in nanomechanical systems bear little resemblance to reactions in crossed molecular beams. In an extended solid system subject to relatively slow mechanical motions, reactive molecular components do not encounter one another with well-defined energies and momenta; broad, thermal distributions dominate. Likewise, during the course of the encounter, the reactive components do not form an isolated system with locally conserved energy, momentum, and angular momentum; the reactive system remains coupled to a thermal bath. The notion of a scattering angle is meaningless, and reaction-induced vibrational excitations are quickly thermalized.

In these respects, reactions in nanomechanical systems resemble the solution-phase reactions familiar to chemists. In such systems, the chief concern is with overall reaction rates, not with the details of trajectories. As a consequence, the detailed shape of the PES (crucial to the details of reactive scattering) is of reduced importance. Reaction rate data only weakly constrain PES properties. Given a few properties of the PES, thermally activated reaction rates are calculated using classical or quantum mechanical transition state theories based on statistical mechanics; these are discussed in Chapter 6.

4.2.3. Generalized trajectories

a. Classical dynamical models. Even in the absence of reactive transformations, the motions of typical nanomechanical systems cannot be completely described in terms of vibrations. Like protein chains or molecules in a liquid, these systems typically permit large displacements subject to complex, interacting constraints. Their motions must be described by more general methods.

It is common practice to model the trajectories of such systems by integrating the classical equations of motion, deriving the forces on each atom from approximate potential functions (e.g., molecular mechanics potentials). Where statistical properties are desired, a common approach is to integrate the equations of motion for a substantial time, taking a time average of the quantities of interest. Time steps are typically $\sim 10^{-15}$ s, and current computers have been used to follow the dynamics of $\sim 10^3$ atom systems for $\sim 10^{-9}$ s. These techniques are suitable for describing the short-term dynamics of nanomechanical components.

b. Quantum corrections. Quantum effects on molecular trajectories can be approximated by an adaptation of classical molecular mechanics techniques in which each atom is represented by a circular chain of atoms with suitable interactions; this has been applied to large molecules, such as the protein ferrocytochrome c (Zheng et al., 1988). The effects are, as one would expect, largest for high-frequency vibrations, such as bond stretching, bending, and torsional motions involving hydrogen atoms. From a statistical perspective, thermal motions result in a certain probability density function for the positions of atoms with respect to their surroundings, and quantum effects broaden this probability density function.

c. Dynamical regimes and corresponding models. The dynamical behavior of nanomechanical systems can be partitioned into three categories: (1) thermally excited vibrational motions within potential wells, (2) thermally excited transitions between potential wells, and (3) other motions, driven by nonthermal energy sources. Motions in category 1 can be treated either classically or quantum mechanically; Chapter 5 compares the statistical distributions resulting from classical and quantum models. Chapter 6 examines motions in category 2, again comparing classical and quantum approaches within a statistical framework. Most motions encountered in category 3 are of low frequency and hence little influenced by quantum effects at ordinary temperatures. They can usually be modeled by applying classical dynamics to molecular mechanics potentials, or (on somewhat larger scales) to bounded continuum models (Section 9.3.3).

4.3. Statistical mechanics

Statistical mechanics (also known as statistical or molecular thermodynamics) is most commonly applied to relate macroscopic thermodynamic properties to statistical descriptions of the behavior of large numbers of molecules. En route to describing properties of bulk matter, statistical mechanics frequently describes probability distributions for underlying molecular variables, such as position and velocity. In the present context, it is these probabilistic descriptions of the behavior of individual molecular objects that are of primary value.

Statistical mechanics can frequently provide estimates of the statistical behavior of nanomechanical systems without the computational cost of running a detailed dynamical simulation for long periods of time. Today, it is expensive to do a simulation of 10^3 atoms for as long as 10^{-9} s, yet a nanomechanism that fails even once per millisecond may be unacceptable. Estimating failure rates by observing more than 10^6 expensive simulations would be impractical. Statistical mechanics, in contrast, can provide estimates for the frequencies of extremely rare events using efficient analytical methods.

Statistical mechanics is commonly used to calculate quantities such as pressure, entropy, and free energy based on averages taken over many molecules in thermal equilibrium. But can temperature, for example, be used to describe a single molecule? There is no fundamental difference between (1) the statistical distribution of a dynamical quantity computed for many similar molecules at a particular instant (which is the usual concern of statistical mechanics), (2) the statistical distribution computed for a single representative molecule over a long period of time, and (3) the probability distribution for a single molecule at a single time. Accordingly, the concepts of pressure, entropy, free energy, and the like can be used to reason about (for example) the expected mean efficiency of a single nanomachine in a single operational cycle. The only caveat stems from the assumption of equilibrium; this is discussed in Section 4.3.4, and the subtle relationship between measurement and equilibrium is discussed in Section 4.3.5.

4.3.1. Detailed dynamics vs. statistical mechanics

By omitting dynamical details, statistical mechanics provides a simplification that can assist both calculation and understanding. In the operation of real nanomechanical systems, the initial conditions are never known with the precision assumed in classical dynamical models, and seldom with the precision assumed in quantum dynamical models. Instead, the motions and displacements resulting from thermal excitation are random variables subject to some distribution. Rather than introducing arbitrary assumptions, statistical mechanics takes these uncertainties as fundamental, yielding inherently probabilistic descriptions of system behavior.

The nanomechanical engineer's task, then, is to devise systems in which all probable behaviors (or all but exceedingly *im*probable behaviors) are compatible with successful system operation. This typically requires designs that present low energy barriers to desired behaviors, and high energy barriers to undesired behaviors, in which the high energy barriers often are a consequence of the use of stiff components.

Even if one were to assume classical, deterministic behavior and nearly perfect information regarding initial conditions, a typical nanomechanical system would soon require a statistical description. Consider the trajectory of a particle rebounding from an atom: because atoms are not flat, a small perturbation typically causes a particle to rebound at a different angle, leading to a larger perturbation at its next point of impact. In a typical system, trajectories that are initially almost identical will rapidly diverge until they have no similarity. This divergence is characteristic of chaos. Further, real nanomechanical systems are in mechanical or radiative contact with an environment at some nonzero temperature, providing a constant source of unpredictable thermal excitations.

4.3.2. Basic results in equilibrium statistical mechanics

Statistical mechanics takes its simplest form for systems at thermodynamic equilibrium. Since this is often a good approximation for real systems, some basic results are worth summarizing. [This discussion largely follows (Knox, 1971).]

a. Quantum statistical mechanics. In quantum statistical mechanics, it is convenient to consider a system that is in thermal equilibrium with a heat bath, yet is assumed to have a set of bath-independent quantum states $i = 0, 1, 2,\ldots$. In equilibrium statistical mechanics, a complete description of a system consists of a specification of the probability, $P(i)$, for each state i. This takes the form

$$P(j) = \frac{\exp[-\mathscr{E}(j)/kT]}{\sum_{i=0}^{\infty} \exp[-\mathscr{E}(i)/kT]} \tag{4.1}$$

where $\mathscr{E}(j)$ represents the energy of state j. The probability that the system is in state i is proportional to the Boltzmann factor, $\exp[-\mathscr{E}(state)/kT]$, and all states of a given energy are thus equally probable.

A quantity of special importance is the denominator of the above expression,

$$q = \sum_{i=0}^{\infty} \exp[-\mathscr{E}(i)/kT] \tag{4.2}$$

where q is a temperature-dependent pure number termed the *partition function* of the system (note that its magnitude depends on the choice of the zero of the energy scale). The partition function can be related to the variables of classical thermodynamics. The mean energy of the system is given by an expression involving a constant-volume derivative of the partition function

$$\overline{\mathscr{E}} = kT^2 \left(\partial \ln q / \partial T\right)_V \tag{4.3}$$

as is the °entropy

$$\mathscr{S} = \left[\partial(kT \ln q)/\partial T\right]_V \tag{4.4}$$

The °Helmholtz free energy is

$$\mathscr{F} = -kT \ln q = \mathscr{E} - T\mathscr{S} \tag{4.5}$$

and the pressure is given by a constant-temperature derivative

$$p = -(\partial \mathscr{F}/\partial V)_T = kT(\partial \ln q / \partial V)_T \tag{4.6}$$

b. Classical statistical mechanics. Paralleling the quantum case, in classical statistical mechanics it is common to consider a system that is in thermal contact with a heat bath, yet is treated as having a bath-independent energy function, $\mathscr{E}(state)$. For a mechanical system, a state is defined as a point in the phase space defined by the set of position coordinates $q_1, q_2, q_3,\ldots, q_{3N}$ (the *position*), and the corresponding momentum coordinates p_1, p_2, p_3,..., p_{3N} (the *momentum*), where N is the number of atoms. Here, a complete description consists of a specification of the probability density function (PDF) over phase space. The fundamental result is

$$f_{state}(state) = \frac{\exp[-\mathscr{E}(state)/kT]}{\int \cdots_{p,q} \int \exp[-\mathscr{E}(state)/kT] dp_1 dq_1 dp_2 dq_2 \ldots dp_{3N} dq_{3N}} \quad (4.7)$$

where the PDF, $f_{state}(state)$, is the probability of occupancy per unit volume of phase space associated with each point in that space.

The denominator of the above expression (times a factor demanded by the correspondence principle*) defines the *classical partition function* for a solid

$$q_c = (2\pi\hbar)^{-3N} \int \cdots_{p,q} \int \exp[-\mathscr{E}/kT] dp_1 dq_1 dp_2 dq_2 \ldots dp_{3N} dq_{3N} \quad (4.8)$$

which (in the classical approximation) can be used as the value of the partition function in Eqs. (4.3) through (4.6).

In nanomechanical design, a common concern is the probability that a thermally excited system will be found in a particular configuration at a particular time. Molecular mechanical systems can usually be described in terms of motion on a single potential energy surface, and the total energy can be divided into potential energy and kinetic energy terms

$$\mathscr{E}(state) = \mathscr{V}(position) + \mathscr{T}(momentum) \quad (4.9)$$

With this division, Eq. (4.7) can be factored, and by integrating over the momentum coordinates of the phase space, a PDF over the position coordinates alone (the classical equilibrium PDF in configuration space) can be obtained

$$f_{position}(position) = \frac{\exp[-\mathscr{V}(position)/kT]}{\int \cdots_{q} \int \exp[-\mathscr{V}(position)/kT] dq_1 dq_2 \ldots dq_{3N}} \quad (4.10)$$

Note that the probability density associated with a position in configuration space is (save for a normalization factor) dependent purely on its potential energy. The distribution of momenta and kinetic energies is independent of the configuration, and hence the mean kinetic energy of a classical system at thermal equilibrium remains unchanged in all configurations. (This uniform behavior of the momentum components simplifies the description of nanomechanical systems near thermal equilibrium, making the configuration space picture of Section 4.3.3 more useful.)

Classical statistical mechanics is frequently a useful approximation. At ordinary temperatures its significant failings occur chiefly in stiff, high-frequency modes; these usually have little effect on the dynamics of nanometer-scale diamondoid components (which are by these standards soft, slow, and ponderous). Chapter 5 examines positional PDFs describing various elementary nanomechanical systems, comparing the results of quantum and classical models; its

---

* Which requires that classical and quantum mechanics give the same answers in the limit of large quantum numbers.

results indicate the limits within which classical statistical mechanics yields accurate results regarding positional uncertainty in nanomechanical engineering systems.

4.3.3. The configuration-space picture

Although it yields no new physical information, it can be helpful to regard a classical mechanical system containing N atoms as a single moving point in a configuration space of $3N$ dimensions, in which each of the three Cartesian coordinates of each atom corresponds to one dimension. Adding a single, orthogonal, "vertical" dimension to represent potential energy yields a potential energy surface in configuration space (Chapter 3). The configuration point can then be imagined as sliding over an undulating, frictionless surface subject to a gravitational force—depending on initial conditions and the shape of the surface, the point may oscillate in a potential well, move along a valley, pass from one well to another through a °col between peaks, and so forth.

To make this dynamical picture work out properly, the configuration-space coordinates corresponding to an atom must vary in proportion to the Cartesian space coordinates of the atom multiplied by $m^{-1/2}$, where m is the mass of the atom. The kinetic energy of the coordinate point is then an isotropic function, depending only on the square of the speed.

In configuration space, a linear, elastic system corresponds to a point moving in a single potential well (neglecting translational and rotational degrees of freedom). For a two-atom system (approximating the bond as a linear spring and neglecting rotational degrees of freedom), this is a simple parabola. For a noncolinear N-atom system, the potential well retains a parabolic form along any line through the equilibrium point, and the isopotential surfaces are concentric $3N - 6$ dimensional ellipsoids; each of the $3N - 6$ axes of an ellipsoid represents the line of motion of a normal mode. A nonlinear system might permit (for example) the interchange of two atoms given a sufficiently great thermal excitation; the corresponding region of the potential surface has two wells joined by a pass.

a. Probability gas in configuration space. For a classical system at thermal equilibrium with a heat bath, statistical mechanics asserts that the probability density of the configuration point is an inverse exponential function of the energy, here represented by the height. Thus, the probability density varies across the configuration-space landscape much as the atmospheric density varies across a real landscape (Figure 4.1). The configuration-space point is like a gas consisting of a single molecule, with a well-defined mean density, flux, and so forth, at every point.

As in an equilibrated atmosphere (unlike Earth's convecting atmosphere), the mean kinetic energy, and hence the temperature, is independent of the height of the land. The equilibrium ratio of the total probability in two connected valleys depends on their effective volumes; these depend on size and altitude, which correspond to the entropy and energy of the corresponding states. The rate at which probability diffuses through a col between the valleys depends on the height and width of the pass, and on the mean speed and overall probability density of the configuration point. All of these factors appear in transition-state theory (Chapter 6).

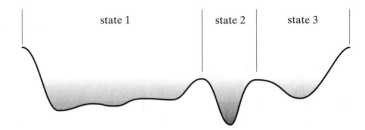

Figure 4.1. A one-dimensional PES shaded to illustrate the *probability gas* visualization of the equilibrium distribution in configuration space. The state boundaries are drawn in accord with the rules in Section 4.3.3b, but omitting those that correspond to energy barriers that are small compared to kT (these would otherwise subdivide state 1 into three regions).

In statistical mechanics, the principle of *detailed balancing* asserts that, at equilibrium, the mean rate of transitions from state 1 to state 2 equals that from state 2 to state 1 for all pairs of states. For states defined as regions in configuration space, this has an intuitive interpretation. At equilibrium, gas molecules cross any arbitrarily defined surface element at equal rates from both sides; this is likewise true for the configuration-point gas, and for each surface element of the boundary separating any two states.

b. States in configuration space. The configuration space picture suggests one natural way to define what is meant by "distinct states" of a solid or liquid system (Stillinger and Weber, 1984).* From each point on the energy landscape, there exists a path of steepest descent, and that path always ends in a point or

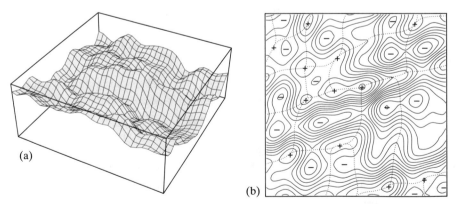

Figure 4.2. Definition of states in terms of potential energy minima. Panel (a) shows a potential energy surface defined over a two-dimensional configuration space; Panel (b) shows the same surface as a contour map partitioned into regions corresponding to local potential energy minima (marked by minus signs).

---

* *State* is an overworked word. Here, it refers to a region of configuration space; in Section 4.3.2b, it referred to a single point in phase space (= configuration space × momentum space). When a thermally excited nanomechanical system is described as being in some particular state of practical significance, this cannot refer to a point, and typically refers to a region in configuration space of the sort discussed here.

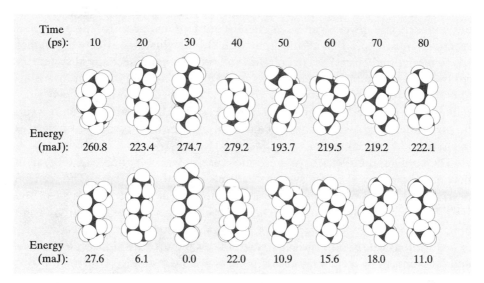

Time
(ps):   10   20   30   40   50   60   70   80

Energy
(maJ): 260.8   223.4   274.7   279.2   193.7   219.5   219.2   222.1

Energy
(maJ):   27.6   6.1   0.0   22.0   10.9   15.6   18.0   11.0

Figure 4.3. Molecular dynamics simulation of *n*-octane at 400 K (MM2/CSC). The upper row illustrates molecular configurations at 10 ps intervals, starting after 10 ps of equilibration at the target energy. Each configuration is labeled with its potential energy relative to the minimum-energy configuration for the molecule, and corresponds to a point on a potential energy surface analogous to that shown in Figure 4.2. Lower row: structures and energies of the corresponding state-defining minima, which in the third state happened to be the global minimum. Note that energy differences between conformers frequently are small. Treating all atoms as distinguishable, the number of distinct conformers for octane is on the order of $3^7 = 2187$ (but this is reduced by excluded-volume effects). °Conformationally mobile structures like *n*-octane are unsuitable for most nanomechanical applications.

region of locally minimal energy. Thus, points correspond to minima, and each local energy minimum can be taken to mark a distinct state, as indicated in Figure 4.2. Figure 4.3 illustrates configurations and corresponding state-defining minima for *n*-octane in a molecular dynamics simulation.

Minima separated by barriers small compared to kT can often be regarded as a single minimum, because the barriers are crossed so easily. A system acting as a good rotary bearing (Section 10.4), for example, will exhibit a chain or loop of minima having essentially the same depth and separated by barriers of negligible height. For all practical purposes, a chain or loop of this sort constitutes a single potential well. In a properly designed nanomechanical system, cols between potential wells are either functional parts of the design, or are high enough to block any substantial probability flux.

*c. **Constraints and thermal excitation.***   Even without being able to visualize interconnected, approximately ellipsoidal potential wells, or ring-shaped valleys in a high-dimensional space, one can get a sense of the strongly constrained nature of the motion of the configuration point in such systems. A similar description of a chemically reacting solution-phase system lacks such well-defined features. Each possible covalent combination of atoms forms a separate valley, and the valleys intertwine in a manner that brings each into close proximity with many others. Which cols are available—determining which reactions can occur

at appreciable rates—depends on the relative heights and widths of the numerous connections between valleys. A chemist attempting to control the pattern of
reactions must exploit small differences in the heights and effective volumes
(the energies and entropies) of cols and valleys. In a nanomechanical system, in
contrast, a reaction might occur between two groups brought together by a gear-
like rotary motion. The available configuration space consists of two ring-
shaped valleys linked by a single col of modest height. Unwanted reactions can
be prevented by gross mechanical constraints, rather than by small differences
in energy and entropy.

The configuration-space picture is most useful when applied to a subsystem
that is coupled to a larger system that acts as a thermal bath. This can be
described in terms of the potential surface that would arise if the rest of the
system were motionless and fully relaxed for all subsystem configurations, combined with a time-varying perturbing potential representing the effects of thermal vibrations external to the subsystem. In this picture, the landscape vibrates,
and total energy of the configuration point is not a constant.

4.3.4. Equilibrium vs. nonequilibrium processes

The relationships cited in Section 4.3.2 describe equilibrium systems, but a func-
tioning nanomechanical system, with motions driven by input power, is not at
equilibrium. How useful, then, is equilibrium statistical mechanics?

Equilibrium statistical mechanics is seldom applied to a true equilibrium
system. Matter under ordinary conditions of temperature and density, for example, has crystalline iron as its equilibrium state—if one allows true equilibration of all the nuclear degrees of freedom. In practice, the necessary reactions
(e.g., fusion) proceed so slowly that they can be ignored. States (or dynamical
domains) that are sufficiently °metastable can be treated as stable in calculating
"equilibrium" properties.

In macromechanical systems, mechanical motion and thermal motion are
quite distinct. In nanomechanical systems, this distinction usually remains clear.
Thermal velocities may exceed the imposed mechanical velocities (which in
most instances considered are quite small), but thermal velocities (unlike mechanical) have a mean value of zero and are independent of any driving force.

a. Deviations from equilibrium. In mechanical systems, frictional forces
convert mechanical energy into °thermal energy in a spatially inhomogeneous
fashion, causing thermal gradients. In estimating the fluctuations resulting from
thermal motion, small regions of matter in such systems can then be approximated as being at thermal equilibrium save for deviations associated with temperature gradients, heat flows, and changing temperatures. When these
deviations are small, equilibrium models give good estimates of the statistics of
thermally induced displacements and motions in nanomechanical systems.

A small thermal gradient is one that produces only a small difference in the
absolute temperature across the diameter of the system under analysis. For a
nanomechanical system with a diameter of 10 nm, a 1% ΔT (at 300 K) requires
a gradient of 3×10^8 K/m. Assuming a thermal conductivity of 10 W/m·K, this
yields a thermal power flux of 3×10^9 W/m^2. This "small" thermal gradient produces a large (i.e., often unacceptable) temperature difference of 300 K in a
modest 10^{-6} m distance.

A small thermal flux is one that produces only a small difference in the equilibrium distribution of thermal vibrations. This equilibrium distribution is characterized by large power fluxes (which cancel, at equilibrium) on the order of (*speed of sound*) × (*thermal energy*)/(*volume*). For a typical material at ordinary temperatures, this is on the order of $(3 \times 10^8 \text{ J/m}^3) \times (10^3 \text{ m/s}) = 3 \times 10^{11} \text{ W/m}^2$. One percent of this value again corresponds to a net thermal power flux of $3 \times 10^9 \text{ W/m}^2$. Since thicker layers result in unacceptable values of ΔT, a "small" power flux of this magnitude would be encountered in a working system only if a $\sim 10^{-6}$ m (or less) thick slab were to dissipate power at $\sim 10^{15} \text{ W/m}^3$ (or more). Despite the extraordinarily high power-*conversion* densities possible in small components (see Section 2.2.2), the power-*dissipation* densities for most complex, multicomponent systems will be small compared to this value.

A small rate of change of temperature is one that produces only a small ΔT during the characteristic vibrational relaxation time. In a crystal, a measure of this relaxation time is the phonon mean free path divided by the speed of sound; a typical value might be $(10^{-8} \text{ m})/(10^3 \text{ m/s}) = 10^{-11}$ s. In a highly inhomogeneous medium (e.g., a typical nanomechanical system), with highly anharmonic interactions (e.g., van der Waals contacts between moving parts), relaxation times typically are shorter. A 1% ΔT (at 300 K) during a single relaxation time then corresponds to a rate of temperature change in excess of 3×10^{11} K/s. For typical volumetric heat capacities ($\sim 10^6 \text{ J/K·m}^3$), this "small" rate of temperature change requires a large power dissipation (as above, $\sim 10^{15} \text{ W/m}^3$ or more).

b. The process of thermalization. Because these "small" thermal gradients, thermal fluxes, and rates of temperature change are all so large, it is often acceptable to divide motions into mechanical and thermal components, describing the thermal motions in terms of the local temperature and the relationships of equilibrium statistical mechanics. The chief exceptions are in descriptions of the processes that convert mechanical motion into thermal excitation, where nonequilibrium vibrational motions are generated and then thermalized.

A typical nonequilibrium event is a transition from one potential well to another that is forced to occur at a particular time by an input of mechanical energy. In the configuration-space picture, the configuration point representing the subsystem passes through a col owing to an imposed change in the shape of the PES, and is thus injected into the new potential well with an unusually high energy and a somewhat-predictable momentum. Regarding the new well as nearly quadratic, the initial state can be viewed as one with a disequilibrium distribution of excitation of normal modes. Relaxation then involves two processes: a tendency toward equilibration of the distribution of modal excitation via anharmonic interactions (given an unusually high total energy) and a tendency toward equilibration of the expected total energy via interactions with a heat bath. For a general anharmonic system in weak contact with a thermal bath, a short-term description of the motion would describe families of trajectories; an intermediate-term description would treat all states of equal energy as equally probable, but would take account of the slowly decaying excess energy; and the long-term description would be in terms of the statistics of fully equilibrated thermal motion.

4.3.5. Entropy and information

a.   Entropy and probability.   The transition from a detailed dynamical description to a statistical description entails discarding information. State variables that are regarded as having definite values in the dynamical description
are regarded as having probability distributions in the statistical description. In
quantum statistical mechanics, distinct states can be enumerated, and the entropy, which can be stated as

$$\mathcal{S} = -k \sum_{states} P(state) \ln[P(state)] \qquad (4.11)$$

is a weighted measure of the number of possible states. This expression yields
the familiar result (the third law of thermodynamics) that a perfect crystal at
absolute zero has zero entropy: the structure of the crystal is known and it is in
the vibrational °ground state with unit probability; $1 \times \ln(1) = 0$, hence $\mathcal{S} = 0$.
This result is equally true for any completely specified structure at absolute zero.

b.   Information can increase free energy.   Probability, however, depends on
the information one possesses. If I flip a coin and peek at it, it may be heads with
probability one for me, while remaining heads with probability 1/2 for you. This
may suggest that the entropy of an irregular solid at absolute zero is positive if
these irregularities are unknown, but zero if they are completely described by
some algorithm or external record (the complete specification just alluded to). If
so, then entropy is not a local property of a physical system.

Studies of entropy and information confirm this view. In concrete terms, a
structure (such as a polymer with a seemingly random sequence of monomers)
can be made to yield more free energy if it is matched against another polymer
with a sequence known to be identical—that is, if one has an external record of
the sequence (Bennett, 1982). Information regarding the sequence increases the
Helmholtz free energy of the polymer, Eq. (4.5), because it eliminates the possibility of many alternative states, and thus lowers $\mathcal{S}$. The free energy extracted
from a system can thus depend on the existence of a record, initially located in
the external environment, that is brought in, read, and returned unchanged.*
This understanding of entropy, although well established in the literature, con-
tradicts the simple perspective presented in most textbooks. (For further discussion, see *entropy* in the glossary.)

---

* More concretely, consider a system that could be in either of two deep, stable potential wells (e.g., a cylinder divided into two compartments by a piston in the middle, with
a gas molecule in one or the other). To convert this into a system with a single well, the
two wells can be merged symmetrically by dropping the barrier between them (by
removing the piston), or asymmetrically (by moving the piston to one end of the cylinder, then removing it). Relative to symmetrical merging, asymmetrical merging can
yield more free energy (the one-molecule gas expands against the moving piston, doing
work), but only if some external record indicates which well is occupied (enabling a
mechanism to choose the right direction for the motion). Proposals to gain this free-
energy increment by observing an initially unknown state can be made quite confusing
but prove futile; if successful, they would enable violations of the second law of thermodynamics. Section 7.6.2 analyzes well-merging processes in more detail.

*c. **Entropy and computation.*** These issues are intimately related to questions regarding the theoretical energy requirements of computation. In the early 1960s, Landauer observed that compressing the logical state of a computer entails compressing its physical state (Landauer, 1961); for example, erasing or overwriting a one-bit storage device with previously unknown contents entails reducing its possible states (one or zero) to a single state (e.g., zero). Such a transformation dissipates a quantity of free energy $\mathcal{F} \geq \ln(2)kT$ per bit erased (Section 7.6.3).

Theorists have examined devices such as Brownian computers with idealized structures (i.e., hard, perfectly rigid parts of arbitrary shape) but subject to realistic dynamics and thermodynamics, as well as computational devices inspired by the molecular machinery of biological systems (Bennett, 1982), idealized, deterministic mechanical systems (Toffoli, 1981; Fredkin and Toffoli, 1982), and quantum mechanical models (Likharev, 1982; Feynman, 1985). These and related studies indicate that logically reversible computations—those where the *output* uniquely specifies the *input*—can be performed in a manner approaching thermodynamic reversibility (that is, the dissipation of free energy per logically reversible operation can be made arbitrarily close to zero). The literature and results in this field have been well reviewed (Bennett, 1982; Landauer, 1988). They provide an improved understanding of the second law of thermodynamics (and the impossibility of a Maxwell's Demon), and are of direct relevance both to mechanical nanocomputers and to many other nanomechanical systems.

4.3.6. Uncertainty in nanomechanical systems

In conventional mechanical systems, the positions and shapes of components are never completely known: manufacturing tolerances, measurement inaccuracies, and environmental vibrations ensure this. When systems work despite these uncertainties, it is because the uncertainties are kept within tolerable bounds, either by reducing uncertainty or by expanding the range of tolerance.

In nanomechanical systems, quantum and statistical mechanics place physical limits on the reduction of uncertainties. For a structure of a given mass and stiffness in equilibrium at a given temperature, positional uncertainties are fixed and irreducible (Chapter 5). Depending on the design of the system, these uncertainties can cause errors at a rate ranging from negligibly low to unacceptably high.

*a. **Quantum and classical parallels.*** There are strong *qualitative* parallels between the uncertainties of quantum mechanics and those of classical statistical mechanics. In both quantum and statistical mechanics, one begins with a potential energy function, $\mathcal{V}(\mathbf{r})$. In quantum mechanics, the Schrödinger equation ensures that a particle of a given energy has a nonzero (though often vanishingly small) probability of penetrating a potential barrier of any finite height and thickness, and of being found in any region of space. In classical statistical mechanics at finite temperature, Boltzmann's law yields the same qualitative result by assigning a nonzero (though often vanishingly small) probability to states of arbitrarily high energy. In quantum mechanics, a harmonic oscillator has a positional uncertainty characterized by a Gaussian probability distribution; classical mechanics at a finite temperature yields a result of the same form.

Quantum uncertainties measured as the product of the uncertainties in conjugate variables have an irreducible minimum,* for example,

$$\Delta x \Delta p \geq \hbar/2 \qquad (4.12)$$

but the uncertainty in either variable can be reduced to an arbitrary degree by a suitable measurement. Classical systems permit similar measurements, but present the illusion that the other variable can simultaneously be specified with arbitrary accuracy as well. This difference in the reducibility of uncertainty is, however, irrelevant in the context of equilibrium statistical mechanics. A measurement that reduces uncertainty also reduces entropy (Section 4.3.5) and hence alters the equilibrium of the system *from the perspective of the observer,* even if in the absence of a physical disturbance of the system itself. Thus, within the equilibrium statistical-mechanical description, uncertainties are irreducible by definition.

In a nanomechanical system, each component has many vibrational degrees of freedom, each subject to thermal excitation. Any attempt to use nanomechanical components within a system to represent the results of measurements performed on other nanomechanical components within a system can succeed in encoding only a small fraction of the information needed to represent the total vibrational state of the system. While one can imagine a device that uses measurement to reduce the uncertainties associated with one or a few critical components, systems will in general be dominated by components subject to statistical uncertainties that are in practice irreducible and that will (as indicated by Section 4.3.4) usually be well described by equilibrium statistical mechanics.

4.3.7. Mean-force potentials

Consider a piston that controls the volume of a cylinder containing an ideal gas. The work done in compressing the gas under isothermal conditions causes no change in the potential energy of the gas or of the piston, and yet so long as the system remains isothermal, the force exerted by the gas on the piston can be treated as the gradient of a potential energy function. This function defines a *mean-force potential,* since it describes the time-average forces experienced by the piston.

The use of a mean-force potential enables the statistical behavior of a portion of an equilibrium (or near-equilibrium) system to be described without including all the coordinates of the full PES. In the piston-and-gas example, the use of a mean-force potential enables the statistical distribution of piston positions and velocities to be calculated directly, without reference to the motions of each of the gas molecules (Section 5.6.2). In general, the mean-force potential defined over the retained coordinates is affected by the omitted coordinates to the extent that the position of the system with respect to the retained coordinates alters the permitted *range of motion* in the omitted coordinates. In the piston-and-gas example, the range of motion in the coordinates corresponding to the gas molecules is directly determined by the position of the piston; softer constraints on motion (e.g., time-varying °elastic restoring forces) have analogous

---

* The uncertainty principle stated here is implicit in all the later derivations based on quantum mechanical principles, for example, those of Chapter 5.

effects. The use of a mean-force potential treats work of compression, which reduces entropy, as equivalent to work done against elastic forces, which increases potential energy.

In quantitative terms, the work done against a mean-force potential (in the classical model) is just $\Delta \mathcal{V} - T\Delta \mathcal{S}$; this equals $\Delta \mathcal{F}$, given that under isothermal conditions the change in kinetic energy $\Delta \mathcal{T} = 0$. A mean-force potential can be applied where imposed motions are slow (permitting thermal equilibration), or where motions—however fast—are spontaneous and thermal, and hence result from the equilibrium state itself.

4.4. PES revisited: accuracy requirements

Both dynamical and statistical mechanical models of molecular behavior rely on potential energy surfaces that are (in all cases of nanomechanical interest) approximations known to deviate from reality. The scientific literature on potential energy surfaces describes efforts to improve the correspondence between experiment and theory, and hence focuses on the imperfections and limitations of existing models. In order to understand the utility of existing models from an engineering perspective, it is useful to consider the sensitivity of different physical phenomena to the existing inaccuracies.

4.4.1. Physical accuracy

In chemical physics, experiments are designed to provide stringent tests of theoretical models of molecular systems (including their potential energy surfaces), and the theoretical models attempt to predict everything that can be experimentally observed. As discussed in Chapter 3, physicists have made extensive use of molecular beam experiments in which molecules are prepared with precise momenta (and sometimes with control of vibrational states, rotational states, and polarization); they are then allowed to scatter (sometimes with a reactive exchange of atoms) and outcomes are observed and analyzed in terms of scattering angle (etc.).

The quantum interference effects that can be observed in such experiments provide a delicate test of potential energy surface models. Scattering events that involve bond formation and cleavage can traverse energy barriers of over 100 maJ, yet the interference effects are sensitive to much smaller energy differences. A characteristic molecular collision time is $>10^{-13}$ s (the time required to travel 10^{-10} m at 10^3 m/s); changing the potential energy along one of the interfering paths by 1 maJ or less shifts the phase of that path by a radian and causes a substantial change in the interference pattern. Since different trajectories explore different parts of the configuration space, describing interference effects correctly can require accuracy of this magnitude across the entire dynamically accessible potential energy surface.

4.4.2. Chemical accuracy

a. Predicting solution-phase reaction rates and equilibria is hard. In solution chemistry, a standard challenge is to predict chemical equilibria, absolute reaction rates, and the relative rates of competing reactions. Synthetic chemistry can be viewed as an engineering discipline aimed at constructing molecules. In this task, rates and equilibria are of central importance: if a reaction equilibrates several species, then the yield of the desired product will depend on the equilibrium concentration ratios; alternatively, if a reaction can proceed along any of several effectively irreversible paths, then the yield of the desired product will depend on the ratio of the reaction rates. In some reactions, yields are affected by both rates and equilibria.

If entropic factors are equal, then the equilibrium ratio of two species is an exponential (Boltzmann) function of the difference in potential energy between the species. Likewise, if entropic factors (and certain dynamical factors) are equal, then the ratio of the rates of two competing reactions is an exponential function of the difference in potential energy between the two transition states. At 300 K, a difference of 1 maJ changes rates and equilibria by a factor of 1.27, a 10 maJ difference results in a factor of 11, and a 100 maJ difference results in a factor of 3.1×10^{10}. To a practicing chemist, an energy difference of 10 maJ *between* two competing species or transition states which changes the yield of a reaction from 8% to 90% usually matters more than would a 100 maJ shift in all energies (causing no change in the course of the reaction) or a 100 maJ shift in the transition-state energies which slows the reaction-completion time from a microsecond to an hour. In discussions of molecular energies, the term "chemical accuracy" usually implies errors of somewhat less than 10 maJ per molecule in describing the energies of chemical species (corresponding to potential wells) and transition states (corresponding to features of cols). Aside from entropic effects dependent on the breadth of potential wells and cols, reaction rates typically exhibit only modest sensitivity to the shape—as distinct from the well depths and °barrier heights—of the potential energy surface.

b.   *Predicting the geometries of flexible structures is hard.* Potential energy functions are also used to predict molecular structures; the MM2 PES has good success for a wide range of small organic molecules. A more challenging test is protein modeling, where the shape and stability of the folded protein molecule depend on a delicate balance of free energy terms in which van der Waals interactions, overlap interactions, hydrogen bonding, torsional strains, and entropic factors all play crucial roles. The net stability of a folded structure typically is ~50 to 100 maJ, or ~0.01 maJ per atom. Although present molecular mechanics potentials are good enough to have found extensive use in protein modeling and design, their errors are large compared to the free energy of folding. Further, energy minimization usually yields structures that differ substantially from those experimentally determined by x-ray diffraction, even when the experimental structures are taken as a starting point.

Even for small organic molecules, a slightly inaccurate molecular mechanics model can predict structures that are totally wrong. Most molecules of concern in organic chemistry can exist in any of a number of conformations, differing by rotations about bonds; a simple example is *n*-octane, Figure 4.3. The relative energies of molecular conformations are sensitive to torsional energies (Section 3.3.2c) and to weak, nonbonded interactions (Section 3.3.2e). Conformational equilibria, like other chemical equilibria, are greatly altered by energy differences of ~10 maJ. A crystal structure (a common source of geometric data) represents one of many possible molecular packing arrangements, again sensitive to weak forces. A small modeling error can result in a predicted crystal structure containing the wrong conformation packed in a lattice of the wrong symmetry.

4.4.3. Accurate energies and nanomechanical design

Nanomechanical systems of the sort considered here are organic structures that resemble (or exceed) proteins in size, and some are used to perform chemical reactions. It is thus important to consider the sensitivity of nanomechanical designs to errors in potential energy surfaces.

a. Sensitive and insensitive designs. It will be possible to design nanomechanical systems that are exquisitely sensitive to the properties of a potential energy surface. For example, successful operation might be made to depend on a ratio of competing transition rates, as in conventional chemistry, or even on interference phenomena in angle-resolved scattering in crossed molecular beams. In general, any measurable physical property can be made essential to the correct operation of a suitable Rube Goldberg device, and hence any known class of unpredictable discrepancies between model and experiment can be used to design a class of devices that cannot with confidence be predicted to work. The design of such sensitive devices is closely related to good design practice in instrumentation and scientific experimentation, but is the opposite of good design practice in conventional engineering.

The robustness, predictability, durability, and performance of nanomechanical designs is usually maximized by making them from strong, stiff materials. Those considered here resemble diamond, consisting of highly°polycyclic, three-dimensional networks of atoms linked by covalent bonds. Predictions of the stability and geometry of these diamondoid structures are far less sensitive to small errors in potential energy surfaces than are similar predictions for folded protein structures or conformationally mobile organic molecules. Figure 4.4 illustrates the results of a molecular dynamics simulation of a polycyclic octane structure (cubane) under conditions like those of Figure 4.3: no conformational freedom is available, hence the deformations are purely vibrational.

b. Stiff molecules can resist "perturbing" modeling errors. Increasing structural stiffness tends to reduce errors in predicted geometry caused by inaccuracies in the PES. Consider a multiatom design that, when modeled with a potential $\mathcal{V}_{model}$, is predicted to have desirable geometrical properties. This function, however, differs from the correct (but unknown) potential, $\mathcal{V}_{correct}$. The transition from model to reality may now be regarded as adding to the model system the perturbing forces defined by the gradient of the difference potential, $\mathcal{V}_{diff} = \mathcal{V}_{model} - \mathcal{V}_{correct}$. A structure of greater rigidity suffers smaller deformations when subject to these perturbing forces; accordingly, a more rigid molecular design is more likely to survive the transition from model to reality without harmful distortion.

Good design practice can increase tolerance for the errors that remain. Nonetheless, errors in predicting molecular geometry can lead to designs that fail to work, in part because small local discrepancies in the predicted equilibrium geometry can have cumulative effects across large structures. Appendix A discusses how this problem can be accommodated during preliminary design and later overcome through experimentation.

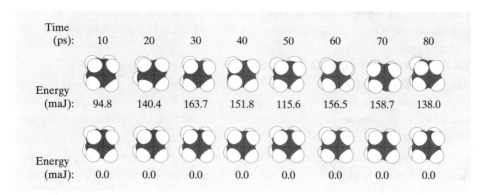

Figure 4.4. Molecular dynamics simulation of pentacyclo[4.2.0.0^{2,5}.0^{3,8}.0.^{4,7}]octane (cubane) at 400 K (MM2/CSC); conditions as in Figure 4.3. The polycyclic structure of cubane, unlike the open-ended chain of *n*-octane, is representative of structures suitable for nanomechanical systems. It is stiff and lacks alternative conformations, hence its shape is insensitive to small errors in the PES, and thermal excitation results in relatively small deformations. All vibrational states correspond to the same minimum, as shown in the lower row. (The length and opacity of formal IUPAC names for polycyclic structures, like that given above for cubane, grows rapidly with molecular size; no attempt will be made to give formal names to structures like those in Figure 1.1.)

*c.   **Mechanosynthesis is relatively insensitive to modeling errors.***   Errors in modeling chemical reactions play a different role. If the correct sequence of reactions is performed, then the correct structure will result, regardless of small errors in describing reaction rates or transition-state geometries during the process (Section 6.2). Distinct structures correspond to distinct states on a PES determined by invariant physical constants, and the manner of construction has no effect on the nature of the product. In conventional manufacturing, small variations in fabrication steps can have cumulative effects on properties such as product geometry, but no parallel problem arises in molecular manufacturing. Like switching events in digital logic, reactions either occur or they don't, and the result is either correct, or it isn't: ambiguous situations are unstable and transient.

In comparison to conventional chemical synthesis, molecular manufacturing processes based on positional control of synthetic reactions are less sensitive to small energy differences. Individual reaction steps can be driven to completion either by ensuring that energy differences greatly favor the product state, or by repeating trials until a molecular measurement verifies successful completion. To achieve regio- and stereospecificity, reactions can be guided not by differences in reaction rates and equilibria as in conventional synthesis, but by rigid control of °reagent access to different sites. In chemical terms, this can create enormous ratios of effective reagent concentration between molecular sites that might otherwise be equally reactive. With positional control of synthesis, the main issues are achieving a high enough reaction rate (fast reactions best exploit the speed of nanomechanical systems) and avoiding (or being able to reverse) unwanted reactions immediately adjacent to the target site. These and related matters are discussed in more detail in Chapter 8.

*d.   **Designs can eventually be tested and corrected.***   At present, nanomechanical design must rely on modeling because the tools necessary for construction and testing are unavailable. Note, however, that errors in specific designs (as distinct from systematic errors regarding what sorts of designs are workable) will be of little significance until the systems they describe can be constructed and used. Then, of course, testing will become possible, and errors can be corrected.

4.5.   Conclusions

Models of molecular dynamics (within the Born–Oppenheimer approximation) begin with a potential energy surface, but differ in their treatment of it. Nonstatistical classical methods proceed by assuming a set of initial conditions and then integrating the classical equations of motion of each atom, subject to the forces implied by the PES; statistical results describing (for example) a thermal distribution can be accumulated by examining long simulations. Statistical mechanics methods compute a statistical distribution, commonly that associated with thermal equilibrium: quantum methods compute a distribution of quantum states; classical methods compute a distribution of positions and velocities. Many nanomechanical systems can be analyzed with good accuracy using relationships based on classical equilibrium statistical mechanics, sometimes with quantum corrections. These relationships can be used to define a mean-force potential (or free-energy potential), enabling systems near equilibrium to be described in terms of fewer coordinates. A consideration of PES accuracy requirements in light of dynamical principles and nanomechanical concerns shows that the mechanical stiffness of diamondoid structures and mechanochemical devices often greatly reduces their sensitivity to PES modeling errors, relative to more familiar molecular systems.

4.6.   Further reading

Molecular dynamics is a broad field with an enormous literature. Basic statistical mechanics is well described in many textbooks, for example (Knox, 1971). Trajectory-based molecular dynamics, being more model dependent and computation intensive, is in a greater state of ferment. A good overview of the dynamics of reactive molecular collisions is (Levine and Bernstein, 1987), which includes many references. Many studies calculate trajectories by integrating the classical equations of motion subject to a molecular mechanics PES. Mitchell and McCammon (1991) review methods for estimating free energy differences between structures using dynamical methods based on molecular mechanics, and improved techniques are discussed in Straatsma and McCammon (1991). These works focus on calculations of the free energy of flexible biomolecules in water, which present difficulties not found in more rigid structures.

5

Positional Uncertainty

5.1. Overview

Positional uncertainty presents a fundamental challenge in nanomechanical engineering, determining many design decisions and limiting the range of workable systems. Positional uncertainty in compliant systems is a result of both thermal motion and quantum mechanical uncertainty, but thermal motion proves more important in almost all mechanical systems at room temperature. This chapter describes positional uncertainty in a set of structures characterized by a PES with a single potential well.

Section 5.2 discusses the concept of positional uncertainty and its importance in engineering. Section 5.3 presents classical and quantum mechanical analyses of displacements in thermally excited harmonic oscillators; Section 5.4 and 5.5 present similar analyses of longitudinal and transverse displacements in rods. These studies of elastic systems provide the basis for describing positional uncertainty in the structures considered in later chapters. Indeed, for analyzing mechanical systems at room temperature, Eq. (5.4) usually provides an adequate approximation. In practice, the rest of the discussion in these sections chiefly serves to quantify the errors in using Eq. (5.4), to show why its range of applicability is greater than its derivation would suggest (Section 5.8.1), and to provide more elaborate expressions that remain accurate at lower temperatures.

The following two sections discuss systems in which positional uncertainty results from fluctuations in entropic springs. The prototype of an entropic spring is a loaded piston in a gas-filled cylinder, analyzed from three perspectives in Section 5.6. A flexible, tensioned rod in a housing that subjects it to transverse elastic restoring forces is a more complex system, but provides a good description of devices useful in nanomechanical signal transmission; this is analyzed in Section 5.7. These results on entropic springs, though useful in analytical work of the sort presented in Part II, are not applied to any of the specific systems described there.

5.2. Positional uncertainty in engineering

In the design of nanomechanical systems, positional uncertainties stemming from thermal excitation and quantum mechanical principles are a fundamental concern. On the scale of conventional mechanical engineering, neither quantum

uncertainties nor thermal excitations are significant; the closest macroscale analogues of these effects arise in systems excited by broad-band acoustic noise, yet
issues arise there (e.g., fatigue and damping) that are alien to the molecular domain. Molecules are not subject to fatigue (for a more precise discussion, see
Section 6.7.1a). Damping, which degrades mechanical energy to heat, cannot
degrade the vibrations of heat itself.

In an engineering context, problems involving positional uncertainty can
frequently be formulated in terms of a °probability density function (PDF) for a
coordinate describing a part of a system. Mechanical designs typically require
that each part should, under specified conditions, occupy a particular position
at a particular time to within specified tolerances. Errors occur at some finite
rate owing to the finite probability mass in the tail of the PDF that extends beyond the tolerance band. Good approximations to the positional PDFs of typical systems are thus of fundamental value in nanomechanical engineering.

5.3. Thermally excited harmonic oscillators

Many parts of nanomechanical systems are adequately approximated by linear models, in which restoring forces are proportional to displacements. The
prototype of such systems is the harmonic oscillator, consisting of a single
mass with a single degree of freedom subject to a linear restoring force (measured by the stiffness k_s). Analytical results for the simple harmonic oscillator
can be adapted and extended to systems with multiple degrees of freedom, as
in the subsequent discussion of rods with internal vibrational modes.

5.3.1. Classical treatment

In classical statistical mechanics, the probability density function for the position coordinate x of a particle subject to the potential energy function $\mathcal{V}(x)$ is
[from Eq. (4.10)]

$$f_x(x) = \exp[-\mathcal{V}(x)/kT] \Big/ \int_{-\infty}^{\infty} \exp[-\mathcal{V}(x)/kT]\,dx \tag{5.1}$$

For the harmonic potential,

$$\mathcal{V}(x) = \tfrac{1}{2}k_s x^2 \tag{5.2}$$

the resulting probability density function is Gaussian:

$$f_x(x) = \frac{\exp(-k_s x^2/2kT)}{\displaystyle\int_{-\infty}^{\infty} \exp(-k_s x^2/2kT)\,dx} = \frac{\exp(-k_s x^2/2kT)}{\sqrt{2\pi}}\sqrt{\frac{k_s}{kT}}$$

$$= \frac{1}{\sqrt{2\pi}\,\sigma_{\text{class}}} \exp(-x^2\,2\sigma_{\text{class}}^2) \tag{5.3}$$

yielding the classical value for the positional variance (= standard deviation
squared = mean square displacement):

$$\sigma_{\text{class}}^2 = kT/k_s \tag{5.4}$$

This relationship has broader applicability than its derivation may suggest.

a. The irrelevance of external bombardment.   The probability density function for the position and velocity of a harmonic oscillator is independent of the nature of its coupling to the surrounding thermal bath. This may occur through vibration of its attachment point, through absorption and emission of thermal radiation, or through bombardment by molecules in a surrounding gas. None of these influences alters the potential energy function, hence none alters the Boltzmann distribution for the oscillator.

This may seem counterintuitive, given a mental image of the dynamical effects of molecular bombardment. At equilibrium, however, an impinging gas molecule is as likely to absorb energy as to deliver it, and so molecular bombardment has no net effect on the amplitude of vibration. How a system is coupled to a thermal bath can affect its detailed dynamics (e.g., the smoothness or irregularity of its trajectory, the decay time for oscillations of unusual amplitude, etc.), but not the statistical distribution of dynamical quantities. This principle holds true for systems in general, and makes the study of positional uncertainty dependent only on potential energy functions.

5.3.2. Quantum mechanical treatment

In quantum statistical mechanics, the classical integral over x (more generally, an integral over phase space, yielding a probability density function for both position and momentum) is replaced by a sum, and the probability density function is replaced by a probability distribution over a series of states i:

$$P(i) = \exp[-\mathscr{E}(i)/kT] \Big/ \sum\nolimits_{i=0}^{\infty} \exp[-\mathscr{E}(i)/kT] \tag{5.5}$$

From elementary quantum mechanics, the vibrational states $n = 0, 1, 2, 3, ...$ of a harmonic oscillator have energies that depend on the frequency

$$\mathscr{E}(n) = \left(n + \tfrac{1}{2}\right)\hbar\,\omega; \quad \omega = \sqrt{k_s/m} \tag{5.6}$$

The probability of finding the oscillator in vibrational state n is

$$P(n) = \exp\left[-\left(n+\tfrac{1}{2}\right)\hbar\,\omega\big/kT\right] \Big/ \sum\nolimits_{n=0}^{\infty} \exp\left[-\left(n+\tfrac{1}{2}\right)\hbar\,\omega\big/kT\right] \tag{5.7}$$

Rearranging and summing the series,

$$P(n) = \exp(-n\hbar\omega/kT) \Big/ \sum\nolimits_{n=0}^{\infty} \left[\exp(-\hbar\omega/kT)\right]^n \tag{5.8}$$

$$\sum\nolimits_{n=0}^{\infty} y^n = 1 + \frac{y}{1-y}, \quad |y| < 1 \tag{5.9}$$

$$P(n) = \exp(-n\hbar\omega/kT)\left[1 - \exp(-\hbar\omega/kT)\right] \tag{5.10}$$

The variance (mean square displacement) of the oscillator can be derived from its mean energy:

$$\overline{\mathscr{E}} = \sum\nolimits_{n=0}^{\infty} P(n)\mathscr{E}_n = \sum\nolimits_{n=0}^{\infty} \exp\left(-n\frac{\hbar\omega}{kT}\right)\left[1 - \exp\left(-\frac{\hbar\omega}{kT}\right)\right]\left(n+\tfrac{1}{2}\right)\hbar\,\omega \tag{5.11}$$

Rearranging and summing both series,

$$\overline{\mathscr{E}} = \hbar\omega\left[1 - \exp\left(-\frac{\hbar\omega}{kT}\right)\right]\left\{\frac{1}{2}\sum_{n=0}^{\infty}\left[\exp\left(-\frac{\hbar\omega}{kT}\right)\right]^n + \sum_{n=0}^{\infty}n\left[\exp\left(-\frac{\hbar\omega}{kT}\right)\right]^n\right\} \quad (5.12)$$

$$\sum_{n=0}^{\infty}ny^n = \frac{y}{(1-y)^2}, \quad y < 1 \quad (5.13)$$

$$\overline{\mathscr{E}} = \hbar\omega\left\{\frac{1}{2} + \left[\exp(\hbar\omega/kT) - 1\right]^{-1}\right\} \quad (5.14)$$

In a harmonic oscillator, the total energy equals twice the mean potential energy, which is proportional to the mean square displacement:

$$\frac{1}{2}\overline{\mathscr{E}} = \overline{\mathscr{V}} = \frac{1}{2}k_s\overline{x^2} = \frac{1}{2}k_s\sigma^2 \quad (5.15)$$

$$\sigma^2 = \frac{\hbar\omega}{k_s}\left\{\frac{1}{2} + \left[\exp(\hbar\omega/kT) - 1\right]^{-1}\right\} \quad (5.16)$$

Describing the frequency in terms of the fundamental mechanical parameters and then rearranging yields an equation between dimensionless quantities (see Figure 5.1):

$$\sigma^2\frac{\sqrt{k_s m}}{\hbar} = \frac{1}{2} + \left[\exp(\hbar\omega/kT) - 1\right]^{-1} \quad (5.17)$$

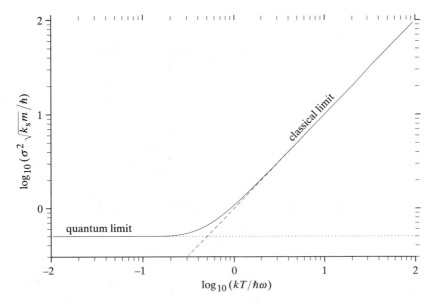

Figure 5.1. A dimensionless measure of variance vs. a dimensionless measure of temperature, Eq. (5.17).

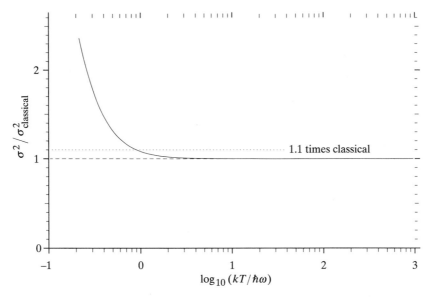

Figure 5.2. The ratio of the actual variance to that predicted by a classical model, vs. a dimensionless measure of temperature, Eq. (5.18).

It often is desirable to determine whether the classical approximation is adequate for describing positional uncertainties in a system of engineering interest. A useful measure is the ratio of the total to the classical variance, which equals the ratio of the total to the classical energy:

$$\frac{\sigma^2}{\sigma^2_{\text{class}}} = \frac{\hbar\omega}{kT}\left\{\tfrac{1}{2}+\left[\exp(\hbar\omega/kT)-1\right]^{-1}\right\} = \frac{\overline{\mathscr{E}}}{kT} \qquad (5.18)$$

This function of the parameter $kT/\hbar\omega$ is graphed in Figure 5.2; where $kT/\hbar\omega$ substantially exceeds unity, the classical variance provides a good approximation.

With less elegance (but more immediately accessible physical significance), these dimensionless quantities can be unfolded into a set of graphs of root mean square displacement (σ) as a function of temperature, spring constant, and mass (Figure 5.3). For perspective, note that the stiffness of nonbonded interactions between atoms at their equilibrium separation typically is about 0.1 N/m; that of covalent bond angle-bending, about 30 N/m; and that of covalent bond stretching, about 400 N/m. Many small components in nanomechanisms have masses in the hundreds to thousands of daltons.

5.4. Elastic extension of thermally excited rods

Various molecular structures resemble rods; these include DNA helices, microtubules, ladder polymers, and several classes of nanomechanical components. Even flexible molecular chains resemble rods (for purposes of the present analysis) when held in an extended conformation by longitudinal tension or by lateral constraints. A rod is the simplest example of an extended object and can serve as a model for various mechanical systems.

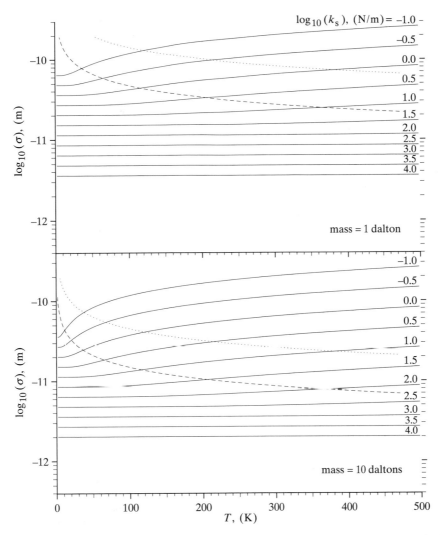

Figure 5.3.   This set of graphs plots the logarithm of the root mean square displacement of harmonic oscillators of varying mass and spring constant as a function of temperature, from Eq. (5.16). Note that the length at the top of the graphs corresponds roughly to an atomic diameter. Above the dashed lines, the classical approximation is accurate to within 10%; above the dotted lines, it is accurate to within 1%.   *(continued)*

A rod clamped at one end and free at the other undergoes both transverse and longitudinal vibrations. This section analyzes the positional variance of the free end resulting from longitudinal vibrational modes; in rods of large aspect ratio, transverse vibrations can make a significant contribution to the longitudinal positional variance of the free end (analyzed in Section 5.6). The following first considers a classical model for a continuous elastic rod, then both an approximate and an exact quantum mechanical model for a rod consisting of a series of identical masses and springs, ending with an empirical approximation to this exact model.

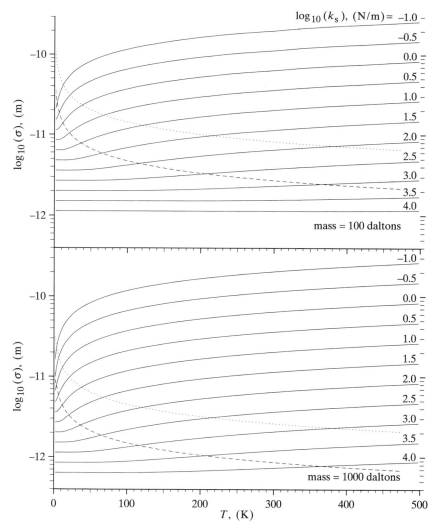

Figure 5.3. *(continued)* A 1 nm^3 volume containing 1000 daltons has $\rho \approx 1661$ kg/m^3.

5.4.1. Classical continuum treatment

A uniform rod of length ℓ clamped at one end and free at the other supports longitudinal vibrational modes ($n = 0, 1, 2, 3, ...$) with wavelengths

$$\lambda_n = \frac{4\ell}{2n+1} \tag{5.19}$$

The speed of sound v_s in the rod can be calculated from the linear modulus E_ℓ and linear density ρ_ℓ yielding the modal frequencies ω_n.

$$v_s = \sqrt{E_\ell / \rho_\ell}; \quad \omega_0 = \frac{\pi}{2\ell}\sqrt{E_\ell / \rho_\ell}; \quad \omega_n = \omega_0(2n+1) \tag{5.20}$$

Each longitudinal mode can be regarded as a harmonic oscillator with a certain frequency, °effective stiffness, and °effective mass. The effective stiffness relates the square of the amplitude (at the rod end, where the variance is to be computed) to the potential energy at maximum displacement during a vibrational cycle, which equals the maximum kinetic energy:

$$\tfrac{1}{2}k_n A_n^2 = \max(\mathcal{V}) = \max(\mathcal{T})$$

$$= \int_0^\ell \tfrac{1}{2}\rho_\ell [v(x)]^2\, dx$$

$$= \int_0^\ell \tfrac{1}{2}\rho_\ell (\omega_n A_n)^2 \sin^2\left[\frac{(2n+1)}{2\ell}\pi x\right] dx$$

$$= \tfrac{1}{4}\rho_\ell \ell (\omega_n A_n)^2 \tag{5.21}$$

This yields the effective stiffness k_n of mode n:

$$k_n = \frac{E_\ell}{\ell}\frac{\pi^2}{8}(2n+1)^2 \tag{5.22}$$

Combining this with the classical expression for positional variance in a harmonic oscillator as a function of temperature and stiffness, Eq. (5.4), the positional variance at the free end associated with mode n is:

$$\sigma^2_{n,\text{class}} = kT\frac{\ell}{E_\ell}\frac{8}{\pi^2}\frac{1}{(2n+1)^2} \tag{5.23}$$

The total variance is the sum of the modal variances:

$$\sigma^2_{\text{class}} = \sum_{n=0}^{\infty}\sigma^2_{n,\text{class}} = kT\frac{\ell}{E_\ell}\frac{8}{\pi^2}\sum_{n=0}^{\infty}\frac{1}{(2n+1)^2} \tag{5.24}$$

Applying the identity

$$\sum_{n=0}^{\infty}\frac{1}{(2n+1)^2} = \frac{\pi^2}{8} \tag{5.25}$$

yields the classical variance in the position of the free end for a rod of uniformly distributed mass and elasticity:

$$\sigma^2_{\text{class}} = kT\,\ell/E_\ell \tag{5.26}$$

This is exactly kT/k_s, where k_s is the stretching stiffness of the rod as a whole. This is the variance for a simple harmonic oscillator of the same stiffness, and hence of a "rod" having a single mass and a single mode—the opposite extreme from a rod of uniformly distributed mass and elasticity. The significance of this equality is discussed in Section 5.7.

5.4.2. Quantum mechanical treatments

a. *Discrete rod models.* For a continuous rod, a quantum mechanical treatment yields a divergent series for the positional variance, owing to the zero-point vibrations of an infinity of high-frequency modes; the continuum model is thus unacceptable even as an approximation. The following will work with a more realistic rod model (Figure 5.4) consisting of a series of N identical springs and masses, supporting N longitudinal vibrational modes. Introducing complexities

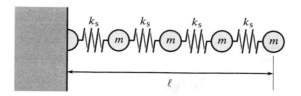

Figure 5.4. Diagram of a discrete rod, showing masses, springs, and the length coordinate (measuring between atomic centers).

such as differing masses would alter the detailed dynamics of rod vibrations, but would usually have little effect on the positional variance. More drastic, however, is the approximation that entire planes of atoms perpendicular to the rod axis can be lumped together and treated as single masses, neglecting the degrees of freedom introduced by the physical extent and flexibility of each plane.

A limiting-case analysis illustrates the essential physics. In one case (Figure 5.5), the atoms in each plane are tightly coupled to one another, each plane shares a single longitudinal degree of freedom, and the approximation under consideration is correct (by construction). In the other limiting case, a "rod" consists of an *uncoupled* bundle of j component rods, each one atom wide and m atoms long; this increases the total number of modes is by a factor of j. If a component rod has a positional standard deviation σ_c, then the position of the end of the bundle (interpreted as the mean position of the component ends) has a standard deviation

$$\sigma_b = \sigma_c / \sqrt{j} \qquad (5.27)$$

and the variance of the bundle end position is thus inversely proportional to j. As we will see, this is exactly the variance that results from treating the bundle as a single unit with $N = m$, hence the uncoupled and the tightly coupled results are identical. N may thus be taken as the number of atomic planes along the length of the rod; a conservatively generous estimate (given that real interatomic spacings are greater than 0.1 nm) is

$$N = 10^{10}\, \ell \qquad (5.28)$$

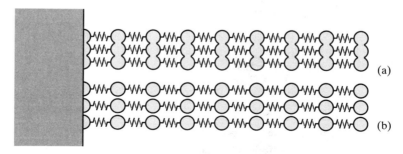

Figure 5.5. Limiting-case rods. Part (a) diagrams a rod in the limiting case in which masses and spring constants are lumped on a plane-by-plane basis; (b) diagrams the limiting case in which the rod is treated as a bundle of decoupled simple rods.

b. Semicontinuum quantum mechanical approximation. A simple approximation to a discrete system uses the continuum results but truncates the sum of the modal variances at $n = N - 1$. The continuum expressions for the modal effective stiffnesses and frequencies, together with Eq. (5.17), yield the positional variance associated with mode n in the continuum approximation:

$$\sigma_n^2 \frac{\sqrt{E_\ell \rho_\ell}}{\hbar} = \frac{4}{\pi(2n+1)} \left\{ \frac{1}{2} + \left[\exp\left(\frac{\hbar \omega_0}{kT} (2n+1) \right) - 1 \right]^{-1} \right\} \qquad (5.29)$$

Summing over the N modes in a discrete rod yields the semicontinuum approximation to the total positional variance of the free end of a clamped rod:

$$\sigma^2 \frac{\sqrt{E_\ell \rho_\ell}}{\hbar} = \sum_{n=0}^{N-1} \frac{4}{\pi(2n+1)} \left\{ \frac{1}{2} + \left[\exp\left(\frac{\hbar \omega_0}{kT} (2n+1) \right) - 1 \right]^{-1} \right\} \qquad (5.30)$$

c. Exact quantum mechanical treatment. The simple model above has serious defects when vibrational wavelengths approach the interatomic spacing of the rod ($\lambda \approx d_a$), because the dispersive wave-propagation characteristics of the discrete structure become important. These effects are large for the single mode of a "rod" with $N = 1$, and continue to be large for a substantial fraction of the modes in rods where $N \gg 1$. Defining the discrete-mass variables in terms of the continuum variables,

$$d_a = \frac{\ell}{N}; \qquad m_a = \frac{\ell \rho_\ell}{N}; \qquad k_s = \frac{E_\ell N}{\ell} \qquad (5.31)$$

Keeping the intuitive definition of the rod length as the distance from the attachment point to the last mass,

$$\lambda_n = \frac{4\ell}{(2n+1)} \left(1 + \frac{1}{2N} \right) \qquad (5.32)$$

The correct, dispersive relationship of frequency to wavelength (Ashcroft and Mermin, 1976) is

$$\omega_n = 2\sqrt{\frac{k_s}{m_a}} \sin\left(\frac{\pi d}{\lambda_n} \right) = \frac{2}{\ell} \sqrt{\frac{E_\ell}{\rho_\ell}} N \sin\left(\frac{2n+1}{2N+1} \frac{\pi}{2} \right) \qquad (5.33)$$

The effective mass m_n of a mode n is the sum over the (equal) masses of the square of the local modal amplitude, divided by the square of the modal amplitude at the free end of the rod:

$$m_n = m_a \left[\sin^2\left(\frac{2n+1}{2N+1} \pi N \right) \right]^{-1} \sum_{i=1}^{N} \sin^2\left(\frac{2n+1}{2N+1} \pi i \right) \qquad (5.34)$$

which simplifies to

$$m_n = \rho_\ell \ell \left[\sin^2\left(\frac{2n+1}{2N+1} \pi N \right) \right]^{-1} \left(\frac{1}{2} + \frac{1}{4N} \right) \qquad (5.35)$$

Given Eq. (5.16) and the above values for frequency and effective mass, the modal variance

$$\sigma_n^2 = \frac{\hbar}{m_n \omega_n}\left\{\frac{1}{2}+\left[\exp\left(\frac{\hbar \omega_n}{kT}\right)-1\right]^{-1}\right\} \tag{5.36}$$

which leads to the following dimensionless expression for the exact total variance expressed in terms of ω_0 (the frequency of the fundamental mode given by the continuum approximation):

$$\sigma^2 \frac{\sqrt{E_\ell \rho_\ell}}{\hbar} = \frac{2}{2N+1}\sum_{n=0}^{N-1}\left[\sin^2\left(\frac{2n+1}{2N+1}\pi N\right)\Big/\sin\left(\frac{2n+1}{2N+1}\frac{\pi}{2}\right)\right.$$
$$\left. \times\left(\frac{1}{2}+\left\{\exp\left[\frac{4N}{\pi}\left(\frac{\hbar \omega_0}{kT}\right)\sin\left(\frac{2n+1}{2N+1}\frac{\pi}{2}\right)\right]-1\right\}^{-1}\right)\right] \tag{5.37}$$

This equation is graphed in Figure 5.6; note that the result for $n = 0$, while identical to that given for the harmonic oscillator in Figure 5.1, is shifted to the left on this graph by the expression of results in terms of ω_0 [Eq. (5.20)] rather than in terms of the true frequency ω.

Note that division of a rod into j component rods with

$$E'_\ell = E_\ell/j; \qquad \rho'_\ell = \rho_\ell/j \tag{5.38}$$

yields

$$\sigma' = \sigma\sqrt{j} \tag{5.39}$$

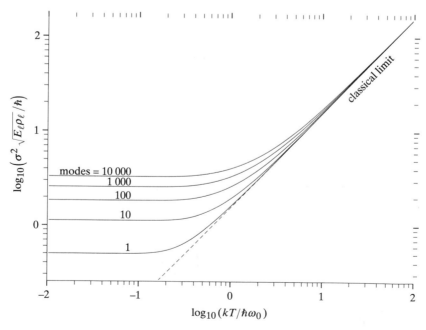

Figure 5.6. A dimensionless measure of variance vs. a dimensionless measure of temperature, for rods supporting varying numbers of longitudinal modes, Eq. (5.37).

and the standard deviation of the bundle end position (i.e., of the mean displacement of the component ends) is

$$\sigma_b = \sqrt{j\sigma'^2}\big/ j = \sigma \qquad (5.40)$$

d. An engineering approximation. For engineering purposes, it is advantageous to have a simple analytical expression rather than a sum over a variable number of terms. In seeking such an expression, we can begin with the relatively simple semicontinuum approximation, Eq. (5.30). This sum can be rearranged into two terms

$$\sigma^2 \frac{\sqrt{E_\ell \rho_\ell}}{\hbar} = \frac{2}{\pi}\sum_{n=0}^{N-1}(2n+1)^{-1} + \frac{4}{\pi}\sum_{n=0}^{N-1}\left((2n+1)\left\{\exp\left[\frac{\hbar\omega_0}{kT}(2n+1)\right]-1\right\}\right)^{-1} \qquad (5.41)$$

and the second term, which dominates in the classical limit, can be simplified by considering the classical limit ($\hbar\omega_0/kT \ll 1$):

$$\sigma^2 \frac{\sqrt{E_\ell \rho_\ell}}{\hbar} = \frac{2}{\pi}\sum_{n=0}^{N-1}(2n+1)^{-1} + \frac{4}{\pi}\frac{kT}{\hbar\omega_0}\sum_{n=0}^{N-1}(2n+1)^{-2} \qquad (5.42)$$

The first, logarithmically divergent sum can be replaced by an integral approximation corrected by a constant. Summing the second series to $N = \infty$ yields a simple expression (also transformed to eliminate ω_0) exhibiting better classical-limit behavior than the original expression:

$$\sigma^2 = \frac{\hbar}{\pi\sqrt{E_\ell\rho_\ell}}[0.54+\log(2N+1)]+kT\frac{\ell}{E_\ell} \qquad (5.43)$$

This expression is accurate in the classical limit ($\hbar\omega_0/kT \to 0$), and provides a good approximation in the quantum limit ($kT/\hbar\omega_0 \to 0$). Its deviations are strictly in a conservative direction (overestimating variance), but can amount to tens of percent for values of $kT/\hbar\omega_0 \approx 1$. A similar expression giving a more accurate result in this middle range can be obtained by multiplying the classical term by an empirically chosen function:

$$\sigma^2 = \frac{\hbar[0.54+\log(2N+1)]}{\pi\sqrt{E_\ell\rho_\ell}}+kT\frac{\ell}{E_\ell}\exp\left[-\left(0.7-\frac{0.39}{\sqrt{N}}\right)\frac{\pi\hbar}{2\ell kT}\sqrt{\frac{E_\ell}{\rho_\ell}}\right] \qquad (5.44)$$

Figure 5.7 compares the results of these two approximations to the results given by Eq. (5.37). Both are shown to give conservative estimates of the variance, and the latter equation is shown to give estimates that are within a few percent of the correct values. It can serve as an adequate approximation for most engineering purposes.

Figures 5.8 and 5.9 plot quantum mechanical variances in forms that are useful for making quick estimates. Figure 5.8 plots positional standard deviations for rods with diamondlike properties at 300 K; Figure 5.9 plots ratios of quantum to classical results for a range of properties and temperatures.

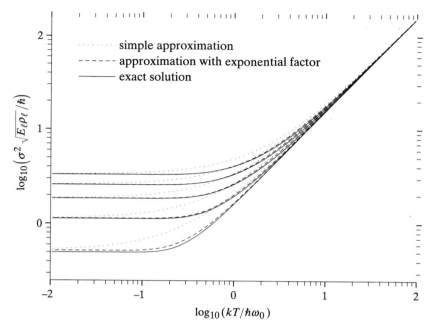

Figure 5.7. A plot of exact variances as in Figure 5.6, with the approximations of Eqs. (5.43) and (5.44) shown for comparison.

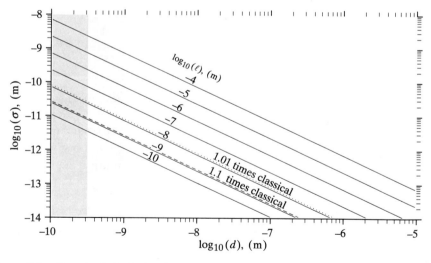

Figure 5.8. Standard deviation in elastic longitudinal displacement for the end of a thermally excited rod, plotted vs. diameter and length, assuming mechanical properties resembling those of bulk diamond ($\rho = 3500 \text{ kg/m}^3$, $E = 10^{12} \text{ N/m}^2$), $N/\ell = 10^{10}$, and $T = 300$ K. Entropic effects, neglected here, are included in Figure 5.16. Note that quantum effects make a major contribution to the positional uncertainty only for $\ell \leq 1$ nm. At sufficiently small dimensions, neglect of atomic-scale structural detail becomes unacceptable even as an approximation (e.g., the dark gray region to the left describes nonexistent subatomic diameters); at larger dimensions, this approximation is excellent.

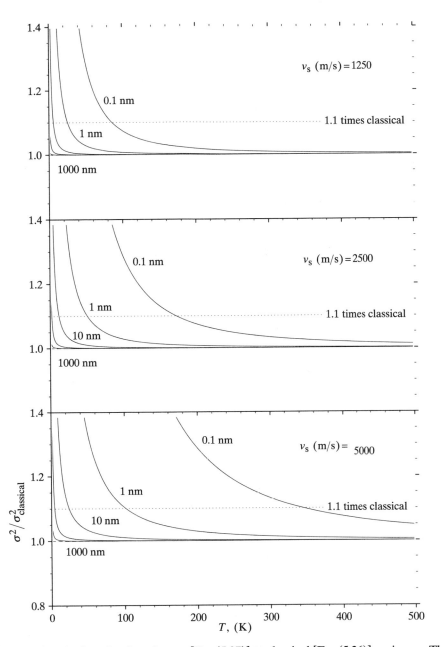

Figure 5.9.   A plot of ratios of exact [Eq. (5.37)] to classical [Eq. (5.26)] variances. The final graph in the sequence assumes an unrealistically high acoustic speed (40 km/s, vs. ~17 km/s for diamond), and thus represents an upper bound on the magnitude of quantum effects in real structures under familiar physical conditions.

(continued)

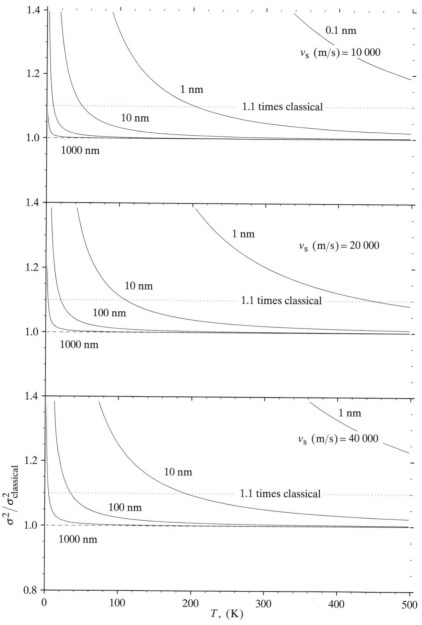

Figure 5.9. *(Continued)*

5.5. Elastic bending of thermally excited rods

The analysis of the transverse positional variance resulting from transverse vibrational modes substantially parallels the longitudinal case. The following first considers a classical model for a continuous elastic rod, then a semicontinuum quantum mechanical model, a fit to the continuum limit of this model, and a conservative approximation to the combined effects of bending and °shear deformation.

5.5.1. Classical treatment

a. Continuum model. Knowledge of modal frequencies and modal effective masses suffices to characterize the system. Following Timoshenko et al. (1974), the bending vibrations in a plane of symmetry of a uniform rod clamped at one end and free at the other (i.e., of a cantilever beam) have normal modes with shapes described by

$$y_n(x) = C_1 \sin(\kappa_n x) + C_2 \cos(\kappa_n x) + C_3 \sinh(\kappa_n x) + C_4 \cosh(\kappa_n x) \quad (5.45)$$

where $y(x)$ is the transverse displacement from the equilibrium position at point x. The constants C_1, C_2, C_3, and C_4 are determined by boundary and normalization conditions. The boundary conditions require that

$$\cos \mathcal{R}_n \cosh \mathcal{R}_n = -1; \quad \mathcal{R}_n = \kappa_n \ell \approx (n + \tfrac{1}{2})\pi \quad (5.46)$$

and the modal frequencies are given by the relationship

$$\omega_n = (\mathcal{R}_n / \ell)^2 \sqrt{k_b / \rho_\ell} \quad (5.47)$$

where k_b is the bending stiffness of the rod in J·m/rad^2, and the elastic energy per unit length resulting from bending

$$\frac{\partial \mathcal{E}_b}{\partial x} = \frac{1}{2} k_b \left(\frac{\partial^2 y}{\partial x^2} \right)^2 \quad (5.48)$$

The modal effective mass (taking the free-end displacement as the generalized position coordinate) can be found by integrating the square of the local normalized amplitude [requiring $y(\ell) = 1$] with respect to the mass:

$$
\begin{aligned}
A &= (\cos \mathcal{R}_n + \cosh \mathcal{R}_n)[\cos(\mathcal{R}_n x/\ell) - \cosh(\mathcal{R}_n x/\ell)] \\
B &= (\sin \mathcal{R}_n - \sinh \mathcal{R}_n)[\sin(\mathcal{R}_n x/\ell) - \sinh(\mathcal{R}_n x/\ell)] \\
m_n &= \int_0^\ell \frac{(A+B)^2}{(\sin \mathcal{R}_n \sinh \mathcal{R}_n)^2} \rho_\ell \, dx = \frac{\rho_\ell \ell}{4}
\end{aligned}
\quad (5.49)
$$

for all values of n. The modal frequency and the modal effective mass define the modal effective transverse stiffness resulting from bending

$$k_{t,b,n} = \frac{k_b}{4\ell^3} \mathcal{R}_n^4 \quad (5.50)$$

which (as in the longitudinal analysis) yields an expression for the total classical transverse variance resulting from bending °compliance:

$$\sigma^2_{t,b,class} = 4kT \frac{\ell^3}{k_b} \sum_{n=0}^{\infty} \frac{1}{\mathcal{R}_n^4} \quad (5.51)$$

Applying the identity

$$\sum_{n=0}^{\infty} \frac{1}{\mathcal{R}_n^4} = \frac{1}{12} \quad (5.52)$$

yields the classical expression for the transverse variance in the position of the free end of a rod of uniformly distributed mass and bending stiffness, neglecting compliance owing to shear and the effects of discreteness (see Section 5.5.1b):

$$\sigma^2_{t,b,class} = kT\ell^3/3k_b \tag{5.53}$$

But elementary theory of flexure shows that the bending stiffness of the end of a continuum cantilever rod is

$$k_{t,b} = 3k_b/\ell^3 \tag{5.54}$$

hence the classical variance is simply $kT/k_{t,b}$. As in the longitudinal case, the vibrational modes prove irrelevant to the classical analysis (see Section 5.4.1); knowledge of the overall stiffness suffices.

For a tube, the bending stiffness is

$$k_b = E\frac{\pi}{4}\left(r_2^4 - r_1^4\right) \tag{5.55}$$

and the transverse stiffness at the far end of an anchored tube is

$$k_{t,b} = E\frac{3\pi}{4\ell^3}\left(r_2^4 - r_1^4\right) \tag{5.56}$$

where r_1 and r_2 are the inner and outer tube radii, and E is °Young's modulus.

b.   Discrete model.   In bending (unlike stretching) the transition from a continuum to a lumped system changes the overall stiffness (not merely the modal stiffnesses). For a rod considered as a series of masses and angular springs, as shown in Figure 5.10, the rod-bending stiffness is related to the intermass spacings and angular stiffnesses:

$$k_b = k_\theta d_a \tag{5.57}$$

For rods in which the number of segments, N, is finite, the compliance of the rod can be described as the sum of contributions from a series of bending joints

$$\frac{1}{k_{t,b}} = \frac{\ell^3}{k_b}\sum_{n=0}^{N-1}\frac{1}{N}\left(1-\frac{n}{N}\right)^2 \approx \frac{\ell^3}{k_b}\left(\frac{3N+4}{9N-2}\right) \tag{5.58}$$

where the approximation is exact for $N = 1, 2$, and ∞; for intermediate values it is always high (i.e., conservative) and in error by less than 1%.

The preceding analysis takes account of transverse compliance caused by bending, but rods of multiatom width also exhibit compliance through shear.

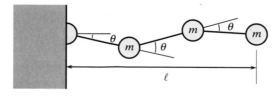

Figure 5.10.   A discrete rod, illustrating the angles entering into calculations of bending stiffness effects.

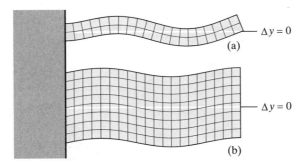

Figure 5.11. Modal deformation in pure bending (a) *vs.* pure shear (b). Bending dominates in high aspect-ratio rods; shear dominates in rods of low aspect ratio. Both deformations are here shown in idealized form.

For typical materials, shear makes a substantial contribution to the total transverse compliance, k_t, when rod diameter $d \approx \ell$; in vibration, it has significant effects on modal shapes and stiffnesses when d is of the same order as the associated wavelength. Figure 5.11 illustrates normal modes from pure bending (zero shear compliance) and pure shear (zero bending compliance) for $n = 2$, with the idealization that plane cross sections remain plane despite shear. A standard engineering approximation treats shear as an independent, additive source of compliance, with transverse stiffness effects from shear taking the same analytical form as longitudinal stiffness effects but substituting the (linear) shear modulus G_ℓ for the (linear) stretching modulus E_ℓ. Combining the resulting shear compliance, $1/k_{t,s} = \ell/G_\ell$, with Eq. (5.58) yields the engineering approximation:

$$\sigma_{t,\text{class}}^2 = kT\left[\frac{\ell^3}{k_b}\left(\frac{3N+4}{9N-2} \right) + \frac{\ell}{G_\ell} \right] \qquad (5.59)$$

Aside from quantum corrections, this approximation breaks down when ℓ/d becomes small, yet bending compliance remains important. In this regime, out-of-plane deformation of the free surface becomes a significant source of additional compliance. To treat objects of such low aspect-ratio as rods is, at best, a crude approximation.

5.5.2. Semicontinuum quantum mechanical treatment

The preceding classical treatment can serve as the basis for a quantum mechanical analysis. As in the earlier analysis of longitudinal vibrations, use of the continuum approximation neglects the dispersive properties of a discrete medium, which become significant as the characteristic wavelengths of normal modes approach interatomic distances. As will be seen, however, zero-point vibrations in the higher transverse modes (unlike those of the higher longitudinal modes) make only a modest contribution to the positional variance. Accordingly, dispersion is neglected in the following analysis; this neglect should have little effect for $N \geq 10$.

We begin with a model including only the effects of bending compliance, deferring discussion of shear compliance. (Figure 5.11 illustrates the difference between these forms of compliance.) The above values of modal effective mass

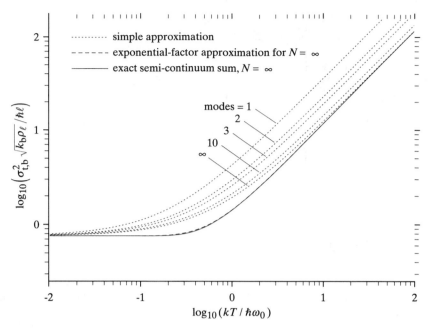

Figure 5.12. Dimensionless transverse variance for rods, neglecting shear compliance. The simple approximation is based on Eq. (5.64); the exponential-factor approximation on Eq. (5.65), and the exact semicontinuum sum on Eq. (5.60).

and effective stiffness yield the semicontinuum approximation to the transverse bending variance for a uniform rod. In dimensionless terms analogous to those of the longitudinal-mode analysis:

$$\sigma_{t,b}^2 \frac{\sqrt{k_b \rho_\ell}}{\hbar \ell} = \sum_{n=0}^{N-1} \frac{4}{\mathcal{R}_n^2} \left(\frac{1}{2} + \left\{ \exp\left[\frac{\hbar \omega_0}{kT} \left(\frac{\mathcal{R}_n}{\mathcal{R}_0} \right)^2 \right] - 1 \right\}^{-1} \right) \qquad (5.60)$$

(see Figure 5.12).

5.5.3. Engineering approximations

Expressing the variance directly,

$$\sigma_{t,b}^2 = \frac{\hbar \ell}{\sqrt{k_b \rho_\ell}} \sum_{n=0}^{N-1} \frac{4}{\mathcal{R}_n^2} \left\{ \frac{1}{2} + \left[\exp\left(\frac{\hbar}{kT} \frac{\mathcal{R}_n^2}{\ell^2} \sqrt{\frac{k_b}{\rho_\ell}} \right) - 1 \right]^{-1} \right\} \qquad (5.61)$$

Multiplying out and taking the classical limit of the second term yields

$$\sigma_{t,b}^2 = \frac{\hbar \ell}{\sqrt{k_b \rho_\ell}} \sum_{n=0}^{N-1} \frac{2}{\mathcal{R}_n^2} + \sum_{n=0}^{N-1} \frac{4}{\mathcal{R}_n^4} kT \frac{\ell^3}{k_b} \qquad (5.62)$$

In the limit of large N, the second term yields the classical continuum result, and may be replaced by the modified classical expression derived above to take account of the reduced bending stiffness of discrete rods where N is small. For the first term (describing zero-point contributions to the variance) solving the equation that defines values of $\mathcal{R}_n$ yields the convergent series

$$\sum_{n=0}^{\infty} \frac{2}{\mathcal{R}_n^2} = 0.5688 + 0.0908 + 0.0324 + 0.0165 + \cdots \approx 0.7588 \tag{5.63}$$

Substituting the infinite series limit yields the approximation

$$\sigma_{t,b}^2 = 0.76 \frac{\hbar \ell}{\sqrt{k_b \rho_\ell}} + kT \frac{\ell^3}{k_b}\left(\frac{3N+4}{9N-2}\right) \tag{5.64}$$

Given the size of the first term in the above series and the shortcomings of the continuum model for rods of low N, it is useful to consider the case $N-1$. In a rod consisting of a single mass and a single point of bending, the continuum model overestimates the effective stiffness by a factor of three and underestimates the effective mass by a factor of four, making its estimate of the quantum mechanical positional variance conservative by a factor of $(4/3)^{1/2} \approx 1.15$.

This approximation, like that of Eq. (5.43), significantly overestimates the positional variance in the transition region between the quantum and classical limits (Figure 5.12). A more accurate result for the important case of large N may be obtained by evaluating the limit of the original sum, Eq. (5.60), as $N \to \infty$ and fitting an empirical expression to the results:

$$\sigma_{t,b}^2 = 0.76 \frac{\hbar \ell}{\sqrt{k_b \rho_\ell}} + kT \frac{\ell^3}{3k_b} \exp\left(-\frac{1.97}{\ell^2}\sqrt{\frac{k_b}{\rho_\ell}}\frac{\hbar}{kT}\right) \tag{5.65}$$

This approximation is always high, but never by more than 1%. Figure 5.12 compares these approximations to Eq. (5.60).

5.5.4. Shear and bending in the quantum limit

The approximations developed for longitudinal positional variance, Eq. (5.43) and Eq. (5.44), have direct analogs for the positional variance that would occur in hypothetical rods having shear but no bending compliance. The more accurate of the two takes the form

$$\sigma_{t,s}^2 = \frac{\hbar[0.54 + \log(2N+1)]}{\pi\sqrt{G_\ell \rho_\ell}} + kT \frac{\ell}{G_\ell}\exp\left[-\left(0.7 - \frac{0.39}{\sqrt{N}}\right)\frac{\pi \hbar}{2\ell kT}\sqrt{\frac{G_\ell}{\rho_\ell}}\right] \tag{5.66}$$

In the classical limit, the effects of bending and shear compliance are simply additive, yielding expressions like Eq. (5.59). In the quantum limit, effective modal masses and frequencies play a role, and a precise analysis would have to include the effects of bending and shear on a mode-by-mode basis. An upper bound on their combined effect can be had more simply. Consider the variance of a harmonic oscillator in the quantum limit

$$\sigma^2 = \hbar/(2\sqrt{k_s m}) \tag{5.67}$$

If we consider its stiffness to have two sources, k_{s1} and k_{s2}, the variance becomes

$$\sigma^2 = \frac{\hbar}{2\sqrt{m}}\sqrt{k_{s1}^{-1} + k_{s2}^{-1}} \tag{5.68}$$

The expression

$$\sigma_{\text{est}}^2 = \sigma_1^2 + \sigma_2^2 = \frac{\hbar}{2\sqrt{k_{s1}m}} + \frac{\hbar}{2\sqrt{k_{s2}m}} \tag{5.69}$$

overestimates the actual variance by a factor

$$1 \le \frac{\sigma_{\text{est}}^2}{\sigma^2} = \frac{\sqrt{k_{s1}/k_{s2}}+1}{\sqrt{(k_{s1}/k_{s2})+1}} \le \sqrt{2} \tag{5.70}$$

Accordingly, it should be conservative to estimate the variance resulting from a vibrational mode subject to both bending and shear as the sum of the variances of a hypothetical mode constrained purely by bending forces and of one constrained by pure shear forces. (The differences between modal shapes in pure bending and those in pure shear would complicate a more precise analysis.) Thus, treating both classical and quantum variances as additive, expressions for the total transverse variance at the end of a rod take the form

$$\sigma_t^2 = \sigma_{t,b}^2 + \sigma_{t,s}^2 \tag{5.71}$$

with the choice of expressions for the shear and bending contributions depending on the desired accuracy, the magnitude of quantum effects, and the value of N. Figure 5.13 uses Eqs. (5.64) and (5.66) to graph the standard deviation of the transverse displacement at room temperature for rods with mechanical properties approximating those of bulk diamond. As can be seen, under these conditions, in the regime where shear and bending compliance are both important, quantum effects on positional uncertainty are minor for rods of nanometer or greater size; accordingly, Eq. (5.59) provides a good approximation.

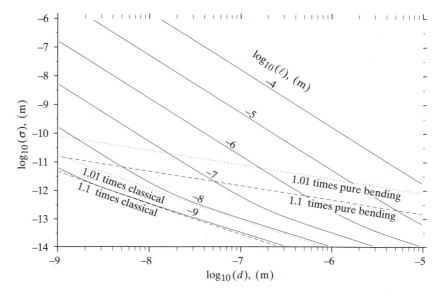

Figure 5.13. Transverse positional standard deviation for rods at 300 K, including shear compliance. The bending component is based on Eq. (5.64); the shear component is based on Eq. (5.66), assuming mechanical properties resembling those of bulk diamond, as in Figure 5.8; similar remarks apply.

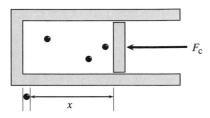

Figure 5.14. Piston, cylinder, and gas molecules. Note that the position coordinate x measures not the distance of the piston from the bottom of the cylinder, but the range of motion available to a gas molecule. A hard-sphere, hard-surface model is assumed here and in the text.

5.6. Piston displacement in a gas-filled cylinder

Earlier sections have considered linear, elastic systems in which the motion can be divided into normal modes, treated as independent harmonic oscillators. A different approach is necessary for nonlinear systems in which the displacement of one component affects the range of motion possible to another, that is, for *entropic springs*. Here, the simplest example is not a mass and spring, but a loaded piston in a cylinder containing an ideal gas.

Figure 5.14 illustrates the system and the defining coordinates. The diameter of the cylinder proves irrelevant, and the displacement of the piston is chosen such that a zero displacement corresponds to zero freedom of movement for the gas molecule in the x direction. The assumption of an ideal gas entails ignoring all forces between gas molecules, and accordingly ignoring the reduction in available volume that a molecule experiences as a result of the bulk of other molecules that may be present. Classical positional uncertainty in this system will be analyzed from three (ultimately equivalent) perspectives.

5.6.1. Weighting in terms of potential energy and available states

Let the compressing force on the piston be a constant, F_c. For an empty cylinder, the potential energy of the piston is then $F_c x$, and applying the Boltzmann weighting to this potential yields the exponential PDF

$$f_x(x) = \frac{\exp(-F_c x/kT)}{\int_0^\infty \exp(-F_c x/kT)\,dx} = \frac{F_c}{kT}\exp\left(-\frac{F_c x}{kT}\right); \quad x \geq 0 \tag{5.72}$$

The addition of N gas molecules does not change the potential energy, but does increase the number of states associated with each piston position by a factor proportional to the space available to each molecule, that is, by a factor proportional to x^N. Introducing this factor to account for the number of states yields the Erlang PDF

$$f_x(x) = \frac{x^N \exp(-F_c x/kT)}{\int_0^\infty x^N \exp(-F_c x/kT)\,dx} \tag{5.73}$$

$$= \frac{1}{N!}\left(\frac{F_c}{kT}\right)^{N+1} x^n \exp\left(-\frac{F_c x}{kT}\right); \quad x \geq 0$$

with mean and variance

$$\bar{x}=(N+1)\frac{kT}{F_c}; \quad \sigma_x^2=(N+1)\left(\frac{kT}{F_c}\right)^2 \tag{5.74}$$

In this approach, the configurational states of the system are treated as known, and the Boltzmann-weighted probabilities are then integrated over these states.

5.6.2. Weighting in terms of a mean-force potential

Alternatively, one can treat the gas as a nonlinear spring described by the ideal gas equation (here in molecular rather than the more familiar molar units)

$$pV = NkT \tag{5.75}$$

yielding the time-average force owing to pressure

$$F_p = -NkT/x \tag{5.76}$$

Since the gas is now treated as a spring external to the piston, its configurational states are now ignored (as discussed in Section 4.3.7). Treating F_p on the same basis as F_c, the work-energy as a function of position is

$$\int F_c - (NkT/x)\,dx = F_c x - NkT\log x + C \tag{5.77}$$

hence the Boltzmann-weighted positional PDF is

$$f_x(x)=\frac{\exp\left(-\dfrac{F_c x}{kT}-n\log x+\dfrac{C}{kT}\right)}{\displaystyle\int_0^\infty \exp\left(-\dfrac{F_c x}{kT}-n\log x+\dfrac{C}{kT}\right)dx}=\frac{x^N \exp(-F_c x/kT)}{\displaystyle\int_0^\infty x^N \exp(-F_c x/kT)\,dx} \tag{5.78}$$

Note that the nonlinearity of the stiffness invalidates the simple relationship

$$\sigma_x^2 = kT/k_s \tag{5.79}$$

Evaluating k_s at the mean and most probable values of x

$$\frac{kT}{k_{s,mean}}=\frac{(N+1)^2}{N}\left(\frac{kT}{F_c}\right)^2; \quad \frac{kT}{k_{s,max\text{-}prob}}=N\left(\frac{kT}{F_c}\right)^2 \tag{5.80}$$

yields differing results, both of which differ from the true variance, Eq. (5.74). Both expressions, however, approach the correct value in the limit of large N, where the standard deviation in position is small enough for a linear approximation to hold for fluctuations of ordinary magnitude.

5.6.3. Weighting in terms of the Helmholtz free energy

In a third approach, we begin with the Helmholtz free energy,

$$\mathcal{F} = \mathcal{E} - T\mathcal{S} \tag{5.81}$$

The internal energy $\mathcal{E}$ is the sum of the potential energy $F_c x$, the kinetic energy of the gas ($3/2\,NkT$, assuming a monatomic gas), and of the piston ($3kT$, assuming freedom to slide, rotate, and rattle). The translational entropy of an ideal gas

(Knox, 1971) is

$$\mathscr{S}_{\text{trans}} = Nk\left(\frac{5}{2} + \frac{3}{2}\ln\frac{mkT}{2\pi\hbar^2} - \ln\frac{N}{V}\right) = Nk\left(C - \ln\frac{N}{x}\right) \tag{5.82}$$

hence the free energy of the system as a function of piston position is

$$\mathscr{F} = \mathscr{E} - T\mathscr{S} = F_c x + \left(3 + \frac{3}{2}N\right)kT - NkT\left(C - \ln\frac{N}{x}\right) \tag{5.83}$$

Taking the Boltzmann-weighted PDF in terms of the free energy again yields

$$\begin{aligned}
f_X(x) &= \frac{\exp\left(-\dfrac{F_c x}{kT} + 3 + \dfrac{3}{2}N - NC + N\ln\dfrac{X}{N}\right)}{\displaystyle\int_0^\infty \exp\left(-\dfrac{F_c x}{kT} + 3 + \dfrac{3}{2}N - NC + N\ln\dfrac{X}{N}\right)dx} \\
&= \frac{x^N \exp(-F_c x/kT)}{\displaystyle\int_0^\infty x^N \exp(-F_c x/kT)\,dx}
\end{aligned} \tag{5.84}$$

5.6.4. Comparison and quantum effects

These three perspectives on the problem are closely related: The variation in available states as a function of piston position considered in the first corresponds to the variation in entropy considered in the third. The second approach considers movement of the piston as doing work on the gas (under reversible, isothermal conditions); work done under these conditions equals the change in $\mathscr{F}$ considered in the third. The choice of perspective in such problems is a matter of convenience.

Physical displacements frequently couple to changes in the range of motion available in some degree of freedom, hence altering its entropy and doing work against an entropic spring. These systems share the basic properties of the piston and cylinder system just considered, including a $(kT)^2$ dependence of positional uncertainty in the classical regime, as opposed to the kT dependence of classical uncertainty stemming from conventional springs. In the quantum limit, systems that would exhibit entropic spring effects at higher temperatures will display a small residual compliance stemming from the compression of zero-point probability distributions. This is not an entropic effect, since (in the quantum limit) all modes are consistently in their ground state with an invariant entropy of zero; nonetheless, it is a compressive effect that is a smooth extension of the behavior in the classical regime.

5.7. Longitudinal variance from transverse deformation

Transverse vibrations in a rod cause longitudinal shortening by forcing it to deviate from a straight line. This longitudinal-transverse coupling causes longitudinal positional variance, and provides another example of an entropic spring. Although the following analysis is not used in this volume, sheathed flexible rods like those described have applications in a wide range of nanomechanical systems.

5.7.1. General approach

Rods sliding in channels can be used to couple mechanical displacements occurring at one location to displacements at a relatively distant location. Thus, it is of interest to consider the longitudinal positional variance of rods in systems where rod motion occurs in a channel that imposes transverse restoring forces (modeled as a stiffness per unit length, k_ℓ) via overlap repulsion, and where boundary conditions may impose a mean tension γ_ℓ on the rod.

The effect of shear compliance on transverse vibration is usually small in systems where longitudinal-transverse coupling is significant. Further, in the approximation that the rod behaves as an isotropic, elastically linear material, shear deformation has no effect on length. Shear is accordingly neglected.

The discrete structure of the rod imposes a limit to the number of modes and modifies the coupling constants as λ_n approaches $2\Delta\ell$ $(= \lambda_{min})$. The following analysis adopts a semicontinuum model that takes account of limitations on modes while neglecting changes in coupling constants. This approximation can break down in systems where modes with $\lambda \approx \lambda_{min}$ dominate the variance. Where the restoring force for these modes is dominated by k_ℓ, the approximation is conservative; where the restoring force is dominated by γ_ℓ, the approximation is accurate; where the restoring force is dominated by k_b, the approximation is too low. In this case, the maximum correction factor for the variance (for $\lambda = \lambda_{min}$) is $(\pi/2)^4 \approx 6.09$, falling to 1.52 for $\lambda = 2\lambda_{min}$ and to 1.02 for $\lambda = 10\lambda_{min}$.

The nonlinearity of overlap repulsions makes the use of the constant stiffness k_ℓ a rough approximation. Since stiffness increases with displacement, this approximation is conservative, underestimating the constraining forces. Discussion of quantum mechanical effects is deferred to Section 5.7.3.

5.7.2. Coupling and variance

The following analysis considers a weighting in terms of potential energy and available states for each of a set of normal-mode deformations, summing the resulting variances to yield the total variance. The use of normal modes here implies nothing about vibrations, but merely provides a convenient set of orthogonal functions with which to describe all possible rod configurations.

For the sinusoidal deformation characteristic of mode n, there exists a constant $\mathscr{C}_n$ (dependent on rod parameters) relating the contraction in the length of the rod to the potential energy of the deformation

$$\Delta \ell_n = \mathscr{C}_n \mathscr{E}_n \qquad (5.85)$$

The classical variance in rod length resulting from modal longitudinal-transverse coupling can be determined from the PDF for the potential energy (which in turn can be derived from the Gaussian PDF for the amplitude),

$$f_{\mathscr{E}_n}(\mathscr{E}_n) = \exp(-\mathscr{E}_n/kT)/\sqrt{\pi kT \mathscr{E}_n} \qquad (5.86)$$

and for the contraction in length

$$f_{\Delta\ell_n}(\Delta\ell_n) = \exp(-\Delta\ell_n/\mathscr{C}_n kT)/\sqrt{\pi \mathscr{C}_n kT \Delta\ell_n} \qquad (5.87)$$

From this one can derive the variance in the potential energy

$$\sigma_{\mathscr{E}_n}^2 = \overline{\mathscr{E}_n^2} - \left(\overline{\mathscr{E}_n}\right)^2$$

$$= \int_0^\infty \frac{\mathscr{E}_n^2 \exp(-\mathscr{E}_n/kT)}{\sqrt{\pi kT \, \mathscr{E}_n}} \, d\mathscr{E}_n - \left[\int_0^\infty \frac{\mathscr{E}_n \exp(-\mathscr{E}_n/kT)}{\sqrt{\pi kT \, \mathscr{E}_n}} \, d\mathscr{E}_n\right]^2$$

$$= \tfrac{1}{2}(kT)^2 \tag{5.88}$$

and hence the variance in length resulting from longitudinal-transverse coupling in mode n.

$$\sigma_{\ell,\iota,n}^2 = \tfrac{1}{2}(\mathscr{C}_n kT)^2 \tag{5.89}$$

The mean contraction is

$$\overline{\Delta\ell_n} = \tfrac{1}{2}\mathscr{C}_n kT \tag{5.90}$$

5.7.3.   Rods with tension and transverse constraints

Consider a long, continuous rod with a bending modulus k_{b}, subject to a mean tension γ_ℓ and a transverse stiffness per unit length k_ℓ. The energy per unit length associated with a sinusoidal deformation can be derived by integrating the contributions from each of these restoring-force terms. For a rod of length ℓ supporting modes with amplitudes A_n and $\lambda = 2\ell/n$ ($n = 1,2,3,\ldots$),

$$\frac{\mathscr{E}_n}{\ell} = \frac{A_n^2}{4}\left[k_{\mathrm{b}}\left(\frac{n\pi}{\ell}\right)^4 + \gamma_\ell\left(\frac{n\pi}{\ell}\right)^2 + k_\ell\right] \tag{5.91}$$

The fractional change in length is

$$\frac{\Delta\ell_n}{\ell} = \left(\frac{n\pi A_n}{2\ell}\right)^2 \tag{5.92}$$

hence

$$\mathscr{C}_n = \frac{\Delta\ell_n}{\mathscr{E}_n} = \left[k_{\mathrm{b}}\left(\frac{n\pi}{\ell}\right)^2 + \gamma_\ell + k_\ell\left(\frac{n\pi}{\ell}\right)^{-2}\right]^{-1} \tag{5.93}$$

and the total (classical) variance resulting from transverse vibrations in one of the two possible polarizations equals the sum of the modal variances

$$\sigma_{\ell,\mathrm{t}}^2 = \sum_{n=1}^N \tfrac{1}{2}(\mathscr{C}_n kT)^2$$

$$= \sum_{n=1}^N \tfrac{1}{2}(kT)^2\left[k_{\mathrm{b}}\left(\frac{n\pi}{\ell}\right)^2 + \gamma_\ell + k_\ell\left(\frac{n\pi}{\ell}\right)^{-2}\right]^{-2}; \qquad N = \ell/\Delta\ell \tag{5.94}$$

The total variance is the sum of contributions from both polarizations; if these are equivalent, the total is twice the above value.

Stretching the rod compresses the range of motion of the transverse modes, doing work and reducing modal entropies; from a mean-force potential perspective, transverse modes introduce a source of (nonlinear) longitudinal compliance. This perspective clarifies the nature of the quantum effects: Each transverse mode can be regarded as a harmonic oscillator with a restoring force modulated by the degree of rod extension. In the quantum-mechanical limit, the transverse positional variance is proportional not to the transverse compliance (as in the classical regime)

$$\sigma^2_{\text{classical limit}} = kT/k_{\text{s}} \qquad (5.95)$$

but to its square root

$$\sigma^2_{\text{quantum limit}} = \hbar/\left(2\sqrt{k_{\text{s}}m}\right) \qquad (5.96)$$

hence as quantum effects become significant, mean transverse amplitudes are less easily compressed by increases in restoring forces. Since the longitudinal compliance just described results from compression of transverse modes as a result of increasing tension, it is lower in the quantum regime than in the classical regime, for a given transverse amplitude. Quantum effects on longitudinal positional variance are neglected here.

To permit a graphical summary, the sum over the modes can be expressed in terms of two parameters, k_{b}/k_ℓ and γ_ℓ/k_ℓ

$$\left(\frac{\sigma^2_{\ell,\text{t}}}{\ell}\right)\frac{k_{\text{b}}\sqrt{\gamma_\ell k_\ell}}{(kT)^2} = \lim_{\ell\to\infty}\sum_{n=1}^{\ell/\Delta\ell}\frac{1}{2\ell}\left(\frac{n\pi}{\ell}\right)^4\frac{k_{\text{b}}}{k_\ell}\sqrt{\frac{\gamma_\ell}{k_\ell}}\left[\frac{k_{\text{b}}}{k_\ell}\left(\frac{n\pi}{\ell}\right)^4 + \frac{\gamma_\ell}{k_\ell}\left(\frac{n\pi}{\ell}\right)^2 + 1\right]^{-2} \qquad (5.97)$$

Because the numerators and denominators of these parameters can both vary over many orders of magnitude, even for submicron systems, the ranges covered by Figure 5.15 are large.

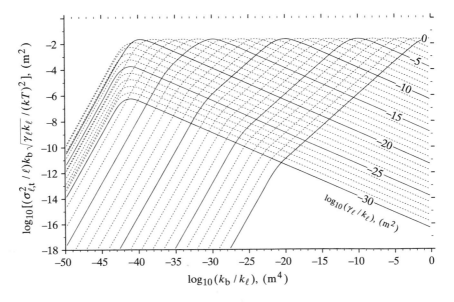

Figure 5.15. A measure of longitudinal variance vs. ratios of restoring forces, based on Eq. (5.97).

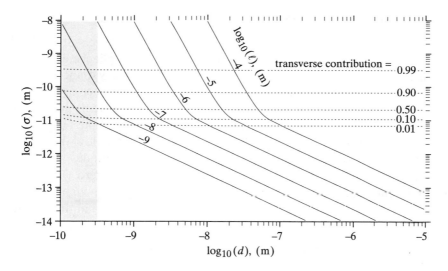

Figure 5.16. Total standard deviation in longitudinal displacement for the end of a thermally excited rod, including elastic and entropic contributions. Plotted vs. diameter and length, assuming mechanical properties resembling those of bulk diamond (as in Figure 5.8); similar remarks apply.

5.7.4.   Rods with freely sliding ends and no transverse constraint

The absence of transverse constraint forces and applied tension increases positional variance. For rods of finite length, angular freedom at the ends again increases the positional variance. Thus, these conditions are appropriate for setting upper bounds on the longitudinal positional variance resulting from longitudinal-transverse coupling in rods where the ends are constrained to the axis (or, equivalently, where the coordinate measured is the distance between the ends) but are otherwise free. Under these conditions, Eq. (5.94) simplifies to

$$\sigma_{\ell,t}^2 = \frac{1}{2}\left(\frac{kT}{k_b}\right)^2\left(\frac{\ell}{\pi}\right)^4 \sum_{n=1}^{N}\frac{1}{n^4} \approx \frac{1}{2}\left(\frac{kT}{k_b}\right)^2\left(\frac{\ell}{\pi}\right)^4 1.082 \qquad (5.98)$$

in which the latter expression represents the rapidly approached limit as N becomes large. Again, contributions from both polarizations must be included. For convenience, Figure 5.16 displays the total longitudinal positional uncertainty resulting from the combined effects of this mechanism (including both equivalent polarizations) and of longitudinal modes as analyzed in Section 5.4.2c, all at 300 K for rods with properties like those of bulk diamond.

It should be noted that the PDF for the value of $\Delta\ell$ resulting from longitudinal-transverse coupling involving a single mode is not Gaussian, but takes the form of Eq. (5.87). The PDF for the case just considered roughly approximates this, since it is dominated by the contribution of a single mode. Where many modes contribute, however, their sum approximates a Gaussian; the differences chiefly affect the tails of the distribution.

5.8.   Elasticity, entropy, and vibrational modes

Elastic and entropic springs show distinct behaviors. In elastic systems in the classical regime, positional variance is proportional to compliance and to temperature. Quantum effects in elastic systems increase the variance over the classical value, and contribute a greater fraction of the variance as stiffnesses increase and as masses and temperatures fall. In entropic systems in the classical regime, positional variance is proportional to the square of the temperature. Quantum effects in entropic systems reduce the variance below the classical value; in the quantum limit of large $\hbar\omega/kT$, all vibrational modes are in their ground state, the entropy is zero, and the entropic variance is zero. (Despite the foregoing, the characteristics that yield entropic-spring behavior at elevated temperatures contribute a compliance that does not go to zero in the quantum limit.) Distinguishing these cases leads to substantial simplifications in analyzing elastic systems within the classical-mechanics approximation.

5.8.1.   Neglect of vibrational modes in classical elastic springs

In analyzing the PDFs of elastic springs with respect to some set of coordinates in the classical limit, modal analysis can be bypassed: only the potential energy as a function of these coordinates is significant. In the quantum regime, an analysis of vibrational modes serves to capture the effects of zero-point energy. In the classical regime, an analysis of vibrational modes is merely a form of entropic bookkeeping; it provides a means of describing all possible system configurations, a step that is useful in analyzing entropic springs but irrelevant in analyzing elastic springs. The simple results Eq. (5.26) and Eq. (5.53) are direct consequences of this.

In the derivation of Eq. (5.26), rod configurations with an end displacement $A_{\text{end}} = \Sigma A_n$ were described as a sum of terms

$$\sum_{n=0}^{\infty} A_n \sin[(2n+1)\pi x/2\ell] \qquad (5.99)$$

with many possible combinations of A_n for a given value of A_{end}. One could equally well describe configurations as a sum of terms

$$A_{\text{end}}x + \sum_{n=0}^{\infty} A_n' \sin[(n+1)\pi x/\ell] \qquad (5.100)$$

This form separates the end displacement from orthogonal rod deformations; the relationship of end position to mean energy, system entropy, etc., would be the same if these deformations were embodied in an entirely separate object. Note that this recasting does not apply in the quantum regime because the deformation $A_{\text{end}}x$ does not correspond to a vibrational mode.

In general, for purely elastic systems in the classical limit, the PDF associated with a set of coordinates can be evaluated by applying a Boltzmann weighting to the potential energy as a function of those coordinates. For linear systems, the result is a Gaussian distribution (a result that holds in the quantum regime as well). Note that, in the absence of imposed accelerations, inertial mass plays no role in determining equilibrium PDFs for the position coordinates of a system, though it does affect the dynamics of fluctuations. For such systems, variance scales linearly with temperature and compliance; accordingly, variance is inversely proportional to linear dimensions, given uniform scaling of an elastic structure.

5.8.2.   Conservative scaling of variance with temperature

Considering quantum effects in elastic systems, it is always conservative to scale up an accurate (or conservative) variance in proportion to temperature or compliance; it is not conservative to scale down in the same manner. At room temperature, objects made of materials as stiff and light as diamond have positional uncertainties dominated by classical effects so long as dimensions exceed one nanometer (Figure 5.8).

For purely entropic systems in the classical regime, variance scales as the square of temperature. Entropic effects in the deformation of structural elements become important as aspect ratios become large (Figure 5.16). For such systems, the variance can be treated as the sum of entropic and elastic contributions.

5.9.   Conclusions

Harmonic oscillators and elastic rods have positional uncertainties described by Gaussian PDFs. At ordinary temperatures in all but very light, stiff systems (e.g., single atomic masses and bond-stretching displacements), the classical value for the positional variance given by Eq. (5.4) is accurate enough for all but the most precise work. Elsewhere in this volume, the greater size of components and the uncertainties in their stiffness values render the use of more elaborate expressions pointless.

Entropic springs have positional uncertainties that scale differently with temperature. Displacement of an entropic spring does no work against elastic restoring forces, but instead does work against a mean-force potential, imposing position-dependent constraints on motions in other degrees of freedom. Descriptions of two systems have been presented.

6

Transitions, Errors, and Damage

6.1. Overview

In the configuration-space picture, a transition in a nanomechanical system corresponds to a motion of the °representative point from one potential well to another. Errors occur when a system executing a pattern of motion transiently enters an undesired potential well; damage occurs when a system permanently exits the set of desired potential wells. Section 6.2 describes standard models used to describe transitions in molecular systems; these are used in later sections to analyze errors and thermomechanical damage. Section 6.3 describes methods for modeling subsystems within machines in terms of time-dependent potential energy surfaces, then uses this approach to model error rates in placing molecular objects into potential wells.

The balance of the chapter examines damage mechanisms. These can be categorized by the energy source that enables the system to surmount the energy barrier that separates working from damaged states. The two broad categories are:

- Internal energy sources, including thermal excitation, mechanical stress, and electromagnetic fields
- External energy sources, including electromagnetic fields and energetic photons and charged particles

Section 6.4 examines the effects of thermal excitation and mechanical stress, drawing on theoretical models of bond cleavage and experimental models of chemical reactivity. Section 6.5 examines photochemical damage and conditions for avoiding it by means of opaque shielding; Section 6.6 examines ionizing radiation damage, presenting models based on experimental studies of damage to biological molecular machines. Section 6.7 summarizes the engineering consequences of these damage mechanisms for the reliability of nanomechanical systems, given the assumption that single-point damage always causes component failure.

Owing to scaling laws (Chapter 2), the magnetic fields that can be generated within nanoscale systems are modest. Where chemical transitions are concerned, these will (at most) influence rates and equilibria in electronic phenomena such as °triplet-singlet intersystem crossing. In typical molecular structures, even the most intense magnetic fields generated in terrestrial physics laboratories cause no damaging transitions; further discussion of chemical effects of magnetic fields is accordingly omitted here. Electric fields, in contrast, can cause electrical breakdown, charge transfer, and molecular damage; Sections 6.4.7 and 11.6.2 touch on these topics.

6.2.   Transitions between potential wells

6.2.1.   Transition state theories

The previous chapter examined the effects of thermal excitation and quantum uncertainty in systems that can be described as points oscillating in a single potential well. This section examines thermal and quantum effects in systems characterized by a PES having two or more potential wells. Within each potential well, the location of the state point can be described by a PDF resembling those derived in the previous chapter. This PDF differs, however, in extending over a col that leads to another potential well; as a consequence a system in one well has a finite probability per unit time of moving to another. *Transition state theories* (TSTs) use a set of approximations to compute these transition rates.

Transition state theories find extensive use in studies of chemical kinetics. They are of little value where conditions are far from microscopic equilibrium (for example, during the first vibration-cycle after a molecule has been photo-chemically excited) but are not impaired by macroscopic disequilibrium (for example, during the first seconds or nanoseconds after two reagents have been mixed). The discussion in Section 4.3.4 of the utility of equilibrium statistical mechanics in nonequilibrium systems is applicable here.

Classical and quantum mechanical TSTs of varying complexity and accuracy have been developed. In the present context, there is little incentive to seek moderate improvements in accuracy at the cost of great increases in complexity. In evaluating transitions in nanomechanical designs, it usually suffices to establish either (1) that an undesirable transition does (or does not) occur at a sufficiently low rate in a given structure, or (2) that a desirable transition does (or does not) occur at a sufficiently high rate with a given reagent or other molecular device. Since transition rates at 300 K commonly vary from $>10^{12}$ s^{-1} to $<10^{-50}$ s^{-1}, an approximate value frequently suffices to distinguish a workable from an unworkable design. The solution-chemist's concern with 10-fold differences in the rates of competing reactions seldom arises.

6.2.2.   Classical transition state theories

Transition state theories (both classical and quantum mechanical) start with a description of the col surrounding the *transition state* of a PES. Within the col between the wells is a saddle point; the surrounding region is locally describable by a quadratic function. With a suitable choice of coordinate axes, the curvature of the surface is negative along one axis (which defines the direction of the *reaction coordinate*) and positive along the other axes. In the most common definition, the transition state corresponds to a surface perpendicular to the reaction coordinate, which passes through the saddle point and divides points corresponding to one well from those corresponding to the other. Some versions of transition state theory define different transition surfaces. (Note that this surface has one fewer spatial dimension than the potential energy surface; the next section speaks of *volumes* that correspond to regions of the PES and *transition surfaces* that separate those regions.)

In solution- and gas-phase chemistry, molecules are free to translate and rotate, and molecular number-densities are important variables. In nanomechanical systems, mechanical constraints can eliminate both translational and rotational degrees of freedom, and most processes are effectively intramolecular, making number densities irrelevant to transition rates. Accordingly, the following treatment is restricted to systems in which the potential wells are localized and distinctly bounded, permitting only vibrational motions within (and transitions between) wells.

a. Standard classical TST.   Elementary transition state theory expresses the transition rate, k_{12} (s^{-1}), as

$$k_{12} = \frac{kT}{2\pi\hbar}\frac{q^{\ddagger}}{q_1}\exp\left(-\Delta\mathcal{V}^{\ddagger}/kT\right) \tag{6.1}$$

In the classical theory, q_1 is the classical °partition function (Section 4.3.2) describing the system when confined to the region of potential well 1, and $q^{\ddagger}$ is the classical partition function describing a hypothetical system constrained to move along the transition surface. In calculating the transition-state partition function, the position and momentum coordinates corresponding to motion along the reaction coordinate are dropped from the integral in Eq. (4.8). The quantity $\Delta\mathcal{V}^{\ddagger}$ represents the barrier height, that is, the potential energy difference between the bottom of well 1 and the saddle point of the col (the superscript ‡ denotes a variable describing a transition state).

Although correct within the classical approximation, Eq. (6.1) does not directly suggest a physical visualization that can aid engineering design. Clear expositions of transition state theory can be found (e.g., Knox, 1971). Other standard expositions, however, speak in terms of fictitious concentrations of fleeting, arbitrarily defined "activated complexes"; these seem more confusing than enlightening. Finally, the presence of $\hbar$ in a classical theory suggests that a detour has been taken somewhere along the path from classical concepts to classical consequences. An alternative approach may help to build a more intuitive understanding, at least of the classical model.

b. The probability-gas visualization.   In a purely classical model, the physics behind transition state theory can be visualized in terms of the diffusion of a probability gas (as introduced in Section 4.3.3). When the state of a system is uncertain (e.g., at equilibrium), motion of the representative point must be treated statistically. Rather than thinking in terms of the behavior of a single point over an indefinitely long period of time, one can instead think in terms of the positions and velocities of an indefinitely large number of independently moving points at a single instant. In mass-weighted coordinates (Section 4.3.3), these points behave like an ideal gas of uniform molecular weight and zero collision cross section: they move with the familiar isotropic Gaussian velocity distribution at every point in the space, and their density at every point varies in proportion to the Boltzmann factor, $\exp[-\mathcal{V}(position)/kT]$. With a single "gas molecule," the density of the gas corresponds to the probability density. The diffusion of the gas through a col corresponds to the dynamics described by transition-state theory.

Figure 6.1(a) shows a top view of a two-dimensional square-well potential like that shown in 6.1(b). Since $\exp(-\mathcal{V}/kT)$ is a constant within the enclosed region, the distribution of the probability gas is uniform at equilibrium. Although there are no subsidiary wells, no proper col, and hence no transition state as ordinarily defined, one can divide the region into two nearly circular wells by defining a suitable transition surface, shown as the dotted line in 6.1(a). At equilibrium, the region on the right (state 2) contains more probability gas than the region on the left (state 1): in a chemical system, state 2 would be described as "having a lower free energy owing to entropic effects."

The rate of transitions from 1 to 2 (given initial occupancy of 1) is easy to calculate. State 1 has a certain volume (area, in two dimensions) and the surface between the states has a certain area (width, in two dimensions). The transition rate is simply the volumetric rate at which probability gas exits the region of state 1 through the available patch of transition surface divided by the total volume of gas in state 1:

$$k_{12} = v_{\text{mean}} S_{\text{trans}}/V_1 \tag{6.2}$$

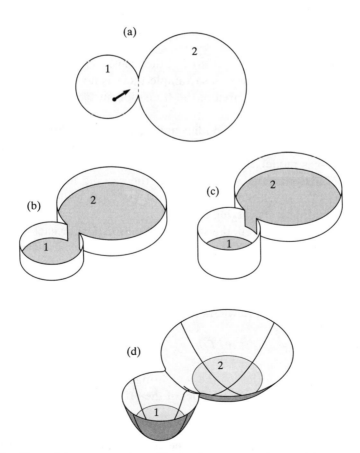

Figure 6.1. Potential energy surfaces illustrating concepts in transition state theory.

where v_{mean} is the mean speed of particles of probability gas along a given direction (that is, the average of the absolute value of the velocity along that direction, divided by two to discount particles traveling in the wrong direction). This expression generalizes to any number of dimensions: An N dimensional volume is always bounded by an $N-1$ dimensional surface; each patch of surface transmits gas at a rate equaling its area times the mean speed perpendicular to its surface, and the mean speed (in the above sense) is independent of the number of dimensions. In mass-weighted coordinates,

$$v_{\text{mean}} = \int_0^\infty \frac{v}{\sqrt{2\pi kT}} \exp\left(-v^2/2kT\right) = \sqrt{kT/2\pi} \tag{6.3}$$

In a slightly more complicated model, Figure 6.1(c), one region has a lower potential energy than the other. At equilibrium, the distribution of gas between states 1 and 2 is determined by the product of their volumes and the Boltzmann factor $\exp(-\mathcal{V}/kT)$ for each region: this product defines a density-weighted volume, or *effective volume*. For suitable choices of V_1/V_2, $\Delta\mathcal{V}$, and T, state 1 will be favored and would be described as "having a lower free energy, despite its lower entropy." The calculation of transition rates from 1 to 2 proceeds as before, but with a Boltzmann factor to account for the lower gas density at the "altitude" of the transition surface between the regions:

$$k_{12} = \frac{v_{\text{mean}}}{V_{\text{state 1}}} S_{\text{col}} \exp\left(-\Delta\mathcal{V}^\ddagger/kT\right) \tag{6.4}$$

This factor reduces the *effective area* of the transition surface relative to the volume of the well. Note that in this example there is no barrier in the reverse direction, hence the effective area of the transition surface relative to well 2 is unchanged.

Figure 6.1(d) represents a smoother and more realistic PES to which quadratic approximations and classical transition state theory can be applied. In the probability-gas model, the effective volume of each well now must reflect the integral of a Boltzmann weighting over the varying floor height, and the effective area of the transition surface (with respect to a well) must likewise be weighted to account for its varying height.

In a general harmonic well, the stiffness k_s may differ for each coordinate.* Along any given coordinate, the effective width of the well is (without mass weighting)

$$w_{\text{effective}} = \sqrt{2\pi}\sigma = \sqrt{2\pi kT/k_s} \tag{6.5}$$

or, with mass weighting as elsewhere in this section,

$$w_{\text{effective}} = \sqrt{2\pi mkT/k_s} = \sqrt{2\pi kT}\frac{1}{\omega} = \sqrt{kT/2\pi}\frac{1}{f} \tag{6.6}$$

allowing expression in terms of the weighting-independent frequency f. The effective volume of a well is the product of the effective widths in each of the N

---

* The configuration coordinates used to describe a well are assumed to be aligned with the principal axes of the hyperellipsoids describing equipotential surfaces; motion along one of these coordinates corresponds to a normal mode.

dimensions

$$V_{\text{effective}} = \prod_{i=1}^{N} \sqrt{kT/2\pi}\, \frac{1}{f_i} = (kT/2\pi)^{N/2} \prod_{i=1}^{N} \frac{1}{f_i} \tag{6.7}$$

and a similar expression of lower dimensionality yields the effective area of the transition surface. Combining these, the transition rate expression becomes

$$k_{12} = v_{\text{mean}} \frac{S_{\text{col}}}{V_{\text{state 1}}} \exp\left(-\frac{\Delta\mathcal{V}^{\ddagger}}{kT}\right)$$

$$= \sqrt{\frac{kT}{2\pi}}\, \frac{\left(\dfrac{kT}{2\pi}\right)^{\frac{N-1}{2}} \displaystyle\prod_{i=1}^{N-1} \dfrac{1}{f_i^{\ddagger}}}{\left(\dfrac{kT}{2\pi}\right)^{\frac{N}{2}} \displaystyle\prod_{i=1}^{N} \dfrac{1}{f_i}} \exp\left(-\frac{\Delta\mathcal{V}^{\ddagger}}{kT}\right)$$

$$= \frac{\displaystyle\prod_{i=1}^{N} f_i}{\displaystyle\prod_{i=1}^{N-1} f_i^{\ddagger}} \exp\left(-\frac{\Delta\mathcal{V}^{\ddagger}}{kT}\right) \tag{6.8}$$

By applying the expression for the classical partition function of a harmonic oscillator,

$$q_{\text{class}} = kT/\hbar\omega \tag{6.9}$$

and the observation that the partition function for a linear elastic system is the product of the partition functions for each of its normal modes, an expression identical to Eq. (6.1) can be derived from Eq. (6.8). This illustrates the equivalence between the probability-gas model (with its geometric visualization) and the more abstract formalism of classical transition state theory.

Equation (6.8) has a simple interpretation in the special case where the reaction coordinate leading over the col is an extension of one of the normal modes of the well (of frequency f_{rc}) and the other modal frequencies are identical in both the well and transition state. A system of this sort is effectively one dimensional, and the ratio of products in Eq. (6.8) reduces to f_{rc}. The representative point can then be described as striking the barrier f_{rc} times per second, with a success rate in surmounting the barrier of $\exp(-\Delta\mathcal{V}^{\ddagger}/kT)$. Where the other modal frequencies are not identical, they can be viewed as affecting the entropy (and thus the free energy) as a function of the reaction coordinate; this coordinate-dependent entropy can be used to define a mean-force potential (Section 4.3.7) for motion along the reaction coordinate. A one-dimensional TST expression in terms of this potential is identical to Eq. (6.8).

c. ***Shortcomings and variational theories.*** Within the classical, equilibrium approximation, the TST described above is exact if the relevant integrals are done with an exact potential energy surface, and *if transitions are defined simply as motions that cross the transition surface separating the states.* On some PESs, however, a large fraction of the trajectories that lead from a low energy in one well to a low energy in another (assuming, as always, coupling to a heat bath) oscillate across the transition surface several times. Other trajectories may cross the surface and be promptly reflected back into the originating well. These crossings are transitions according to the above definition, but not in any practical chemical sense. Accordingly, classical TSTs provide only an upper bound to the (classical) transition rate of practical interest. The difference is sometimes accommodated by an ad hoc "transmission factor."

Variational transition state theories (which lack ad hoc factors) adjust the geometry of the transition surface to minimize the transition rate. The more sophisticated forms adopt different transition surfaces for trajectories of different energies. These methods yield lower transition rates that are still upper bounds on the rate of practical interest. A review of TSTs, including variational TSTs, can be found in Bérces and Márta (1988). Standard transition state theory usually fits exact classical calculations with reasonable accuracy.

6.2.3. Quantum transition state theories

Quantum mechanical corrections to classical transition state theory take three forms: (1) use of a barrier height that accounts for the zero-point energies of the well and transition state, (2) replacement of classical with quantum mechanical partition functions, and (3) use of a correction to account for tunneling through the barrier in the region of the transition state (Bérces and Márta, 1988). These are addressed separately.

It is common to correct barrier heights by adding a zero-point energy term to the energy of the well minimum. The zero-point energy of a linear elastic system is the sum of the zero-point energies of its (real-valued) normal modes; the effective minimum energy of a state is then

$$\mathscr{V}' = \mathscr{V} + \sum_{i=1}^{N} \frac{1}{2} \hbar \omega_i \tag{6.10}$$

Application of Eq. (4.2) to the harmonic oscillator yields a standard form for the quantum-mechanical partition function,

$$q_{\text{quant,zpe}} = \left[1 - \exp(-\hbar \omega / kT) \right]^{-1} \tag{6.11}$$

which takes the vibrational ground state as the zero of energy and is compatible with corrected barrier heights. Alternatively, one can omit the zero-point energy correction from the barrier height, and instead use a partition function

$$q_{\text{quant}} = \exp(-\hbar \omega / 2kT) \left[1 - \exp(-\hbar \omega / kT) \right]^{-1} \tag{6.12}$$

which takes the well minimum as the zero of energy. The choice is a matter of notational convenience, and expressions using this form more closely correspond to the classical equations.

The partition function for a linear system having N normal modes is the product of the partition functions of the modes; using Eq. (6.12),

$$q_{\text{quant}} = \prod_{i=1}^{N} \exp(-\hbar \omega / 2kT) \left[1 - \exp(-\hbar \omega / kT) \right]^{-1} \tag{6.13}$$

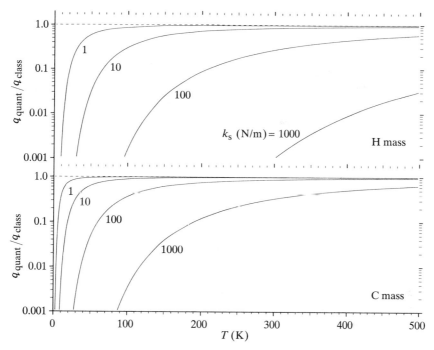

Figure 6.2.   Ratios of quantum mechanical to classical partition functions for harmonic oscillators, [Eq. (6.11)]/[Eq. (6.9)], as a function of temperature and effective stiffness for two values of the effective mass at the well minimum. Since the TST expression contains one more modal factor in the denominator than in the numerator, *decreases* in the above quantum-to-classical ratios tend to *increase* transition rates.

As in the classical theory, transition state partition functions arc computcd by omitting factors associated with motion along the reaction coordinate. Figure 6.2 shows how the ratio of classical and quantum-mechanical partition functions varies with temperature and stiffness.

The standard Wigner approximation to the tunneling correction factor is

$$\Gamma^* = 1 + \frac{1}{24}\left|\hbar\omega_{rc}/kT\right|^2 \tag{6.14}$$

where

$$\omega_{rc} = \sqrt{k_{s,rc}/m_{eff,rc}} \tag{6.15}$$

is an imaginary frequency associated with motion along the reaction coordinate [m_{eff} is the effective mass along this coordinate, and the stiffness along the reaction coordinate $k_{s,rc}$ is negative (Bell, 1959; Bigeleisen, 1949)]. Figure 6.3 shows how Γ^* varies with temperature and stiffness. (Section 6.4.4b notes the inapplicability of Γ^* at low temperatures.)

Combining the preceding substitutions and corrections yields the approximate quantum-mechanical TST expression

$$k_{12} = \Gamma^* \frac{kT}{2\pi\hbar} \frac{q^{\ddagger}_{quant}}{q_{quant,1}} \exp\left(-\Delta\mathcal{V}^{\ddagger}/kT\right) \tag{6.16}$$

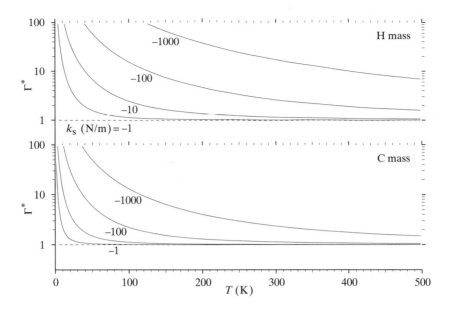

Figure 6.3. The Wigner tunneling correction factor, Eq. (6.14), as a function of temperature and effective stiffness for two values of the effective mass for motion along the reaction coordinate.

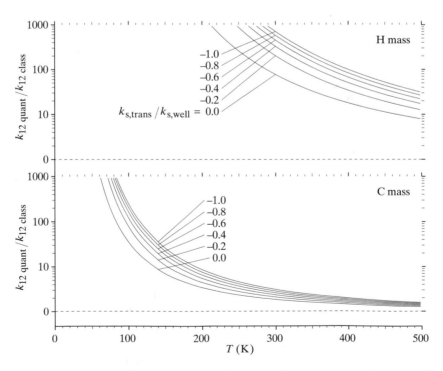

Figure 6.4. The ratio of quantum to classical transition rates, assuming a one-dimensional model with the stated ratios of stiffnesses along the reaction coordinate. The stiffness at the well minimum is 500 N/m (typical of the stretching stiffness of a covalent bond). Note that a stiffness ratio of zero implies an infinitely thick barrier and an absence of tunneling.

The ratio of this quantum mechanical rate constant to the classical rate constant is plotted in Figure 6.4 for examples involving the motion of H and C atoms. More sophisticated quantum-mechanical TSTs are reviewed in Bérces and Márta (1988).

6.2.4. Tunneling

Tunneling processes dominate the transition rate in all systems at sufficiently low temperatures and in many electronic systems at room temperature. When this occurs, transition state theories based on the Wigner tunneling correction can radically underestimate the transition rate. A different model must be used.

Estimates of tunneling transition rates for a particle in a well bounded by a barrier are derived by multiplying the barrier transmittance T by a suitable frequency factor, such as $\omega/2\pi$ (for a harmonic well) or the round-trip traversal time (for a square well). In the classical picture, the particle strikes the barrier with a certain frequency and probability of penetration.

Quantum theory yields exact expressions for T for various cases (such as rectangular barriers) but real systems seldom correspond to one of these. Nanomechanical and electromechanical designs of the sort described in this volume require approximations of broader utility. As with transition state theory, the typical objective is to discriminate between those systems that do and do not have tunneling rates low enough to be acceptable (e.g., low enough for an insulator to transmit little current or for a stressed bond to have a long expected lifetime). A standard method applicable to many systems of low T exploits the WKB (Wentzel-Kramers-Brillouin) approximation.

a. The WKB tunneling approximation. A useful approximation to the transmittance of a potential barrier $\mathcal{V}(x)$ for a normally incident particle of energy $\mathcal{E}$ and effective mass m_{eff} is

$$T \approx \exp\left[-2\int_{x_0}^{x_1}\sqrt{\frac{2m_{\text{eff}}}{\hbar^2}[\mathcal{V}(x)-\mathcal{E}]}\right] \tag{6.17}$$

where the limits of integration define the region in which $\mathcal{V}(x) > \mathcal{E}$ (Figure 6.5). The approximation requires that the exponential decay of the wave function in the region of negative energy be the dominant phenomenon, making $T \ll 1$; it can be applied when tunneling occurs into a continuum of energy states, such as that characterized by free motion of a particle. Where several potential barriers occur in close succession, resonance effects can yield more complex relationships between energies and transmittances.

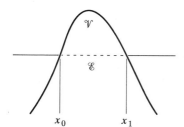

Figure 6.5. Total energy and potential energy of a tunneling particle, showing the limits of integration used in the WKB approximation.

6.3. Placement errors

6.3.1. Time-dependent PES models

Section 6.2 implicitly assumes that the PES is unchanging. From a perspective that includes all particles, this is always correct; yet it frequently is convenient to consider mechanical subsystems separately, accounting for the effects of other subsystems by imposing changing boundary conditions on the subsystem under analysis. These boundary conditions involve changes in the PES as a function of time. Detailed dynamical simulations (e.g., Landman et al., 1990) sometimes model the mechanical motion of an adjacent structure by imposing definite trajectories on boundary atoms, and model thermal equilibration by adjusting the mean kinetic energies of a subset of the atoms in thermal contact with the rest. The more abstract models discussed in this section describe a time-dependent PES directly; like standard TSTs, they introduce a thermal-equilibrium assumption by means of a basically statistical description.

Various processes in nanomechanical systems can be described as attempts to place something in the proper potential well. In systems that perform measurement (Chapter 11), the proper potential well corresponds to an accurate measurement of a physical parameter; in nanomechanical computer systems (Chapter 12), the proper potential well corresponds to the correct state of a logic device; in molecular manufacturing (Chapters 13 and 14), the proper potential well corresponds to the desired result of an assembly step. The potential wells and trajectories that specify the process of placement will in the general case involve motions describable only in terms of a PES of high dimensionality; in practice, typical examples of each of the above cases can be described rather well in terms of thermal displacements of an atom (or small group of atoms) in one or two dimensions, subject to an increasing constraint imposed by a mechanical motion in a second or third dimension.

For a concreteness, picture a probe atom at the tip of an elastic beam descending through vacuum toward a surface (Figure 6.6), described by the coordinates x, y, z, and z'. In this model, z' is the equilibrium value of z in the absence of thermal excitation and perturbing forces from the surface, or alternatively, the limit of z as the stiffness components $k_{sx} = k_{sy} = k_{sz} \to \infty$. In a nanomechanical system, the value of z' is determined by the mechanical configuration of the rest of the system. The potential $\mathcal{V}(x,y,z)_{z=z'}$, can be regarded as a z'-dependent two-dimensional potential $\mathcal{V}(x,y)$.

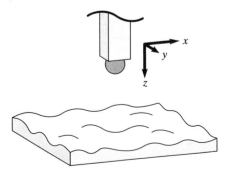

Figure 6.6. Model system consisting of a probe atom positioned by an elastic beam and subject to perturbations both from thermal excitation and from interactions as it encounters the surface below.

This potential is harmonic at large distances from the surface, owing to the beam stiffness; as the probe approaches the surface, corrugations are superimposed on the harmonic potential, growing in magnitude as it grows closer. (At contact, the probe may be forced into a hole by pressure from the arm, or attracted to a site by electrostatic interactions, or bonded to a reactive group.) We seek a potential of the form $\mathcal{V}^\dagger(x,y,z')$ in order to describe time-dependent two-dimensional systems $\mathcal{V}^\dagger[x,y,z'(t)]$. This is possible if we assume that the probe is mechanically stable in z (i.e., that the z stiffness, summing the beam stiffness and the probe-surface interaction stiffness, remains positive); without this assumption, transitions involving z excitations in a time-dependent three-dimensional potential must be considered explicitly.

How does $\mathcal{V}^\dagger$ differ from $\mathcal{V}$? At zero K, it equals

$$\mathcal{V}_0^\dagger(x,y,z') = \mathcal{V}(x,y,z_e) + \frac{1}{2}k_{sz}(z'-z_e)^2 \qquad (6.18)$$

where z_e is the equilibrium value of z for the given value of x, y, and z'. More generally, one can apply a mean-force potential in which an entropic term accounts for variations in the compression of vibrations in the z direction. Making the harmonic approximation for vibrations about z_e and applying the classical vibrational partition function, the potential at finite temperatures is

$$\mathcal{V}^\dagger(x,y,z') = \mathcal{V}(x,y,z_e) + \frac{1}{2}k_{sz}(z'-z_e)^2 - kT\left[1 + \ln(kT/\hbar\omega)\right] \qquad (6.19)$$

6.3.2. Error models

A detailed model would use Monte Carlo methods to estimate probabilities, integrating the equations of motion for a system. This method could take account of complex potentials and rapid motions in systems far from thermal equilibration, but would be computationally expensive and would offer little general insight.

Transition-state theories can provide a basis for a relatively detailed model of placement errors. In this approach, one would define regions corresponding to growing potential wells, and moving transition surfaces corresponding to growing barriers. Initial assignments of probabilities to wells would follow the Boltzmann distribution, and integration of transition rates would trace the evolution of those probabilities over time. To each transition rate predicted by a fixed-potential TST would be added a rate corresponding to the product of the probability density along each transition coordinate and the rate of motion of the associated transition surface in the time-dependent potential. This approach would be tractable for many problems.

A simpler model can be constructed starting with the observation that, in most systems of engineering interest, barriers between states are initially absent and eventually so high as to preclude significant transition rates. This leads to consideration of models in which equilibration among regions of configuration space switches from complete to nonexistent, omitting consideration of the narrow range of barrier heights in which transition rates are neither fast nor negligible compared to the characteristic time scale of the placement process. As usual, the goal is not to make a perfect prediction of physical behavior, but to distinguish workable systems, identifying them with a methodology that yields a low rate of false positives without an overwhelming rate of false negatives.

6.3.3. Switched-coupling error models

The probability of a state (or set of states) is effectively frozen* when the lowest barrier between it and other states meets the approximate condition

$$f_{\text{TST}}\exp\left[-\Delta\mathcal{V}^{\dagger}(t)/kT\right]\approx\frac{1}{kT}\frac{\partial}{\partial t}\Delta\mathcal{V}^{\dagger}(t) \tag{6.20}$$

assuming that the barrier height $\Delta\mathcal{V}^{\dagger}(t)$ and the equilibrium ratio of probabilities are both smoothly increasing. (In this expression, f_{TST} is a frequency factor taken from transition state theory.) This can be termed the time of *kinetic decoupling* for the two wells; one useful approximation treats this as a discrete time at which equilibration is switched off. A one-dimensional classical model illustrates the consequences of switching equilibration off at different times.

a. The sinusoidal-well model. The simplest time-dependent PES for modeling placement errors combines a fixed, one-dimensional harmonic potential with a time-varying sinusoidal potential

$$\mathcal{V}^{\dagger}(t)=\frac{1}{2}k_{\text{s}}x^2+\left[1-\cos(2\pi x/d_{\text{err}})\right]At \tag{6.21}$$

This model captures several basic features of placement errors in nanomechanical systems: the harmonic well is aligned to maximize the probability of

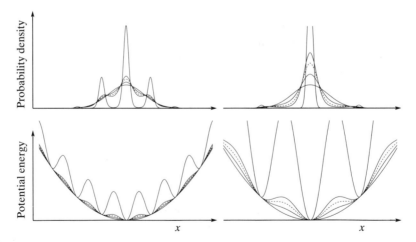

Figure 6.7. The lower panels illustrate time-dependent potentials described by Eq. (6.21), with higher curves representing later times; the upper panels illustrate the corresponding Boltzmann probability density functions at a particular temperature. In both panels, the dotted curves correspond to the potential at t_{crit} (Section 6.3.3d); other curves are at 0.5, 1.5 and 10 times t_{crit}. Note the decrease in probability mass in the outlying peaks as the well spacing increases.

---

* A process that equilibrates different states can be described as *freezing* when the transitions that establish equilibrium effectively cease, that is, when they become rare on the time scale under consideration. Often this occurs because the temperature falls; here it occurs because the barrier height increases. The result is a *frozen distribution*.

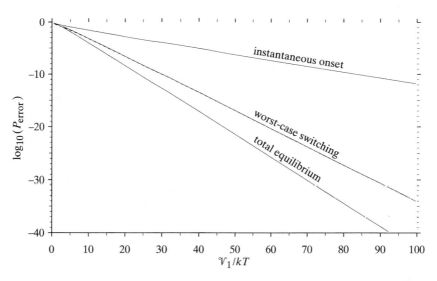

Figure 6.8. Error rates per placement operation resulting from the application of different equilibration assumptions to the sinusoidal-well model.

finding the system in the target potential well in the center; other wells, corresponding to error states, appear at some distance d_{err} from the target well; well depths grow over time at a controlled rate. The bottom panel of Figure 6.7 illustrates the shape of this potential at varying times for two values of d_{err}.

b.   The total-equilibrium limit.   If equilibration continues until the wells that will eventually hold significant probability are all sharp compared to the initial harmonic well, then the probability mass associated with each is determined only by its depth, which is the height of the harmonic well at a minimum of the sinusoid. In this model, the probability of error is given by the ratio of sums

$$P_{err} = 2 \sum_{1}^{\infty} \exp\left(-k_s (n d_{err})^2 / 2kT\right) \Big/ \sum_{-\infty}^{\infty} \exp\left(-k_s (n d_{err})^2 / 2kT\right) \qquad (6.22)$$

Figure 6.8 shows the probability of error in the total equilibrium limit, as a function of $\mathcal{V}_1$. The generalization of this model is standard chemical equilibrium, in which wells can have varying free energies (that is, depths and effective volumes) and have the corresponding Boltzmann-weighted probabilities of occupancy.

c.   The instantaneous-onset limit.   If the rate parameter A is sufficiently large, the barriers imposed by the sinusoidal potential become large in a time that is short compared to the transit time of a single well, and probabilities are frozen in their prebarrier distribution. The probability of an error is the integral

$$P_{err} = 2 \int_{d_{err}/2}^{\infty} \sqrt{kT/2\pi k_s} \, \exp\left(-k_s x^2 / 2kT\right) dx \qquad (6.23)$$

of the initial harmonic Boltzmann distribution in the regions beyond the eventual location of the peaks that will bound the target well. The upper curve in Figure 6.8 illustrates how this probability varies as a function of $\mathcal{V}_1$, the energy difference between the target state and the lowest-energy error state. The generalization of this instantaneous limit in a multidimensional system would start with an ellipsoidal Gaussian distribution and impose on this the state boundaries of the final potential surface (Section 4.3.3b), taking the integrated probability outside the boundaries of the correct state as the probability of error.

For a nanomechanical system to be well-approximated by the instantaneous limit, its mechanical speeds must be large compared to its thermal speeds. For combinations of sufficiently large effective masses and sufficiently high mechanical speeds, this may hold true, yet it is unlikely to be a common case. The instantaneous limit maximizes error rates, hence high speeds (which also increase energy dissipation) are usually best avoided.

d. The worst-case decoupling model. Within the bounds of the kinetic-decoupling model (allowing free equilibration until some time after energy barriers appear), there exists a decoupling time that maximizes the error rate. In the sinusoidal-well model, the first barriers appear at a time

$$t_{crit} = 0.2332\, d_{err}^2 / A \tag{6.24}$$

then grow in height and move inward toward the origin. Figure 6.7 illustrates two time-varying potentials and their corresponding PDFs, showing t_{crit}. Dividing the PDF at the time and position of the initial barrier appearance yields the dashed curve in Figure 6.8; varying the time (and resulting PDF and barrier position) to maximize the error probability yields the solid curve slightly above it.

This model, which permits barrier crossing only so long as it increases errors, yet gives error rates lower than in the instantaneous-onset model, seems a good choice for making conservative estimates of error rates in many circumstances. The conservatism breaks down when error rates are large, that is, when $d_{err} < \sim 2(kT/k_s)^{1/2}$, but this is outside the usual range of interest. A more common failure is excessive conservatism in systems where a large well may appear *after* being cut off by a high barrier: this model will then assign an unrealistically high probability to that well. With the notion of state boundaries that change over time, the worst-case switching model readily generalizes to multidimensional systems.

6.4. Thermomechanical damage

6.4.1. Overview

Thermal excitation, sometimes aided by mechanical stress, can cause permanent damage to devices built with molecular precision. Damaged states (as distinct from transiently disturbed states) occur when a transition occurs into a stable potential well corresponding to a nonfunctional device structure. Causes of damage include:

- Reactions like those between small molecules
- Rearrangements like those within small molecules
- Reactions like those at solid surfaces
- Rearrangements like those within solid objects
- Bond cleavage accelerated by mechanical stress

The first four points can be approached via analogies to known chemical systems. Several of the following sections follow this strategy, building on empirical knowledge and theoretical scaling relationships that describe how transition rates vary with temperature. Here (as in Chapter 8) it seems wise to begin by comparing the general conditions of solution-phase and °machine-phase chemical processes, as an aid to forming the proper generalizations from present chemical experience.

As shown in Section 6.6, damage caused by a process with a transition lifetime $\geq 10^{20}$ s ($\approx 3 \times 10^{12}$ years) is small compared to that caused by ionizing radiation under typical conditions (assuming, as usual, that complete device failure will result from a single damaging transition). This rate can accordingly be regarded as small in practical terms, and is used here as a standard of comparison.

It can be useful to draw a distinction between two classes of reactions: those that begin with the breakage of a bond (and may then move on to further transformations), and those that proceed with simultaneous bond breakage and bond formation, permitting lower transition-state energies. In processing metals, ceramics, and semiconductors, temperatures high enough to promote rapid bond breakage are commonplace. In solution chemistry, however, and in nanomachines operating at room temperature, only low-energy (or highly strained) bonds can break at high rates (Sections 6.4.3 and 6.4.4), and the dominant chemical processes therefore involve simultaneous bond breakage and formation. Many organic molecules have no available reaction pathways that permit this; they can accordingly be quite stable at ordinary temperatures when surrounded by similarly stable structures or vacuum.

6.4.2. Machine- vs. solution-phase stability

Machine-phase chemistry comprises systems in which all potentially reactive moieties follow controlled trajectories (Section 1.2.2a). Although borderline cases can be imagined, the contrast is clear between systems of small, diffusing molecules in the liquid phase and °eutactic systems of molecular machinery as envisioned here. In solution-phase chemistry, potentially reactive moieties encounter each other in relative positions and orientations constrained by only local molecular geometry, not by control of molecular trajectories. In the solution phase, access to a reactive site may be blocked (e.g., by attached hydrocarbon chains); in the machine phase, even an exposed reactive site may never encounter another molecule.

a. Examples. The stability of a structure often depends on its physical environment. For example, a molecular structure including the fragment:

6.1

would be unstable in solution, readily forming dimeric products such as:

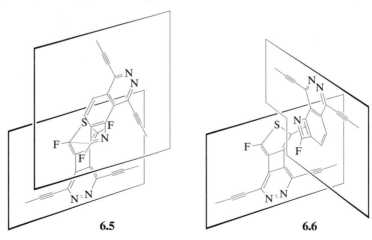

6.2

among others. The formation of dimers, however, requires that two molecules of the same kind encounter one another; this is inevitable in solution, but not in molecular machines.* Structure **6.1** has been proposed for a role in a nanomechanical computer (Drexler, 1988b) in which such encounters would not occur.

In that design, however, **6.1** would encounter other structures including the fragment **6.3**, which in solution might yield the Diels–Alder adduct **6.4**:

6.3

6.4

and various other products. To approach the Diels–Alder transition state, the rings would have to encounter one another in a nearly parallel orientation, as in **6.5**. In the application proposed, however, a surrounding matrix would mechanically constrain the rings, forcing them to approach in a perpendicular orientation, as in **6.6**, thereby precluding this reaction.

6.5   **6.6**

---

* The four-membered ring in **6.1** resembles cyclobutadiene, which is stable at room temperature if intermolecular collisions are blocked (Cram et al., 1991).

b. *Machine-phase damage mechanisms.* To generalize from these examples:

- Structures that are unstable in solution environments owing to intermolecular reactions between identical molecules can be stable in eutactic systems because collisions do not occur.
- Structures that react on collision in solution environments can fail to react on collision in eutactic systems owing to constraints on orientation.

In solution chemistry, exposed moieties always encounter solvent molecules and often encounter dissolved oxygen, dissolved water, trace contaminants, and container materials, each potentially reactive. In machine-phase chemistry, potentially reactive moieties are constrained, encountering other structures only by design (Chapters 11, 13, and 14 discuss maintenance of controlled environments). In some nanosystems (discussed in Chapters 8, 13, and 14), certain structures and encounter conditions are designed to foster reactions; °reagent devices of this sort are considered elsewhere. Other structures and encounter conditions, however, are designed to discourage reactions, and reactions then constitute damage.

In nonreagent nanomechanical systems, contacts between structures serve mechanical functions: bearing surfaces, °cam riders, gear teeth and the like are typical examples. Chapter 11 addresses the stability of the resulting interfaces. Where no mechanical function is served, no contact is necessary, and intermolecular reactions can be excluded from possibility.

Nanomechanical systems are, however, subject to increased reactivity owing to mechanical stresses not found in solution chemistry. These stresses can be used to facilitate desired reactions (Sections 8.3.3c and 8.5) but can also facilitate damage. Stresses are subject to engineering control, and estimation of their chemical effects is important to engineering design. These stress-induced reactions, however, are special cases of intramolecular damage processes involving intramolecular reactions of sorts familiar in conventional chemistry. These reactions are little affected by control of trajectories.

c. *Design and intramolecular stability.* To provide a thorough discussion of intramolecular reactions would require a thorough discussion of much of chemistry: this is impossible in a single volume, and still further from possibility in a section of a chapter. Even a discussion of organic intramolecular reactions would include nearly the whole of organic chemistry, since the ends of long, flexible molecules can react with each other essentially like separate molecules.

The pursuit of engineering objectives, however, greatly restricts the range of structures to be considered. For example, these objectives usually favor designs in which structures are either intrinsically stiff or held in an extended conformation by tension or by a surrounding matrix. Designs with these characteristics can exclude this class of reactions.

Likewise, the pursuit of engineering objectives often favors the selection of structures with strong, stable patterns of bonding. Errors in discriminating between stable and unstable structures during the preliminary phases of nanomechanical engineering will later be eliminated by more thorough analysis and (eventually) experimental testing. The aim of the present discussion of thermomechanical damage is to outline some principles of importance in (1) generating reasonable designs and (2) using the chemical literature and computational experiments to evaluate specific designs in more detail.

With regard to reading the chemical literature, a possible source of cognitive bias is worth noting: In studying solution chemistry, one tends to focus on the active reagent molecules, not on the comparatively inert solvent molecules. Indeed, in a typical textbook of organic chemistry, roughly half the chemical species appear to the left of an arrow representing a transformation to another species, A $\to$ B; if A were stable under the specified reaction conditions, it would rarely appear in the equation, save as a notation of reaction conditions. A superficial reading of the chemical literature can thus foster the illusion that almost any set of molecules is likely to react at an appreciable rate. In nanomechanical systems, however, many structures play roles more like those of solvents: these structures can move with respect to one another (and transport reagents) without reacting. If need be, the interactions between nanomechanical surfaces can be made to resemble those among, for example, hexane, benzene, and methyl ether more closely than those among, for example, sodium hydroxide, bromoethane, hydrochloric acid, and cyclopentadiene. Reactivity depends on surface structure and interaction geometry, and will be subject to engineering control guided by experimentation.

6.4.3. Thermal bond cleavage

Direct cleavage of a bond by thermal excitation, forming a pair of °radicals, is usually slow at ordinary temperatures. Among the major exceptions are organic peroxides, diacyl peroxides, and azo compounds, which find use as initiators of radical reactions:

$$R\text{--}O\text{--}O\text{--}R' \longrightarrow R\text{--}O\bullet + \bullet O\text{--}R' \tag{6.25}$$

$$\text{diacyl peroxide} \longrightarrow R\bullet + 2\,CO_2 + \bullet R' \tag{6.26}$$

$$R\text{--}N{=}N\text{--}R' \longrightarrow R\bullet + N_2 + \bullet R' \tag{6.27}$$

a. Thermal bond cleavage rates.   The rate of bond cleavage k_{cl} cannot be estimated using standard transition state theory when molecules separate into gas-phase radicals: the effective transition state is at infinity, and the transition-state frequencies are zero. Variational theories can give sensible answers, but a rough approximation, good in a wide range of circumstances (and low by a small factor in this instance), is provided by a one-dimensional theory. In this,

$$k_{cl} = \frac{kT}{2\pi\hbar} \frac{1}{q_{long}} \exp(-\Delta\mathcal{V}/kT) \tag{6.28}$$

where q_{long} is the partition function for longitudinal vibrations of the bond,

$$q_{long} = \exp\left(-\frac{\hbar}{kT}\sqrt{\frac{k_s}{\mu}}\right)\left[1 - \exp\left(-\frac{\hbar}{kT}\sqrt{\frac{k_s}{\mu}}\right)\right]^{-1} \tag{6.29}$$

approximating the effective mass of the system as the °reduced mass

$$\mu = \frac{m_1 m_2}{m_1 + m_2} \qquad (6.30)$$

and the effective stiffness as the bond stiffness, k_s. The tunneling correction Γ^* is unity owing to the effectively infinite barrier thickness. Rate increases owing to the increasing effective area of the escape channel with increasing bond length are neglected. Figure 6.9 shows values for the characteristic bond-cleavage time, $\tau_{cl} = k_{cl}^{-1}$, based on Eq. (6.28). As shown in Section 6.7, there is seldom reason to seek bond lifetimes $> 10^{20}$ s; of the bonds selected for listing in Table 3.8, only the O—O bond of organic peroxides falls short of this lifetime without °strain or other adverse influences on bond stability.

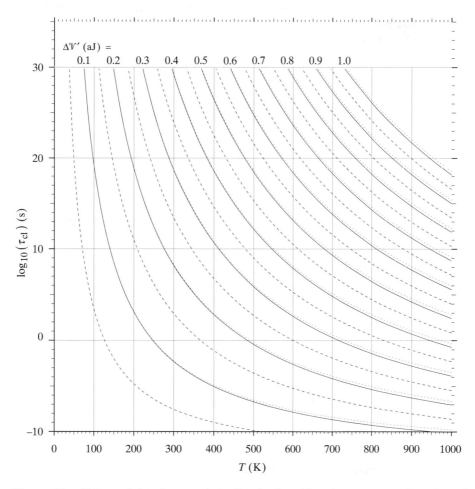

Figure 6.9. Values of the characteristic time for bond breakage, τ_{cl}, as a function of temperature and the bond energy $\Delta\mathcal{V}$, as given by Eq. (6.28). The solid and dashed lines represent systems in which the partition function q_{long} is unity (corresponding to the high-frequency limit); the dotted lines just above the solid lines represent systems in which this partition function is calculated for $\omega = 7 \times 10^{13}$ rad/s, a relatively low frequency like that of a Si—Si bond.

b. *Rate modifications in liquid and solid media.* In liquids, the observed rate of thermal bond cleavage is substantially reduced by the dynamic *cage effect:* radicals surrounded by solvent molecules are confined for several molecular vibration times, permitting opportunities for immediate (geminate) recombination. In solids, geminate recombination is often far more probable than in liquids owing to the near absence of diffusion: the radicals are trapped in solid cages. In the interior of a stiff diamondoid solid, and in the absence of intense stress, separation of radicals commonly requires the simultaneous cleavage of additional bonds, multiplying the effective magnitude of the barrier height and drastically reducing failure rates; this can be termed a *solid-cage effect.*

Attachment to a °rigid structure can have dramatic effects even for surface moieties. For example, while an ordinary organic peroxide, Eq. (6.31), is subject to irreversible cleavage in solution (once separated, the two radicals are unlikely to recombine), an analogous peroxide structure attached to a rigid structure will (in the absence of neighboring reactive moieties) reliably recombine on a 10^{-13} s time scale Eq. (6.32).

$$\text{O–O} \longrightarrow \text{O•} \quad + \quad \text{•O} \tag{6.31}$$

$$\text{O–O} \quad \underset{\sim 10^{-13}\,\text{s}}{\overset{\text{slow}}{\rightleftharpoons}} \quad \text{O••O} \tag{6.32}$$

At interior sites, a physical model for thermally activated structural damage is self-diffusion in diamondlike crystals of germanium and silicon, believed to occur largely through a vacancy-hopping mechanism (Reiss and Fuller, 1959). This process requires bond cleavage, as would a damage mechanism in a diamondoid solid; unlike an atom in the interior of a well-designed diamondoid solid, however, an atom adjacent to a vacancy has available to it both room into which to move and unsatisfied valencies with which to bond. Despite these facilitating structural features, the activation energies (which approximate the barrier energies discussed above) for vacancy diffusion in germanium and silicon (Lide, 1990) have been estimated to be 503 and 745 maJ, about 1.6 and 2.0 times their respective bond energies.

A better physical model for thermal damage in the interior of a stable diamondoid solid might be the creation of a new vacancy within a covalent crystal. For this, activation energies are far greater.

6.4.4. Thermomechanical bond cleavage

Tensile stress destabilizes bonds, increasing the rate of thermal cleavage and sometimes opening a tunneling path to cleavage. Large angular strains are common in organic chemistry, for example, in small rings. Substantial tensile strains are less common in organic chemistry but frequently are important in nanomechanical systems, where engineering performance often depends on the imposed stress, providing an incentive to push toward the allowable limits of stress.

a.   Thermomechanical bond cleavage rates.   Cleavage rates for bonds under simple tensile stress at ordinary temperatures can be approximated with a one-dimensional quantum TST like that in Section 6.2.3, but with frequencies and barrier heights calculated with the aid of a potential energy function describing bond extension. Combining the Morse function Eq. (3.10) with the potential energy resulting from an applied force F yields

$$\mathscr{V}_{\text{stressed}} = D_e \left(\{ 1 - \exp[-\beta(\ell - \ell_0)] \}^2 - 1 \right) - F(\ell - \ell_0) \qquad (6.33)$$

Some examples are plotted in Figure 6.10.

The Morse function overestimates the bonding energy in the high extension region, underestimating the energy gradient and the tensile strength of the bond (Section 3.3.3a); it is thus conservative for engineering analyses where bond stability is required for a workable design. Equation (6.33) assumes that the applied force is independent of the displacement. Where applied forces are associated with large positive stiffnesses (e.g., for bonds in diamondoid structures) this analysis, by ignoring the solid-cage effect, grossly overestimates cleavage rates.

Manipulating the Eq. (6.33) yields the classical barrier height (Kauzman and Eyring, 1940)

$$\Delta \mathscr{V} = D_e \sqrt{1 - 2F/\beta D_e} + \frac{F}{\beta} \ln \left(\frac{1 - \sqrt{1 - 2F/\beta D_e}}{1 + \sqrt{1 - 2F/\beta D_e}} \right) \qquad (6.34)$$

and frequencies at the well minimum and barrier maximum that yield both a one-dimensional partition function (making the conservative approximation of

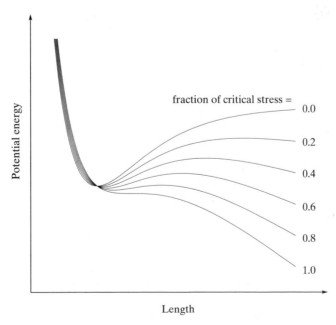

Figure 6.10.   Morse curves modified by tensile stresses ranging from 0.0 to 1.0 times the critical stress for barrier elimination.

a parabolic well),

$$q_{\text{quant}} = \exp(-\hbar\omega/kT)[1-\exp(-\hbar\omega/kT)]^{-1} \qquad (6.35)$$

where the frequency

$$\omega = \sqrt{\frac{D_e\beta}{\mu}\left(\beta - 2F/D_e + \sqrt{\beta^2 - 2\beta F/D_e}\right)} \qquad (6.36)$$

and an imaginary frequency that can be used to calculate the Wigner tunneling correction [Eq. (6.14)] is

$$\omega_{\text{rc}}^{\ddagger} = \sqrt{\frac{D_e\beta}{\mu}\left(\beta - 2F/D_e - \sqrt{\beta^2 - 2\beta F/D_e}\right)} \qquad (6.37)$$

The zero-point energy estimate uses the harmonic approximation, yielding a (conservative) underestimate of the barrier height. All factors incorporating bond frequencies neglect the increase in effective mass and effective stiffness resulting from atoms not directly participating in the bond.

b. Tunneling cleavage rates.   At sufficiently low temperatures, the Wigner tunneling correction becomes inadequate and methods like those discussed in Section 6.2.4 must be applied. At zero K, only an estimate of tunneling from the ground state is necessary; the model used here makes the preceding assumptions regarding the potential energy function and effective mass, approximating the transition rate as the product of the vibrational frequency in the well and the WKB transmission probability, Eq. (6.17). Intermediate cryogenic temperatures would require a more complex treatment, giving intermediate values of the bond lifetime.

c. Allowable stresses in covalent bonds.   Figure 6.11 graphs the characteristic bond cleavage time τ_{cl} vs. applied stress for various bond types at 0, 300, and 500 K, with bond stiffnesses and dissociation energies from Table 3.8. It shows that the criterion $\tau_{\text{cl}} > 10^{20}$ s is met for a C—C bond at 300 K if the stress is $\leq \sim 1.2$ nN/bond. Replacing each stressed C—C bond by a pair of C—C bonds (modeled by doubling the mass, energy, and stiffness) yields the second curve from the right; note that the threshold stress for this system is ~ 6 nN, substantially more than twice the single-bond threshold stress. This is a consequence of the additivity of the barrier energies for individual bonds combined with the exponential dependence of the cleavage rate on the total barrier. In diamondoid structures, failure commonly requires the simultaneous cleavage of several bonds, hence the acceptable working stress per bond can be substantially higher than for a structure relying on just one bond.

Allowable stresses are strongly temperature dependent: Although this volume is concerned with systems at room temperature, a wider range of structures and stresses can be used at cryogenic temperatures. Conversely, as temperatures increase, workable designs become increasingly constrained, ultimately dwindling to none.

These calculations can be compared to the theoretical tensile strength of diamond estimated by Lawn as $\sim 1.9 \times 10^{11}$ N/m^2 (Field, 1979). Diamond normally cleaves along a (111) plane, which has 1.8×10^{19} bonds/m^2; the strength just cited corresponds to ~10.6 nN/bond, roughly twice the theoretical limiting strength implied by the more conservative calculations of this section. Lawn also calculates a shear strength of $\sim 1.2 \times 10^{11}$ N/m^2 (Field, 1979), or 6.7 nN/bond. At ~1300 K, diamond undergoes plastic flow under mean shear stresses as low as 0.18 nN/bond (Brookes et al., 1988), but this process depends on the breakage of bonds in dislocations (where stresses are far higher than the mean) and occurs at temperatures that rapidly cleave C—C bonds in normal hydrocarbons.

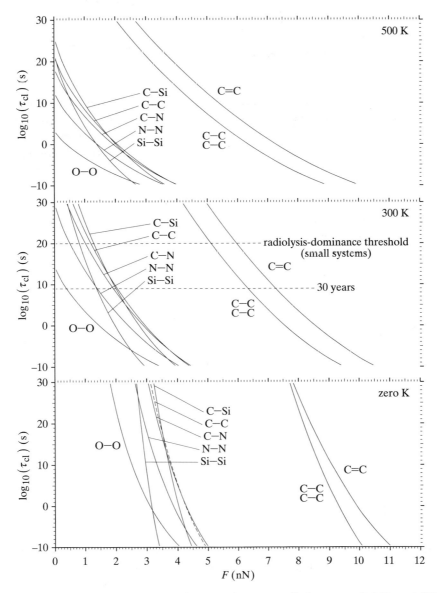

Figure 6.11. Characteristic bond cleavage time vs. applied stress at 0, 300, and 500 K. Calculations at 300 and 500 K are based on quantum transition state theory (Section 6.2.3); those at zero K are based on the WKB tunneling approximation (Section 6.2.4). The second curve from the right represents a pair of single C—C bonds mechanically constrained to cleave in a concerted process.

6.4.5. Other chemical damage mechanisms

a. Elementary, nonelementary, and solvent-dependent reactions. As suggested by Section 6.4.2c, intramolecular reactions provide important models for damage mechanisms of significance in nanomechanical systems. Common intramolecular reactions include elimination (loss of part of a molecule) and rearrangements (structural transformations that leave overall composition unchanged). Often, however, such transformations occur by intermolecular processes, or by intramolecular processes that are solvent dependent.

A simple example is an elimination reaction sometimes written as

$$ \text{(6.38)} $$

This does not occur at an appreciable rate among molecules isolated in vacuum, because the reaction does not occur as shown. The above equation represents a reaction (or compound reaction), but not an elementary reaction, which is to say a single molecular transformation. Not all reactions are elementary, and mistaking a compound reaction for an elementary reaction can lead to mistaken conclusions.

The steps yielding ethylene in Eq. (6.38) can proceed by at least two distinct mechanisms comprising three elementary reactions. The first mechanism is a single-step E2 process, such as:

$$ \text{(6.39)} $$

The second is a two-step E1 process, such as:

$$ \text{(6.40)} $$

$$ \text{(6.41)} $$

Neither reaction mechanism actually yields an HBr molecule as shown in Eq. (6.38), and both require the participation of other molecules. The *2* in E2 refers to the requirement for participation of two molecules in the slow, rate-limiting step; the 1 in E1 indicates that the rate-limiting step is unimolecular. Even the first step of the E1 reaction is solvent dependent: the separation of a positive and negative ion from 0.3 nm to 3 nm in vacuum requires ~700 maJ of energy, but in a medium with the dielectric constant of water, only ~9 maJ. For similar reasons, Na^+ and Cl^- ions readily leave a salt crystal in water, causing swift dissolution despite the lack of any significant tendency for salt to evaporate

Table 6.1. Temperatures for 30-minute volatilization half-lives in the absence of air (Schnabel, 1981).

Polymer	T_{vol} (C)	T_{vol} (K)
Polytetrafluoroethylene	510	~780
Polybutadiene	410	~680
Polypropylene	400	~670
Polyacrylonitrile	390	~660
Polystyrene	360	~630
Polyisobutene	350	~620
Poly(ethylene oxide)	350	~620
Polymethylacrylate	330	~600
Polymethylmethacrylate	330	~600
Polyvinylacetate	270	~540
Polyvinylchloride	260	~530

into the vacuumlike emptiness of air. In general, processes generating free ions are unusual in vacuum at 300 K.

b. Evidence from pyrolysis. The one-dimensional TST-derived relationships described in Section 6.4.3 can be used to estimate intramolecular reaction rates at 300 K from reaction rates at higher temperatures (the ratio of partition functions is constant in the classical harmonic approximation, and is treated as constant here). If rates scale as in Eq. (6.28), then a species requiring ≥ 1 s at ≥ 750 K (≈ 480 °C) to undergo thorough pyrolysis has a transformation lifetime $\tau_{trans} \geq 10^{20}$ s/molecule at 300 K. Thermal exposure times of 1 to 100 s are common in pyrolysis experiments in organic chemistry (Brown, 1980).

These experiments (discussed in Section 6.4.5d), like the polymer degradation experiments discussed in the next section, give evidence regarding the suitability of various classes of structures for use as major components of long-lifetime nanomechanical systems. As should be expected, some are adequate and others are too unstable.

c. Polymer pyrolysis. In the pyrolysis of polymers, 50% weight loss during 30 min at 610 K (≈ 340 °C; centigrade is temporarily adopted here in reporting results from the chemical literature) suggests a rate for fragmentation reactions of $\leq 10^{-20}$ s^{-1} per monomer at 300 K. Table 6.1 lists the characteristic temperatures for various polymers; data regarding heat resistant polymers can be found in Critchley et al. (1983). Note that many exceed 610 K despite the availability of intermolecular processes for degradation; polyvinylchloride, for example, is thought to undergo elimination of HCl via a radical chain reaction mechanism. In estimating the failure rate of nanomechanical systems, most analyses in this volume assume that the first chemical transformation is sufficient to cause failure (Section 6.7.1a), making chain reactions irrelevant. Polymer degradation studies are accordingly better for establishing lower bounds on achievable stability than they are at establishing upper bounds.

These empirical observations can also be criticized for being sensitive to the cleavage of polymer backbones, but perhaps not to rearrangements of side chains. Rearrangements are better examined in small molecules.

d. Unimolecular elimination, fragmentation, and rearrangement. The rapid pyrolysis of small organic molecules in the gas phase provides a useful model for long-term damage to machine-phase systems at room temperature. These processes can provide a guide to the relationship between stability and local molecular structure. This section offers examples of reactions and sketches some imperfect generalizations regarding molecular stability.

Many structures suffer elimination reactions that proceed at unacceptably high rates at room temperature. In reactions such as

$$\cdots \longrightarrow [\cdots]^{\ddagger} \longrightarrow \bigcirc + N_2 \qquad (6.42)$$

$$\cdots \longrightarrow [\cdots]^{int} \longrightarrow \cdots + N_2 \qquad (6.43)$$

bonds rearrange to make the extremely stable N_2 molecule. Reaction (6.42) proceeds quickly at 200 K, but reaction (6.43) does so only at ≥ 370 K (Carey and Sundberg, 1983b); the difference in reaction mechanism stems from differences in orbital symmetry. Other processes of this sort yield CO, CO_2 and H_2. Under pyrolytic conditions, hydrogen halides can be eliminated from alkyl halides [possibly as shown in Eq. (6.44), or in a surface reaction].

$$\cdots \longrightarrow [\cdots]^{\ddagger} \longrightarrow \cdots + HX \qquad (6.44)$$

$$\cdots \longrightarrow \bigcirc + H_2 \qquad (6.45)$$

$$\cdots \longrightarrow \cdots + HCl \qquad (6.46)$$

An extensive review of eliminations under gas-phase pyrolytic conditions appears in Brown (1980), from which reactions (6.44) through (6.46) are taken. Chapter 7 of Carey and Sundberg (1983b) provides a brief overview of elimination reactions and discusses some classes of unimolecular rearrangements of chemical interest. Examples include Cope rearrangements

$$\cdots \longrightarrow [\cdots]^{\ddagger} \longrightarrow \cdots \qquad (6.47)$$

and unimolecular ene reactions

$$(6.48)$$

Brown (1980) surveys reactions occurring during gas-phase pyrolysis, describing hundreds of examples of reactions from numerous classes. It is possible to draw some conclusions from this body of data (though without formulating generalizations that neatly categorize molecules according to stability). Brown's stated selection criteria favor the inclusion of reactions that are (1) useful in organic synthesis and (2) proceed at temperatures >350 °C. The criterion of usefulness favors the inclusion of molecules having a specific reactivity, that is, a specific weakness relative to more inert molecular structures (such as aliphatic hydrocarbons of low strain). Further, the objective of maximizing reaction specificity tends to favor use of the lowest temperatures at which reaction rates are adequate for practical work. Of the structures in Brown (1980), selected for exhibiting useful pyrolytic reactions, the fraction having listed reaction temperatures >500 °C (suggesting adequate stability for nanomechanical use at room temperature) is roughly half.

For pyrolytic reactions in general, the structural characteristics favoring reaction temperatures ≤480 °C are too complex to enumerate, but include inherently weak covalent bonds, resonant stabilization of reaction products, high strain, and others. A generalization does hold for the set of unimolecular rearrangements reviewed in Brown (1980): listed reaction temperatures are >480 °C unless the molecule either (1) contains two or more pi bonds, or (2) contains both a strained, three-membered ring and one or more pi bonds, or (3) contains a strained, four-membered ring that includes a single pi bond (many molecules with one or more of these features nonetheless have listed reaction temperatures >480 °C). Exceptions to this generalization (including any number of compounds with inherently weak bonds) can be found elsewhere in the literature.

It should be noted that surface-catalyzed and intermolecular gas-phase reaction pathways may occur in some of the systems reviewed in Brown (1980). Removing these pathways can only improve stability. (Rearrangements occurring under various conditions are reviewed in the volumes of Thyagarajan, 1968–1971.)

The thermal rearrangement of hydrocarbons, which has been practiced on a large scale in the petroleum industry to increase the octane rating of gasoline, is relevant to estimating the failure rates of molecules having no special destabilizing characteristics. The temperatures applied are in the 780 to 840 K range (~500 to 570 °C), with exposure times of many seconds; this implies $\tau_{trans} > 10^{20}$ s at 300 K.

The study of pyrolytic reactions in small molecules provides useful indications of stability in nanomechanical systems at 300 K, but fails to account for two expected differences. The first is the presence of mechanically imposed tensile and shear stresses: these are destabilizing and are, at best, only partially modeled by the methods of Section 6.4.4. Since activation energies for rearrangements frequently fall within the range of activation energies for the cleavage of structurally useful bonds, it seem safe to assume that failure from rearrangement under moderate applied stresses will in these instances meet the $\tau_{trans} > 10^{20}$ s at 300 K criterion.

The second difference is that potentially unstable substructures found in nanomechanical systems commonly are constrained by diamondoid matrices. As we have seen, thermal bond cleavage can be strongly inhibited by the solid-cage effect. Likewise, many rearrangements involve atomic displacements that are feasible in flexible structures, but not in more diamondlike structures. An example is the conrotary thermal ring opening of cyclobutene, Eq. (6.49) [with orbitals and transition geometry shown in Eq. (6.50)] contrasted with the expected behavior of a constrained analogue, Eq. (6.51).

$$(6.49)$$

$$(6.50)$$

$$(6.51)$$

A useful exercise is to examine a list of rearrangement reactions and observe how few could occur if most bonds to hydrogen atoms on a molecular surface were instead bonds to carbon atoms in a surrounding matrix of diamondlike rigidity. Diamond itself is stable to 1800 K in an inert atmosphere ($\tau_{trans} > 10^{85}$ s at 300 K), despite the well-known energetic advantage of rearrangement to graphite.

For the nanomechanical design process, it would be useful to develop an automated classification system capable of reliably labeling structures as adequately stable, inadequately stable, or of unknown adequacy, with the set of structures labeled as adequately stable being large enough to permit effective nanomechanical design. Such a design tool could in substantial measure be validated using the existing data on small-molecule pyrolysis. Lacking a tool of this sort, one must proceed using informal chemical reasoning (drawing on the accumulated generalizations of organic chemistry) and analogy based on model compounds (drawing on the yet more massive accumulation of data in organic chemistry), supplemented by computational methods. The strategy pursued in the present work is to favor structures in which strong bonds form frameworks of diamondlike rigidity that lack substructures of known instability. This strategy is more likely to exclude workable designs (yielding false negatives) than to include unworkable designs (yielding false positives). As development progresses, both classes of error will be reduced.

e. Stability of reagent devices. Molecular manufacturing systems include devices containing moieties that play the role of reagents in organic synthesis. These evidently cannot be designed for high stability in a general sense, hence their stability against *unwanted* chemical transformations must in practice be examined on a case-by-case basis.

Some broad observations can be made: Reagent moieties make up only a small fraction (10^{-3} to 10^{-6}) of the mass of a molecular manufacturing system as presently conceived; this somewhat reduces their quantitative significance as targets for damage. Further, they are subject to rapid cycles of synthesis and use, rather than serving as structural elements that must be stable for the life of the system; this can reduce their significance as targets for damage. Finally, the reagent devices used in a molecular manufacturing system can be selected from a wide range of candidate structures, including analogues of familiar reagents, catalysts, and reactive intermediates; this generates a large set of options with varying stability characteristics.

The stability of a reagent device depends on its design, which depends in part on its application. Chapter 8 discusses reagent devices in the context of mechanochemical operations, giving special attention to the stability and use of structures that would in solution chemistry be regarded as nonisolable reactive intermediates.

6.4.6. The stability of surfaces

Surfaces present special problems only if one regards a bulk phase as the norm. The present work, however, relies chiefly on concepts and models based on the chemistry of small molecules. These molecules are, in effect, all surface. Accordingly, the concepts and models need no modification to make them applicable to surfaces of the sort contemplated here. From this perspective, surfaces are the norm, and it is (for example) the special stability of the diamond *interior* relative to small, surface-dominated hydrocarbons that has been worthy of remark.

It is well known in materials science that surface diffusion occurs more readily than diffusion in bulk materials, that surfaces frequently undergo spontaneous °reconstruction to form arrangements unlike those in the interior, that surfaces are associated with modified electronic properties in semiconductors, and so forth. In judging the significance of these concerns in a particular instance, the nature of the material is crucial.

Metals, for example, frequently exhibit high surface reactivity, diffusion, or both. Nanometer-scale pits made on gold surfaces have been observed to refill through surface diffusion in minutes to hours at room temperature (Emch et al., 1988), but the surface atomic layer of gold is unusual: it has a slightly different lattice spacing from that of the bulk material, presumably facilitating motion. Although some metals and metal surfaces have adequate stability for use in eutactic nanosystems, this volume does not propose using metals in structural roles in nanomechanical components. Section 6.5 discusses the use of aluminum films for photochemical shielding, but films of the sort required are familiar in present technology and can tolerate atomic rearrangements. Section 11.6.1 discusses metals as conductors in electromechanical systems, but even here, alternative conductors are feasible.

Unlike metals, semiconductors are covalently bonded solids somewhat resembling the diamondoid components discussed in Part II. Clean semiconductor surfaces in vacuum consistently exhibit high reactivity and often undergo reconstruction. This unstable behavior, however, disrupts surfaces that a chemist would regard as dense arrays of free radicals or highly strained bonds. Nanomachines need not have such surfaces. Polished, *hydrogenated* diamond surfaces, in contrast, do not undergo surface reconstruction until heating (at ≥ 1275 K) has removed the hydrogen (Hamza et al., 1988). Until then, diamond is an unusually stable hydrocarbon. A wide range of other diamondoid structures with chemically reasonable surfaces should likewise exhibit excellent stability.

6.4.7. Thermal ionization and charge separation

The production of an unwanted ion pair in a nanomechanical system could easily disrupt its function; the threshold energies for ionization are therefore of interest. Ionization of a typical organic molecule in vacuum requires >1.3 aJ, and producing an electron-hole pair within diamond requires ~0.864 aJ. Structures of this sort undergo thermal ionization at negligible rates at ordinary temperatures, owing to the large energy required to produce charge separation. Mechanical energy applied to moving nanomechanical parts can do charge-separation work, but only if initial charge separation has occurred, placing opposite charges on adjacent moving surfaces.* The ease of producing ions by thermal processes in polar solvents, however, shows that ionization can readily occur in some structures. Indeed, with a suitable arrangement of polar groups around a pair of sites, one of low ionization energy and the other of high electron affinity, ionization can be spontaneous and stable. Ensuring a stable charge distribution will preclude the use of certain structures, but alternatives will usually be feasible. For example, if bearing interfaces with facing arrays of polar groups [e.g., Figure 10.15(d)] were to pose a problem, redesigned bearings with less-polar surfaces [e.g., Figure 10.15(a)] could in most instances be substituted with little loss of system-level performance. Unwanted thermomechanical ionization can be limited to negligible rates in well-designed nanomechanical systems at 300 K.

6.5. Photochemical damage

6.5.1. Energetic photons

Photochemical processes can excite molecules to energies that are effectively unavailable in equilibrium systems at ambient temperatures. Photons at visible and ultraviolet wavelengths (for example, in sunlight) have energies that are

---

* Based on difficulties arising in silicon micromachine research, P. Barth has raised concerns regarding electrostatic stiction and repulsion caused by charging of moving parts. Micromachines have irregular surface structures with sliding interfaces subject to high-energy processes including bond breakage, wear, and tribological charge separation; these phenomena can be avoided in nanomechanical systems. Dissimilar insulating surfaces brought into contact can transfer charge by a combination of thermal activation and tunneling, but only if the energies of occupied electronic states in one object approach those of unoccupied states in the other; avoiding this can impose design constraints (an experimental study of contact electrification forces is reported in Horn and Smith, 1992).

Table 6.2. Wavelengths, frequencies, and energies

Band name	Wavelength range (nm)	Maximum frequency (Hz)	Maximum energy (aJ)
Visible	700 – 400	7.5×10^{14}	0.50
UV-A	400 – 320	9.4×10^{14}	0.62
UV-B	320 – 280	1.1×10^{15}	0.71
UV-C	280 – 200	1.5×10^{15}	1.00

characteristic of far higher temperatures (for example, the ~5800 K of the solar photosphere) and they can deliver that energy to a single molecular site.

Where photochemical damage in the ambient terrestrial environment is concerned, the electromagnetic spectrum has traditionally been divided as shown in Table 6.2. Ultraviolet exposure is limited by atmospheric opacity: From the visible to the UV-B, the atmosphere transmits solar radiation; within the UV-C, absorption by oxygen and ozone effectively blocks solar radiation; at energies beyond the UV-C (the vacuum ultraviolet range), air is opaque. Consequently, both solar and local radiation sources are hazardous in the visible to UV-B range, only local sources are hazardous in the UV-C, and exposure in the vacuum ultraviolet is (within the atmosphere) usually negligible. At yet higher energies, the UV spectrum merges into the x-ray spectrum and photons become both penetrating and ionizing; the resulting damage is discussed in Section 6.6.

6.5.2. Overview of photochemical processes

Photochemical processes at a given wavelength begin with the absorption of a photon, which requires a chemical species able to absorb at that wavelength. Ordinary °alkanes, alcohols, °ethers, and °amines absorb at wavelengths <230 nm (Robinson, 1974), deep in the UV-C. In the absence of oxygen and other photochemically sensitive molecules, lack of absorption renders these substances photochemically stable (by ordinary standards) under ambient UV conditions. Systems of pi electrons characteristically absorb at longer wavelengths (organic dyes, which absorb at visible wavelengths, contain large °conjugated pi systems). Quantum mechanical selection rules constrain electronic transitions and absorption cross sections. Multiphoton absorption processes relax constraints on photon energies but require intensities seldom encountered in the absence of lasers.

For a process to do permanent damage to a nanomachine, bonds must be rearranged or cleaved, or a charge must be displaced and trapped. For a given molecular structure, each process has an energy threshold. Energy thresholds for ionization have been discussed in Section 6.4.7; typical energies for molecular ionization are in the vacuum ultraviolet range, and the production of an electron-hole pair in diamond requires a photon of λ <230 nm. Energy thresholds for rearrangements vary greatly with molecular structure and can be quite low. In laboratory photochemistry, rearrangements often involve conjugated pi systems. Broad classes of structures resist low-energy rearrangement.

Bond cleavage requires an energy equal to or greater than the dissociation energy of the bond; in practice, it often requires significantly more energy, since the photochemical process that overcomes the negative potential energy of the bond also deposits energy in vibrational and electronic excitations. Many bonds of interest in constructing nanomechanical systems—including carbon-carbon bonds—are subject to cleavage at UV-A and UV-B energies.

Photochemical bond cleavage is more common in low-pressure gases than in condensed matter, where competing processes more efficiently dissipate excitation energy. Further, in condensed matter, photochemical (like thermal) bond cleavage frequently is reversed by geminate recombination. Bond cleavage can be characterized by quantum yield, the ratio of bonds cleaved to photons absorbed. For °carbonyl-rich polymers, such as poly(methyl-isopropyl ketone), quantum yields are as high as 0.22 at $\lambda = 253.7$ nm; for polystyrene, the quantum yield is 9×10^{-5} (Ranby and Rabek, 1975).

In diffusive chemistry, the consequences of photochemical reactions can be complex. Free radicals can initiate chain reactions involving various reactive species, and all species have unconstrained opportunities for collision. In a structured solid phase, the opportunity to build more constrained systems enables the construction of systems with simpler behavior, including reduced sensitivity to photochemical damage.

6.5.3. Design for photochemical stability

In designing nanomechanical systems, photochemical damage can be limited or avoided in three alternative ways:

1. by keeping the entire system away from UV light,
2. by providing the system with a UV-opaque surface layer, or
3. by requiring that all components be photochemically stable.

Approach (1) is simple and workable for many purposes, but limits the nature of operating environments. Approach (2) is analyzed in the next section. Approach (3) would be preferable for some applications, but adding the constraint of photochemical stability would substantially increase the complexity of molecular systems design and eliminate device designs that would otherwise be attractive. The discussion in the present work assumes the use of approach (1) or (2), thereby avoiding photochemical stability constraints in device structures.

It is nonetheless worth briefly considering how one might design systems within the constraints of approach (3). One methodology would focus on absorption processes, attempting to avoid all use of structures that absorb in the UV-A and UV-B bands, but this may not prove practical. An alternative methodology would examine all potential modes of absorption and attempt to ensure that the absorbing structures can tolerate the resulting photochemical excitation. Tolerance for excitation is more achievable in a structured solid phase than in a liquid phase, owing to greater control of reaction opportunities and radical recombination processes. One approach would be to seek structures in which mechanical constraints force swift recombination of any cleaved bond: complications include photoexcitation to triplet states (delaying recombination in a manner that is rare in thermal processes) and rearrangements that trap the system in a ground-state potential well other than that desired. In summary, it may prove possible to develop a library of nanomechanical components that meet the stringent conditions required for UV-B exposure tolerance, but for the present it is easier to assume the use of shielding.

Table 6.3.   UV optical properties of aluminum (Gray, 1972).

Wavelength (nm)	n	k
120	0.057	1.15
160	0.080	1.73
200	0.110	2.20
320	0.280	3.56
400	0.400	4.45

6.5.4.   Photochemical shielding

Most metals block UV radiation of ordinary wavelengths, and aluminum is among the most opaque in thin-film form. The optical transmittance of a metal layer is determined (Gray, 1972) by its index of refraction, n, its extinction coefficient, k, and (secondarily) by the indexes of refraction of the adjacent media, n_1 and n_2. Table 6.3 lists values of n and k for aluminum at UV wavelengths. When an absorbing layer is thick enough, one can neglect interference among reflected waves within the layer. In this single-path approximation, appropriate for shielding calculations, the transmittance is given by:

$$T = \frac{16 n_1 n_2 \left(n^2 + k^2\right)}{\left[\left(n+n_1\right)^2 + k^2\right]\left[\left(n+n_2\right)^2 + k^2\right]} \exp\left(-4\pi k d / \lambda\right) \qquad (6.52)$$

where d is the thickness of the metal layer. Figure 6.12 graphs the transmittance of aluminum layers as a function of their thickness, at several UV wavelengths.

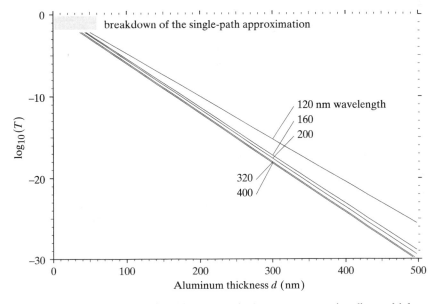

Figure 6.12.   Transmittance (fraction of optical power transmitted) vs. thickness of aluminum for various wavelengths, $n_1 = n_2 = 1.5$. Values computed from Eq. (6.52) and Table 6.3.

Note that transmittance is nearly independent of wavelength from the UV-A to the edge of the vacuum ultraviolet.

A typical shield thickness is readily estimated. Assume that the mean time between (photochemical) failures is required to be 30 years ($\approx 10^9$ s) for a system with an area of one square micron exposed to terrestrial sunlight. Assume that the mean power density at effective wavelengths is 5 W/m^2, that all shield-penetrating photons are absorbed in the system, that cleavage of one bond causes the system to fail, and that the quantum efficiency of bond cleavage is 10^{-2}. With these assumptions, the mean photon flux is $\sim 10^{19}$ m^{-2}s^{-2}, the 30-year dose to the system surface is 10^{16} photons, and the transmittance must be limited to 10^{-14} or less, which can be achieved with a shield thickness of just under 250 nm. Including a shield of this thickness imposes a large volumetric penalty on a system with an overall diameter of 2 μ, but only a modest penalty on a system with a diameter of 10 μ.

6.6. Radiation damage

6.6.1. Radiation and radiation dosage

Forms of ionizing radiation include high-energy photons and charged particles. At a nanometer scale, most of the damage done by these forms of radiation is mediated by high-energy secondary electrons. Experiments show that the energy carried by charged particles (including secondary electrons) is chiefly deposited in chains of excitation events with spacings ranging from a several hundred nanometers at MeV energies, to ~10 nm at 5 keV, to a virtual continuum at low energies; the local chemical effects are like those of UV radiation in the 10 to 30 eV range (Williams, 1972).

Ionizing radiation is measured both in rads (1 rad = 100 ergs absorbed per gram = 10^{-2} J/kg) and in roentgens (defined in terms of ionization produced in air by x-ray and gamma radiation; 1 roentgen deposits ~87 ergs in a gram of air). For many forms of ionizing radiation impinging upon light-element targets, the quantity of energy absorbed by a small volume of matter is roughly proportional to its mass. Background radiation in the terrestrial environment seldom exceeds 0.5 rad/year.

6.6.2. Classical radiation target theory

Existing experimental evidence relates radiation dosage to damage in molecular machinery, where that molecular machinery takes the form of proteins serving as enzymes. Classical radiation target theory (Lea, 1946) is a standard technique for estimating the molecular mass of enzymes from the loss of enzymatic activity as a function of radiation dose applied to dried enzyme in vacuum (Beauregard and Potier, 1985; Kepner and Macey, 1968). In the target-theory model, a single hit suffices to inactivate an enzyme molecule, and the probability that a molecule will be hit is proportional to its mass (for radiation doses small enough to make multiple hits improbable). Studies of the inactivation of enzymes of known molecular weight indicate that roughly 10.6 aJ of absorbed radiation is required to produce one inactivating hit (Kepner and Macey, 1968), yielding ~10^{15} hits/kg·rad.

How good is this relationship for estimating the rate of destruction of nano-mechanical systems owing to radiation damage? As a rough guide, it should be fairly reliable because it rests directly on experimental evidence. It must, however, be somewhat inaccurate owing to the physical differences between the targets. Proteins can tolerate small changes in side-chain structure at many sites without loss of function, as shown by protein engineering experience (Ponder and Richards, 1987b; Bowie et al., 1990); this suggests a significant ability to absorb structure-altering events while remaining active, which in turn suggests that nanomachines of more tightly constrained design tend to be *more* sensitive to radiation damage. Weighing on the other side, however, is the greater radiation tolerance of diamondoid structures: where geminate recombination of radicals is strongly favored by mechanical constraints, only a small fraction of excitation events cause a permanent change. This suggests that typical nanomachines tend to be *less* sensitive to radiation damage than are proteins, which readily undergo backbone cleavage. In summary, nanomachines have a greater fraction of sites at which excitations of bond-cleaving magnitude cause no permanent alteration, and proteins have a greater fraction of sites at which permanent alterations can be tolerated. On the whole, it seems reasonable to assume that diamondoid nanomachines, like proteins, suffer $\sim 10^{15}$ inactivating hits per kilogram per rad. For this estimate to err by being substantially too low would require that enzymes have an implausibly large probability of surviving random structural damage. In this model, the probability that an initially functional component has not been destroyed by radiation damage is

$$P_{\text{func}} \approx \exp\left(-10^{15} Dm\right) \tag{6.53}$$

where D is the dose in rads and m the component mass in kilograms.

6.6.3. Effects of track structure

Aside from structural differences, typical nanomechanical systems are far larger than typical enzymes. This affects scaling relationships because radiation hits are distributed not at random, but along particle tracks. In the range of sizes for which the diameter of a mechanism is large compared to the spacing of excitation events along a typical particle track (while remaining small compared to the length of the track), damage becomes proportional not to mass but to area, and hence scales not as m but as $m^{2/3}$. In typical radiation environments, components of ordinary density with a diameter of 100 nm or greater should benefit from this favorable breakdown in linearity. A rough model representing this shift from m to $m^{2/3}$ scaling in the vicinity of 100 nm diameters is

$$P_{\text{func}} \approx \exp\left(\left\{ -\left(10^{15} Dm\right)^{-1} - \left[10^{11} D(m/\rho)^{2/3}\right]^{-1} \right\}^{-1} \right) \tag{6.54}$$

where ρ is the density of the component in kg/m^3. This model (graphed in Figure 6.13) neglects various factors affecting track structure (though ionizing-event spacing is taken to scale inversely with density). It uses an arbitrary functional form to represent the transition between regimes, and should be taken as only a rough guide to damage rates for large components.

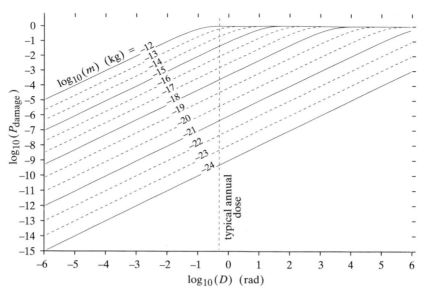

Figure 6.13. Probability of component failure owing to ionizing radiation damage vs. radiation dose, according to Eq. (6.54), for components of differing masses; assumes failure after a single hit and a component density equaling that of water. Background radiation in the terrestrial environment seldom delivers more than 0.5 rad/year, marked as "typical annual dose." Note that at the assumed density, the graphed curves correspond to component volumes ranging from 1 nm^3 to 10^{12} nm$^3 = (10\,\mu)^3$.

6.6.4. Radiation shielding

To shield against ionizing radiation usually requires macroscopic thicknesses of dense material, ranging from a fraction of a millimeter for medium-energy alpha particles to meters for gamma and cosmic rays. Earth's atmosphere has an areal mass density of $\sim 10^4$ kg/m^2, yet showers of secondary particles from cosmic rays deliver a significant annual dose at sea level.

Any nonzero exposure to ionizing radiation adversely affects either the reliability or the feasible performance of nanoscale systems, yet even with extensive macroscale engineering, radiation exposure seems inescapable. Thick layers of nonradioactive shielding materials, although able to reduce radiation exposure by large factors, cannot prevent sporadic showers of secondary particles resulting from high-energy cosmic-ray neutrinos. These particles can penetrate planetary thicknesses of rock, and the resulting showers of secondaries have been observed in deep mines, moving upward (Hendricks et al., 1969; Krishnaswamy et al., 1969).

6.7. Component and system lifetimes

6.7.1. Component lifetimes

a. The single-point failure assumption.   Under the conditions of machine-phase chemistry, some sources of molecular damage (e.g., reactions with water and free oxygen) are excluded, and most others are subject to control during the design phase. Damage *rates* can accordingly be quite low. A single chemical transformation, however, can cause failure. Since a typical system may consist of 10^6 to 10^{12} atoms, damage *sensitivity* can be quite high.

Components of sufficient size can avoid this sensitivity: macroscale machines are proof of this. The exclusive use of large-scale components, however, would sacrifice many of the scaling-law advantages of systems of minimal size. Accordingly, to simplify analysis and ensure conservatism, the present work in most instances assumes that components fail if they experience a single chemical transformation anywhere in their structures. This is the *single-point failure assumption*. (Chapter 14 discusses larger, damage-tolerant structures.)

In macroscale systems, cumulative wear and fatigue often determine component lifetimes. Wear involves removal of material from a surface, which requires chemical transformations. Fatigue is more subtle, but entails changes in chemical bonding associated with displacements of atoms (e.g., the motion of dislocations). Note, however, that cumulative wear and fatigue play no role in the component lifetimes when the single-point failure assumption is applied: the first chemical transformation step is regarded as fatal, making cumulative processes irrelevant. The elementary mechanisms responsible for cumulative wear and fatigue in macroscale systems can cause damage in nanomechanical systems, but in diamondoid structures they are mostly subsumed by the thermomechanical damage mechanisms discussed in Section 6.4. If damage is avoided, cumulative effects do not occur, and so wear and fatigue as such are absent. Failure rates in as-yet undamaged components are accordingly independent of age.

b. Choosing reliability criteria.   What will be the main causes of chemical transformation and resulting failure? By a suitable choice of design, almost any mechanism could be made dominant. By permitting exposure to photochemically active UV wavelengths, photochemical damage could be made dominant; by using organic peroxides or highly strained bonds as structural elements, bond cleavage could be made dominant; other choices could make rearrangements or interfacial reactions dominant. All of these damage mechanisms, however, are subject to rate laws that are exponentially dependent on parameters that are subject to engineering control. Under ambient terrestrial conditions, all can be reduced to low levels by careful design, augmented by experimental testing and redesign if necessary.

Ionizing radiation is less subject to control. Although (imperfect) shielding is possible, it would often be awkward: major reductions in radiation exposure require shielding thicknesses measured in meters, in contrast to the hundreds of nanometers that suffice at UV-A, -B, and -C wavelengths. The present work assumes that ambient terrestrial radiation sets a practical lower bound on molecular damage rates.

This practical lower bound can be used to choose design criteria for other damage rates. Various engineering objectives can be served by using structures of marginal stability; examples include high mechanical speeds (associated with high bond stresses), compact designs (associated with exploitation of relaxed design rules on bond types and geometries), and so forth. Reliability, however, is itself a major engineering objective. Given the assumption of a fixed, structure-independent failure rate resulting from ionizing radiation, it is reasonable to constrain the controllable damage mechanisms to be comparatively small. This is the *radiation-damage dominance criterion*.

In the terrestrial environment, the level of background radiation rarely exceeds ~0.5 rad/year, which corresponds to $\sim 1.5 \times 10^7$ hits/kg·s (Section 6.6). The single-point failure assumption implies that a hit may be equated to a cleaved bond or any other damaging transition. If we demand that all other damage rates be an order of magnitude lower than radiation damage rates, then an acceptable mean failure rate per bond in small structures (100 nm diameter or less) is $\sim 10^{-20}$ s^{-1}, given typical numbers of bonds per kilogram. The 250 nm photochemical shielding thickness calculated in Section 6.5.4 meets this criterion in full sunlight at Earth's surface. For thermally activated processes, Eq. (6.28) (with $q_{long} = 1$) indicates that failure rates remain acceptable so long as all barrier heights ($\approx$ activation energies) exceed ~313 maJ at 300 K, or ~366 maJ at 350 K.

In the absence of strong stresses or other destabilizing influences, most covalent bonds of interest (in particular, those between C and H, N, O, F, Si, P, S, and Cl) substantially exceed these thresholds (Table 3.8); all of those mentioned yield cleavage rates below 10^{-33} s^{-1}. If most bonds in a system have failure rates this low, then a smaller population can have higher failure rates while still meeting the 10^{-20} s^{-1} criterion for the mean failure rate. If this population is 1% of the total, then its barrier heights (by the above method) need only exceed 294 maJ at 300 K and 344 maJ at 350 K. For perspective, the former condition is *almost* met by isolated organic peroxide linkages (bond strength ~259 maJ, Table 3.8), which are usually regarded as quite reactive.

c. Assumptions and damage rates.   The single-point failure assumption and the radiation-damage dominance criterion are both subject to criticism and improvement, but neither is a hypothesis. Single-point failure is the most conservative rule of calculation, and can be relaxed if a detailed model of a component demonstrates that it is more robust. Radiation-damage dominance is a design objective that can be discarded whenever alternative objectives prove superior. Both are stated in order to guide design and analysis within the context of the present work.

So long as components meet the above 10^{-20} failure/bond·s criterion, the uncorrelated radiation-damage lifetime model, Eq. (6.53), can with reasonable accuracy be treated as the overall component lifetime model. Systems meeting the slightly more stringent 10^{-22} failures/bond·s criterion can be described by Eq. (6.54) over the range of component volumes graphed in Figure 6.13.

6.7.2. System lifetimes

For a given level of reliability, finite damage rates and the single-point failure assumption set upper bounds to component size. Figure 6.13 indicates that components of 10^9 nm^3 ($= 1$ μ^3) have annual failure rates of several percent if they meet the radiation-damage dominance criterion. Among macroscopic systems, the mean time to failure is quite brief unless damage can be tolerated.

Where component failures cannot be excluded but systems must be reliable, standard engineering practice resorts to redundancy. Although the existence of radiation tracks violates the assumption that component failures are randomly distributed in space and introduces geometrical complications, the essential effects of redundancy can be illustrated by a random-damage model.

Assume that a system consists of N sets of components, $s_1, s_2, s_3, ..., s_N$, with each set s_i containing n_i components of type c_i, and assume that the system remains functional so long as at least one component in each set remains functional. In other words, any component can substitute for the other components *within its set*, but components in different sets play different roles (e.g., in manufacturing systems like that discussed in Section 14.3). If $n_i = 1$ for each set, the system is nonredundant, and the probability that it remains functional is the product of the probabilities that each of components remains functional:

$$P_{\text{func}}(\text{system}) = \prod_{i=1}^{N} P_{\text{func}}(c_i) = \prod_{i=1}^{N} \exp\left(-10^{15} Dm_i\right) \qquad (6.55)$$

where m_i is the mass of a component of type c_i, and the final expression is based on Eq. (6.53). With redundancy, however, the probability that the system remains operational depends on the probability that at least one member of each set s_i remains functional:

$$P_{\text{func}}(s_i) = 1 - \left[1 - P_{\text{func}}(c_i)\right]^{n_i} \qquad (6.56)$$

hence

$$P_{\text{func}}(\text{system2}) = \prod_{i=1}^{N} P_{\text{func}}(s_i) = \prod_{i=1}^{N} \left\{1 - \left[1 - \exp\left(-10^{15} Dm_i\right)\right]^{n_i}\right\} \qquad (6.57)$$

To understand how reliability depends on redundancy and component lifetimes, it is useful to consider an idealized system in which all components have the same mass m and all sets have the same degree of redundancy n:

$$P_{\text{func}}(\text{system3}) = \left\{1 - \left[1 - \exp\left(-10^{15} Dm_i\right)\right]^{n_i}\right\}^{N} \qquad (6.58)$$

$$\approx \exp\left\{-N\left[1 - \exp\left(-10^{15} Dm_i\right)\right]^{n_i}\right\} \qquad (6.59)$$

in which the approximation is valid so long as the probability remains low that all of the components in a given set have failed. Choosing a value of D and m such that $10^{15}Dm = 0.1$ (implying a $\sim.9$ probability that a given component is functional), a nonredundant system with $N = 1000$ sets will be functional with a probability of only $\sim 10^{-44}$. If $n = 5$, however, the system will be functional with probability $\sim.992$, and if $n = 10$, the probability of system failure shrinks to $<10^{-7}$.

Calculations show that substantial levels of redundancy can make failure extremely improbable, even in macroscopic systems, so long as each component has a reasonably large probability of remaining functional. For example, if the components have a mass of $\sim 10^{-18}$ kg (a diameter of ~ 100 nm, at ordinary densities), then $10^{15}Dm = 0.1$ after >100 years of background radiation (Figure 6.13). If these components are combined to form a 1000 kg system containing $N = 4 \times 10^{19}$ sets of $n = 25$ components, Eq. (6.54) predicts a failure probability of only $\sim 10^{-6}$.

Where indefinitely prolonged system life is required, the standard engineering answer is a combination of redundancy and replacement of damaged components. This is feasible in nanoscale systems, but at the cost of substantial increases in system complexity.

6.8. Conclusions

Transition state theories based on statistical mechanics can adequately describe thermally driven transition rates in most processes of nanomechanical interest, including mechanosynthesis, thermomechanical bond cleavage, and nonchemical transitions between potential wells. A simpler model suffices to set bounds on error rates in placing objects in potential wells (a concept that requires treatment of time-dependent potential energy functions). In general, thermally driven transition rates are an exponentially decreasing function of barrier heights, and thermally driven placement errors are an exponentially decreasing function of differences in well depths. Placement errors can be made rare either by increasing the spacing between alternative wells, or by increasing the stiffness of the placement mechanism. Thermomechanical damage can be made rare by choosing strong covalent structures that lack unusually low-energy pathways leading to elimination or rearrangement reactions, and by limiting applied mechanical stress (exclusion of stray reactive molecules from the environment is assumed). Impinging energy can also cause damage: photochemical reactions can be prevented by excluding light, but reactions induced by ionizing radiation are inescapable, setting practical lower bounds on damage rates and component lifetime. Later chapters assume that damage from readily controllable mechanisms is limited to levels low enough that ionizing radiation damage is dominant.

Some open problems.   An open problem, the solution of which would substantially aid nanomechanical design by nonchemists, is the development of software able to check nanomechanical structures for instability (e.g., as an adjunct to a molecular mechanics package). A realistic system would include a set of criteria for identifying clearly stable structures, a set of criteria for identifying clearly unstable structures, and an ability to report which substructures in a design are either unusable or questionable. A highly conservative rule set can easily be defined, but recognizing increasingly broad classes of stable structures (without making errors of inclusion) presents increasing challenges. For example, estimating rates of cleavage of sets of strained bonds (including shear forces not considered in Section 6.4.4a) must in general reflect nonlocal elastic deformation of the structure.

A further open problem relates to photochemical stability: What structures of nanomechanical interest are compatible with what ranges of photon energies? Progress in this area may relax the constraint of photochemical shielding.

7

Energy Dissipation

7.1. Overview

Energy dissipation imposes constraints on the design of nanomechanical devices, particularly when they are aggregated to form macroscale systems. Owing to limits on feasible cooling capacity, energy dissipation in macroscale systems limits the feasible rate of operations, for example, in massively parallel computers. It could likewise reduce the attractiveness of nanomechanical systems relative to alternatives, for example, in performing chemical transformations that could be achieved through diffusive, solution-phase chemistry.

Despite its importance, energy dissipation seldom imposes *qualitative* limits, that is, constraints on the *kinds* of operations that can be performed on a nanoscale, as opposed to constraints on the speed and efficiency with which they can be performed. In studies of the potential of molecular nanotechnology, energy dissipation is often important to estimate, yet seldom crucial to estimate precisely. Moderate overestimates of dissipation yield conservative estimates of performance, while seldom falsely implying that a design is infeasible.

The present chapter surveys mechanisms of energy dissipation that appear significant, estimating or bounding their magnitudes. Mechanisms of energy dissipation specific to metals are ignored since the systems under consideration are chiefly dielectric; mechanisms occurring during plastic deformation are ignored (except as analogies) since, under the rules adopted in Chapter 6, any degree of plastic deformation is counted as catastrophic damage.

Several kinds of dissipation mechanisms are described. These are:

- *Acoustic radiation from forced oscillations,* which transports mechanical energy to remote regions where it is thermalized
- *Phonon scattering,* in which mechanical motions disturb °phonon distributions by reflection
- *°Thermoelastic effects and phonon viscosity,* in which elastic deformations disturb phonon distributions via anharmonic effects and shear
- *Compression of potential wells,* in which nonisothermal processes result in thermodynamic irreversibility
- *Transitions among time-dependent potential wells,* in which the merging of initially separate wells dissipates free energy by a combination of free expansion and forced oscillation.

Different mechanisms result in power dissipation rates that scale differently with speed. For acoustic radiation from an oscillating force, $P \propto v^2$; for radiation from an oscillating torque, $P \propto v^4$. Phonon scattering, thermoelastic and phonon viscosity effects, and nonisothermal compression of potential wells all (to a good approximation) exhibit $P \propto v^2$. Transitions among time-dependent potential wells, in contrast, are better described by $P \propto v$. Of these mechanisms, all but the last exhibit a speed-dependent energy dissipation per operation (or per unit displacement) which approaches zero as $v \to 0$.

Note that qualifiers such as *typically, frequently, many systems,* and the like are used throughout this chapter to indicate situations and parameters characteristic of systems of practical interest. This usage reflects the results of design and analysis from Part II of the present work. Parameter values used in sample calculations are usually chosen to yield significant but moderate energy dissipation by one or more mechanisms.

7.2. Radiation from forced oscillations

7.2.1. Overview

Time-varying forces in an extended material system can excite mechanical vibrations that are eventually thermalized. The energy dissipated in this fashion is distinct from the energy transiently stored in local elastic deformations, which is (unless subject to other dissipation mechanisms) recovered in the course of a cycle. This mode of energy dissipation is affected by the structure of the system in a substantial region surrounding the device in question. In estimating magnitudes, it is natural to begin by modeling the structure as a uniform elastic medium and the vibrations as acoustic waves in that medium. At the frequencies and energies of interest here, quantum effects are small.

The accuracy of this approximation varies, but it is good when the wavelength of the acoustic radiation is long compared to the scale of inhomogeneities in the mechanical system. Many systems considered in later chapters have structures that include an extended matrix or housing that supports numerous nanoscale moving parts. To estimate dissipation via acoustic radiation, it is reasonable to treat such a system as uniform on a scale of tens of nanometers or more. If a structural matrix of diamondlike stiffness is roughly 1/10 the total mass (larger fractions lower losses), then the speed of sound across the system is $\sim(1/10)^{1/2}$ times the speed of sound in diamond, or ~ 5000 m/s. For $\lambda = 100$ nm, $\omega \approx 3 \times 10^{11}$ rad/s. At higher frequencies, the acoustic radiation model should still yield results of the right order, so long as estimates are based on mean properties of the structure within a wavelength of the device. (Phonon scattering processes depend on material properties on a nanometer scale.)

Treating dissipation as simple acoustic radiation still leaves a complex problem. Only a few cases are treated here, and then by approximation. Acoustic radiation in fluids is commonly described; expressions for the power radiated by a pulsating sphere and a piston in a wall appear in Gray (1972); Nabarro (1987) adapts an expression for a pulsating cylinder in a fluid to describe analogous radiation losses in a solid. Of more interest in the present context is radiation resulting from a sinusoidally varying force, torque, or pressure at a point (or small region) in an elastic medium. General expressions for radiation from a time-varying force applied at a point within a solid medium appear in the physics literature (e.g., Hudson, 1980). Sections 7.2.3, 7.2.4, and 7.2.5 derive approximations for a sinusoidally varying force, torque, and pressure. Section 7.2.6 discusses radiation from traveling disturbances in a medium, taking dislocations in crystals as a model.

7.2.2. Acoustic waves and the equal-speed approximation

An isotropic elastic medium supports transverse waves of velocity

$$v_{s,t} = \sqrt{G/\rho} \tag{7.1}$$

(where G is the °shear modulus and ρ the density), together with longitudinal waves of velocity

$$v_{s,\ell} = \sqrt{\frac{E}{\rho} \frac{1-v}{(1+v)(1-2v)}} \tag{7.2}$$

where E is Young's modulus and v is °Poisson's ratio, which [except in unusual structures of negative v (Lakes, 1987)] falls in the range $0 \leq v \leq 0.5$. In an isotropic medium,

$$E = 2G(1+v) \tag{7.3}$$

hence $v_{s,\ell} > v_{s,t}$. For diamond, $v \approx 0.1$, and $v_{s,\ell} \approx 1.5 v_{s,t}$.

Given the approximations involved in treating a nanomechanical system as a uniform medium, it is not unreasonable to add the approximation that $v_{s,\ell} = v_{s,t}$ for waves radiated from the origin. This equal-speed approximation in effect assumes anisotropic elastic properties that simplify the mathematics, rather than making the mathematics fit an arbitrarily assumed isotropy. The equal-speed approximation also underlies the standard Debye model of heat capacity and phonon distributions. It is used extensively in the phonon-drag models of Section 7.3.

7.2.3. Oscillating force at a point

Many mechanical systems cause disturbances that can be approximated by an oscillating force applied to a point. Among these are unbalanced rotors, reciprocating power-driven mechanisms, and vibrating, elastically restrained masses.

a.   A model.   With the equal-speed approximation (Section 7.2.2), propagating disturbances can be a function of radius alone: the restoring forces between uniformly displaced spherical shells are then uniform over each shell, leading to uniform accelerations and continued spherical uniformity of displacements. The linearized dynamical equation has the form

$$\frac{\partial}{\partial r}\left(4\pi r^2 M \frac{\partial}{\partial r} y(r,t)\right) = 4\pi r^2 \rho \frac{\partial^2}{\partial t^2} y(r,t) \tag{7.4}$$

where the function $y(r,t)$ specifies a displacement along the line of the force, and M is a modulus of elasticity, uniform in magnitude but differing in nature from the axis aligned with the force [where it is equivalent to $E(1-v)/(1+v)(1-2v)$] to the plane perpendicular to that axis (where it is equivalent to G).

The oscillating force of amplitude F_{max} is introduced through the boundary condition

$$4\pi r^2 M \frac{\partial}{\partial r} y(r,t) \Big|_{r=0} = F_{max} \sin(\omega t) \qquad (7.5)$$

and solutions corresponding to outbound waves are required. These constraints yield

$$y(r,t) = -\frac{F_{max}}{4\pi Mr} \sin \omega \left(t - r\sqrt{\rho/M} \right) \qquad (7.6)$$

The instantaneous power at a given radius is the force transmitted times the velocity

$$P = \left(4\pi r^2 M \frac{\partial}{\partial r} y(r,t) \right) \frac{\partial}{\partial t} y(r,t) \qquad (7.7)$$

which has a time-average value equaling the isotropic mean radiated power

$$P_{rad} = F_{max}^2 \omega^2 \sqrt{\rho} \frac{1}{8\pi M^{3/2}} \qquad (7.8)$$

b.   Damping of an embedded harmonic oscillator.   A harmonic oscillator like that in Figure 7.1 is damped by acoustic radiation. At a given amplitude, the force is related to the energy and effective stiffness by

$$F_{max} = \sqrt{2 k_s \mathcal{E}} \qquad (7.9)$$

Equating the net radiated power to the time-average value, Eq. (7.8), yields an exponential decay of the oscillation energy with a time constant (in seconds) of

$$\tau_{osc} \approx \frac{4\pi m M^{3/2}}{k_s^2 \sqrt{\rho}} \qquad (7.10)$$

Alternatively, the fractional energy loss per cycle can be expressed as

$$f \approx \frac{1}{2} \sqrt{\rho/m} \left(k_s/M \right)^{3/2} \qquad (7.11)$$

for $f \ll 1$.

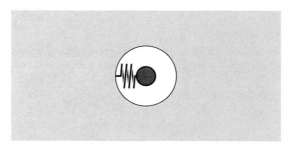

Figure 7.1.   Model of a mechanical harmonic oscillator embedded in a medium. The oscillator can be treated as a point source of force so long as its dimensions are small compared to the wavelength of the sound emitted at its characteristic frequency.

Many systems of low stiffness will be constrained by nonbonded interactions with strong anharmonicity. In the limiting case, the stiffness at equilibrium is small, and the oscillation can be viewed as a series of collisions with bounding walls. Energy loss is then better modeled using thermal accommodation coefficients (Section 7.5.1).

7.2.4. Oscillating torque at a point

Torsional harmonic oscillators are directly analogous to the linear oscillators just discussed, and can be modeled as sinusoidally varying torques applied to a point. The potential energy of an imperfect bearing varies with the rotational angle, causing a varying torque. For a bearing in uniform rotation, the resulting torque can be treated as a sum of sinusoidally varying components, each causing acoustic radiation.

a. A model. An oscillating torque in a uniform, isotropic medium radiates pure shear waves, hence such media serve as a convenient approximation for real systems. The analysis roughly parallels that given in Section 7.2.3. Again, spherical shells can be treated as undergoing rigid motion (here rotation rather than displacement), reducing the problem to a linearized equation with a single spatial dimension. The linearized dynamical equation is

$$\frac{\partial}{\partial r}\left(\frac{8}{3}\pi r^4 G \frac{\partial}{\partial r} y_\theta(r,t)\right) = \frac{8}{3}\pi r^4 \rho \frac{\partial^2}{\partial t^2} y_\theta(r,t) \qquad (7.12)$$

where the function $y_\theta(r,t)$ specifies an angular displacement about the axis of the applied torque, which sets the boundary condition at the origin:

$$\left.\frac{8}{3}\pi r^4 G \frac{\partial}{\partial r} y_\theta(r,t)\right|_{r=0} = T_{max}\cos(\omega t) \qquad (7.13)$$

Together with the requirement for outbound waves, this yields the solution

$$y_\theta(r,t) = \frac{T_{max}\omega\sqrt{\rho}}{8\pi G^{3/2}}\left[\frac{1}{r^2}\sin\omega\left(t - r\sqrt{\frac{\rho}{G}}\right) + \frac{1}{r^3\omega}\sqrt{\frac{G}{\rho}}\cos\omega\left(t - r\sqrt{\frac{\rho}{G}}\right)\right] \qquad (7.14)$$

(which includes a near-field component). This solution implies a time-average radiated power

$$P_{rad} = T_{max}^2\omega^4\frac{\rho^{3/2}}{48\pi G^{5/2}} \qquad (7.15)$$

which is steeply dependent on frequency.

b. Damping of an embedded torsional harmonic oscillator. Paralleling the development in Section 7.2.3b, a torsional harmonic oscillator characterized by an angular spring constant k_θ (J/rad^2) and a moment of inertia I (kg·m^2) has a time constant for radiative decay of oscillation energy

$$\tau_{osc} \approx \frac{24\pi I^2 G^{5/2}}{k_\theta^3 \rho^{3/2}} \qquad (7.16)$$

and a fractional energy loss per cycle

$$f \approx \frac{1}{12}(\rho/I)^{3/2}(k_\theta/G)^{5/2} \qquad (7.17)$$

7.2.5. Oscillating pressure in a volume

A component sliding through a channel with corrugated walls exerts a varying pressure on its surroundings. The force applied in one direction is balanced by the force applied in the opposite direction, distinguishing this from the case described in Section 7.2.3. This and related systems can be modeled as a sinusoidally varying pressure in a spherical cavity.

a. A model.   An oscillating pressure in a spherical cavity in an isotropic, homogeneous medium radiates spherical, longitudinal waves. In the near field, however, hoop stresses transverse to the wave can play a dominant role in the balance of forces. The materials of greatest engineering interest have low values of v; for example, diamond has $v \approx 0.1$; the analysis can be simplified and rendered somewhat more conservative by assuming $v = 0$ and treating the effective modulus M as equal to °Young's modulus E. With these approximations, the linearized dynamical equation is

$$\frac{\partial}{\partial r}\left(4\pi r^2 M \frac{\partial}{\partial r} \mathrm{y}(r,t)\right) - 8\pi M\, \mathrm{y}(r,t) = 4\pi r^2 \rho \frac{\partial^2}{\partial t^2} \mathrm{y}(r,t) \qquad (7.18)$$

where the function $\mathrm{y}(r,t)$ specifies a radial displacement. Because the effects of the applied pressure are opposed and contained by surrounding layers of the medium in a way impossible for forces or torques (which must be transmitted), a radius of application r_0 must be defined for the applied pressure p and the associated total force $F\,(= \pi r^2 p)$. The boundary condition imposed by the oscillating force is then

$$4\pi r^2 M \frac{\partial}{\partial r}\mathrm{y}(r,t)\bigg|_{r=r_0} = F_{\max}\sin(\omega t) \qquad (7.19)$$

Together with the requirement for outbound waves, this yields the solution

$$\mathrm{y}(r,t) = \frac{F_{\max}}{4\pi M}\left(\frac{\rho\omega^2 r_0^2}{M} + \frac{4M}{\rho\omega^2 r_0^2}\right)^{-1/2}$$
$$\times \left[\frac{1}{r}\sin\omega\left(t - r\sqrt{\frac{\rho}{M}}\right) - \frac{1}{\omega r^2}\sqrt{\frac{M}{\rho}}\cos\omega\left(t - r\sqrt{\frac{\rho}{M}}\right)\right] \qquad (7.20)$$

This again includes a near-field component. This solution implies a time-average radiated power

$$P_{\mathrm{rad}} = F_{\max}^2 \omega^2 \sqrt{\rho}\, \frac{1}{16\pi M^{3/2}}\left(\frac{\rho\omega^2 r_0^2}{2M} + \frac{2M}{\rho\omega^2 r_0^2}\right)^{-1} \qquad (7.21)$$

The trailing factor, in parentheses, strongly reduces the radiated power when the radius of the driven region is small compared to the radiated wavelength.

7.2.6. Moving disturbances

a. Dislocations as a model.   Dislocations provide a model for nanometer-scale mechanical disturbances moving through a medium, exhibiting many distinct energy dissipation mechanisms. The major role of dislocation motion in determining the strength of bulk materials has encouraged extensive analysis and experimentation; recent reviews include Nabarro (1987) and Alshits and Indenbom (1986). Several of the following sections draw directly or indirectly on this body of analysis.

b.   Subsonic disturbances.   Among the sources of moving mechanical disturbance are objects sliding or rolling on a surface and alignment bands (Section 7.3.5a) in sliding interfaces. At any given point, the motion imposed by a moving disturbance takes the form of an imposed oscillation. Nonetheless, in a uniform environment, a uniform disturbance moving at a uniform, subsonic speed radiates no acoustic power. In a real system, inhomogeneities and variations in speed and in force as a function of time cause forced-oscillation radiation, but these mechanisms can be considered separately.

c.   Supersonic disturbances.   Material motions of subsonic speed can cause supersonic patterns of disturbance. The chief mechanism of interest here is the motion of bands of atomic alignment (closely analogous to dislocations) in sliding interfaces.

Figure 7.2 illustrates the geometry for two rows of atomic bumps, with spatial frequencies k_1 and k_2 (rad/m). Panels (a), (b), and (c) show three successive configurations as surface 2 moves over surface 1: the arrow to the left shows the motion of an atom in surface 2; the arrow to the right shows the motion of an alignment band. Taking v as the velocity of 2 with respect to 1, and k_2 as the spatial frequency of surface 2, it can be seen that the spatial frequency of the alignment bands is $|k_{\mathrm{bands}}| = |k_2 - k_1|$, and that the velocity ratio R is

$$R = \frac{v_{\mathrm{bands}}}{v} = \frac{|k_1|}{|k_2 - k_1|} \tag{7.22}$$

which can attain arbitrarily high values as $|k_1 - k_2| \to 0$.

More generally, each surface can be viewed as having many sets of rows, with sets being described by wave vectors that are not necessarily collinear with each other, or with the sliding velocity vector. Interpreting k_1 and k_2 as vectors with signs chosen to minimize $|k_2 - k_1|$, each pair of opposed row-sets defines a set of alignment bands in which $k_{\mathrm{bands}} = k_2 - k_1$. From a geometrical construction, it

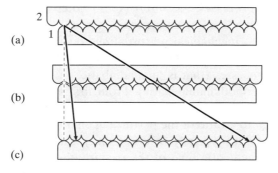

Figure 7.2.   The motion of bands of atomic alignment as two surfaces with differing row spacings slide over each other. Panels (a), (b), and (c) represent three successive positions, the left-hand arrow traces the motion of an atom in surface 1; the right-hand arrow traces the motion of an alignment band between the surfaces.

can be seen that

$$R = \frac{v_{\text{bands}}}{v} = \left| (k_2 - k_1) \frac{|k_1|}{|k_2 - k_1|^2} \cos\theta + \frac{k_1 \times (k_2 \times k_1)}{|k_1 \times (k_2 \times k_1)|} \sin\theta \right| \qquad (7.23)$$

$$\leq \frac{|k_1|}{|k_2 - k_1|} + 1$$

where θ measures the angle between the velocity vector v and the vector k_1.

In the limiting case, $|k_2 - k_1| = 0$, $R = \infty$, and the interface as a whole period-
ically enters and leaves the aligned state, radiating sound like an oscillating piston [Eq. (7.21), considering the power per unit area in the limit of large r_0]. This
limiting case sets an upper bound to the power dissipation of supersonic alignment bands. Nanomechanical bearings of several kinds contain sliding interfaces
(Chapter 10). The present work adopts the design constraint that the alignment-
band speeds remain subsonic, thereby avoiding this mode of energy dissipation.

The limiting case just described can be modeled as a compliant interface in
which a sinusoidal variation in the equilibrium separation occurs at a frequency
$\omega = kv$. The time-average radiated power is then

$$P_{\text{rad}} = A^2 \omega^2 \sqrt{M\rho} \left(M\rho \omega^2 / k_a^2 + 4 \right)^{-1} S \qquad (7.24)$$

where S is the area of the interface, A is the amplitude of the variation in equilibrium separation (the limit of the actual amplitude as $\omega \to 0$), both media are
assumed identical, and the calculated value includes power radiated from both
sides of the surface. Typically, unless ω is unusually high or the interfacial stiffness k_a is unusually low, the approximation

$$P_{\text{rad}} \approx A^2 \omega^2 \sqrt{M\rho} \frac{S}{4} \qquad (7.25)$$

is accurate (it is always conservative). For $M = 10^{11}$ N/m^2, $\rho = 2000$ kg/m^3, $k =$
2×10^{10} rad/m, and $A = 0.05$ nm, the radiated acoustic power is $\sim 4 \times 10^6$ W/m^2
at a speed of 1 m/s, and $\sim 4 \times 10^2$ W/m^2 at a speed of 1 cm/s. Again, most sliding interfaces need not suffer losses by this mechanism.

d. Nonadiabatic processes. J. Soreff (1991) notes that, if one surface of a
sliding interface is modeled as an array of atomic-scale harmonic oscillators,
these are exposed to mechanical perturbations resulting from the passage of
atomic features on the other surface and are subject to excitation at some rate
by nonadiabatic processes (that is, "nonadiabatic" in the quantum-mechanical
rather than the thermodynamic sense). From first-order perturbation theory
(Kogan and Galitskiy, 1963), the probability of an encounter causing an excitation is proportional to a ratio of the perturbing energy to the oscillator quantal
energy ($\hbar\omega$), a quantity typically of order unity, times the factor $\exp(-2\omega\tau)$,
where τ is the characteristic time of the perturbation. Since $\omega\tau \approx v_s/v$, Soreff
observes that $\exp(-2\omega\tau)$ typically is extremely small. For example, in a material
with $v_s = 10^4$ m/s, a system with v as high as 10^2 m/s has a transition probability
on the order of 10^{-85}. In a typical system, a single excitation dissipates ~ 1 maJ.

7.3. Phonons and phonon scattering

7.3.1. Phonon momentum and pressure

Thermal phonons in a crystal resemble blackbody radiation in a cavity, and both resemble a gas. As discussed further in Sections 7.3.3, 7.4.1, and 7.4.2, the phonon gas is responsible for energy dissipation by several mechanisms analogous to those in ordinary gases. Here, we consider drag resulting from scattered and reflected phonons.

In calculating drag owing to scattering, phonons can be treated (Lothe, 1962) as having a momentum equal to their quasi-momentum (i.e., crystal momentum), of magnitude

$$|\mathbf{p}| = \hbar k = \hbar \omega / v_s = \mathcal{E}/v_s \tag{7.26}$$

where k in this section is to be taken as the *magnitude* of the wave vector (in rad/m) and v_s is the speed of sound (here again approximated as constant for all frequencies and modes). With the substitution of c for v_s, Eq. (7.26) also describes the momentum of photons in vacuum.

A phonon-reflecting surface in an isotropic medium with an energy density ε experiences a pressure

$$p = \varepsilon/3 \tag{7.27}$$

Note that this pressure is exerted on a (hypothetical) surface able to move with respect to the medium, but *not* on features, such as a free surface of a crystal, that can move only by virtue of elastic deformation of the medium. Accordingly, phonon pressure makes no contribution to the thermal coefficient of expansion, which for an ideal harmonic crystal is zero (Ashcroft and Mermin, 1976).

7.3.2. The Debye model of the phonon energy density

Phonon scattering drag depends on the phonon energy density and (more generally) on the energy distribution vs. wave vector. The commonly used Debye model of the distribution (discussed at greater length in Ashcroft and Mermin, 1976) assumes that all waves propagate at a uniform speed v_s. It gives the total phonon energy density as an integral over a spherical volume in k-space,

$$\varepsilon = \frac{3\hbar v_s}{2\pi^2} \int_0^{k_D} k^3 \left[\exp(\hbar k v_s/kT) - 1\right]^{-1} dk \tag{7.28}$$

where k is to be interpreted as the magnitude of the wave vector, and k_D (the Debye radius) is a function of n_v, the atomic number density (m^{-3}):

$$k_D = (6\pi n_v)^{1/3} \tag{7.29}$$

The Debye temperature

$$T_D = \hbar k_D v_s / k \tag{7.30}$$

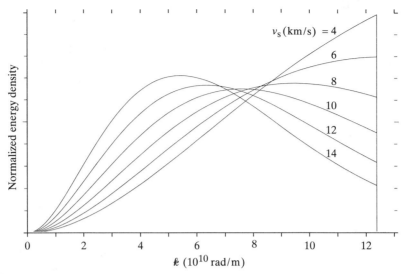

Figure 7.3. Phonon energy density per unit interval of k in the Debye model for $n_v = 100/nm^3$ and $T = 300$ K, normalized to constant total energy. The maximum value of $k = k_D = 1.24 \times 10^{10}$ rad/m.

is a measure of the temperature at which the highest-frequency modes of a solid are excited. For $T \ll T_D$, $\varepsilon \propto T^4$, as in blackbody radiation. To yield the correct value of the phonon energy density for $T \ll T_D$, v_s in the preceding expressions must (in an isotropic material) be taken as

$$v_s = \left(\tfrac{1}{3} v_{s,\ell}^{-3} + \tfrac{2}{3} v_{s,t}^{-3} \right)^{-1/3} \tag{7.31}$$

which has a maximum value of $1.084\, v_{s,t}$ in the limiting case of $E = 2\,G$. For a hypothetical isotropic substance with $v_{s,t}$ and $v_{s,\ell}$ equal to those of diamond along an axis of cubic symmetry, these relations give $v_s = 1.38 \times 10^4$ m/s and $T_D = 1570$ K [vs. a value for diamond itself of 2230 K (Gray, 1972)]. Phonon energy distributions and magnitudes according to the Debye model are plotted in Figures 7.3 and 7.4.

The Debye model has several shortcomings in describing real crystals, to say nothing of nanomechanical systems treated as continuous media; in the present context, its chief defects arise from its neglect of waves of low group velocity. Near the limiting value of k, acoustic dispersion (ignored in the Debye model) results in group velocities approaching zero. The Debye model also neglects so-called optical modes of vibration, which typically have low group velocities. The effect of these shortcomings can be significant, but typically is small for $T \ll T_D$ (Alshits and Indenbom, 1986).

7.3.3. Phonon scattering drag

A scattering center moving with velocity v experiences drag from the "phonon wind" resulting from its velocity (Alshits and Indenbom, 1986); this can be treated as analogous to scattering of photons in a vacuum. In the simplest situation, a scattering center has both an isotropic cross section in the rest frame and an isotropic emission pattern in its own frame. Drag can then be calculated from

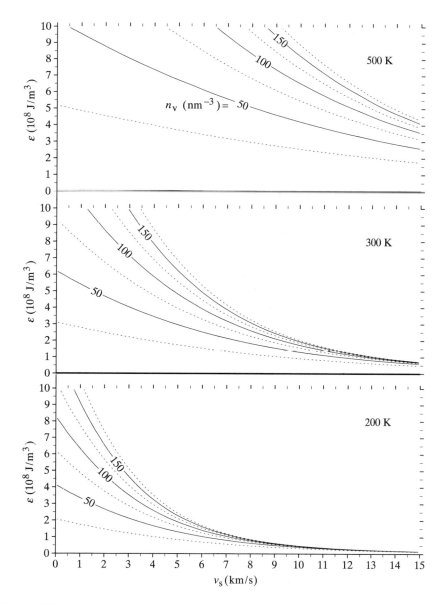

Figure 7.4. Phonon energy density in the Debye model vs. the *effective* speed of sound, v_s, and the atomic number density, n. For perspective, in diamond the effective speed $v_s \approx 13.8$ km/s, Eq. (7.31), and $n \approx 176/\text{nm}^3$.

the anisotropic momentum distributions and encounter frequencies of phonons seen in the moving frame.

Phonons approaching at an angle θ from the forward direction are Doppler shifted, changing their frequency and energy by a factor $[1 + v\cos(\theta)/v_s]$; the rate of encounter for phonons from this direction is altered by the same factor. Phonons approaching from the side experience an aberrational shift in direction through an angle $-v\sin(\theta)/v_s$ (for $v \ll v_s$). Combining these factors, discarding terms of order v^2/v_s^2 and higher, and integrating yields an approximation to the phonon-scattering drag and power dissipation for the scattering center:

$$F_{\text{drag}} \approx -\frac{4}{3}\varepsilon\sigma_{\text{therm}}\frac{v}{v_s}, \qquad P_{\text{drag}} \approx \frac{4}{3}\varepsilon\sigma_{\text{therm}}\frac{v^2}{v_s} \qquad (7.32)$$

where σ_{therm} is a thermally weighted scattering cross section (in m^2) derived from a scattering cross section $\sigma(k)$ that is a function of the magnitude of the wave vector. For the Debye model of the phonon distribution,

$$\sigma_{\text{therm}} = \frac{\int_0^{k_D} \sigma(k)k^3 \left[\exp\left(\frac{\hbar k v_s}{kT}\right)-1\right]^{-1} dk}{\int_0^{k_D} k^3 \left[\exp\left(\frac{\hbar k v_s}{kT}\right)-1\right]^{-1} dk} \qquad (7.33)$$

7.3.4. Scattering from harmonic oscillators

Various nanomechanical components can be treated as moving scattering centers. A roller bearing moving across a surface, a °follower moving over a °cam, an object sliding in a tube: each results in the motion of a small region of contact with respect to a medium. The effect of the contact can be modeled as an embedded harmonic oscillator of the sort described in Section 7.2.3b, excited by incident phonons and radiating to a degree that can be approximated by Eq. (7.8). (Dissipation from broad sliding contacts of regular structure can be modeled as described in Sections 7.3.5, 7.3.6, and 7.4.3.)

In the limit of large mass and low stiffness, the motion of the mass is small and the oscillating force is proportional to the stiffness, making $\sigma \propto k_s^2$. In the limit of low mass and large stiffness, the deformation of the spring is small and the force is proportional to the mass, making $\sigma \propto m^2$. In general, far from resonance,

$$\sigma \approx \frac{1}{2\pi}\left(\frac{k_s}{M}\right)^2 \left(\frac{k_s}{m\omega^2}+1\right)^{-2} \qquad (7.34)$$

Resonant scattering occurs when $k v_s \approx (k_s/m)^{1/2}$, with resonant cross sections limited by radiation damping. Values of σ_{therm} can be estimated by numerical integration of the damped harmonic oscillator cross section over the Debye phonon distribution. The results for representative values of material parameters at 300 K are graphed in Figure 7.5, for oscillators with isotropic effective stiffnesses and masses. For oscillators with differing values along three principal axes, the cross section is the mean of those that would be exhibited by three isotropic oscillators with these values.

A sliding scattering center typically is coupled to the medium by nonbonded interactions. The relationships summarized in Figure 3.8 suggest that, regardless of how low the equilibrium stiffness may be, thermal excitation of small scattering centers at 300 K frequently explores regions in which the stiffness is of the order of 10 N/m. Thus, anharmonicity and thermal excitation place an effective lower bound on the effective mean stiffness, and hence on the phonon scattering cross section.

It is useful to examine the magnitude of drag for a typical case. A sliding contact with a stiffness of ~30 N/m has $\sigma \approx 10^{-20}$ m^2 in a moderately stiff medium. With $v_s \approx 10^4$ m/s and $\varepsilon \approx 2 \times 10^8$ J/m^3, the power dissipation [Eq. (7.32)] is

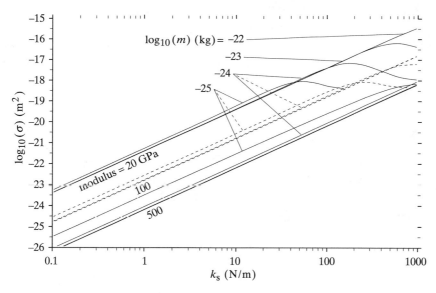

Figure 7.5. Values of σ_{therm} at 300 K for a range of values of m and k_s and three values of the modulus M. For perspective, 10^{-25} kg is the approximate mass of 5 carbon atoms, and the Young's modulus of diamond is $\sim$1000 GPa ($= 10^{12}$ N/m^2). The assumed density is 2000 kg/m^2 (vs. $\sim$3500 for diamond), with $n_v = 100$/nm^3.

$\sim 3 \times 10^{-16}$ W at 1 m/s and $\sim 3 \times 10^{-20}$ W at 1 cm/s, or (equivalently) 3×10^{-25} and 3×10^{-27} J/nm traveled.

7.3.5. Scattering from alignment bands in bearings

a. Alignment bands in bearings. Sliding interfaces between regular surfaces contain alignment bands that are closely analogous to dislocations. Phonon-scattering drag plays a major role in dislocation dynamics and has accordingly received substantial attention (Alshits and Indenbom, 1986; Lothe, 1962; Nabarro, 1987). Causes of scattering include both the mechanical inhomogeneity of the dislocation core and *flutter,* in which phonons excite oscillations in dislocations, which then reradiate power.

In typical materials, dislocations are narrow, causing severe disruption of crystalline alignment over $\sim$5 atomic spacings (Lothe, 1962), and inducing local variations in stress that are significant compared to the modulus E. Alignment bands in sliding-interface bearings, however, typically are broad and induce only small variations in stress. These variations in stress, however, can yield significant variations in the stiffness of the interface owing to the strong anharmonicity of nonbonded interactions (Section 3.3.2e). Alignment bands are sufficiently similar to dislocations that analogues of both flutter scattering and inhomogeneity scattering occur, yet are different enough to invalidate the approximations that have been used to model them. Suitable approximations are developed in Sections 7.3.5b–7.3.5e.

b. Common features of the models. Alignment bands in interfaces can be modeled as sinusoidally varying disturbances moving at a speed v_{bands}, related to the sliding speed v by Eq. (7.22) or the final expression in Eq. (7.23), in two limiting cases. The nature of the disturbance varies with the mode of scattering.

Both flutter scattering and stiffness scattering are here described by approximate models, intended to provide only upper bounds.

In many systems of engineering interest, the shear stresses transmitted across the interface are small compared to the normal stresses, and the shear stiffnesses are likewise small compared to the normal stiffnesses. Shear stresses and stiffnesses are accordingly neglected in the following discussion, although their treatment would be entirely analogous.

As noted by Soreff (1991), the assumption that $v_{s,\ell} = v_{s,t}$ permits waves to be resolved into components with x, y, and z polarizations, where the polarization axes may be chosen for convenience, without regard to the direction of propagation. For scattering from band stiffness variations, only polarizations perpendicular to the interface are relevant; for flutter scattering, only polarizations parallel to the interface and oriented in the band-shifting direction are relevant.

c. Band-stiffness scattering. The interface can be treated as a compliant sheet with a stiffness per unit area k_a and a transmission coefficient (Section 7.3.5e) of T_{trans} (this coefficient includes the factor of 1/3 resulting from the effectiveness of only one of three polarizations). The mechanical inhomogeneity of the alignment bands can be approximately described as a sinusoidal variation in stiffness of amplitude $\Delta k_a/2$, which (across interfaces with small values of T_{trans}) causes variations in the transmission coefficient on the order of $\Delta T_{trans} = 1.7 T_{trans} \Delta k_a / k_a$ (Section 7.3.5e).

Neither specular reflection nor simple transmission of phonons contributes to the drag, but a fraction of incident phonons $\leq \Delta T_{trans}$ is subject to diffractive scattering from the bands owing to the spatial variation in the transmission (and accordingly reflection) coefficient. For normally incident phonons of $k \gg k_{bands}$, the diffraction angle is small, and for $k < k_{bands}$, it is zero. In these cases, the scattering contribution to drag is relatively small or nonexistent. (Owing to the angular variation in the transmission coefficient, the actual results are strongly influenced by the diffraction of phonons at grazing incidence.) The present estimate nonetheless assumes isotropic scattering for all k, tending to overestimate the drag.

The incident power on a single side of a surface is $v_s \varepsilon/4$, and hence the average collision cross section for a flat surface of area S (counting both sides and all angles of incidence) is $S/2$. (The quantity $\varepsilon/4$ appears frequently in normalization expressions and can be termed the *effective energy density*.) Combining these factors yields the bound

$$P_{drag} < 0.85 \varepsilon T_{trans} \frac{\Delta k_a}{k_a} R^2 \frac{v^2}{v_s} S \tag{7.35}$$

where v is the sliding speed of the interface and a factor of order unity [analogous to the 4/3 in Eq. (7.32)] has been dropped. (Note that this and similar expressions do not hold when the value of $k_a \approx 0$.)

As discussed in Chapter 10, $\Delta k_a/k_a$ can be made small in properly designed bearings of certain classes, and values of T_{trans} (Section 7.3.5e) can easily be less than 10^{-3}. A not-atypical value for R is 10. For these values, with $\Delta k_a/k_a \approx 0.1$, $v_s = 10^4$ m/s, and $\varepsilon = 2 \times 10^8$ J/m^3, P_{drag} from this mechanism is bounded by

~200 W/m^2 at $v = 1$ m/s, and ~0.02 W/m^2 at 1 cm/s. The latter values correspond to 2×10^{-25} and 2×10^{-27} J/nm^2 per nanometer of travel.

d. Band-flutter scattering. Alignment bands also cause deformations of the equilibrium shape of each surface of the interface with amplitude A and spatial frequency k_{bands}; this results in sinusoidally varying slopes with a maximum magnitude of Ak_{bands}. A shear wave of the proper polarization causes the bands to move by a distance that is a multiple R [see Eqs. (7.22) and (7.23)] of the particle displacements caused by the shear wave itself. The ratio of the amplitude of the equilibrium displacement of the interface to the amplitude of the incident shear wave is $Ak_{bands}R$. These displacements are like those imposed by an incident wave of perpendicular polarization and scaled amplitude; after this transformation, the interface can again be regarded as a moving diffraction grating.

Taking the mean square value of the slope over the interface introduces a factor of 1/2 in the scattered power; consideration of radiation from both surfaces introduces a compensating factor of 2. A time-reversed equilibrium system is an equilibrium system, hence in the limit of slow band motion, power scattered from parallel to perpendicular polarizations by band flutter must equal power scattered from perpendicular to parallel. This introduces a further factor of 2 in the drag expression.

With these bounding approximations, the analysis proceeds essentially as before, yielding

$$P_{drag} < \varepsilon T_{trans} \left(Ak_{bands} R \right)^2 R^2 \frac{v^2}{v_s} S \qquad (7.36)$$

Equation (7.22) implies that the product $k_{bands}R = k_1 = 2\pi/d$, where d is the spacing of atomic rows in either surface. [In the general case described by Eq. (7.23), this remains a reasonable approximation.] This yields the expression

$$P_{drag} < \varepsilon T_{trans} \left(\frac{2\pi A}{d} \right)^2 R^2 \frac{v^2}{v_s} S \qquad (7.37)$$

As with stiffness variations, A/d can be made small in properly designed bearings of certain classes; a reasonable value is 10^{-2}. For values of other variables as previously assumed, the drag power from this mechanism is bounded by ~10 W/m^2 at 1 m/s.

e. The transmission coefficient. As Sections 7.3.5c and 7.3.5d illustrate, the phonon transmission coefficient T_{trans} greatly affects drag at sliding interfaces. In a simple one-dimensional model of longitudinal vibrations propagating along a rod with a linear modulus E_ℓ interrupted by spring of stiffness k_s, the transmission coefficient is

$$T_{trans,rod} = \left[1 + \left(E_\ell k / 2k_s \right)^2 \right]^{-1} \qquad (7.38)$$

where k is the spatial frequency (rad/m).

A detailed analysis by Soreff (1991) shows that in a medium in which all speeds of sound are equal, the transmission coefficient at a planar interface takes the same form,

$$T_{trans,perp} = \left[1 + \left(Mk_z / 2k_a \right)^2 \right]^{-1} \qquad (7.39)$$

in which k_z is the z-component of the wave vector of an incident wave of perpendicular polarization, and M is the single modulus. The overall mean power transmission coefficient can then be estimated by an integral over one hemisphere of the allowed volume of k-space, weighting contributions from different wave vectors in accord with the Debye model of the distribution of phonon energy and including a factor of 1/3 to account for the transmission of incident power in only one of the three possible polarizations:

$$T_{trans} = \frac{\dfrac{4}{3}\displaystyle\int_0^1 \frac{k'^3}{\exp(k'/T')-1} \int_0^{\pi/2} \frac{\sin(2\theta)}{\left(d'k'\cos\theta\right)^2+4}\,d\theta\,dk'}{\displaystyle\int_0^1 \frac{k'^3}{\exp(k'/T')-1}\,dk'} \tag{7.40}$$

where $T' = T/T_D$, and d' is a dimensionless measure of the stiffness of the interface, $d' = k_D M/k_a$. Values of T_{trans} are plotted in Figure 7.6 with respect to $d_n = n^{-1/3}M/k_a\ (\propto d')$ for a range of values of interest in the present context. The parameter d_n can be interpreted as the thickness of a slab of the medium, in atomic layers (assuming a simple cubic lattice), that has a stiffness per unit area equaling that of the interface itself.

Equation (7.40) does not lend itself to easy evaluation or use in analytical models. A reasonable engineering approximation is

$$T_{trans} \approx z/1+3z; \quad z = 0.6 d_n^{-1.7}\left(1+0.075 T'^{-1.8}\right) \tag{7.41}$$

or

$$T_{trans} \approx z, \quad z \ll 1 \tag{7.42}$$

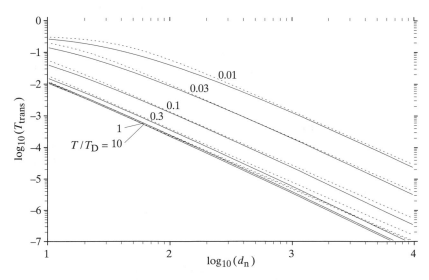

Figure 7.6. The transmission coefficient for a compliant interface, Eq. (7.40), plotted for several values of a dimensionless measure of temperature, T/T_D, and a dimensionless measure of interface compliance, d_n. Dotted curves represent the approximation given by Eq. (7.41).

Equation (7.41) is plotted in Figure 7.6. Both expressions consistently overestimate transmission (and hence drag), yielding conservative results for most purposes; their forms have been chosen to give a good fit for parameters in the anticipated region of engineering interest, rather than to exhibit correct behavior at the simple physical limits (e.g., $T' \to 0$).

f. Curved interfaces and dissimilar media.   In the model described by Eq. (7.40), when d_n is large and T' is not small, normally incident phonons are reflected almost perfectly and grazing-incidence phonons make a large contribution to T_{trans}. The efficient transmission of grazing-incidence phonons results from a resonant process that depends on (1) the prolonged interaction associated with grazing-incidence collisions, and (2) the matching acoustic speeds of the media on either side of the interface. In the limit of small angles, the transmission probability approaches unity.

Short or curved interfaces disrupt this process, reducing T_{trans}. A difference in acoustic speed between the two media does likewise, causing the grazing-incidence transmission probability to fall to zero on one side (owing to total internal reflection) and to approach zero in the limit of small angles on the other side (as shown by the angular variation of reflectivity in analogous optical systems). Sharply curved interfaces are common in nanomechanical bearings, and differences in acoustic speed are likewise common; indeed, such differences can be a design objective. In critical applications, practically significant differences in acoustic speed can be ensured even between chemically identical structures by building them from different isotopes (e.g., ^{12}C vs. ^{13}C). An analysis taking account of curved interfaces or differentiated media would be desirable, but models based on Eq. (7.40) provide an upper bound on transmission-dependent drag processes. This suffices for present purposes.

7.3.6. Shear-reflection drag

The asymmetry of Doppler shifts among phonons transmitted through a sliding interface shows that a (symmetrical) equilibrium distribution is transformed into a nonequilibrium distribution, implying an increase in the free energy of the phonons at the expense of the energy of sliding. Analyzing this energy loss mechanism, however, is difficult. A study by Soreff (1991) of sound propagation through the model interface of Section 7.3.5e yields an expression for the wave-vector resolved transmission coefficient of an interface as a function of the Mach number, M_s, and the dimensionless measures of interfacial stiffness and phonon spatial frequency described in Section 7.3.5e:

$$T_{\text{trans,perp}} = 4r_z \left[\left(1 + r_z\right)^2 + \left(d' k' r_z \sin \psi \cos \phi\right)^2 \right]^{-1} \tag{7.43}$$

where

$$r_z = \frac{\sqrt{\sin^2 \psi \cos^2 \phi - 2M_s \cos \psi + M_s^2 \cos^2 \psi}}{\sin \psi \cos \phi} \tag{7.44}$$

In this expression, the coordinates are chosen such that sliding motion occurs along the x axis with the z axis perpendicular to the interface; ψ measures the angle between the wave vector and the x axis, and ϕ measures the angle between the x-$\pmb{k}'$ plane and the x-z plane. The square root in Eq. (7.44) is to be taken as positive, with formally imaginary values taken as zero. Soreff observes that phonons crossing the interface experience no change in the component of their wave vector parallel to the interface and hence no change in that component of momentum; only velocity-dependent asymmetries in the transmission coefficient cause momentum transfer and hence drag. (Soreff has verified the correctness of equating crystal momentum and ordinary momentum in this instance by a detailed analysis of transverse forces in the phonon-deformed interface.)

The drag per unit area per unit phonon effective energy density can be expressed as the ratio of integrals in Eq. (7.45). This closely resembles the expression for the transmission coefficient save for a change of coordinates, use of the velocity-dependent expression for transmission, and the introduction of a $\cos(\psi)$ factor in the numerator to account for contributions to the x momentum. At low Mach numbers, contributions from the leading and trailing regions of $\pmb{k}$-space nearly cancel and are associated with rapidly varying length scales; these characteristics make numerical analysis difficult. Soreff reports that several different approaches for developing analytical approximations or bounds fail to yield useful results.

On physical grounds, one expects that at low Mach numbers the drag is approximately proportional to the energy density, to the zero-velocity transmission coefficient, and to the Mach number itself. This encourages consideration of the expression

$$D_{\text{sr}} = \frac{-\dfrac{8}{3\pi}\displaystyle\int_0^1 \frac{k'^3}{\exp(k'/T')-1}\int_0^{\pi/2}\int_0^{\pi}\frac{r_z\sin^2\psi\cos\psi\cos\phi}{\left(1+r_z\right)^2+\left(d'k'r_z\sin\psi\cos\phi\right)^2}\,d\psi\,d\phi\,dk'}{M_sT_{\text{trans}}\displaystyle\int_0^1\frac{k'^3}{\exp(k'/T')-1}\,dk'} \tag{7.45}$$

in hopes that D_{st} is a slowly varying quantity. A numerical investigation suggests that D_{st} is of order unity and does indeed vary only moderately across a range of parameters in which the drag varies by more than seven orders of magnitude. Use of the approximation $D_{\text{st}}=1$ appears conservative for the systems of interest in the present context, frequently overestimating drag by a factor of 10 or more (a more thorough numerical investigation would be desirable). The considerations raised in Section 7.3.5f often introduce a further measure of conservatism in the use of this approximation.

The expression for the power dissipation from shear-reflection drag includes a factor of two to account for phonons approaching an interface from each side:

$$P_{\text{drag}} \approx \frac{\varepsilon}{2}T_{\text{trans}}M_sD_{\text{sr}}vS = \frac{\varepsilon}{2}T_{\text{trans}}D_{\text{sr}}\frac{v^2}{v_s}S \tag{7.46}$$

The magnitude of drag from this mechanism relative to those described by Eqs. (7.35) and (7.37) varies with the design parameters. For bearings in which alignment-band drag has been minimized, it can be dominant. With the assumptions of Section 7.3.5c, the drag power is ~ 10 W/m^2 at 1 m/s.

7.3.7. Interfacial phonon-phonon scattering

Nonlinear interactions permit phonons to scatter from one another, and the restoring forces in a nonbonded interface are substantially nonlinear. In particular, phonons in one medium can scatter from corrugations in the interface induced by phonons in the other medium; interfacial sliding will give these corrugations a drift velocity, enabling them to transmit net energy to the scattered phonons and so cause drag. A preliminary evaluation suggests that this effect is small, but a more thorough investigation would be desirable.

7.4. Thermoelastic damping and phonon viscosity

The phonon gas causes energy loss by mechanisms analogous to those occurring in the compression and shear of ordinary gases. These mechanisms are termed *thermoelastic damping* and *phonon viscosity*.

7.4.1. Thermoelastic damping

When a typical solid is compressed, the energy of its normal modes increases. In the absence of equilibration, phonon energies likewise increase and the solid becomes hotter. Since this process involves changes in the dimensions of the solid rather than motions with respect to the lattice, no work is done against the pressure of the phonon gas directly (as in the compression of an ordinary gas or a photon gas). Instead, phonon energies increase because of anharmonic effects: interatomic potentials become stiffer as distances shrink. A widely used measure of anharmonicity is the Grüneisen number,

$$\gamma_G = \beta K / C_{vol} \qquad (7.47)$$

where β is the volume coefficient of thermal expansion (K^{-1}), K is the bulk modulus (N/m^2), and C_{vol} is the heat capacity per unit volume ($J/K \cdot m^3$). (C_{vol} equals the molar heat capacity at constant volume divided by the molar volume.) Values of γ_G for many ordinary materials fall in the range 1.5 to 2.5 and have little temperature dependence near 300 K. For diamond, $\beta = 3.5 \times 10^{-6} \, K^{-1}$, $K \approx 4.4 \times 10^{11} \, N/m^2$, and $C_{vol} \approx 1.7 \times 10^6 \, J/K \cdot m^3$ at 300 K, yielding $\gamma_G \approx 0.9$.

Thermoelastic damping arises from the difference between the adiabatic and the isothermal work of compression. Starting with an expression for small values of ΔV and ΔT (Lothe, 1962), and applying thermodynamic identities,

$$\Delta W = \frac{1}{2}(K_{adiabatic} - K)\frac{\Delta V^2}{V} \qquad (7.48)$$

$$= \frac{1}{2}\gamma_G^2 T C_{vol}\frac{\Delta V^2}{V}$$

$$= \frac{1}{2}\beta^2 \frac{T}{C_{vol}}\Delta p^2 V \qquad (7.49)$$

A worst-case thermodynamic cycle involves adiabatic compression of a volume (increasing the temperature) followed by nonequilibrium cooling (producing entropy), followed by adiabatic expansion and nonisothermal warming, causing an overall energy dissipation of $2\Delta W$, from Eq. (7.49). For diamond, this amounts to $\sim 2.2 \times 10^{-24} (\Delta p)^2 \, J/nm^3 \cdot$cycle, where p is here measured in nN/nm^2 (= GPa).

Thermoelastic damping falls to zero if the cycle is either adiabatic or isothermal, and nanomechanical systems frequently approach the isothermal limit. The ratio of the energy dissipated in a nearly isothermal cycle to that dissipated in the worst-case cycle equals the ratio of the mean displacement-weighted temperature increments. A component undergoing smooth mechanical cycling with a period t_{cycle} and a characteristic time for thermal equilibration τ_{therm} experiences a temperature rise during the cycle on the order of $\tau_{\text{therm}}/t_{\text{cycle}}$ times that of the adiabatic case, implying

$$\Delta W_{\text{cycle}} \approx 2\beta^2 \frac{T}{C_{\text{vol}}}(\Delta p)^2 \frac{\tau_{\text{therm}}}{t_{\text{cycle}}} V \tag{7.50}$$

For a component in good thermal contact with its environment, τ_{therm} is on the order of

$$\tau_{\text{therm}} \approx \max\left(\frac{C_{\text{vol}}}{K_{\text{T}}}\ell^2, \frac{\ell}{v_{\text{s}}}\right) \tag{7.51}$$

where K_{T} is the thermal conductivity (W/m·K) and ℓ is a characteristic dimension. Values of K_{T} for glasses and nonporous ceramics typically are between 1 and 10 W/m·K, with the value for diamond being ~700 (Gray, 1972). For $K_{\text{T}} = 10$, $\ell = 10$ nm, and $C_{\text{vol}} = 2 \times 10^6$ J/K·m^3, $\tau_{\text{therm}} \approx 10^{-11}$ s; ΔW_{cycle} accordingly is multiplied by a factor of ~10^{-2} at 1 GHz and ~10^{-5} at 1 MHz relative to the values given by the worst-case expression, Eq. (7.49). These calculations are in terms of an isotropic pressure; a compressive load of the same formal magnitude but applied along a single axis will have a lesser effect. This correction will be neglected here, tending to favor overestimates of energy dissipation.

7.4.2. Phonon viscosity

Shear deformation causes no volume change and hence no thermoelastic losses. Shear does, however, cause compression along one axis and extension along another: phonons traveling along one axis are increased in energy; those along the other, reduced. Within a factor of 3/2, an analogy between this and the thermoelastic effect yields an estimate of the difference between the adiabatic and isothermal shear moduli (Lothe, 1962), resulting in an effective viscosity

$$\eta_{\text{phonon}} = \tau_{\text{relax}} \frac{3}{2}\gamma_{\text{G}}^2 TC_{\text{vol}} \tag{7.52}$$

The analysis of the energy dissipation proceeds essentially as for thermoelastic damping, but with the substitution of the phonon relaxation time τ_{relax} for τ_{thermal}, and the shear stress γ for the pressure p, yielding

$$\Delta W_{\text{cycle}} \approx \frac{3}{2}\beta^2 \frac{T}{C_{\text{vol}}}(\Delta\gamma)^2 \frac{\tau_{\text{relax}}}{t_{\text{cycle}}} V \tag{7.53}$$

The time τ_{relax} measures the rate of equilibration of phonon energy between different directions in the solid, which can be accomplished by elastic scattering such as that occurring at the boundaries of a solid body or at internal inhomogeneities. In nanomechanical systems, scattering typically limits phonon mean free

Table 7.1. Values of the volumetric thermal coefficient of expansion β for several strong solids in the neighborhood of 300 K (Gray, 1972).

Material	β $(10^{-6}\,\mathrm{K}^{-1})$
Diamond	3.5
Silicon	7.5
Silicon carbide	11.1
Sapphire	15.6
Silica, crystalline (quartz)	36.0
Silica, vitreous	1.2

paths to nanometer distances, resulting in values of $\tau_{\mathrm{relax}} \approx 10^{-13}$ s for $v_s \approx 10^4$ m/s. Because of this small time constant, phonon viscosity losses typically are small compared to thermoelastic losses, save in systems undergoing very high frequency motion or nearly pure shear.

7.4.3. Application to moving parts and alignment bands

Alignment bands, like sliding and rolling components, impose moving regions of stress on the surrounding medium. These regions can be characterized by their volume $V \approx \ell^3$ (for contact regions) or $\approx d\ell^2$ (for bands extending over a distance d), where ℓ is a measure of the scale of the region (e.g., the wavelength of the alignment bands). For motions of velocity v, $\tau_{\mathrm{cycle}} \approx \ell/v$. This leads to an estimate of the magnitude of the thermoelastic drag of a set of bands:

$$P_{\mathrm{drag}} \approx \beta^2 \frac{T}{K_{\mathrm{T}}} \ell (\Delta p)^2 R^2 v^2 S \qquad (7.54)$$

based on the assumption that phonon mean free paths are shorter than ℓ. With $\beta = 3.5 \times 10^{-6}\,\mathrm{K}^{-1}$, $K_{\mathrm{T}} = 10$ J/m·K, $\ell = 10$ nm, $R = 10$, $T = 300$ K, and $\Delta p = 10^8$ N/m^2, $P_{\mathrm{drag}} \approx 4$ W/m^2 at $v = 1$ m/s, or 0.04 W/m^2 at 1 cm/s. As with stiffness and displacement, good design can in many instances yield low values of Δp.

These estimates of thermoelastic dissipation have used a value of β appropriate for diamond. Table 7.1 lists values for several other strong solids that can serve as models for the materials of nanomechanical components. The large difference between SiO_2 as quartz and as vitreous silica indicates that β is sensitive to patterns of bonding and hence is subject to substantial design control in the products of molecular manufacturing, including nanomechanical components.

7.5. Compression of potential wells

7.5.1. Square well compression

Many useful nanomechanical systems contain components that move over a relatively flat potential energy surface within a bounded region of variable size. Thermodynamically, the motion of the component within this region is like the motion of a gas molecule in a container; changes in the size of the region correspond to compression and expansion. At any finite speed, compression is a nonisothermal process, heating the gas, raising its pressure, and so increasing the work of compression and causing energy dissipation. The better the thermal contact between the component and its environment, the lower the dissipation. A conservative model assumes contact only between the "gas molecule" component and two moving pistons (Figure 7.7), treating the component as a one-dimensional gas consisting of one molecule.

a. Accommodation coefficients. Thermal contact between a gas and a solid is usually described by a thermal accommodation coefficient α that measures the extent to which the excess energy of an impinging gas molecule is lost in a single collision with a wall:

$$(T_1 - T_s)\alpha = T_1 - T_2 \tag{7.55}$$

where the temperature of the incident molecules is T_1, that of the surface is T_s, and that of the outbound molecules is T_2. Separate accommodation coefficients can be defined for the energy of translation, rotation, and vibration. The most accurate measurements have been made for monatomic gases, in which all the energy is translational. As just defined, α is a function of three temperatures; in the limit as T_1, T_2, and T_s become equal, α becomes a function of a single temperature. A value of the latter sort (an equilibrium accommodation coefficient) can be used with reasonable accuracy so long as none of the three temperatures differs greatly from the reference temperature. Theory and experimental data for thermal accommodation coefficients are reviewed in Goodman (1980), Goodman and Wachman (1976), and Saxena and Joshi (1989).

Save for light gases (helium, neon) impinging on a clean surface with massive atoms (tungsten), most tabulated values of α range from 0.25 to ~1.0. Surface contamination tends to increase accommodation; stable structures with similar effects could be provided in many systems.

b. A square-well temperature-increment model. A simple model for the temperature rise assumes that the statistics of the velocity of a moving component of mass m are those of a system at equilibrium at some temperature T_{rest}, with a root mean square velocity (along the axis) of $v_{g,rms} = (kT_g/m)^{1/2}$. If each piston moves at a speed $(f/2)(kT_g/m)^{1/2}$ relative to the wall, then

$$v_{1,rms} = v_{g,rms}\left(1 + f/\sqrt{\pi} + f^2/4\right)$$
$$v_{2,rms} = v_{g,rms}\left(1 - f/\sqrt{\pi} + f^2/4\right) \tag{7.56}$$

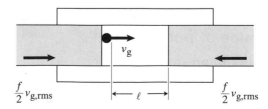

Figure 7.7. Two-piston model of a compressed square well; cylinder walls are assumed adiabatic.

where $v_{1,\text{rms}}$ and $v_{2,\text{rms}}$ are measured in the frame of reference of the scattering piston (pistons are assumed to interact only with particles moving in their direction in the rest frame, neglecting molecules overtaken; this approximation is good for small f). Since mean square velocity is proportional to temperature, Eq. (7.55) can be converted to velocity terms, yielding the condition

$$\left(v_{1,\text{rms}}^2 - kT_{\text{s}}/m\right)\alpha = v_{1,\text{rms}}^2 - v_{2,\text{rms}}^2 \tag{7.57}$$

where kT/m is the mean square thermal speed, and hence

$$R_{\text{temp}} = \frac{T_{\text{g}}}{T_{\text{s}}} = \left[1 + \frac{f^2}{4} - f\sqrt{\frac{2}{\pi}\left(\frac{2}{\alpha} - 1\right)}\right]^{-1} \tag{7.58}$$

A useful approximation, good for small f and moderate to large α, is

$$R_{\text{temp}} \approx 1 + f\sqrt{\frac{2}{\pi}\left(\frac{2}{\alpha} - 1\right)} \tag{7.59}$$

c. Energy losses. The cases of greatest interest in the present context are those in which $f \ll 1$, and $T_{\text{g}} \approx T_{\text{s}}$. The work done in isothermally compressing a freely moving particle from a range of motion ℓ_1 to a range ℓ_2 is

$$W = -\int_{\ell_1}^{\ell_2} \frac{kT}{\ell}\,d\ell = kT \ln\frac{\ell_1}{\ell_2} \tag{7.60}$$

and for a system undergoing compression at a uniform speed, resulting in a constant value of ΔT_{comp}, the free energy lost (with the above approximations) is

$$\Delta W \approx k\Delta T_{\text{comp}} \ln\frac{\ell_1}{\ell_2} = kT_{\text{s}}f\sqrt{\frac{2}{\pi}\left(\frac{2}{\alpha} - 1\right)}\ln\frac{\ell_1}{\ell_2} \tag{7.61}$$

$$\approx \sqrt{\frac{2mkT_{\text{s}}}{\pi}}\left(\frac{2}{\alpha} - 1\right)\ln\frac{\ell_1}{\ell_2}v_{\text{total}} \tag{7.62}$$

where v_{total} is the speed of one piston with respect to the other. For a sliding component with $m = 2 \times 10^{-25}$ kg, a compression ratio of 10, $\alpha = 0.5$, and $T_{\text{s}} = 300$ K, the energy lost is $\sim 1.6 \times 10^{-22}$ J at 1 m/s, and $\sim 1.6 \times 10^{-24}$ J at 1 cm/s. So long as f remains small, Eqs. (7.61) and (7.62) are equally applicable to nonisothermal expansion losses.

d. Large molecules. For relatively massive moving components, the literature values of α for ordinary gas molecules offer only a poor guide. As molecular motions become slow, collisions become more nearly elastic, and energy transfer decreases, but as molecules become large, the slowing of their *free* motion is offset by the effect of their increased van der Waals attraction energy (Goodman, 1980): the final approach to a surface is accelerated, and the loss of a small portion of the resulting *increment* in kinetic energy can result in a negative *total* energy relative to the free state. This results in adsorption, complete thermal accommodation, and (in the present context) elimination of further nonisothermal-compression losses until the pistons begin to press the molecule from both sides. (Energy losses resulting from the fall into the van der Waals potential well can be described within the framework of Section 7.6.)

7.5.2. Harmonic well compression

Nanomechanical systems sometimes contain components confined to approximately harmonic wells of time-varying stiffness. This is, for example, a reasonable description of the final compression of a single-molecule gas when it is subject to repulsive forces from both pistons. Compression corresponds to an increase in k_s, reducing the effective volume (Section 6.2.2b) available to the oscillator. Thermal exchange with the medium in these instances can be modeled as acoustic radiation from a harmonic oscillator with an energy equaling the excess thermal energy; near equilibrium, the same coefficients describe the absorption of energy by an oscillator undergoing expansion.

a. A harmonic-well temperature-increment model. Assume that the compression process is slow compared to the vibrational period, and that the temperature increment ΔT_{comp} is small compared to the equilibrium temperature T. The total work done in compressing the system by increasing the stiffness from $k_{s,1}$ to $k_{s,2}$ is

$$W = kT \ln \sqrt{k_{s,2}/k_{s,1}} \tag{7.63}$$

Equating the derivative of W with respect to k_s to the net radiated acoustic power, Eq. (7.8), associated with the excess energy $k\Delta T_{\text{comp}}$ yields the expression

$$\Delta T_{\text{comp}} = 2\pi m M^{3/2} \rho^{-1/2} T \frac{\partial k_s}{\partial t} k_s^{-3} \tag{7.64}$$

where the constants are as defined in Section 7.2.2.

b. Energy dissipation models. The energy dissipated is the integral of the difference in work resulting from ΔT_{comp}, or

$$\Delta W = \int_{k_{s,1}}^{k_{s,2}} \frac{k\Delta T_{\text{comp}}}{2k_s} dk_s = \frac{\pi m M^{3/2}}{3\sqrt{\rho}} kT \frac{\partial k_s}{\partial t} \left(k_{s,1}^{-3} - k_{s,2}^{-3} \right) \tag{7.65}$$

assuming that the stiffness increases linearly with time. In systems where k_s results from nonbonded repulsions, Eq. (3.18) implies that $k_s \approx 3.5 \times 10^{10} F_{\text{load}}$ (N/m); a roughly linear increase of F_{load} is not uncommon.

The value of this expression is strongly sensitive to the value of $k_{s,1}$, the stiffness at the onset of compression. This frequently is on the order of the stiffness of an unloaded nonbonded contact between two objects. A model for this, in turn, is the contact between two planes in solid graphite. With an interlayer spacing of 0.335 nm and a modulus of 1.0×10^{10} N/m^2 (Kelly, 1973), the stiffness of this contact is $\sim 3 \times 10^{19}$ N/m^3, or 30 N/m per nm^2. The contact area between a blocky component and a surface typically is on the order of

$$S \approx (m/\rho_c)^{2/3} \tag{7.66}$$

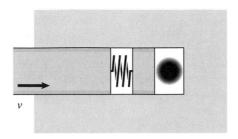

Figure 7.8. Model of compression by an elastic system.

and hence

$$k_{s,1} \approx 3 \times 10^{19} (m/\rho_c)^{2/3} \tag{7.67}$$

In a typical situation (Figure 7.8), a system with a finite stiffness external to the interfaces under consideration, $k_{s,ext}$, is loaded by a steady displacement at a rate v_{ext}. Using Eq. (7.65) and assuming substantial compression ratios,

$$\Delta W \approx kT \frac{\pi m M^{3/2}}{3\sqrt{\rho}} 3.5 \times 10^{10} \frac{k_{s,ext} v_{ext}}{k_{s,1}^3} \tag{7.68}$$

where the approximation assumes a $k_{s,ext}$ substantially less than $k_{s,1}$. Assuming that $T = 300$ K, $m = 10^{-24}$ kg, $M = 5 \times 10^{11}$ N/m^2, $\rho = \rho_c = 2000$ kg/m^3, $k_{s,ext} = 10$ N/m, and $k_{s,1} = 19$ N/m, $\Delta W \approx 2 \times 10^{-21}$ J at $v_{ext} = 1$ m/s, and $\approx 2 \times 10^{-23}$ J at 1 cm/s.

These estimates indicate magnitudes of energy dissipation that are large relative to those identified thus far. Of the parameters subject to design control, the most significant is M; the value assumed above is for a diamondlike material, which strongly suppresses radiative coupling and increases ΔT. In the immediate neighborhood of a component undergoing fast compression, a better choice is a structure with properties like those of an engineering polymer, $M \approx 3 \times 10^9$ N/m^2, $\rho \approx 1000$ kg/m^3. This decreases ΔW by a factor ~7×10^{-4}, yielding ~1×10^{-24} J at $v = 1$ m/s, and ~1×10^{-26} J at 1 cm/s.

It would be of interest to test the predictions of Eqs. (7.62) and (7.68) against direct molecular dynamics simulations. Studies along these lines could reduce uncertainties regarding accommodation coefficients and velocity distributions in the square-well regime and could probe energy losses in the transition to a harmonic well compression regime. For this purpose, the routines for integrating the equations of motion must conserve energy with considerable accuracy.

7.5.3. Multidimensional systems

Sections 7.5.1 and 7.5.2 assume one-dimensional motion of the compressed component. Real systems have two additional degrees of freedom in translation and further degrees of freedom in torsion and internal vibration. Usually, excitation of a single translational degree of freedom by nonisothermal compression results in rapid equilibration with the other degrees of freedom, approaching equipartition of energy. Each of these degrees of freedom opens new channels for equilibration with the environment, reducing ΔT_{comp}; commonly, some have higher frequencies and more effective radiative coupling than the single mode considered above. Consequently, energy dissipation is lower than that implied by a one-dimensional model.

*a. **Multidimensional square-wells.*** In a simple gas-in-cylinder model, the one-dimensional approximation is equivalent to assuming that the walls of the cylinder are adiabatic, permitting energy loss only to the pistons. A realistic, multidimensional model would include equilibration through collisions with the walls.

*b. **Multidimensional harmonic wells.*** In a realistic, approximately harmonic well, longitudinal vibrations are coupled (at a minimum) to transverse vibrations, providing two additional vibrational modes for equilibration through acoustic radiation. Further, vibrational frequencies in these transverse modes can often be made substantially higher than those in the longitudinal mode: a component can be tightly clamped against transverse motion while performing a function requiring relatively low longitudinal stiffness. This can greatly reduce ΔT_{comp}.

*c. **Mixed systems.*** A component may be subject to strong harmonic constraints against transverse motion, yet move in an approximately square-well potential in longitudinal motion. Again the modes are coupled, and again the transverse modes provide relatively effective coupling to the heat bath of the surrounding medium.

7.6.   Transitions among time-dependent wells

7.6.1.   Overview

Each of the energy dissipation mechanisms discussed in Sections 7.2 through 7.5 approaches zero energy per operation (or per unit displacement) as the speed of the system approaches zero. The merging of potential wells, in contrast, can dissipate energy no matter how slowly it is performed. Since well merging can dissipate on the order of kT or more per cycle, it can impose a significant energy cost on systems operation.

7.6.2.   Energy dissipation in merging wells

Energy dissipation in merging wells occurs when transitions between states move a disequilibrium system closer to equilibrium: the dissipation equals the reduction in free energy. In analyzing time-dependent wells, we begin by describing wells as distinct states linked by cols which are divided by transition surfaces (Section 6.2.2). For concreteness, we will consider the case of systems with two wells linked by a single col, using the harmonic approximations made in Section 6.2.

In the cases of interest, transition rates are low compared to vibration times, and the partition function of each state can be evaluated separately under the assumption of internal equilibrium. Losses owing to nonisothermal compression within a well can be estimated by adapting the methods of Section 7.5.2; this mechanism can be treated as independent given either that the transition rate is negligible, or that the temperature rise has little effect on the transition rate. The latter condition is satisfied if

$$\Delta T_{comp}/T \ll kT/\Delta \mathcal{V}^{\ddagger} \tag{7.69}$$

a. Free energy. The free energy for a system of two wells can conveniently be written as the sum of the free energy of each well (assuming unit probability of occupancy), weighted by the actual probabilities of occupancy, plus an entropic term reflecting the uncertainty regarding which well is occupied:

$$\mathcal{F}_{1,2} = P_1 \mathcal{F}_1 + (1 - P_1)\mathcal{F}_2 + kT\left[P_1 \ln(P_1) + (1 - P_1)\ln(1 - P_1)\right] \tag{7.70}$$

or, applying Eq. (4.5) and simplifying,

$$\mathcal{F}_{1,2} = -kT\left[P_1 \ln(q_1/P_1) + (1 - P_1)\ln(q_2/1 - P_1)\right] \tag{7.71}$$

where q_1 and q_2 are the partition functions of the wells considered as isolated, occupied systems.

b. Free-energy changes. For a fixed value of P_1, the free energy of the two-well system $\mathcal{F}_{1,2}$ depends on the shape of the PES. This changes over time, but so long as equilibrium is maintained, $\Delta \mathcal{F}_{1,2}$ exactly equals the work done on the system by whatever mechanisms are responsible for the change in the PES. The total free energy is then conserved, and no dissipation results.

For a fixed PES, a finite change ΔP_1 results in a change in free energy without work being done by (or on) the external mechanisms. Accordingly, $\Delta \mathcal{F}_{1,2}$ can only be negative and corresponds to an increment of energy dissipation. The rate of decrease of free energy, holding the PES constant, is thus

$$\left(\frac{\partial \mathcal{F}_{1,2}}{\partial t}\right)_{PES} = \frac{\partial P_1}{\partial t}\frac{\partial \mathcal{F}_{1,2}}{\partial P_1} = -R_{12}\left(\mathcal{F}_1 - \mathcal{F}_2 + kT\ln\frac{P_1}{1 - P_1}\right)$$
$$= -R_{12}kT\ln\frac{P_1 q_2}{(1 - P_1)q_1} \tag{7.72}$$

Note that at equilibrium

$$\mathcal{F}_1 - \mathcal{F}_2 = -kT\ln\left(P_{1,eq}/1 - P_{1,eq}\right) \tag{7.73}$$

hence

$$P_{1,equil} = \exp\left(-\frac{\mathcal{F}_1 - \mathcal{F}_2}{kT}\right)\left[\exp\left(-\frac{\mathcal{F}_1 - \mathcal{F}_2}{kT}\right) + 1\right]^{-1} \tag{7.74}$$

and the flow of probability from well to well occurs without dissipation.

The net rate of transitions from 1 to 2, R_{12}, can be expressed in terms of the unidirectional transitions rates k_{12} and k_{21} calculated from an appropriate version of transition state theory (Section 6.2):

$$R_{12} = -\partial P_1/\partial t = P_1 k_{12} - (1 - P_1)k_{21} \tag{7.75}$$

The total energy dissipation in a process between times t_1 and t_2 is

$$\Delta \mathscr{F}_{diss} = -\int_{t_1}^{t_2} R_{12}(t)\left(\mathscr{F}_1(t) - \mathscr{F}_2(t) + kT \ln\left(P_1(t)/1 - P_1(t)\right)\right) dt,$$

$$(7.76)$$

$$\text{where} \quad P_1(t) = P_{1,init} - \int_{t_1}^{t} R_{12}(\tau)\, d\tau$$

c.   Switched-coupling models.   As in calculations of error rates (Section 6.3), one useful approximation treats the transition rate as switching between rapid and negligible. As two wells merge and the barrier between them shrinks, energy dissipation typically is low when transition rates are negligible, and low again when rates are high enough to ensure the maintenance of near equilibrium values of P_1; between these regimes is a period in which transition rates first become significant and then establish and maintain near equilibrium. The switched coupling model approximates that period as a single time, t_{cpl}, hence

$$\Delta \mathscr{F}_{diss} = \mathscr{F}_{1,2}\left(t_{cpl}, P_{1,equil}\right) - \mathscr{F}_{1,2}\left(t_{cpl}, P_{1,init}\right) \quad (7.77)$$

where the equilibrium probability $P_{1,equil}$ is given by Eq. (4.74) with values of $\mathscr{F}_1$ and $\mathscr{F}_2$ evaluated at t_{equil}. A value of t_{cpl} can always be chosen such that Eq. (7.77) yields the same value as Eq. (7.76); in practice, simple rules for choosing t_{cpl} yield approximations to this value.

7.6.3.   Free expansion and symmetrical well merging

A simple and important case is the symmetrical merging of two identical wells [e.g., Figure 7.9(a)], where one is initially occupied with $P_1 = 1$. At all times, $\mathscr{F}_1 = \mathscr{F}_2$, the losses resulting from transitions are purely entropic, and the switched-coupling model yields

$$\Delta \mathscr{F}_{diss} = -kT \ln 2 \quad (7.78)$$

which is simply $-T\mathscr{S}$, where $\mathscr{S}$ is the entropy increase resulting from a volume-doubling free expansion of a single-molecule gas; this equals the work required to restore the initial condition by isothermally compressing a single-molecule

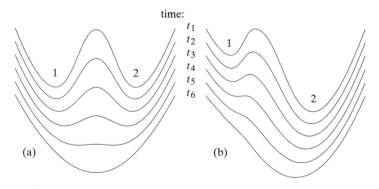

Figure 7.9.   Merging of symmetrical (a) and asymmetrical (b) potential wells, showing potentials at a series of times.

gas to half its initial volume. In computational systems, the symmetrical merging of two wells corresponds to the erasure of one bit of information; information erasure is the only abstract computational process that *of necessity* results in an increase in entropy and a corresponding loss of free energy (Bennett, 1982; Landauer, 1988; see also Section 4.3.5).

7.6.4. Asymmetrical well merging

Simple approximations suffice when merging wells are sufficiently different in free energy, as in Figure 7.9(b). If $(\mathscr{F}_1 - \mathscr{F}_2) \gg kT$ during the time of equilibration, and $P_{1,\text{init}} = 1$, then

$$\Delta\mathscr{F}_{\text{diss}} \approx \mathscr{F}_2(t_{\text{equil}}) - \mathscr{F}_1(t_{\text{equil}}) \tag{7.79}$$

In this case, equilibration changes P_1 from 1 to $\sim\exp[-(\mathscr{F}_1 - \mathscr{F}_2)/kT] \approx 0$, hence the entropy of uncertain well-occupancy is almost unchanged. If the difference in free energy between the wells chiefly results from a difference in potential energy (rather than within-well entropy), then the system can be viewed as falling into a deeper well, with subsequent thermalization of the energy of the fall. Processes of this sort occur, for example, when an elementary step in a chemical reaction is exothermic, or when a mechanical process places an elastic component under load, then allows it to slip past an obstacle. $\Delta\mathscr{F}_{\text{diss}}$ can be large compared to kT.

If $(\mathscr{F}_1 - \mathscr{F}_2) \gg kT$ during the time of equilibration, and $P_{1,\text{init}} = 0$, then

$$\Delta\mathscr{F}_{\text{diss}} \approx -kT\exp\left[-(\mathscr{F}_1 - \mathscr{F}_2)/kT\right] \tag{7.80}$$

In this case, $\Delta\mathscr{F}_{\text{diss}} \ll kT$, and is dominated by the entropy of free expansion into the small effective volume of the upper well.

7.6.5. Optimal well merging under uncertainty

The value of $\Delta\mathscr{F}_{12} = \mathscr{F}_1 - \mathscr{F}_2$ for a pair of merging wells often is subject to design control, since well depths can be modulated by various influences. If so, and if the value of P_1 is known, then the energy dissipation resulting from well merging can be kept near zero by ensuring that $\Delta\mathscr{F}_{12}$ nearly satisfies Eq. (7.74) at $t = t_{\text{cpl}}$ in the switched coupling model, or during the critical period surrounding t_{cpl} in the model described by Eq. (7.77). Note that in a cyclic process where $P_1 \neq 0$ or 1, this uncertainty indicates that some other step in the cycle has increased the entropy and dissipated a corresponding increment of free energy; optimal well merging merely avoids imposing an additional loss.

7.7. Conclusions

Multiple energy dissipation mechanisms operate in nanomechanical systems, each strongly influenced by design parameters. Most have a magnitude per operation or per unit displacement that is proportional to the speed of the system; merging of potential wells (including information erasure in computational systems) is the chief exception, since the energy per operation typically is speed independent, or nearly so. With the exception of the last, each of these mechanisms causes an energy dissipation small compared to kT for nanoscale systems with reasonable choices of physical parameters (one criterion for a reasonable choice, of course, is that it result in acceptable energy dissipation). The approximations and bounds described and developed in this chapter are applied and occasionally elaborated in the chapters of Part II.

Some open problems. Many of these mechanisms are associated with open problems. The theory of phonon transmission through sliding interfaces, in particular, is challenging and in need of further work. For example, in the regime of lowest drag—at separations where the mean interfacial stiffness goes to zero— the phonon transmission model described here becomes unrealistic. Further, the interfaces of greatest engineering interest are not between identical half-spaces; they may be curved, or juxtapose dissimilar materials, or couple a small component to large environment. Finally, drag resulting from interfacial phonon-phonon scattering deserves a thorough treatment. A survey of the devices discussed in Chapters 10–14 reveals many situations of practical interest for which the models presented in this chapter provide only rough estimates of energy dissipation. In addition, many of these models can be improved by means of better approximations or more thorough analysis.

8

Mechanosynthesis

8.1. Overview

Mechanosynthesis and nanomechanical design are interdependent subjects. Advanced mechanosynthesis will employ advanced nanomachines, but advanced nanomachines will themselves be products of advanced mechanosynthesis. This circular relationship must be broken both in exposition and (eventually) in technology development. In development, the circle can be broken by using either conventional synthesis or noneutactic mechanosynthesis to construct first-generation nanomachines, as discussed in Chapters 15 and 16 of Part III. In the following exposition, the circle is broken as follows:

(1) This chapter shows that a wide range of diamondoid structures can be made, provided that accurately controlled mechanical motions can somehow be imposed as a boundary condition in molecular systems;

(2) Part II then shows how diamondoid structures can serve as components of nanomachines able to power and guide accurately controlled mechanical motions, including motions of the sort assumed in this chapter.

8.1.1. Mechanochemistry: terms and concepts

The term *mechanochemistry* was coined by Ostwald decades ago to describe "a branch of chemistry dealing with the chemical and physico-chemical changes of substances of all states of aggregation due to the influence of mechanical energy" (Heinicke, 1984); it has more recently been used to describe both the conversion of mechanical energy to chemical energy in polymers (Parker, 1984) and the conversion of chemical energy to mechanical energy in biological molecular motors; proponents of the first use (Casale and Porter, 1978) urge that processes of the second sort be termed "chemomechanical" in accord with the direction of energy flow. As used here, *mechanochemistry* refers to processes in which mechanical motions *control* chemical reactions, guiding the molecular encounter and often providing °activation energy; this is consistent with Ostwald's broad definition. Since mechanochemical systems can approach thermodynamic reversibility (Section 8.5), the direction of conversion between chemical and mechanical energy cannot be used as a defining characteristic. The present usage is thus related to but substantially distinct from those just cited. The term *mechanosynthesis* is here introduced to describe mechanochemical processes that perform steps in the construction of a complex molecular product.

Mechanochemistry, in the present sense, has features in common with reactions in organic crystals (McBride et al., 1986) and in enzymatic active sites (Creighton, 1984). In both crystals and enzymes, the environment surrounding a molecule can apply substantial stresses to it, or to a reaction intermediate, and can constrain reaction pathways so as to produce a specific product. In crystals, these forces limit motion (McBride et al., 1986). Mechanochemical systems of the sort contemplated here, in contrast, characteristically transport reactive moieties over substantial distances, induce substantial displacements in the course of a reaction, and supplement local thermal and molecular potential energies with work done by external devices.

Mechanochemistry plays several roles in the systems described in Part II. These include the transformation of small molecules, generation of mechanical power, preparation of bound reactive moieties, and mechanosynthesis of complex structures using such moieties. Although the discussion in the present chapter centers on reagent preparation and mechanosynthesis, the principles described are of general applicability.

8.1.2. Scope and approach

Like solution-phase chemistry, °machine-phase chemistry embraces an enormous range of possible reactions. Its potential breadth places the field beyond any hope of thorough exploration at present, and beyond any possibility of summarizing in the space of a chapter. Nonetheless, an attempt must be made to delineate the general capabilities and limitations of mechanochemical processes. How is this problem to be approached? In the absence of opportunities for immediate experimentation, two options suggest themselves:

The first approach relies on detailed theoretical modeling of mechanochemical processes. This yields quantitative descriptions of the sort familiar to those schooled in physics and engineering; several examples are presented in Section 8.5. Theoretical modeling suffers from several difficulties and limitations, however. Chemical reactions cannot be modeled using a molecular mechanics PES, but require either a specialized PES (e.g., Sections 8.5.3, 8.5.4) or the use of *ab initio* molecular orbital methods to calculate the energies of a set of molecular configurations (e.g., Section 8.5.4b). These quantum calculations have been expensive, and are of limited accuracy.* Available techniques in computational quantum chemistry (to say nothing of future techniques) can be used to provide an extensive and detailed understanding of the capabilities of mechanosynthesis, but this will require considerable time, effort, and expense. Today, a survey of the capabilities of mechanosynthesis must rely on other methods.

---

* Note, however, that stiff positional control of reactants can make predictions of the ratios of reaction rates among competing reaction pathways quite insensitive even to large errors in predicted transition-state energies (Section 4.4.3c). Further, the equilibrium results of highly exoergic reactions can be predicted with confidence despite substantial errors in calculating the energies of the reactants and products. Thus, the results of many mechanosynthetic reactions can be predicted with ample confidence despite errors in calculated energies that would be unacceptable in predicting the results of typical solution-phase reactions.

The second approach begins by surveying the demonstrated capabilities of solution-phase organic synthesis, then examines how the conditions of mechanochemistry both add capabilities and impose constraints relative to this model. The resulting picture of the capabilities of mechanosynthesis thus is linked to experimental results in chemistry, rather than being directly founded on theory. This picture contains fewer quantitative descriptions than a physicist or an engineer may expect. If, however, we think that chemists are competent in chemistry (as their achievements show), then we have reason to assume that their methodologies and modes of explanation are well suited to their subject matter. Organic synthesis is a notoriously qualitative field [it is "very little dependent on mathematical calculations," (Hendrickson, 1990)], and much of the discussion in Sections 8.3 and 8.4, taking organic synthesis as its point of departure, accordingly represents a substantial departure from the style of exposition in the preceding chapters. Like much of chemistry, it relies heavily on example and analogy, reasoning from experimental results on model systems supplemented by the general principles of chemical kinetics and thermodynamics.

The following sections approach this topic from several directions: Section 8.2 provides a brief overview of the nature and achievements of solution-phase organic synthesis, offering a perspective from which to judge the expected capabilities of mechanosynthesis. Section 8.3 compares solution-phase synthesis and mechanosynthesis, examining their relative strengths and limitations. This section describes quantitative criteria for reaction reliability in light of the anticipated requirements of molecular manufacturing, and provides a preliminary description of the role of mechanical forces in chemical reactions. Section 8.4 provides an overview of several classes of reactive species from a mechanosynthetic perspective; Section 8.5 then examines several mechanochemical processes in more detail, focusing in particular on the exemplary processes of tensile bond cleavage and hydrogen abstraction. Finally, Section 8.6 presents several approaches to the mechanosynthesis of diamondoid structures, drawing on the preceding sections and on studies of reaction mechanisms in the chemical vapor deposition of diamond.

8.2. Perspectives on solution-phase organic synthesis

8.2.1. The scale and scope of chemistry

The advent of organic synthesis commonly is dated from the 1828 synthesis of urea by Wöhler, at a time when "chemists were just beginning to speculate about the arrangement of atoms within organic molecules" (Brooke, 1985). By the late 1800s, the basic concepts of atoms and molecular geometry were well established (except in the view of ardent Positivists), synthesis was subject to rational planning, and synthetic organic chemicals had become a major industry.

Today, a single nationally based professional group, the American Chemical Society, has ~150 000 members. One year's output of *Chemical Abstracts* now occupies ~2 m of shelf space, containing abstracts of ~5×10^5 articles; the volumes published to date occupy ~30 m of shelf space, with the indexes occupying ~30 m more. After 110 years of publication, *Beilsteins Handbuch der Organischen Chemie* occupies ~20 m of shelf space with over 300 volumes. The chemists' view of the current capabilities of organic synthesis is suggested by the observation that "the problem of synthesis design is simply that the number of possible synthetic routes to any target molecule of interest is enormous" (Hendrickson, 1990). This author estimates tens of millions of paths for a typical five-step synthesis, with vastly differing yields. A further indication is the barely cautious warning that "the idea persists that we can synthesize anything. The major flaw in this view is that it fails to recognize effectiveness and practicality" (Trost, 1985).

During the first century of organic synthesis, chemists proceeded with no guidance from quantum mechanics, relying instead on a growing set of concepts regarding molecular geometry and thermodynamics, together with rules of thumb for estimating the energy differences between similar molecular configurations and the rate differences between similar reactions. These concepts and rules grew from efforts to rationalize experimental results; they are still being extended and remain central to organic synthesis to this day. Many reactions are known to yield particular results; somewhat fewer can be described in mechanistic detail (i.e., which atoms go where as the system passes through its transition state); far fewer have been modeled at the level of potential energy surfaces and molecular dynamics. Detailed theory thus describes only a fraction of the synthetic operations available to the modern chemist.

8.2.2. The prominence of qualitative results in organic synthesis

Organic synthesis is fundamentally qualitative, in much the same sense that topology (as distinct from, say, analytic geometry) is fundamentally qualitative. In synthesis, the chemist seeks to construct molecules with a particular pattern of bonding that can be described (give or take discrete, stereochemical differences) in topological terms. The discovery that substances of type A can be transformed into substances of type B is inherently qualitative and often of great value. The development of such useful species and processes as Grignard reagents, Wittig reactions, and Sharpless epoxidations are examples of contributions of this kind.

In describing an organic synthesis, numbers are used to specify temperatures, quantities of material, and characteristics such as melting points, refractive indexes, and so forth, but these numbers play a peripheral role. One can frequently examine several chapters of a monograph on chemistry while encountering no numbers save for dimensionless integers, such as those used to specify molecular structures and charge states. To be useful in experimental work, most measurements need only discriminate among different molecular species, and accurate discrimination need not require great precision. Many theoretical contributions of great fame and value take the form of nonquantitative rules (some subject to violation) that are expressed in terms of structure and geometry; examples include Markownikoff's rule for regioselectivity of addition to alkenes, the Alder rule for the production of isomers in Diels–Alder reactions, and the Woodward–Hoffmann rules regarding orbital symmetry in °cycloaddition processes.

Returning to the picture of potential energy surfaces in configuration space, the discovery of a new kind of reaction corresponds to the discovery of a set of similar cols between potential wells corresponding to similar molecular species. In standard chemistry, useful cols must be low enough to be thermally accessible at moderate temperatures *and* other cols must be relatively inaccessible (or harmless to traverse). The quantitative results of greatest interest in organic synthesis relate to the height and width of the cols (which determine reaction kinetics) and the height and width of the potential wells (which determine reaction equilibria).

8.2.3. A survey of synthetic achievements

No simple first-principles analysis leads from elementary facts regarding molecular physics to conclusions regarding the range of structures that can be made by diffusive, solution-phase chemical processes. Accordingly, one must examine the actual capabilities that chemists have developed. These include the synthesis of:

- Highly strained molecules, such as cyclopropane rings **8.1**, with 60° bond angles, rather than the optimal 109.5°; cubane **8.2**, with a total strain energy of ~1.15 aJ (Eaton and Castaldi, 1985), about twice the typical C—C bond energy; and [1.1.1]-propellane **8.3** (Wiberg and Walker, 1982), in which the bonding configuration at two of the carbon atoms is inverted relative to the normal tetrahedral geometry—all four bonds are in one hemisphere.

| **8.1** | **8.2** | **8.3** |

- Highly symmetrical molecules, such as dodecahedrane **8.4** (Ternansky et al., 1982) and cyclo[18]carbon **8.5** (Diederich et al., 1989).

| **8.4** | **8.5** |

- Molecules that spontaneously decompose to free highly reactive species, such as carbenes (species containing divalent carbon), Eq. (8.1), and even free carbon atoms (Shevlin and Wolf, 1970), Eq. (8.2). (These species ordinarily exist only briefly before reacting.)

$$\text{(8.1)}$$

$$\text{(8.2)}$$

- Some six hundred organic structures, containing a few tens to a few hundred atoms, identical to molecules synthesized by molecular machinery in living systems (Corey and Cheng, 1989). Some examples include **8.6** [anno-

tinine (Wiesner et al., 1969)], **8.7** [bleomycin A$_2$ (Aoyagi et al., 1982)], and
8.8 [endiandric acid C (Nicolaou et al., 1982)]

- Chains of a hundred or more monomers, joined in a precise sequence to build biologically active protein and DNA molecules containing on the order of 1000 precisely bonded atoms (Caruthers, 1985; Kent, 1988).
- Molecules designed to self-assemble into larger structures (Cram, 1988; Lehn, 1988; Rebek, 1987), and even to catalyze the synthesis of copies of themselves (Tjivikua et al., 1990).

8.3. Solution-phase synthesis and mechanosynthesis

8.3.1. Analytical approach

This section uses the demonstrated mechanisms and capabilities of solution-phase organic synthesis as a basis for understanding the capabilities of mechanosynthesis. It proceeds by examining differences:

- The solution-phase conditions and unit operations *excluded* by mechanosynthetic constraints.
- The new conditions and unit operations *provided* by mechanosynthesis.
- The additional system-level capabilities implied by these differences.
- The additional constraints that must be satisfied to exploit these system-level capabilities.

Taken together with the known capabilities of diffusive organic synthesis, this set of differences gives substantial insight into the capabilities of mechanosynthesis.

a.   A terminological note: exothermic vs. exoergic.   In solution-phase chemistry, when molecular potential energy is transformed into mechanical energy, this promptly appears as heat. Reactions that reduce potential energy are accordingly termed *exothermic;* those that increase it, *endothermic*. In mechanochemical systems, however, potential energy released by a reaction can often be stored elsewhere in the mechanical system, either as potential energy or as kinetic energy in the form of orderly motion. Accordingly, the more general terms *exoergic* and *endoergic* are appropriate. Changes in free energy determine the direction of a reaction, and since changes in entropy are usually small in mechanosynthetic steps (which typically transform one highly constrained structure into another), changes in free energy and in potential energy are often almost equal. Under these conditions, exoergic reactions can produce both heat and work, and most endoergic reactions are driven more by work than by heat.

b. Mechanochemical parameters. Two major themes in the following sections are reaction rates and reaction reliability in mechanochemical systems. Useful variables for an approximate analysis include the actuation time t_{act}, during which motions carry the system toward and then away from a reactive geometry; the transformation time t_{trans}, during which the geometry of the system permits a reaction to occur; the reaction rate k_{react}, which measures the probability per unit time of a reactive transformation during t_{trans}, given that a reaction has not yet occurred; the intersystem crossing rate k_{isc}, which measures the rate of the electronic intersystem crossing "reaction"; and the error rate k_{err}, which measures the rate of unwanted reactions. These parameters will be related to the probability of an error in a mechanochemical operation, P_{err}. It will usually be assumed that $t_{act} \approx 10^{-6}$ s, $t_{trans} \approx 10^{-7}$ s, and that each contribution to P_{err} is to be $\leq 10^{-15}$ (see Sections 8.3.4c to 8.3.4f).

8.3.2. Basic constraints imposed by mechanosynthesis

a. Loss of natural parallelism. Solution-phase synthesis often yields macroscopic quantities of product. In replacing a macroscopic volume of reagent solution with a single device, the number of reacting entities is reduced by a factor on the rough order of 10^{-23}. To compensate for this difference requires many mechanosynthetic devices, each operating at a high frequency. Some consequences of this constraint are discussed in Section 8.3.4c.

b. Limitations on reagents and products. Mechanical control of reactions typically requires that reagent moieties be bound to extended "handle" structures; control of products typically requires that they, too, be bound, excluding the use or production of small, freely moving species. This constraint is less severe than it might seem, however, because free and bound species can often participate in similar reactions. For example, although an S_N2 reaction producing a free ion [Eq. (8.3)] would cause unacceptable disorder, (8.4) is permissible.

$$\text{(8.3)}$$

$$\text{(8.4)}$$

In reality, an ionic process like reaction (8.3) would require a favorable electrostatic environment, and that environment could be used to bind the product ion. In water, small ions (from Li^+ to I^-) are found in solvent cages, and are bound with energies ≥ 330 maJ relative to vacuum (Bockris and Reddy, 1970a). In condensed-phase acid reactions "H^+" is always bound. Reactions characteristic of small free °radicals (e.g., H, F, OH) can frequently be made to proceed by concerted mechanisms in which the radicals are never free. Binding of such species usually lessens reactivity, but Section 8.3.3 describes several effective techniques for speeding reactions in mechanochemical systems. Purely °steric confinement sometimes is feasible, and has little effect on the electronic structure of the reagent.

The requirement for handles (of any kind) places significant steric constraints on feasible reactions. Small, diffusing species can reach many sites (e.g., deep in pockets on a surface) that would be inaccessible to a mechanically positioned reagent moiety. In planning the synthesis of an object, geometric constraints are thus of increased importance. Finally, for reaction processes to be reliable, reagents must have adequate stability against unimolecular decomposition reactions (Section 8.3.4f).

c. Lack of true solvation. Fully °eutactic systems must lack true solvation, since solvents are not eutactic. As discussed in Section 8.3.3b, however, solvent-like eutactic environments can be superior to true solvents in promoting reaction speed and specificity. Mechanical transport can likewise substitute for diffusive transport; note that this sidesteps the constraint that reactants and products be soluble.

d. Limited usefulness of true photochemistry. Although similar results can be achieved using other sources of electronic excitation, photochemical reactions in a strict sense require photons. Optical wavelengths, however, are measured in hundreds of nanometers, and a cubic-wavelength block accordingly contains millions of cubic nanometers. This mismatch relative to the molecular scale limits the utility of photochemistry in mechanosynthesis (but note Section 8.3.3d).

e. Limitation to moderate temperatures. Mechanosynthetic systems are sensitive to damage from thermomechanical degradation (Section 6.4), and the single-point failure assumption (Section 6.7) is adopted here. With careful design, reliable operation at substantially elevated temperatures may be feasible. Nonetheless, temperatures like those used in melt-processing of typical ceramics are almost surely infeasible. The present work assumes that temperatures remain near 300 K, thereby eliminating from consideration a large class of reactions that require elevated temperatures to proceed at an acceptable rate. This constraint could be partially evaded by exploiting highly localized, nonthermal kinetic energy sources to provide activation energy (e.g., impacts driven by stored elastic energy), but this approach is not pursued here.

8.3.3. Basic capabilities provided by mechanosynthesis

a. Large effective concentrations. In a solution-phase reaction A + B → C, etc., the reaction frequency of molecules of type A is directly proportional to the concentration of molecules of type B (so long as molecules of B rarely interact with one another). In bulk chemistry, concentrations are commonly stated in moles per liter (M); here, they are stated as a molecular number density ($1\ nm^{-3} \approx 1.66\ M$). The concentration of water molecules in liquid water is $\sim 33\ nm^{-3}$.

Because their moieties are held in proximity to one another, mechanosynthetic reactions resemble intramolecular reactions; these, in turn, often resemble intermolecular reactions, save for a covalent link between reacting moieties. From the rate for an intramolecular reaction (in s^{-1}) and the rate constant for an analogous intermolecular reaction (in $nm^3 s^{-1}$), one can compute the *effective concentration* of a moiety B with respect to a moiety A in the intramolecular process by computing the (hypothetical) concentration of B-like molecules (in nm^{-3}) that would subject an A-like molecule to the same reaction frequency in the intermolecular process. The concept of effective concentration, although of no intrinsic importance to mechanosynthesis, is quite useful in estimating machine-phase reaction rates from solution-phase data.

Experiment shows that effective concentrations for reactions between moieties on small molecules are not limited to $\sim 10\ nm^{-3}$ (typical of immersion in a liquid); they can be as large as $\sim 10^3$ to $10^{10}\ nm^{-3}$ (Creighton, 1984). Two examples cited by Creighton are:

$$\text{Eff. conc.} = 2.2 \times 10^3\ nm^{-3} \qquad (8.5)$$

$$\text{Eff. conc.} = 3.3 \times 10^9\ nm^{-3} \qquad (8.6)$$

The occurrence of large magnitudes can readily be understood: In the probability-gas picture of classical transition state theory (Section 6.2.2), the rate of transitions is proportional to the density of the gas. For concreteness and ease of visualization, consider an intramolecular reaction in which a single atom is transferred (e.g., atom abstraction). The concentration of an atom at the bottom of the potential well defining its initial state is just the reciprocal of the effective volume; for example, using Eq. (6.5) and a stiffness of 20 N/m in each of three dimensions (vs. ~ 30 N/m typical of bond angle-bending; Section 3.3.2b) yields a number density of $\sim 2 \times 10^4\ nm^{-3}$; this local density is unrelated to (and far larger than) the mean number-density of nuclei in solids at ordinary pressures. Further, many reactions require a relatively precise orientation of the reagent moieties; structural constraints in intramolecular reactions can greatly increase the concentration of properly oriented moieties, multiplying the effective concentration by a large factor. (Intramolecular reactions can also be accelerated by intermoietal strain; this can, however, be viewed as a violation of the intramolecular-intermolecular analogy that motivates the notion of effective concentration, and is more closely analogous to the piezochemical effects described in Section 8.3.3c.)

Binding of reagents so as to ensure high effective concentrations is a major mechanism of enzymatic catalysis (Creighton, 1984). Mechanochemical processes can likewise present reagent moieties to one another in favorable orientations and positions, producing high effective concentrations. Relative to intermolecular reactions in solution (with a typical concentration of $\sim 1 \text{ nm}^{-3}$), this mechanism should routinely yield rate accelerations on the order of 10^4 or more.

b. Eutactic "solvation." Where reactions involve substantial changes in charge separation, solvent effects can alter reaction rates by factors in excess of 10^5. Relative to vacuum or nonpolar environments, a polar solvent greatly decreases the energetic cost of creating dipoles or separated ions by orienting its molecules in a manner that decreases electrostatic field strengths. The feasible reduction in free energy is, however, lessened by the entropic cost of orienting the solvent molecules. A preorganized structure (Cram, 1986), in contrast, can equal the electrostatic effects of a good solvent without imposing entropic costs on the transition state.

Indeed, a preorganized environment can yield larger electrostatic effects. The solvent structure at the transition state of a solution-phase reaction minimizes the free energy of the transition-state system as a whole, including that of the solvent. A different organization of solvent molecules would minimize the free energy of the transition-state reactants alone; this organization typically is more highly ordered (imposing additional entropic costs) and more highly polarized (imposing substantial energetic costs). Preorganized environments can be constructed to mimic this hypothetical, high-free-energy solvent state, paying the free-energy cost in advance and thereby outperforming a good solvent in promoting the desired reaction (a principle exploited by enzymes).

The magnitude of these effects can be roughly estimated. Ionic reactions commonly involve charge displacements on the order of one electron-charge and one bond length, or $\sim 2.5 \times 10^{-29} \text{ C·m}$. Fields of $2 \times 10^9 \text{ V/m}$ (e.g., 1 V across 0.5 nm) are easily achieved in a polarized environment; contact potentials in metallic systems (Section 11.7.2a) can produce fields of this magnitude. The energy difference associated with this field and charge displacement, $\sim 50 \text{ maJ}$, gives a rough measure of the effect that local electrostatics can have on the energies of transition and product states in ionic reactions. This energy suffices to change rates and equilibria by factors of $> 10^5$ at 300 K. The electrostatic energy difference between two orientations of an HCN molecule adjacent to a pair of oriented HCN molecules (Figure 8.1) is $\sim 110 \text{ maJ}$, an energy differential sufficient to change a rate or equilibrium by a factor $> 10^{10}$.

Enzymatic active sites often resemble preorganized solvent structures, reducing the energy of a transition state. Mechanochemical systems based on a more general class of molecular structures can equal or exceed enzymes in this regard. The requirement that all structures be well bound, prohibiting true solvation (Section 8.3.2c), does not appear to be a significant sacrifice.

As enzymes show, molecular flexibility can be achieved without freely moving small molecules. In a mechanochemical context, however, it frequently is desirable to provide relatively rigid support for a reagent structure; one strategy is to surround the structure with a nonbonded (hence solventlike) but rigid shell.

Figure 8.1. Dipole-dipole interaction geometries and energies. In (a), the two parallel HCN molecules (modeling bound nitrile moieties) have been held fixed at a separation of 0.3 nm while the energy of the third is minimized. In (b), the orientation of the third has been reversed. The increase in the MM2/CSC electrostatic (and total) energies from (a) to (b) is ~110 maJ.

*c. **Mechanical forces.*** The term *piezochemistry* is in general use to describe solution-phase chemical processes in which mechanical pressure modifies chemical reactivity.* Derived from the Greek *piezein* ("to press"), the term is here adopted to refer to a wider range of machine-phase chemical processes in which time-dependent mechanical forces (not necessarily a homogeneous, isotropic, slowly varying pressure) modify chemical reactivity.

In conventional piezochemistry, pressures accessible in commercially-available laboratory equipment (e.g., 0.1 to 2 GPa) frequently have substantial effects on reaction rates and equilibria. Transition states in solution-phase chemistry are characterized by an activation volume $\Delta V^{\ddagger}$ defined by Eyring in terms of a constant-temperature partial derivative:

$$\Delta V^{\ddagger} = -kT \left(\frac{\partial}{\partial p} \ln k_{\text{react}} \right)_T \tag{8.7}$$

where k_{react} is the rate constant for the reaction (Section 6.2.1) and p is the pressure. Typical values of $\Delta V^{\ddagger}$ are in the range of -0.01 to -0.10 nm^3; values can be positive, for example, in fragmentation reactions (Isaacs and George, 1987; Jenner, 1985). Many reactions have negative volumes of activation because their transition states combine bond making with bond breaking, and because the shortening of nonbonded distances by partial bond formation exceeds the lengthening of bonded distances by partial bond breakage. The relatively compact transition state is then favored by increased pressure.†

---

* Pressure in piezochemistry plays a fundamentally different role from pressure in gas-phase reactions. In the gas phase, so-called pressure effects on kinetics and equilibria have no direct relationship to the applied force per unit area, being mediated entirely by changes in molecular number density and resulting changes in collision frequencies. In the gas phase, the PES describing an elementary reaction process is independent of pressure, since each collision occurs (locally) in vacuum, free of applied forces. Likewise, so-called pressure effects in liquid and solid-surface environments exposed to reactive gases usually result more from changes in molecular number density than from piezochemical modifications to the PES of the reaction. (Section 6.3.1 outlines the concept of a PES dependent on boundary conditions, such as pressure.)

† In polar, solution-phase systems, the effects of solvent reorganization can be large, and bond-forming but charge-neutralizing reactions can have positive volumes of activation: the polar solvent, freed of strong fields, relaxes and expands.

In the (poor) approximation that $\Delta V^{\ddagger}$ is independent of pressure, the rate increases exponentially with pressure. In reality, increasing pressures alter $\Delta V^{\ddagger}$ by altering the potential energy surface. Equation (8.7) and moderate-pressure values of $\Delta V^{\ddagger}$ give only a rough guide to the magnitudes of the effects that can be expected in higher-pressure piezochemical processes. With this assumption, a pressure of 2 GPa applied to a reacting system with $\Delta V^{\ddagger} = -0.02$ nm^3 results in a rate increase of 1.6×10^4 at 300 K.

Nanomechanical mechanosynthetic devices can be built of diamond and diamondoid structures, and in the continuum approximation (Chapter 2), stress is a scale-independent parameter. Accordingly, mechanosynthetic devices can apply pressures equaling those in macroscale diamond-anvil pressure cells. These reach ≥ 550 GPa (Mao et al., 1989; Xu et al., 1986), corresponding to a ≥ 30 nN compressive load per bond in a (111) diamond plane (neglecting the change in areal bond density with pressure). The effects of such pressures on bonding are substantial: at ~ 150 GPa, H_2 becomes metallic (Hemley and Mao, 1990), as does CsI at ~ 110 GPa (Mao et al., 1989); xenon has likewise been metallized. Even a lesser pressure of 50 GPa (~ 3 nN/bond) can have large chemical effects. In the constant $\Delta V^{\ddagger}$ approximation, $p = 50$ GPa and $\Delta V^{\ddagger} = -0.01$ nm^3 would yield a 500 maJ reduction in activation energy (and a physically unrealistic $> 10^{52}$ speedup); effects in real systems are smaller, but still large. The change in free energy resulting from high pressures can greatly exceed the change resulting from high temperatures (within the conventional laboratory range).

Mechanically applied energy (unlike random thermal vibration) is subject to precise control. This discussion has been cast in terms of pressure and volume because it draws on experiments and theories applicable to solution-phase processes. In mechanochemistry, however, force and displacement are more useful variables, and (for a particular mode of displacement) an activation length

$$\Delta \ell^{\ddagger} = -kT \left(\frac{\partial}{\partial F} \ln k_{\text{react}} \right)_T \tag{8.8}$$

can be defined in terms of the applied force F. A typical magnitude for $\Delta \ell^{\ddagger}$ is ~ -0.1 nm, and feasible compressive loads can extend to > 5 nN, which would once again yield a characteristic energy of ~ 500 maJ. (Section 8.5.2a and 8.5.4 develop more realistic estimates for certain systems.)

Mechanical instabilities can limit compressive loads: for example, if a single-atom tip is pressed against another atom, it has a tendency to slip sideways ("down off the hill") unless this is resisted by an adequate transverse stiffness. The negative stiffness associated with this instability has a magnitude

$$k_{\text{s,instab}} = -\frac{F_{\text{compr}}}{r_1 + r_2} \tag{8.9}$$

in the (conservative) model of hard spheres with radii r_1 and r_2. A typical transverse stiffness $k_{\text{s}\perp}$ for an unloaded tip is 20 N/m (characteristic of angle bending for a single bond; Section 3.3.2b), and a typical atomic radius for a nonbonded contact under substantial loads is ~ 0.1 nm (Figure 3.9), hence the condition

$$k_{\text{s,instab}} + k_{\text{s}\perp} > 0 \tag{8.10}$$

permits $F_{compr} \approx 4$ nN. A load of this magnitude suffices to store >100 maJ of potential energy in overlap repulsion between two unreactive atoms (Figure 3.8), and can store additional energy in the more-compliant deformation modes that lead toward a chemical reaction.

Unlike forces resulting from hydrostatic pressure, forces applied by mechanochemical devices can be highly anisotropic and inhomogeneous on a molecular scale: large loads (including tension, shear, and torsion) can be applied to specific atoms and bonds in a controlled manner. As noted in Section 6.4.4, these forces can cleave otherwise-stable bonds. Further, steric difficulties can be reduced by molecular compression and deformation in conventional piezochemistry (Jenner, 1985). Under mechanochemical conditions, larger effects can be obtained which can significantly offset steric difficulties posed by the mechanosynthetic requirement for bound reagents [for comparison, at 4 GPa the atomic number density in liquid cyclohexane is increased by a factor of ~1.5 (Gray, 1972)]. In general, the availability of controlled forces of bond-breaking magnitude permits piezochemical modulation of reactions greatly exceeding that seen in solution-phase chemistry or in the comparatively low-strength, low-stiffness environment of an enzymatic active site. Section 8.5 discusses piezochemical effects in further detail.

d.   Localized electrochemistry, "photochemistry."   Mechanochemical systems can exploit nonmechanical energy sources, for example, through electrochemistry and energy transfer via electronic excitations. Both of these mechanisms can be controlled more precisely in a mechanochemical environment than in solution or solution-surface systems.

Electrochemistry finds significant use in organic synthesis (Kyriacou, 1981). Electrostatic potentials and tunneling rates can vary sharply on a molecular scale (Bockris and Reddy, 1970b), permitting molecular-scale localization of electrochemical activity. Accordingly, electrochemical processes are well suited to exploitation in a mechanochemical context; they are also subject to modulation by piezochemical means (Swaddle, 1986). In electrochemical cells, pyridine can withstand an electrode potential of 3.3 V without reaction and tetrahydrofuran, −3.2 V, both with respect to a (catalytically active) platinum electrode (Kyriacou, 1981); these potentials correspond to energy differences with a magnitude >500 maJ per unit charge. In field-ion microscopes (which provide one model for an electrode surface), electric fields can reach ~50 V/nm (Nanis, 1984). Chapter 11 will discuss certain aspects of electrical engineering on a nanometer scale, but electrochemical processes, despite their undoubted utility, are not exploited in the following analysis.

Direct photochemistry suffers from problems of localization (which do not preclude its use), but photochemical effects can often be achieved by nonphotochemical means. Photochemical processes begin with the electronic excitation of a molecule by a photon, but this energy often can migrate from molecule to molecule as a discrete *exciton* before inducing a chemical reaction, and this process is highly sensitive to molecular structures and positions. Accordingly, the transfer of excitons can provide a better-controlled means for achieving photochemical ends; potential applications are, however, neglected in this volume.

e.   Broadened options for catalysis.   The structural requirements for mechanochemical reagents discussed in Section 8.3.2b are satisfied by many catalytic structures: some are parts of solid surfaces already, and others (e.g., many
homogeneous transition metal catalysts) have analogues that can be covalently
anchored to a larger structure. The remarks of Section 8.3.2b apply to small catalytic species (e.g., hydrogen and hydroxide ions).

Aside from regeneration treatments (which are infrequent on a molecular
time scale), conventional catalysts operate under steady-state conditions. In a
typical catalytic cycle, reagents are bound to form a complex, the complex rear-
ranges, and a product departs, all in the same medium at constant pressure, temperature, and so forth. If any transition state in this sequence of steps is too high
in energy, its inaccessibility will block the reaction. If any intermediate state is
too low in energy, its stability will block the reaction. If any feasible alternative
reaction (with any reagent or contaminant in the diffusing mixture) leads to a
stable complex, the catalyst will be poisoned. Many successful catalysts have
been developed, but the above conditions are stringent, requiring a delicately
balanced energy profile across a sequence of steps (Crabtree, 1987).

The range of feasible catalytic processes is broadened by the opportunities
for control in mechanochemical processes. The elementary reaction in a catalytic cycle can occur under distinct conditions, lessening the requirement for delicate compromises to avoid large energy barriers or wells. As discussed in
Section 8.5.10c, mechanochemical catalysts can be subjected to manipulations
that modulate bond and transition state energies, typically by many times kT_{300}.
Finally, comprehensive control of the molecular environment enables the
designer to prevent many unwanted reactions (Section 8.3.3f), permitting the
use of more reactive species (Section 8.3.3g).

f.   Avoidance of competing reactions.   In diffusive synthesis, achieving 95%
yield in each of a long series of steps would ordinarily be considered excellent.
At the end of a 100-step process, however, the net product would be ~0.6%; and
at the end of a 2000-step process, $\sim 10^{-43}$%. A million tons of starting reagents
would then reliably yield zero molecules of the desired product. At equilibrium,
a reaction with $\Delta \mathscr{F} \leq -145$ maJ at 300 K leaves less than 10^{-15} of the starting
molecules unreacted. An energy difference of this magnitude is not uncommon,
and a series of reactions with this yield would permit over 10^{10} sequential steps
with high overall yield, in the absence of side reactions.

The complexity of the structures that can be built up by diffusive synthesis is
limited not by an inability to add molecular fragments to a structure, but by the
difficulty of avoiding mistaken additions. This problem is substantial, even in
100-atom structures. Moreover, as structures grow larger and more complex,
they tend to have increasing numbers of functional groups having similar or
identical electronic and steric properties (on a local scale). Reliably directing a
conventional reagent to a specific functional group becomes increasingly difficult, and ultimately impossible.*

---

* Note, however, that convergent, "structure-directed" synthesis strategies (Ashton et
al., 1989) can loosen this constraint by combining larger fragments in a manner analogous to biological self-assembly. Distinctive features of larger-scale structures then
guide the reactions, with no obvious bound on the feasible complexity of the products.

In a well-designed eutactic mechanochemical system, unplanned molecular encounters do not occur, and *most* unwanted reactions accordingly are precluded. One class of exceptions consists of reactions analogous to unimolecular fragmentation and rearrangement; these instabilities are discussed in Section 6.4, and are discussed further in connection with reagent moieties in Sections 8.4 and 8.5. The other class of exceptions consists of reactions that occur in place of desired reactions; these can be termed *misreactions.*

A typical mechanosynthetic step involves the mechanically guided approach of a reagent moiety to a target structure, followed by its reaction at a site on that structure. In general, *un*guided reactions would be possible at several alternative sites, each separated from the target site by some distance (properly, a distance in configuration space). At one extreme, the alternative sites are separated by a distance sufficient to make an unwanted encounter in the guided case energetically infeasible (e.g., requiring that the mechanical system either break a strong bond or undergo an elastic deformation with a large energy cost). At the other extreme, the potential energy surface is such that passage through a single transition state leads to a branching valley, and then to two distinct potential wells, only one of which corresponds to the desired product; in this circumstance, unwanted reactions would be unavoidable. In intermediate cases, transition states leading to desired and undesired products are separated by intermediate distances, and the mechanical stiffness of the guiding mechanism imposes a significant energy cost on the unwanted transition state, relative to the unguided case.

Considering the approach of a reagent moiety to a target structure in three-dimensional space, a reaction pathway can be characterized by the trajectory of some atom in the moiety (such as an atom that participates in the formation of a bond to the surface). Inspection of familiar chemical reactions concurs with elementary expectations in suggesting that reaction pathways leading to alternative products commonly are characterized by trajectories that differ by a bond length ($d_t \approx 0.15$ nm) or more at the competing transition states (for example, see Section 8.5.5a). This is not universally true, but it is the norm; accordingly, avoidance of the exceptional situations imposes only modest limitations on the available sequences of chemical reactions. Where reagents are selective, or have reactivities with a strong orientational dependence, alternative trajectories usually are separated by far greater distances.

A reasonable stiffness for the displacement of an atom in a mechanically guided reagent moiety (measured with respect to a reasonably rigid or well-supported target structure) is 20 N/m (comparable to the ~30 N/m transverse stiffness of a carbon atom with respect to an sp^3 carbon, Section 3.3.2b). Note that the mechanical constraints on a reactive moiety can in many instances include not only the stiffness of its covalent framework, but forces resulting from an unreactive, closely packed surrounding structure that substantially blocks motion in undesired directions (see Figure 8.2). Using the sinusoidal worst-case decoupling model of Section 6.3.3d to estimate error rates with nonsinusoidal potentials, the condition $P_{err} \leq 10^{-15}$ requires that the elastic energy difference between the two locations be $\geq$ ~180 maJ at 300 K. With $k_s = 20$ N/m, this condition is satisfied at transition-state separations $d_t \geq 0.135$ nm. Accordingly, with modest constraints on the chemistry of the reacting species, suppression of

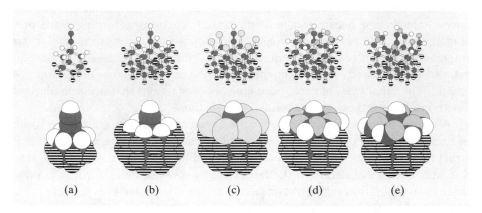

Figure 8.2. Models of a reagent moiety of low intrinsic stiffness with varying degrees of support from surrounding nonbonded contacts (MM2/CSC potential). Each model includes an °alkyne moiety representing (for example) an alkynyl radical of the sort that might be used as a hydrogen °abstraction tool (Section 8.5.4c). In (a), the moiety is supported by an adamantyl group; the bending stiffness of the moiety at its most remote carbon atom, taking the hydrogen atoms attached to the six-membered ring opposite the moiety as fixed, is ~6 N/m. In (b), a more crowded environment increases the stiffness to ~11 N/m (relative to those hydrogen atoms not either bonded to or in contact with the alkyne moiety); (c) replaces H atoms with Cl, increasing the stiffness to ~20 N/m; greater stiffness could be achieved by adding structures outside the Cl ring that press its atoms inward. Structure (d) surrounds the alkyne with a ring of oxygen-linked carbon atoms, yielding a calculated stiffness of ~65 N/m. Structure (e), however, represents a possible failure mode of (d) in which a rearrangement has cleaved six C—O bonds, yielding six aldehyde groups. Although this process is quite exoergic, the transition state may be effectively inaccessible at room temperature.

unwanted reactions by factors of better than 10^{-15} should be routinely achievable. With $k_s = 30$ N/m and $d_t \geq 0.15$ nm, this model yields $P_{err} \leq 10^{-27}$. (Although the differing ratios of accessibility between alternative transition points at different times during the approach complicate the situation, the essential conclusions remain unchanged.)

Section 8.5.5a describes a specific, relatively challenging case: discriminating between the ends of the double bond in an alkene during radical addition. It concludes that error rates $\leq 10^{-15}$ are achievable with stiffnesses of ~10 N/m, or ≤ 6 N/m using a trajectory-biasing technique.

g.   Reliable control of highly reactive reagents.   It might seem that the most useful reagents would be those (e.g., strong free radicals, carbenes) that can react with many other structures. In solution-phase chemistry, however, unwanted reactions are the chief limit to synthesis, and reagents are prized less for their reactivity than for their selectivity. The ideal reagent in solution-phase synthesis is inert in all but a few circumstances, and it need not react swiftly when it reacts at all.

In mechanosynthesis, however, *selectivity* based on the local steric and electronic properties of the reagents themselves can be replaced by nearly perfect *specificity* based on positional control of reagent moieties by a surrounding mechanical system. Accordingly, highly reactive reagents gain utility, including

Table 8.1. Comparison of solution-phase synthesis and mechanosynthesis (a briefer, more general comparison appears in Table 1.2).

Characteristic	Solution-phase synthesis	Mechanosynthesis
Parallelism	Natural	Requires many devices
Reagent structure	Unconstrained (but soluble)	Bound to "handle"
Electrostatic environment	Control of solvent dielectric constant	Control of dielectric constant, fields
Electrochemistry	Useful	Useful, localized
Photochemistry	Useful	Poorly localized (or exciton mediated)
Temperatures	Can be high	Must be moderate
Max. effective concentrations	~100 nm^{-3}	>10^9 nm^{-3}
Available pressures	Up to ~2 GPa (routinely)	>500 GPa
Control of forces	Magnitude of uniform pressure	Location, magnitude, direction, shear, torque
Positional control	None	All three degrees of freedom
Orientational control	None	All three degrees of freedom
Reagent requirements	Selectivity	Stability and reactivity
Reaction site selectivity	Steric, electronic influences	Direct positional control
Max. synthesis complexity	~100 to 1000 steps[a]	>10^{10} steps

[a] Values in this range are seldom achieved.

reagents that are highly unstable to intermolecular reactions among molecules of the same type (e.g., benzynes, reactive dienes, and other fratricidal molecules). The use of reagents with increased reactivity can increase reaction frequencies, adding to (or, more properly, multiplying together with) the influences of high effective concentration, well-designed eutactic "solvation," and applied force.

8.3.4. Preview: molecular manufacturing and reliability constraints

The capabilities of mechanosynthetic processes can be gauged by comparing the limitations discussed in Section 8.3.2 with the strengths discussed in Section 8.3.3, taking the capabilities of diffusive synthesis as a baseline. This comparison, summarized in Table 8.1, suggests that mechanosynthetic operations are (overall) more capable than diffusive operations. These capabilities must, however, be judged in light of their proposed applications. This section previews proposed molecular manufacturing systems and discusses their associated speed and reliability requirements. Molecular manufacturing systems are examined more thoroughly in Chapters 13 and 14 after the analysis of nanomechanical components and systems summarized in Chapters 9 through 12.

a. Molecular manufacturing approaches. Mechanochemical devices are of interest in this volume chiefly as components of mechanosynthetic systems capable of building large, complex structures, including systems of molecular machinery. The necessary positioning and manipulation of reagent moieties can be achieved in any of several ways, but these can be roughly divided into *molecular mill* approaches and *molecular manipulator* approaches. Molecular mills perform simple repetitive motions; molecular manipulators enable complex, programmable motions (see Chapters 13 and 14). Mills are well suited to the task of preparing reagent moieties for manipulators, but can also produce final products directly. In all cases considered, mechanical transport mechanisms replace diffusion.

b. Reagent preparation vs. application. Product synthesis can use reagent moieties that have been prepared from other reagent moieties, starting ultimately with simple feedstock molecules. The preparative steps can occur in an environment significantly different from that of the final synthetic steps.

In *reagent preparation,* the entire surrounding structure can be tailored to facilitate the desired transformation. In this respect, the reaction environment can resemble that of an enzymatic active site, but with the option of exploiting a wider range of structures, more active reagents, piezochemical processes, and so forth. The freedom to tailor the entire reaction environment during reagent preparation is a consequence of the small size and consequent steric exposure of the structures being manipulated.

In the final synthetic step, however, one reacting surface must be a feasible intermediate stage in constructing the product, and a prepared reagent must be applied to that surface.* Construction strategies can be chosen to facilitate the sequence of synthetic reactions, but the freedom to tailor the entire environment solely to facilitate a single reaction is not available. In these *reagent application* steps, highly reactive moieties are of increased utility and moieties compatible with supporting structures of low steric bulk are desirable. These remarks apply both to manipulator-based systems and to the reagent-application stages of a mill-style system engaged in direct synthesis of complex products.

c. Reaction cycle times. One measure of the productivity of a manufacturing system is the time t_{prod} required for it to make a quantity of product equaling its own mass. Mill-style systems are anticipated to contain $\sim 10^6$ atoms per processing unit, with each unit responsible for converting a stream of input molecules into a stream of reagent moieties which are incorporated into identical sites in a stream of product structures. Manipulator-style systems are anticipated to contain $\sim 10^8$ atoms per manipulator, with each unit responsible for performing a series of mechanosynthetic operations on a single product structure. Each processing unit and manipulator operates on products at some frequency f_{synth}. A set of devices will transfer a mean number of atoms n_{synth} per operation; in any one operation, the number transferred may be positive or negative (e.g., in abstraction reaction to prepare a radical site on a workpiece), but a typical value is $n_{synth} \approx 1$. Accordingly, if $f_{synth} = 10^3$ Hz, $t_{prod} = 10^3$ s for a mill-based system and 10^5 s for a manipulator-based system; for $f_{synth} = 10^6$ Hz, the corresponding values of t_{prod} are 1 and 100 s respectively. Values of $f_{synth} \geq 10^3$ Hz are acceptable for many practical applications, and $\sim 10^6$ Hz is used as a reference value in the systems analysis of Chapter 14.

---

* Note, however, that a convergent synthesis can reduce the asymmetry of the reagent application process; see Section 8.6.6.

The exemplar calculations of the following sections assume that the time available for a reaction $t_{trans} = 10^{-7}$ s. This is compatible with $t_{act} \approx 10^{-6}$ s, or with $f_{synth} \approx 10^6$ Hz. These times are all long compared to the characteristic time scale of molecular vibrations, $\sim 10^{-13}$ s, but require reactions with low energy barriers.

d. Constraints on misreaction rates. In molecular manufacturing, two basic classes of error are (1) those that damage one product structure (fabrication errors), and (2) those that damage the manufacturing mechanism (destructive errors). Overall reliability can be increased in each instance by dividing a system (product or manufacturing) into smaller modules and replacing those that fail. Where fabrication errors occur, the module being made can be discarded before being incorporated into a product system. Where destructive errors occur, the manufacturing module can be stopped and its function taken over by one or more backup modules (Section 13.3.6). Simplicity of design favors the use of relatively large modules, which in turn requires good reliability. For simplicity, the systems analysis of Chapter 14 assumes that all fabrication errors are destructive, placing further demands on reliability.

For many purposes, 10^8-atom modules can be considered large. Fabrication errors at a rate of $\leq 10^{-10}$ per reagent application operation, resulting in a fabrication failure rate of $\leq 10^{-2}$, can thus be considered low. If we demand that each module in a manufacturing system process $\geq 10^2$ times its own mass before failing (and assume $n_{synth} \approx 1$), then a destructive-error rate of $\leq 10^{-10}$ per reagent application operation is again acceptable. The following calculations, however, take as a requirement that both classes of error occur at rates $\leq 10^{-15}$ per mechanosynthetic operation; Section 8.3.3f describes conditions on mechanical stiffness and transition-state separation that can satisfy this constraint. An error rate of 10^{-15} is compatible with the assumptions made in Chapters 13 and 14; higher error rates could be tolerated at the cost of increased design complexity.

e. Meeting constraints on omitted reactions in a single trial. Section 8.3.3f describes conditions for keeping the probability of misreactions $\leq P_{err} = 10^{-15}$; a related but distinct problem is to ensure that the desired reaction occurs, with a probability of omission (failure to react) $\leq P_{err}$. As also noted in Section 8.3.3f, a reaction with $\Delta\mathscr{F} \leq -145$ maJ at 300 K (at the time of kinetic decoupling, in terms of the approximation discussed in Section 6.3.3) can proceed with an equilibrium probability of remaining in the starting state $< 10^{-15}$. For a strongly exoergic reaction characterized by a rate constant k_{react}, the probability of omission in a single trial falls to $< 10^{-15}$ when the available reaction time $t_{trans} \geq 35/k_{react}$. A one-dimensional model based on transition state theory (Section 6.2.2), together with the bounds just described, yields a bound on the allowable °barrier height for the reaction

$$\Delta\mathscr{V}^{\ddagger} \leq kT \ln\left(\frac{t_{react}\, f_{TST}}{-\ln(P_{err})}\right) \tag{8.11}$$

where f_{TST} is the transition state theory frequency factor; for mechanochemical reactions with relatively rigid, well-aligned reagent moieties, a reasonable value is $f_{TST} \geq 10^{12}$ s^{-1}. At 300 K, with $t_{react} = 10^{-7}$ and $P_{err} = 10^{-15}$, $\Delta\mathscr{V}^{\ddagger} \leq 33$ maJ is acceptable.

How does this compare to typical solution-phase reactions? In the laboratory, characteristic reaction times (the reciprocals of the reaction rates) vary widely, from $<10^{-9}$ to $>10^6$ s; a not unusual reaction time in organic synthesis is $\sim 10^3$ s at a reactant concentration of ~ 1 nm^{-3}. Relative to this, achieving a reliable reaction in $t_{\text{trans}} = 10^{-7}$ s requires a speedup of $\sim 3 \times 10^{11}$. Increased effective concentration owing to mechanical positioning can easily provide a speedup of $>3 \times 10^4$ (Section 8.3.3a). Achieving the remaining factor of 10^7 requires that the energy barrier be lowered by ~ 70 maJ. Earlier sections have shown that electrostatic effects in eutactic environments can exceed 100 maJ, and that (crudely estimated) piezochemical effects can exceed 500 maJ; shifts from less-active to more-active reagents can likewise have large effects. Achieving adequate reaction speeds does not appear to be a severe constraint. The duration of t_{trans} can be increased by $\sim 10^2$ without undue cost by mechanisms that prolong the encounter process (Section 13.3.1).

Intersystem crossing from the °singlet state to a low-lying °triplet can cause errors, including omitted reactions and misreactions. Singlet transition-state geometries often resemble triplet equilibrium geometries (Salem and Rowland, 1972), with corresponding reductions in singlet-triplet energy gaps. To ensure reliability, it is sufficient to ensure either (1) that the singlet-triplet gap $\Delta \mathcal{V}_{\text{s,t}}$ always exceeds 145 maJ [$\approx kT \ln(10^{15})$ at 300 K], or (2) that as the $\Delta \mathcal{V}_{\text{s,t}}$ increases, it exceeds 145 maJ at some time t_0 (prior to the time t_1 at which the geometry is no longer suitable for correct bond formation), and that the integral of the intersystem crossing rate k_{isc} meets the condition that

$$\int_{t_0}^{t_1} k_{\text{isc}}(t)\, dt \geq \ln(P_{\text{err}}) \qquad (8.12)$$

Meeting a somewhat weaker but more complex condition would also suffice.

f. Avoiding omitted reactions through conditional repetition. The conditions described in Section 8.3.4e were chosen to assure a reliable reaction in a single trial, but this is unnecessary. A *conditional repetition* process uses the outcome of a measurement process (Chapter 11) to determine whether an operation is repeated. For example, consider a reaction process in which the reaction rate and time ensure equilibrium, but in which $\Delta \mathcal{F} = 0$ rather than -145 maJ. This process has a .5 probability of success in any single trial. If a series of trials is terminated whenever success is achieved, then the mean number of trials required to achieve success is 2, and after 50 trials the probability of failure by omission is $<10^{-15}$. A similar process with an exoergicity of 25 maJ would (at 300 K) have a .9976 probability of success in a given trial, requiring a sequence of 6 conditional repetitions to achieve $P_{\text{err}} \leq 10^{-15}$, with the mean number of trials before success ≈ 1.002.

Conditional repetition can be viewed as a variation on the chemical strategy of driving a reaction to completion by steadily removing product from a reacting mixture. A simple and efficient approach to implementing conditional repetition in a molecular mill is presented in Section 13.3.1c.

g. Requirements for reagent stability. To meet the reliability objectives stated in Section 8.3.4d, reagent instability must not cause errors at a rate greater than the rate of misreactions. If the mean time between reactions is

$\leq 100 t_{act} = 10^{-4}$ s, and the frequency factor for the instability is $\leq 10^{13}$ Hz, then limiting the instability-induced error rate per reaction to $\leq 10^{-15}$ requires a barrier height ≥ 230 maJ.

Many potential rearrangement and dissociation reactions can be suppressed in a mechanochemical context by surrounding the reagent moiety with a suitably structured environment. For example, nonbonded interactions that block motions on the pathway to a rearrangement before the structure reaches an alternative, deeper potential well can suppress the instability. Solid cage structures that block the escape of a molecular fragment can likewise suppress dissociative instabilities.

Since each of the failure mechanisms discussed in this and the previous sections is exponentially dependent on energy parameters, exceeding the specified objectives yields large improvements in reliability. For example, with a barrier height ≥ 275 maJ the error rate per reaction is $\sim 10^{-20}$, and sets of $\sim 10^4$ reagent moieties have a mean time to failure (via this instability) of $\sim 10,000$ years.

8.3.5. Summary of the comparison

Solution-phase chemistry has enabled the synthesis of small molecules with an extraordinary range of structures (Section 8.2), but has not yet succeeded in constructing large structures while maintaining eutactic control. The specific comparisons made in Sections 8.3.2–8.3.4 support some general conclusions regarding the relative capabilities of mechanosynthesis:

a.   Versatility of reactions.   Relative to diffusive synthesis, mechanosynthesis imposes several significant constraints on the kinds of reagents that can be effectively employed. It requires that reagent moieties be bound, which can reduce reactivity and impose steric constraints. It requires that reactions be fast at room temperature, limiting the magnitude of acceptable activation energies. It requires that reagent moieties have substantial stability against unimolecular decomposition reactions, precluding the use of some reagents that are acceptable in the diffusive synthesis of small molecules.

Offsetting these limitations, however, are several advantages. Fundamentally, *mechanochemical processes permit the control of more degrees of freedom than do comparable solution-phase processes;* these degrees of freedom include molecular positions, orientations, force, and torques. As a consequence, highly reactive moieties can be guided with precision, enabling the exploitation of reagents that are too indiscriminate for widespread use in solution-phase synthesis. Bound reagent moieties are subject to mechanical manipulation, enabling piezochemical effects to speed reactions and overcome substantial steric barriers through localized compression and deformation of molecular structures. Finally, entirely new modes of reaction become available when reagent moieties can be subjected to forces of bond-breaking magnitude. Overall, these gains in versatility appear to exceed the losses, and hence the range of local structural features that can be constructed by mechanosynthesis should equal or exceed the range feasible with diffusive synthesis.

b.   Specificity of reactions.   In diffusive synthesis, most reactions entail substantial rates of misreaction, and the probability of a misreaction during any given step tends to increase with the size of the product structure. Experience suggests that the cumulative probability of error becomes intolerable for product structures of more than a few hundred to a few thousand atoms.

In mechanosynthesis, reliable exclusion of misreactions can be achieved given (1) a sufficient distance between alternative transition states, and (2) a sufficient mechanical stiffness resisting relative displacements of the reagent moieties. Distances on the order of a bond length combined with stiffnesses comparable to those of bond angle-bending yield $P_{err} < 10^{-15}$. Error rates (per step) are independent of the size and complexity of the product structure, given that the product is either stiff or well supported.

c. Synthetic capabilities. Within the constraints required to achieve reliable, specific reactions, the versatility of the set of chemical transformations available in mechanosynthesis can be expected to equal or exceed the versatility of the set of transformations available in solution-phase synthesis. This versatility is sufficient to suggest that most kinetically stable substructures will prove susceptible to construction (the challenging class of diamondlike structures is considered in more detail in Section 8.6). Transformations that (1) satisfy these reliability constraints, and (2) yield kinetically stable substructures can then be composed into long sequences that maintain eutactic control and yield product structures of 10^{10} or more atoms.

8.4. Reactive species

8.4.1. Overview

This section examines several classes of reagents from a mechanosynthetic perspective. *Numerous classes are omitted, and those included are discussed only briefly.* As noted in Section 8.2, chemistry is a vast subject; introductory textbooks on organic chemistry commonly exceed 1000 pages.

Polycyclic, broadly diamondoid structures are the products of greatest interest in the present context, hence this discussion focuses on the formation of carbon-carbon bonds; much of what is said is applicable to analogous nitrogen- and oxygen-containing compounds. As Section 8.3.3g indicates, highly reactive species are of particular interest. Many of the following species would be regarded as reaction intermediates (rather than reagents) in solution-phase chemistry.

8.4.2. Ionic species

Although ions are important in many solution-phase chemical systems, they are uncommon in the gas phase at room temperature; the electrostatic energy of an ion in vacuum can reduce stability by ≥ 500 maJ, favoring neutralization. In mill-style reagent preparation, reactions can occur in an electrostatically tailored environment; ionic species then can be as stable as those in solution and desired transformations can be driven by local electric fields. Accordingly, the utility of ionic species in these environments is, if anything, enhanced. In manipulator-based reagent application, however, it may frequently be desirable to expose reagent moieties on a tip moving through open space toward a product structure that is not tailored for favorable electrostatics. The electrostatic energy of ionic species is then high (making solution precedents inapplicable), and the utility of ionic species in manipulator-based operations may be relatively limited.*

a. Rearrangements and neutralization. Ionic species vary in their susceptibility to unimolecular rearrangement. °Carbonium ions, for example, are prone to 1,2-shifts with low [or zero (Hehre et al., 1986)] barrier energies:

$$(8.13)$$

A reagent moiety in which this process can occur will likely fail to meet the stability criterion for use in molecular manufacturing processes. If, however, each of the R-groups is part of an extended rigid system, this rearrangement is mechanically infeasible, as in Eq. (8.14). Rearrangement could also be prevented by steric constraints from a surrounding matrix, or (since the rearrangement involves charge migration) by local electrostatics.

$$(8.14)$$

°Carbanions, having filled °orbitals, are somewhat less prone to rearrangement (Bates and Ogle, 1983) and can more readily meet stability criteria. Again, structural, steric, and electrostatic characteristics can in many instances be used to suppress unwanted rearrangements.

Charge neutralization provides another, nonlocal failure mechanism. To avoid this requires a design discipline that takes account of the ionization energies and electron affinities of all sites within a reasonable tunneling distance (several nanometers) of the ionic site, ensuring that charge neutralization is energetically unfavorable by an ample margin (e.g., ≥ 145 maJ).

8.4.3. Unsaturated hydrocarbons

In diffusive chemistry, °unsaturated hydrocarbons (alkenes and alkynes) find extensive use in the construction of carbon frameworks. Their reactions characteristically redistribute electrons from relatively high-energy °pi bonds to relatively low-energy °sigma bonds, thereby increasing the number of covalent linkages in the system while reducing the bond order of existing linkages. This process has a strongly negative $\Delta V^{\ddagger}$ and ΔV_{react}, since the increase in interatomic separation resulting from a reduction in bond order is outweighed by the decrease resulting from the conversion of a nonbonded to a bonded interaction.

---

* Strongly polar tip structures may mitigate this problem.

Accordingly, these reactions are subject to strong piezochemical effects. Typical examples are Diels–Alder reactions such as Eq. (8.15),

$$\text{(8.15)}$$

which have $\Delta V^{\ddagger}$ in the range of ~-0.05 to -0.07 nm^3.

Unsaturated hydrocarbons undergo useful reactions with ionic species, and their reactions with other reagents are touched on in the following sections. Sections 8.5.5–8.5.9 describe certain classes of reactions in somewhat more detail.

From the perspective of solution-phase chemistry, mechanosynthesis has a greater freedom to exploit reactions involving strained (and therefore more reactive) alkenes and alkynes. The cyclic and polycyclic frameworks necessary to enforce strain will be commonplace, hence use of strained species can be routine; their use is desirable and practical for reasons discussed in Sections 8.3.3g and 8.3.4c. Planar alkenes have minimal energy, but molecules as highly pyramidalized as cubene **8.9** (Eaton and Maggini, 1988) and as highly twisted as adamantene **8.10** (Carey and Sundberg, 1983a) have been synthesized.

8.9   **8.10**

The reduced bonding overlap in these species (zero for **8.10**) makes the energetic penalty for unpairing of the pi electrons small enough to permit them to engage in diradical-like reactions under mechanochemical conditions.

Alkynes are of lowest energy when linear, but such highly reactive species as benzyne **8.11**, cyclopentyne **8.12**, and acenaphthyne **8.13**, have been synthesized (Levin, 1985).

8.11   **8.12**   **8.13**

Structures **8.11** to **8.13** have short lifetimes in solution owing to intermolecular reactions, but all have mechanically anchored analogues that are kinetically stable in eutactic environments but highly reactive in synthetic applications. Again, one pi bond is sufficiently weak to permit diradical-like reactivity under suitable conditions (Levin, 1985).

Also of high energy are allenes **8.14**, cumulenes **8.15**, and polyynes **8.16** (Patai, 1980).

8.14   **8.15**

8.16

Cumulenes and polyynes are of particular interest in building diamondoid structures consisting predominantly of carbon (Section 8.6); they bring no unnecessary atoms into the reaction.

Unsaturated hydrocarbons are prone to various rearrangements, subject to constraints of geometry, bond energy, and orbital symmetry. Those shown here are stable, however, and (as with carbonium ions) mechanical constraints from a surrounding structure can inhibit many rearrangements of otherwise-unstable structural moieties.

8.4.4. Carbon radicals

Free radicals result when a covalent bond is broken in a manner that leaves one of the bonding electrons with each fragment. Radicals thus have an unpaired electron spin and (in the approximation that all other electrons remain perfectly paired) a half-occupied orbital. Radicals can be stabilized by delocalization, for example in pi systems, but most are highly reactive.

a. Reactions. Among the characteristic reactions of radicals are *abstraction,* in which the radical encounters a molecule and removes an atom (e.g., hydrogen), leaving a radical site behind,

$$(8.16)$$

addition to an unsaturated hydrocarbon (here, a reactive pyramidalized species), generating an adjacent radical site on the target structure,

$$(8.17)$$

and *radical coupling,* the inverse of bond cleavage.

$$(8.18)$$

Radical addition and coupling have significant values of $\Delta V^{\ddagger}$ [for addition, ~ -0.025 nm^3 (Jenner, 1985)] and ΔV_{react}. Abstraction reactions often have a smaller $\Delta V^{\ddagger}$, and no significant ΔV_{react}. Their susceptibility to piezochemical acceleration is analyzed in Section 8.5.4.

b. Radical coupling and intersystem crossing. Electron spin complicates radical coupling and related reactions. Bond formation demands that opposed spins be paired (a singlet state), but two radicals may instead have aligned spins (a triplet state), placing them on a repulsive PES. Bond formation then requires an electronic transition (triplet $\rightarrow$ singlet intersystem crossing). As the radicals approach, the gap between the triplet and singlet state energies grows, but this *decreases* the rate of intersystem crossing. In delocalized systems, bond formation can occur without intersystem crossing, at the energetic cost of placing some other portion of the system into a triplet state. If intersystem crossing is required during the transformation time, however, then a condition like Eq. (8.12) must be met, but with t_1 representing the time by which bond formation must have occurred (if, that is, the system is to operate correctly). The condition that $\Delta \mathcal{V}_{s,t} \geq 145$ maJ imposes a significant constraint because k_{isc} varies inversely with the electronic energy difference $\Delta \mathcal{V}_{isc}$, which (in the absence of mechanical relaxation) would equal the difference in equilibrium energies $\Delta \mathcal{V}_{s,t}$, and will frequently be of a similar magnitude.

Values of k_{isc} for radical pairs in close proximity vary widely, and intramolecular radical pairs (diradicals) provide a model (Salem and Rowland, 1972). For 1,3 and 1,4 diradicals, k_{isc} has been estimated to be comparable that of the $S_1 \to T_1$ intersystem crossing in °aromatic molecules (Salem and Rowland, 1972), where k_{isc} commonly is between 10^6 and 10^8 s^{-1} (Cowan and Drisko, 1976; Salem and Rowland, 1972). Experiments with 1,3 diradicals in cyclic hydrocarbon structures found more favorable values of k_{isc}, ranging from ~10^7 to >10^{10} s^{-1} (Adam et al., 1987); the differences were attributed to conformational effects on orbital orientation, confirming rules proposed in Salem and Rowland (1972).

The presence of high-Z atoms relaxes the spin restrictions on intersystem crossing by increasing spin–orbit coupling (Cowan and Drisko, 1976; Salem and Rowland, 1972). In an aromatic-molecule model, k_{isc} for the $S_1 \to T_1$ transition in naphthalene is increased by a factor of ~50 by changing the solvent from ethanol to propyl iodide, and bonded heavy atoms have a larger effect: k_{isc} increases from ~6×10^6 s^{-1} in 1-fluoronaphthalene to >6×10^9 s^{-1} in 1-iodonaphthalene (Cowan and Drisko, 1976). The energy-gap condition is met at ordinary temperatures in this system: $\Delta \mathcal{V}_{isc}$ for 1-iodonaphthalene is 209 maJ (Wayne, 1988).

From these examples, it is reasonable to expect that, in the absence of special adverse circumstances, the inclusion of high-Z atoms bonded in close proximity to radical sites can be used to ensure values of $k_{isc} \geq 10^9$ in radical coupling processes where $\Delta \mathcal{V}_{s,t} \geq 145$ maJ. This is consistent with $P_{err} < 10^{-15}$ and $t_{trans} < 10^{-7}$ s. Failure to achieve intersystem crossing rates of this magnitude would increase the required value of t_{trans} for a particular operation, but would have only a modest effect on processing rates in a system as a whole (Section 8.3.4c), and no effect on the set of feasible transformations.

c.   Types of radicals.   Carbon radicals can broadly be divided into pi radicals (e.g., **8.17**) and sigma radicals (e.g., **8.18** to **8.20**), depending on the hybridization of the radical orbital. Of these, sigma radicals are higher in energy and hence more reactive; examples include radicals at pyramidalized sp^3 carbon (e.g., the 1-adamantyl radical) **8.18**, aryl radicals **8.19**, and the alkynyl radical **8.20**.

8.17   **8.18**   **8.19**   **8.20**

The alkynyl radical forms the strongest bonds to hydrogen and has excellent steric properties. *Ab initio* calculations on the ethynyl radical (HCC·), however,

Figure 8.3. A structure with a sterically exposed alkynyl carbon (here in a model alkyne group) having MM2/CSC stiffnesses of 4.5 and 21 N/m in orthogonal bending directions.

predict a low-energy electronic transition, $A\,^2\Pi \leftarrow X\,^2\Sigma$, with an energy of only ~40 maJ (Fogarasi and Boggs, 1983). If alkynyl radicals have a similar state at a similar energy, they will have a significant probability of being found in the wrong electronic state. Interconversion, however, requires no intersystem crossing and should be fast compared to typical values of t_{trans}. Figure 8.3 illustrates an exposed alkyne stiffened by nonbonded contacts.

d. Radical rearrangement. Radicals are much less prone to rearrangement than are carbonium ions. Intramolecular abstraction and addition reactions of radicals are common, where they are mechanically feasible, but shifts analogous to that shown in Eq. (8.13) are almost unknown unless the migrating group is capable of substantial electron delocalization (e.g., an aryl group).

8.4.5. Carbenes

Carbenes are divalent carbon species, formally the result of breaking two covalent bonds. The two nonbonding electrons in a carbene can be in either a singlet or a triplet state; the unpaired electrons in the triplet species behave much like those in radicals. Some carbenes are ground-state singlets in which the singlet-triplet energy gap $\Delta\mathcal{V}_{s,t} \geq 145$ maJ, making the probability of occupancy of the triplet state $\leq 10^{-15}$ at 300 K. Singlet carbenes can react directly to form singlet ground-state molecules; to achieve analogous results with triplet carbenes requires intersystem crossing (Section 8.4.4b).

a. Carbene reactions. Carbenes are of broad synthetic utility.* They can undergo addition to double bonds, yielding cyclopropanes

$$\begin{array}{c} R' \\ | \\ R{-}C{:} \end{array} \qquad \rightarrow \qquad \begin{array}{c} R' \\ R \end{array} \tag{8.19}$$

insertion into C—H bonds,

$$\begin{array}{c} R' \\ | \\ R{-}C{:} \end{array} \qquad \rightarrow \qquad \begin{array}{c} R' \\ R{-}C{-}H \end{array} \tag{8.20}$$

---

* "It is no exaggeration to claim a major role for carbenes in the modern chemist's attitude that he can very probably make anything he wants." (Baron et al., 1973)

and coupling (Neidlein et al., 1986)

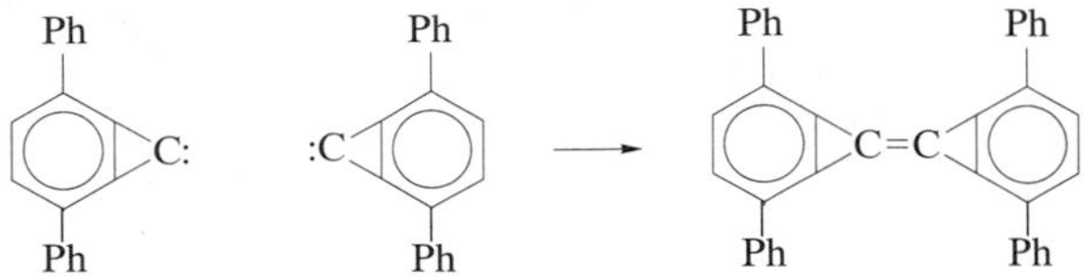

(8.21)

These reactions frequently proceed with energy barriers of ≤ 20 maJ; many have a barrier of zero (Eisenthal et al., 1984; Moss, 1989). Coupling of a singlet carbene and a radical

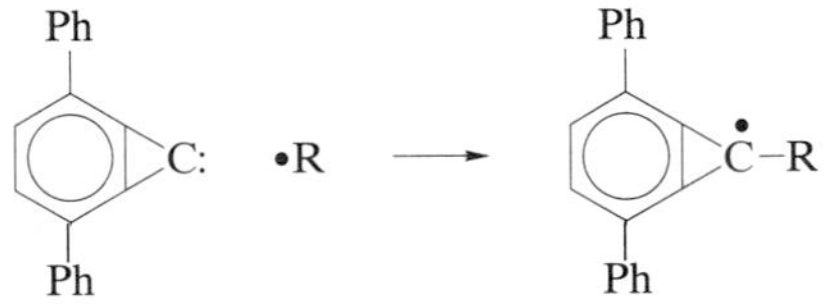

(8.22)

should likewise proceed with little or no barrier.

b.   *Singlet and triplet carbenes.*   The prototypical carbene is methylene, **8.21**, a ground-state triplet with $\Delta\mathcal{V}_{s,t} = -63$ maJ (Schaefer, 1986).

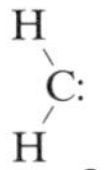

8.21    **8.22**

8.23 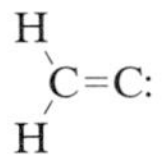   **8.24** 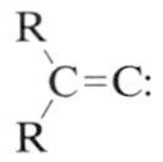   **8.25** 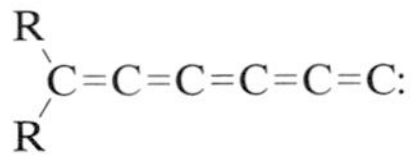   **8.26**

Adamantylidene **8.22** is thought to be a ground-state triplet (Moss and Chang, 1981). Carbenes with better steric properties (i.e., with supporting structures occupying a smaller solid angle) tend to be ground-state singlets: reducing the bond angle at the carbene stabilizes the singlet state, as does a double bond. In cyclopropenylidene **8.23** the predicted value of $\Delta\mathcal{V}_{s,t}$ is ~490 maJ (Lee et al., 1985); in vinylidene **8.24**, a prototype for alkylidenecarbenes **8.25**, $\Delta\mathcal{V}_{s,t} \approx$ 320 maJ (Davis et al., 1977); and in cumulenylidenecarbenes such as **8.26**, $\Delta\mathcal{V}_{s,t}$ >400 maJ (based on studies of odd-numbered carbon chains by Weltner and van Zee, 1989). These gaps are consistent with reliable singlet behavior in a molecular manufacturing context. Since singlet states of carbenes are appreciably more polar than triplets, singlet states can be significantly stabilized by a suitable electrostatic environment. For diphenylcarbene, experiment indicates that the shift from a nonpolar solvent to the highly polar acetonitrile increases $\Delta\mathcal{V}_{s,t}$ by ~ 10 maJ (Eisenthal et al., 1984). A preorganized environment (Section 8.3.3b) having a polarization greater than that induced in a solvent by the singlet dipole (and doing so without imposing an entropic cost) can increase the stabilization. Substituents including N, O, F, and Cl also tend to stabilize the singlet state.

c.   *Carbene rearrangements.*   Carbenes have a substantial tendency to rearrange; alkylcarbenes, for example, readily transform into alkenes:

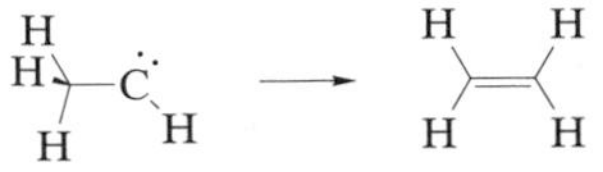

(8.23)

Among unsaturated carbenes, alkadienylidenecarbenes and cumulenylidene-carbenes have no available local rearrangements at the carbene center, and although vinylidene itself readily transforms to ethyne [indeed, it may not be a potential energy minimum (Hehre et al., 1986)], reaction (8.24) is not observed (Sasaki et al., 1983), and related cyclic species, as in reaction (8.25), may also be stable against this process.

$$ \text{(8.24)} $$

$$ \text{(8.25)} $$

8.4.6. Organometallic reagents

Various reagents with metal-carbon bonds are used in organic synthesis; this section briefly discusses only a few classes.

a. *Grignard and organolithium reagents.* Grignard reagents (RMgX) and organolithium reagents (RLi) find extensive use, providing weakly bonded carbon atoms with a high electron density (Carey and Sundberg, 1983b). As is common with organometallic species, their metal orbitals can accept electron pairs from coordinated molecules; this provides options for improving stability and mechanical manipulability. For example, in an ether solution a typical Grignard reagent structure is **8.27**; organometallic species such as **8.28** to **8.30** are typical of species that might be used in mechanosynthesis.

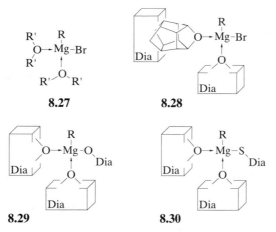

b. *Transition-metal complexes.* Complexes containing a *d*-block transition-metal atom exhibit versatile chemistry; such complexes are prominent in catalysis, including reactions that make and break carbon-carbon bonds. The presence of accessible *d* orbitals in addition to the *p* orbitals available in first-row elements changes chemical interactions in several useful ways: the orbital-symmetry constraints of reactions among first-row elements are relaxed, and bonded structures can have six or more °ligands (rather than four); further, the relatively long bonds (typical M—C lengths are ~0.19 to ~0.24 nm) reduce steric congestion, thereby facilitating multicomponent interactions. (Longer bonds result in coordination shells with areas ~1.6 to 2.6 times larger than those of first-row atoms). Many transition-metal complexes readily change their coordination number and oxidation state in the course of chemical reactions. Electronic differences among transition metals are large; complexation further increases their diversity. Clusters containing multiple metal atoms eventually resemble metal surfaces, which also find widespread use in catalysis.

Transition metals in bulky complexes appear more useful in reagent preparation and small-molecule processing than in sterically constrained reagent-application operations. In enfolded sites, ligand arrangements can be determined by mechanical constraints in the surrounding structure and placed under piezochemical control. Further discussion of the mechanochemical utility of these species is deferred to Section 8.5.10.

8.5. Forcible mechanochemical processes

8.5.1. Overview

Section 8.3.3 delineates some fundamental characteristics of mechanochemical processes, giving special attention to the use of mechanical force. Section 8.4 describes various reactions and reactive species, weighing their utility in a mechanochemical context. The present section examines a selected set of forcible mechanochemical processes in more detail. It starts by expanding on the discussion of piezochemistry in Section 8.3.3c, introducing the issue of thermodynamic reversibility. Tensile bond cleavage and hydrogen abstraction are then presented as model reactions and examined in quantitative detail. Several other reactions (involving alkene, alkyne, radical, carbene, and transition-metal species) are considered, building on results from the cleavage and abstraction models.

8.5.2. General considerations

a. Force and activation energy. Forces in piezochemical processes alter the reaction PES, reducing the activation energy; in some instances, they can eliminate energy barriers entirely, thereby merging initially distinct states. Section 8.3.4e calculates that barrier reductions of ~70 maJ can convert reactions that take 10^3 s in solution into reactions that complete reliably in 10^{-7} s. How much force is required to have such an effect?

Initial motion along a reaction coordinate can resemble the stretching of bonds or the compression of nonbonded contacts. As these motions continue, the resisting forces increase, but (usually) not so rapidly or so far as they would in simple bond cleavage or in the compression of an unreactive molecular substance. The potential energy curve instead levels off, passes through a transition state, and falls into another well. Since the energy stored in a given degree of freedom by a given force is proportional to compliance, the energy stored by a force applied through an unreactive bonded or nonbonded interaction is usually lower than that stored in a reactive system.

As shown by Figure 3.8, a compressive load of ~2 to 3 nN stores 70 maJ in a nonbonded interaction in the MM2 model; the ~30 nN per-bond compressive load in a diamond anvil cell (Section 8.3.3c) is an order of magnitude larger. The energy stored in bonds is more variable, but for a C—C bond in the Morse model (Table 3.8), a tensile force of ~5 nN stores 70 maJ.

b. Applied forces and energy dissipation. When actuation times are relatively long ($\sim 10^{-6}$ s), acoustic radiation from time-varying forces (Section 7.2) is minimal, as are free-energy losses resulting from potential-well compression (Section 7.5), given reasonable values for critical stiffnesses. Likewise, with small displacements (~ 1 nm) and low speeds ($\sim 10^{-3}$ m/s), phonon-scattering losses (Section 7.3) are small. In an elementary reaction process, the most significant potential sources of dissipation are transitions among time-dependent wells (Section 7.6).

Although the issue is distinct from the basic qualitative question of mechanosynthesis (i.e., what can be synthesized?), minimizing energy losses is of practical interest. Losses can broadly be divided into three classes: (1) those that are many times kT, resulting from the merging of an occupied high-energy well with an unoccupied low-energy well; (2) those on the order of kT, resulting from the merging of wells of similar energy; and (3) those of negligible magnitude, resulting from the merging of a low-energy, occupied well with a high-energy, unoccupied well. The simplest way to achieve high reaction reliability is to follow route (1), dissipating ≥ 145 maJ per operation. During forcible mechanochemical processes, however, it will in many instances be possible first to follow route (2) or (3) to a state in which the wells are merged (or rapidly equilibrating over a low barrier), then to use piezochemical effects to transform the PES to a type (1) surface before separation. This yields a process with reliability characteristic of (1), but with energy dissipation characteristic of (2) or (3). Systems capable of altering relative well depths by ≥ 150 maJ in mid-transformation can achieve error rates $< 10^{-15}$ with an energy dissipation $< 0.1 kT_{300}$. Opportunities for this sort of control are discussed in several of the following sections.

8.5.3. Tensile bond cleavage

Cleavage of a bond by tensile stress is perhaps the simplest mechanochemical process, providing an instance of the conversion of mechanical energy to chemical energy and illustrating the relationship between stiffness and thermodynamic reversibility. Further, tensile bond cleavage plays a role in several of the mechanosynthetic processes described in later sections.

As Figure 6.11 suggests, a typical C—C bond has a relatively large strength. As Table 3.8 shows, k_s for such a bond is lower than that for bonds to several other first-row elements, but higher than that for bonds to second-row elements. In many practical applications, the bond to be cleaved is of lower strength and stiffness than a typical C—C bond. Cleavage of the C—C bond will be considered in some detail, however, and can serve as a basis for comparison to other bond cleavage processes.

a. Load and strength. The 300 K bond-lifetime curves in Figure 6.11 indicate the tensile loads required to effect rapid bond cleavage. To achieve a level of reliability characterized by P_{err} requires a barrier meeting the condition given by Eq. (8.11); for a C—C bond in this model, achieving $P_{err} \leq 10^{-15}$ within $t_{trans} = 10^{-7}$ s requires a tensile load of ~ 4.2 nN. The Morse potential underestimates bond tensile strengths, but the problem of achieving sufficient tensile loads for rapid bond cleavage essentially pits the strength of one bond against that of others, hence errors in estimated strengths approximately cancel.

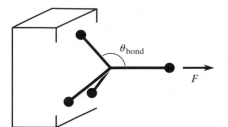

Figure 8.4. Bond angle in a distorted tetrahedral geometry.

As long as a carbon atom occupies a site with tetrahedral symmetry, strain-
ing one bond to the theoretical zero-Kelvin, zero-tunneling breaking point necessarily does the same to the rest. To concentrate a larger load on one bond
requires that the angle θ_{bond} be increased (Figure 8.4), thereby increasing the
alignment of the back bonds with the axis of stress and reducing their loads.
Figure 8.5 illustrates a structure that has a geometry of this sort when at equilibrium without load. With load, however, even an initially tetrahedral geometry distorts in the desired fashion. Increase of θ_{bond} from 109.5° to 115°
reduces back-bond tensile stresses to 0.79 of their undistorted-geometry values
(neglecting the favorable contributions made by angle-bending forces in typical
structures). Breaking of a back bond under these conditions would of necessity be a thermally activated process, and the energy barrier for breaking more
than one bond at a time would be prohibitive. Moreover, in the structures considered here, breaking of a single bond is strongly resisted by angle-bending
restoring forces from the remaining bonds (a ~0.1 nm, bond-breaking deformation is associated with an angle-strain energy of ~300 maJ); as has been discussed, this solid-cage effect invalidates the model used in Section 6.4.4a and
strongly stabilizes structures. These stress and energy differences are more than
adequate to ensure a $>10^{15}$ difference in rates of bond cleavage.

b. Stiffness requirements for low-dissipation cleavage. Transitions between
time-dependent potential wells can cause energy dissipation (Section 7.6.2), and
distinct wells are of necessity separated by regions of negative stiffness in the
potential energy surface. The potential energy for a pair of atoms undergoing
bond cleavage can be described as the sum of the bond energy and the elastic

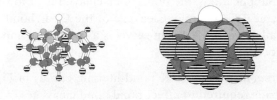

Figure 8.5. A structure having a surface carbon atom with a significantly nontetrahe-
dral geometry ($\theta_{bond} = 116.6°$). In this structure, $k_{sz} = 225$ N/m (MM2/CSC) for vertical
displacement of the central surface carbon atom with respect to the lattice-terminating
hydrogens below (shown in ruled shading); with an approximate correction for compliance of a surrounding diamondoid structure (Section 8.5.3d), $k_{sz} \approx 190$ N/m.

Structural diagram: Mechanical model:

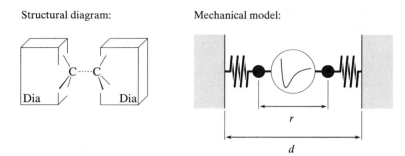

Figure 8.6. Diagrams illustrating tensile bond cleavage and the corresponding coordi-nates. The lengths d_0 and r_0, corresponding to unstrained springs and an unstrained bond, determine the values of the coordinates $\Delta d = d - d_0$ and $\Delta r = r - r_0$.

deformation energy of the structures in which the atoms are embedded:

$$\mathcal{V}_{cleave}(\Delta r, \Delta d) = \mathcal{V}_{bond}(\Delta r) + \tfrac{1}{2}k_{s,struct}(\Delta d - \Delta r)^2 \qquad (8.26)$$

where Figure 8.6 and its caption describe Δd and Δr, and the function $\mathcal{V}_{bond}(\Delta r)$ is the bond potential energy. The elastic deformation energy (neglecting modes orthogonal to the reaction coordinate) is a function of the displacement be-tween the bonded atoms and their equilibrium positions with respect to the sup-porting structure, and corresponds to some positive stiffness $k_{s,struct}$. The bond energy function, however, has a negative stiffness in the bond-breaking separa-tion range (Section 3.3.3a). The Morse potential predicts an extreme value of $-0.125k_s$, where k_s is the stretching stiffness of the bond at its equilibrium length; the Lippincott potential predicts negative stiffnesses of greater magni-tude (-115 rather than -55 N/m, for a standard C—C bond), and hence is more conservative in the present context. It is adopted in the following analysis.

Figure 8.7 illustrates potential energy curves as a function of Δr for various values of Δd for one set of model parameters. As can be seen, a steady increase

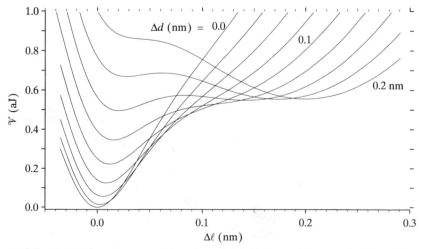

Figure 8.7. Potential energy as a function of CC distance, for several values of support separation. Stiffness of support = 50 N/m.

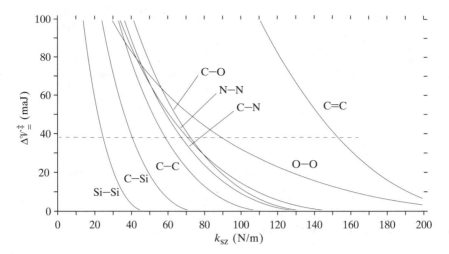

Figure 8.8. Barrier heights vs. stiffness, for various bonds placed under a tensile load which equalizes the well depths (bond parameters from Table 3.8).

in Δd causes the evolution of the system from a single well, to a pair of wells, to a single well again; larger values of $k_{s,struct}$ first reduce and then eliminate the barriers.

At finite temperatures and modest speeds, transitions can occur without causing substantial dissipation, so long as good equilibration occurs between the old well and the new before the energy of the new well has fallen substantially below that of the old. This requires that, during the time when the wells are of nearly equal energy, the mean interval between transitions be short compared to the time required for significant changes in relative well depth to occur. A transition rate of 10^9 s^{-1} ensures low dissipation (small compared to kT) when the characteristic time for the evolution of the wells is ~10^{-7} s. Assuming a frequency factor of 10^{13} s^{-1}, this is achieved for barrier heights $\leq$38 maJ at 300 K.

For estimating energy dissipation, a conservative measure of the barrier height for the process as a whole is the height when the two wells are of equal depth, $\Delta \mathcal{V}_{=}^{\ddagger}$ (assuming substantial values of $k_{s,struct}$, to limit the entropic differences between the two wells). Figure 8.8 plots $\Delta \mathcal{V}_{=}^{\ddagger}$ as a function of $k_{s,struct}$ for several bond types. For processes in which $k_{s,struct} >90$ N/m, and characteristic times are $\geq 10^{-7}$ s, dissipation is small relative to kT for all the single bonds shown; for standard C—C bonds, $k_{s,struct} > 60$ N/m is sufficient.

c. ***Spin, dissipation, and reversibility.*** In the absence of intersystem crossing, bond cleavage yields a singlet diradical. In a well-separated diradical, however, the singlet-triplet energy gap approaches zero, and thermal excitation soon populates the triplet state. If bond cleavage is fast compared to intersystem crossing, this equilibration process results in $\Delta \mathcal{F} = -\ln(2)kT$ (corresponding to the loss of one bit of information). Conversely, if intersystem crossing is fast, the thermal population of the (repulsive) triplet state results in a reduction of the mean-force bond potential energy during cleavage, and no significant dissipation. Note that slow intersystem crossing can cause large energy dissipation in mechanically forced radical coupling, even when the reverse process has a dissipation $<kT$.

d. Atomic stiffness estimation. In the linear, continuum approximation, the z-axis deformation of a surface at a radius r from a z-axis point load is $\propto 1/r$ (Timoshenko and Goodier, 1951). Accordingly, most of the compliance associated with displacement of an atom on a surface results from the compliance of the portion of the structure within a few bond radii.

A carbon atom on a hydrogenated diamond (111) surface can be taken as a model for sp^3 carbons on the surfaces of diamondoid structures. The MM2 model value for the z-axis stiffness k_{sz} of such a carbon atom on a semi-infinite lattice can be accurately approximated by measuring the stiffness in a series of approximately hemispherical, diamondlike clusters of increasing radius (Figure 8.9), holding the lattice-terminating hydrogen atoms fixed. In the continuum model, the compliance of the region outside a hemisphere is $\propto 1/r$; treating the number of carbon atoms as a measure of r^3 and fitting the four cluster-stiffness values with this model yields $k_{sz} = 153 \pm 2$ N/m.

This model holds for small displacements, but at the larger displacements corresponding to peak forces in the bond-cleavage process, bond angles and lengths are significantly distorted; this affects stiffness. Examination of the energy as a function of z-axis displacements shows that the decrease in bond-stretching stiffness resulting from tension is, under loads in the range of interest, more than offset by the increase in k_{sz} resulting from changes in bond angle. Under tensile loads of 3 to 7 nN, the stiffness is increased by a factor of ~ 1.05. The MM2 model is known to have unrealistically low bond-bending stiffnesses (Section 3.3.2g); increasing these stiffness values by a factor of 1.49 (to approximate MM3 results) increases k_{sz} by a factor of 1.14, yielding an overall estimate of $k_{sz} \approx 183$ N/m.

Since the compliances of the two bonded atoms are additive, $k_{s,struct} = k_{sz}/2 \approx 90$ N/m. This significantly exceeds the required stiffness for low-dissipation cleavage of all the single bonds in Figure 8.8, save for O—O. For the C—C bond, available stiffness exceeds the requirement by a factor of 1.5; accordingly, low-dissipation bond cleavage should be a feasible process in a broad range of circumstances.

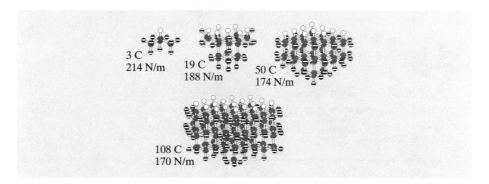

3 C
214 N/m

19 C
188 N/m

50 C
174 N/m

108 C
170 N/m

Figure 8.9. A series of structures modeling a diamond (111) surface, with corresponding values of k_{sz} (MM2/CSC) for vertical displacement of the central surface carbon atom with respect to the lattice-terminating hydrogens below (shown in ruled shading). Note that this carbon atom has poor steric exposure, but that a better-exposed carbon atom need not sacrifice stiffness (e.g., Figure 8.5).

8.5.4. Abstraction

Abstraction reactions can prepare radicals for use either as tools or as activated workpiece sites. Although various species are subject to abstraction, the present discussion focuses on hydrogen. Most exoergic or energetically neutral hydrogen abstraction reactions have activation energies ≤ 100 maJ (Bérces and Márta, 1988). The abstraction reaction

$$H_3C\cdot + HCH_3 \longrightarrow H_3CH + \cdot CH_3 \tag{8.27}$$

can serve as a model for reactions involving more complex hydrocarbons, including diamondoid moieties. The energy barrier for this process is substantial, ~ 100 maJ (Wildman, 1986), making it representative of relatively difficult abstraction reactions.

a.   Abstraction in the extended LEPS model.   The piezochemistry of abstraction reactions can be modeled using the extended LEPS potential, Eq. (3.23). This three-body potential automatically fits the bond energies, lengths, and vibrational frequencies of each of the three possible pairwise associations of atoms. In a symmetrical process such as Eq. (8.27), the two independent Sato parameters can be used to fit the barrier height and the CC separation, $r(CC)$, at the linear, three-body transition state [calculated to be 0.2669 nm for reaction (8.27) using high-order *ab initio* methods (Wildman, 1986)]. Fitting a LEPS function to these values using parameters based on standard bond lengths and energies for methane and ethane (Kerr, 1990) and bond-stretching stiffnesses from MM2 (Table 3.8), yields Sato parameters of 0.132 for the two CH interactions and –0.061 for the CC interaction. (The pure LEPS potential predicts deflection of the hydrogen atom from the axis at small CC separations, but a modest angle-bending stiffness suffices to stabilize a linear geometry.)

With this function in hand, it is straightforward to evaluate the energy barrier $\Delta \mathcal{V}^{\ddagger}$ as a function of the compressive load F_{compr}, as shown in Figure 8.10. In

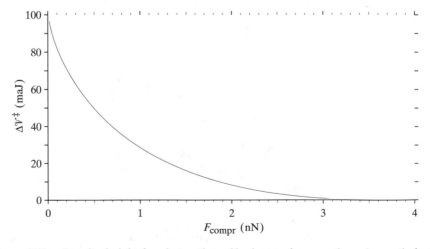

Figure 8.10.   Barrier height for abstraction of hydrogen from methane by methyl, plotted as a function of the compressive load applied to the carbon atoms (linear geometry). This function is based on an extended LEPS potential fitted to *ab initio* transition state parameters for an unloaded system.

this model, $\Delta \mathcal{V}^{\ddagger} = 0$ at $F_{compr} \approx 3.6$ nN, and $r(CC) \approx 0.242$ nm; this load is well within the achievable range (Section 8.3.3c). At ordinary temperatures, however, there is little practical reason to drive $\Delta \mathcal{V}^{\ddagger}$ to zero (indeed, with tunneling, there is little reason to do so even at cryogenic temperatures). The barrier is reduced to $2kT_{300}$ at a load of 2 nN, with $r(CC) \approx 0.251$ nm at the transition state. At a load of 1 nN, $\Delta \mathcal{V}^{\ddagger} \approx 29$ maJ, and $r(CC) \approx 0.258$ nm. Assuming a frequency factor of 10^{13} s^{-1}, this barrier would be consistent with an omitted-reaction probability $\leq 10^{-15}$ in a mechanochemical system with a transformation time of $\sim 4 \times 10^{-9}$ s (if, that is, the reaction were also sufficiently exoergic). These conditions are consistent with achieving high reaction reliability via the conditional-repetition mechanism (Section 8.3.4f).

*b. **Exoergic abstraction reactions.*** A reliable single-step reaction must meet stringent exoergicity requirements (Section 8.3.4e) at the time of kinetic decoupling. The energy differences in Eqs. (8.28) and 8.30) are consistent with $P_{err} < 10^{-15}$ without using cycles of repetition and measurement, while for Eq. (8.29), $P_{err} \approx 10^{-13}$:

$$H-C\equiv C\bullet \quad H-C\!\!\!\bigcirc\!\!\!\bigcirc \quad \longrightarrow \quad H-C\equiv C-H \quad \bullet C\!\!\!\bigcirc\!\!\!\bigcirc \quad + \sim 146 \text{ maJ} \qquad (8.28)$$

$$\bigcirc\!\!\!\bigcirc C\bullet \quad H-\overset{Me}{\underset{Me}{C}}-Me \quad \longrightarrow \quad \bigcirc\!\!\!\bigcirc C-H \quad \bullet\overset{Me}{\underset{Me}{C}}-Me \quad + \sim 123 \text{ maJ} \qquad (8.29)$$

$$\overset{Me}{\underset{Me}{Me-C\bullet}} \quad H-\overset{H}{C}\!\!\!\bigcirc \quad \longrightarrow \quad \overset{Me}{\underset{Me}{Me-C-H}} \quad \overset{H}{\bullet C}\!\!\!\bigcirc \quad + \sim 154 \text{ maJ} \qquad (8.30)$$

The energies in these reactions are computed from differences in C—H bond strengths (Kerr, 1990). See also Musgrave et al. (1992).

Figure 8.11 compares potential energy curves under compressive loads for the methyl-methane abstraction reaction (discussed previously) to a set of similar curves for a strongly exoergic reaction: with greater exoergicity, the barrier tends to disappear under a substantially smaller load.

*c. **Hydrogen abstraction tools.*** The large C—H bond strength of alkynes (~ 915 maJ) enables alkynyl radicals to abstract hydrogen atoms from most exposed nonalkyne sites with a single-step $P_{err} < 10^{-15}$. As shown in Figure 8.2, alkyne moieties can be buttressed by nonbonded contacts to increase their stiffness, greatly reducing the probability of reacting with sites as much as a bond length from the target (Section 8.3.3f). In a typical situation, a supporting structure need only have good stiffness in resisting displacements in one direction, and can permit larger excursions in directions that lack nearby reactive sites. Figure 8.3 illustrates an alkyne moiety that achieves greater steric exposure through selective buttressing.

Moieties with C—H bond energies between those of alkynes and aryls can abstract hydrogen atoms from most sp^3 carbon sites with good reliability. Strained alkenyl radicals such as **8.31** or **8.32** should bind hydrogen with the

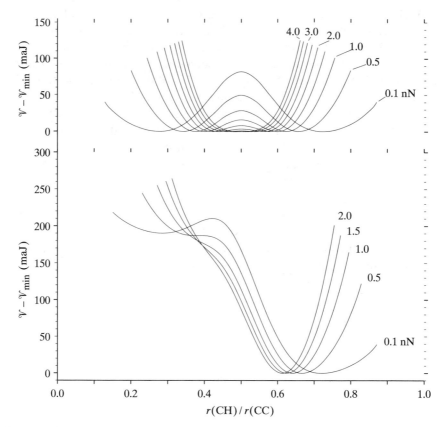

Figure 8.11. Potential energy surfaces for abstraction reactions under various compressive loads, plotted as a function of a reaction coordinate, the ratio of a CH distance to the (variable, optimized) CC distance. The curves in the upper panel are for the abstraction of hydrogen from methane by methyl, based on the extended LEPS model described in the text. The lower curves are representative of low-barrier exoergic processes (bond lengths and energies fit the abstraction of H from methane by ethynyl, but all Sato parameters are an arbitrary 0.15.)

necessary energy ($\geq$790 maJ, vs. ~737 maJ for ethenyl).

These structures can exhibit good stiffness without buttressing from nonbonded contacts. Their steric properties are desirable, and their supporting structures strongly suppress bond-cleavage instabilities that would otherwise be promoted by the radical site.

d. Hydrogen donation tools. Structures capable of resonant stabilization of a radical permit easy abstraction of a hydrogen; these include cyclopentadiene, shown in Eq. (8.31), and various other unsaturated systems. Some atoms bind hydrogen weakly, including tin and many other metals. Moieties like these, with C—H bond energies less than ~530 maJ, can donate hydrogen atoms reliably to carbon radicals where the product is an ordinary sp^3 structure. Section 8.6 accordingly assumes that the abstraction and donation of hydrogen atoms can be performed at will on diamondoid structures.

8.5.5. Alkene and alkyne radical additions

The addition of radicals to alkenes and alkynes often is exoergic by ~160 to 190 maJ; this is sufficient to ensure reaction reliability. Because radical addition converts a (long) nonbonded contact into a (short) bond, compressive loads can couple directly to the reaction coordinate, reducing the activation energy and driving the reaction in the forward direction. Activation energies for alkyl additions to unsaturated species are typically 30 to 60 maJ (Kerr, 1973). Since this barrier is lower than in the methyl-methane hydrogen abstraction reaction (Section 8.5.4), and since piezochemical effects are expected to be larger, modest loads (~1 nN) can be expected to reduce the energy barrier sufficiently to permit fast, reliable reactions with $t_{trans} = 10^{-7}$ s. A semiempirical study (using the MNDO method) of the addition of the phenyl radical to ethene yielded a substantial overestimate of the barrier height (~98 maJ) and a maximal repulsive force along the reaction coordinate of ~1.5 nN (Arnaud and Subra, 1982).

In these reactions, the radical created at the adjacent carbon destabilizes the newly formed C—C bond, permitting the addition reaction to be reversed. The energy of the destabilized bond (~160 to 190 maJ), together with the barrier on the path between bonding and dissociation (~30 to 60 maJ, above), makes many such bonds acceptable with respect to the stability requirements for reagents and reaction intermediates (barriers ≥230 maJ, Section 8.3.4g). Stability problems in other structures can often be remedied either by satisfying the radical (e.g., by hydrogen donation) or by mechanically stabilizing the newly added moiety using other bonds or steric interactions.

The energy dissipation caused by radical addition (and its inverse) depends on an interplay of stiffness and reaction PES like that examined in Section 8.5.3 for bond cleavage. Obtaining the necessary PES data would make an interesting *ab initio* study.

a. Stiffness and misreaction rates in radical addition. Radical addition reactions provide a convenient example of a process having two similar, closely spaced transition states. Structure **8.33** illustrates a transition state in the attack of an aryl radical on an alkene, with asterisks marking equivalent, alternative locations for the position of the attacking radical. Although transition states can be arbitrarily close together, the separation of the pair of locations on a single side of the alkene (moderately more than a bond length) is toward the low end of the range of anticipated typical cases, presenting a relatively challenging situation for achieving reliable positional control.

8.33

Quantum calculations can be used to estimate the transition-state geometry for the addition of a phenyl radical to ethene (Arnaud and Subra, 1982), yielding lengths of 0.135 nm for the breaking double bond and 0.225 nm for the bond undergoing formation, with an angle of 95.4° between them. This yields a distance between alternative positions for the attacking radical of 0.177 nm. (In the full configuration-space picture, the distance between transition states is increased by contributions from other displacements, including differences in the angle of the attacking radical.)

At this distance, an elastic energy difference of >180 maJ can be achieved with a stiffness of ~12 N/m; a stiffness in this range can effectively suppress misreactions (rates $\leq 10^{-15}$). Because the most probable misreactions result from a displacement in a particular direction, however, an advantage can be gained by biasing the equilibrium position of the attacking moiety in the opposite direction during the approach to the reaction site. For example, with a stiffness of 6 N/m, a 0.1 nm bias of this sort results in an elastic energy difference of ~200 maJ between the two transition state geometries, while raising the energy of the favored transition by only 25 maJ. Accordingly, stiffnesses in the 5 to 10 N/m range should prove adequate to suppress misreactions with high reliability in a wide range of synthetic operations.

8.5.6. Pi-bond torsion

Rotation of one of the methylene groups of ethene by 90° breaks the pi bond, yielding a diradical. Analogous structures and transformations can be used to modulate the strength of adjacent sigma bonds. The transition (a) → (b) in Figure 8.12 represents an R—C bond cleavage yielding a twisted alkene; (c) → (d) represents an R—C cleavage yielding a planar alkene. (Figure 8.13 illustrates a specific structure with good steric properties.) The differences in alkene energy around the illustrated cycle can be used to estimate the difference in R-group bond energies between (a) and (b). The (a) → (c) transition involves torsion of a single bond that links centers of roughly twofold and threefold symmetry; the resulting energy difference is usually small in the absence of substantial steric interference between the end groups. The difference in R-group bond energies between (a) and (c) thus approximates the energy difference between twisted and planar alkenes (b) and (d). The analogous energy difference is 453 maJ for ethene (Ichikawa et al., 1985). Alternatively, one can assume that the R-group bond energy in a structure like (a) is unaffected by the presence of an adjacent *twisted* radical (i.e., would be unchanged if the radical site were hydrogenated), then use the bond weakening caused by an *unconstrained* radical to estimate the energy difference; from thermochemical data (Kerr, 1990), this difference is ~410 to 440 maJ for hydrocarbon R-groups. Thus, the bond energy of a typical R-group in (a), ~550 to 650 maJ, drops dramatically when the structure is twisted to the configuration of (c). Indeed, elimination of a more weakly bonded group at this site can become exoergic.

A typical reaction cycle for a mechanism of this kind could involve (1) abstraction of a relatively tightly bound moiety by a radical site like that in (b), yielding a structure like (a); (2) torsion to a state like (c) weakening the new bond; and (3) abstraction of the moiety by another radical, delivering it to a more weakly bound (i.e., high-energy) site and leaving a structure like (d). If

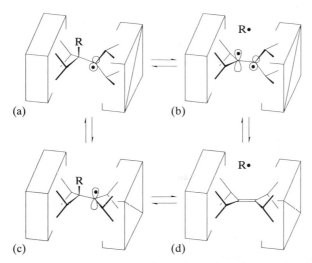

Figure 8.12. Modulation of bond strength by pi-bond torsion; see Section 8.5.6 for discussion.

steps 1 and 2 must be exoergic by 145 maJ, then the net increase in the energy of the transferred moiety can be ≥ 120 maJ. Moreover, by modulation of the torsion angle during (rather than between) reaction steps, the transition states for steps 1 and 2 can be approached forcibly under conditions that make the transitions endoergic or isoergic, with separation under exoergic conditions. This meets the conditions stated in Section 8.5.2b for a low-dissipation process, and the isolated twisted-alkene state (b) never occurs. Alternatively, using the conditional repetition approach (Section 8.3.4f) to remove exoergicity requirements, the net increase in the energy of the transferred atom can be >400 maJ.

Note that abstraction of a moiety by a radical to yield an alkene resembles radical coupling: it requires spin pairing, raising questions of intersystem crossing rates. As discussed in Section 8.4.4b, k_{isc} is usually adequate in the presence

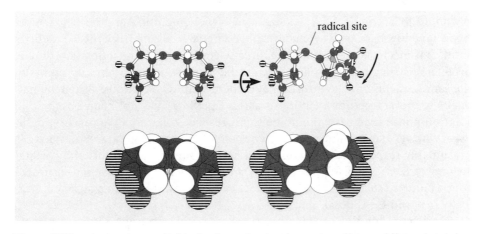

Figure 8.13. A structure suitable for imposing torsion on an alkene while maintaining good steric exposure at one of the carbon sites (pyramidalization of the radical site increases the change in energy).

of a suitably coupled high-Z atom. Bismuth ($Z = 83$), with its ability to form three (albeit weak) covalent bonds, is a candidate for inclusion at a nearby site in the supporting diamondoid structure.

Mechanochemical processes involving pi-bond torsion appear to have broad applications. For example, reactions of dienes (e.g., Diels–Alder and related reactions) can be accelerated by torsions that weaken the initial pair of double bonds. In the reverse direction, the sigma bonds resulting from the reaction can (for reactions yielding noncyclic products) later be cleaved with the aid of radicals generated by torsion of the new double bond. Some other systems having a single pi bond can likewise be manipulated by mechanical torsion; these include some metal–ligand and boron–nitrogen double bonds. In the latter instance, torsion results in a lone pair and an empty orbital, causing a shift in electron density.

8.5.7. Radical displacements

Various mechanochemical processes are analogous to pi-bond torsion in that they modulate the strength of a sigma bond by altering the availability of a radical able to form a competing pi bond. For example, the addition of a radical to an alkyne can facilitate an abstraction reaction as shown in Figure 8.14.

A process of this sort could employ a weakly bonding moiety (or one with modulable bonding, as in Section 8.5.6) as the attacking radical R, generating a strong alkynyl radical by tensile bond cleavage [step (d) → (e)]. A mechanochemical cycle based on these steps can first use an alkynyl radical to abstract a tightly bound atom from one location, then deliver the atom to a moderately bound site R′, and finally regenerate the original alkynyl radical by cleaving a weak bond to R. Given the strong weakening of the bond to hydrogen in (b), each of these steps should proceed rapidly and reliably under moderate mechanical loads. [Note that step (b) → (c) requires spin pairing.] Analogous displacement operations can be used to regenerate alkene- and aryl-derived sigma radicals (e.g., **8.19**, **8.31**, **8.32**).

8.5.8. Carbene additions and insertions

As noted in Section 8.4.5a, the standard carbene addition and insertion reactions have low barriers. The changes in geometry resulting from these reactions show that mechanical loads can be directly coupled to the reaction coordinate, producing strong piezochemical effects; the increase in bond number points to the same conclusion. The insertion of carbenes into C—C bonds, although analogous to metal insertion reactions such as Eq. (8.35), has not been observed in solution-phase chemistry; the ubiquitous presence of alternative reaction pathways with less steric hindrance, lower energy barriers, or greater exoergicity is presumably responsible, since this reaction is exoergic and permitted by orbital symmetry. Section 8.6.4c presents a reaction step which assumes that positional control and mechanical forces can effect carbene insertion into a strained, sterically exposed C—C bond.

Since bond-forming carbene reactions are highly exoergic, single-step reactions can be highly reliable (assuming, as always, that conditions and mechanical constraints are chosen to exclude access to transition states leading to unwanted products). Fast, reliable reactions involving triplet carbenes can require fast

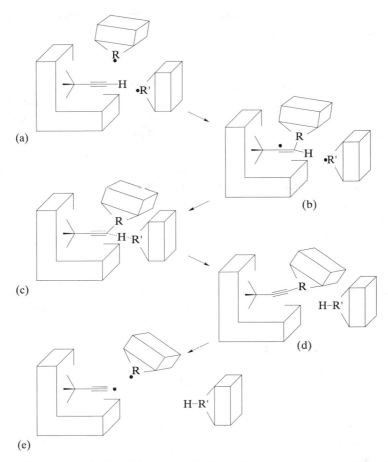

Figure 8.14. Removal of a tightly bound alkynyl hydrogen atom, facilitated by the addition of a radical. A supporting structure provides nonbonded contacts to one side of the alkyne; this enables the application of force to accelerate the radical addition step, (a) → (b). See Section 8.5.7 for discussion.

intersystem crossing (Section 8.4.5). The potential for low-dissipation carbene reactions is at present unclear.

8.5.9. Alkene and alkyne cycloadditions

°Cycloaddition reactions can find broad use in mechanosynthesis, as they have in solution-phase synthesis. The [4 + 2] Diels–Alder reactions [e.g., Eq. (8.15)] form two bonds and a ring simultaneously, and, as has been mentioned (Section 8.4.3), they have large negative volumes of activation at moderate pressures ($\sim$−0.05 to −0.07 nm^3). These reactions are more sensitive to piezochemical effects than are abstraction reactions (Section 8.5.4). Energy barriers vary widely, from <60 to >130 maJ. Reaction (8.15) proceeds at low rates at ordinary temperatures and pressures; piezochemical rate accelerations under modest loads should make them consistent with $t_{trans} = 10^{-7}$ s. Under these conditions, alkenes can be replaced with less-reactive alkynes, yielding less saturated products. Exoergicity [$\sim$280 maJ for Eq. (8.15)] is sufficient to yield stable intermediate products (Section 8.3.4g) in a reliable process.

Cycloaddition reactions are subject to strong orbital-symmetry effects. The [2 + 2] cycloaddition reaction

$$\| \quad \| \quad \longrightarrow \quad \square \qquad\qquad (8.31)$$

is termed *thermally forbidden* (according to the Woodward–Hoffmann rules), because the highest-energy occupied orbital on one molecule fails to mesh properly with the lowest-energy unoccupied orbital on the other (the unoccupied orbital presents two lobes of opposite sign, divided by a node; the occupied orbital has no such node, and creating the required node is akin to breaking a bond). But, as with well-observed "forbidden" spectral lines, the ban is not complete. Piezochemical techniques (including pi-bond torsion and direct compression) can force the formation of one of the new sigma bonds, even at the expense of breaking both pi bonds, thereby creating a pair of radicals that can then combine to form the second sigma bond:

$$\| \quad \| \quad \longrightarrow \quad \underset{\cdot\;\cdot}{\bigcup} \quad \longrightarrow \quad \square \qquad\qquad (8.32)$$

This reaction can also proceed through a polar, asymmetric intermediate, which is likewise subject to strong piezochemical effects. For the dimerization of ethene, the activation energy is ~305 maJ (Huisgen, 1977).

8.5.10. Transition-metal reactions

Section 8.4.6 surveys some of the general advantages of transition-metal compounds as intermediates in mechanosynthetic processes. The present section describes some reactions and mechanochemical issues in more detail.

*a. **Reactions involving transition metals, carbon, and hydrogen.*** Transition metals participate in a wide variety of reactions, including many that make or break carbon-carbon and carbon-hydrogen bonds:

$$L_nM\cdots\underset{H}{\overset{H}{|}} \;\rightleftharpoons\; L_nM\overset{H}{\underset{H}{\big\backslash}} \;\rightleftharpoons\; L_nM\overset{H}{\underset{H}{\diagup}} \qquad\qquad (8.33)$$

$$L_nM\cdots H\underset{R}{\diagdown} \;\rightleftharpoons\; L_nM\overset{H}{\underset{R}{\big\backslash}} \;\rightleftharpoons\; L_nM\overset{H}{\underset{R}{\diagup}} \qquad\qquad (8.34)$$

$$L_nM\;\underset{R'}{\overset{R}{|}} \;\rightleftharpoons\; L_nM\overset{R}{\underset{R'}{\big\backslash}} \;\rightleftharpoons\; L_nM\overset{R}{\underset{R'}{\diagup}} \qquad\qquad (8.35)$$

All illustrated states and transformations in each sequence have been observed, though not necessarily in a single complex (Crabtree, 1987).

The complexes in Eqs. (8.33)–(8.35) have single bonds between metal and carbon, but double-bonded species (metal carbene complexes and alkylidenes **8.34**) and triple-bonded species (metal alkylidynes **8.35**) are also known (Nugent and Mayer, 1988).

$$L_nM{=}C\overset{\diagup R}{\underset{\diagdown R}{}} \qquad\qquad L_nM{\equiv}C{-}R$$
$$\textbf{8.34} \qquad\qquad\qquad\qquad \textbf{8.35}$$

$$L_nM{=}C\text{:} \qquad\qquad L_nM{\equiv}C\bullet$$
$$\textbf{8.36} \qquad\qquad\qquad\qquad \textbf{8.37}$$

Species such as metal alkylidene carbenes, **8.36**, and metal alkylidyne radicals, **8.37**, can presumably exist (given stable ligands and a suitable eutactic environment) and may be of considerable use in synthesis. (Note that high-Z transition metals can accelerate intersystem crossing in reactions in which they participate.)

Among the reactions of metal carbene complexes are the following (Dötz et al., 1983; Hehre et al., 1986; Masters, 1981):

$$L_nM=C \quad \rightleftharpoons \quad L_nM-C^- \quad \rightleftharpoons \quad L_nM \qquad \qquad (8.36)$$

$$
\begin{array}{c}
\text{R'} \quad \text{R''} \\
\| \quad \| \\
L_nM \quad \text{R}
\end{array}
\rightleftharpoons
\begin{array}{c}
\text{R'--R''} \\
| \\
L_nM--R
\end{array}
\rightleftharpoons
\begin{array}{c}
\text{R'=R''} \\
\\
L_nM=R
\end{array}
\qquad (8.37)
$$

(Again, all illustrated states and transformations in each sequence have been observed, though not necessarily in a single complex.) Transition metals can also serve as radical leaving groups in S_H2 reactions at sp^3 carbon atoms, transferring alkyl groups to radicals (Johnson, 1983).

b. Ligands suitable for mechanochemistry. The reactivity of a transition metal atom is strongly affected by the electronic and steric properties of its ligands. These can modify the charge on the metal, the electron densities and energies of various orbitals, and the room available for a new ligand. A ligand can be displaced by a new ligand, or can react with it and dissociate to form a product.

In a molecular manufacturing context, reagent stability is commonly adequate if all components are bound with energies ≥ 230 maJ (Section 8.3.4). Typical M—C bond strengths (Crabtree, 1987) are ~210 to 450 maJ (vs. ~550 maJ for typical C—C bonds); the stronger bonds have adequate stability by themselves, and weaker bonds are acceptable if incorporated into a stabilizing cyclic structure. Stabilization by cyclic structures will often be necessary to prevent unwanted rearrangements in the ligand shell. Since M—H bonds are estimated to have strengths ~100 to 200 maJ greater than M—C bonds (Crabtree, 1987), their strengths should be adequate in the absence of special destabilizing circumstances. Electronegative ligands such as F and Cl should also be relatively stable, particularly in structures that lack adjacent sites resembling anion solvation shells in polar solvents.

The stability and manipulability of ligand structures in a mechanochemical context are greatest when they are bound to strong supporting structures. This is easily arranged for a wide variety of ligands having metal-bonded carbon, nitrogen, oxygen, phosphorus, or sulfur atoms. Of these, phosphorus (in the form of tertiary phosphines, PR_3) has been of particular importance in transition-metal chemistry. Carbon monoxide, another common ligand, lacks an attachment point for such a handle and may accordingly be of limited use. Isonitriles, however, are formally isoelectronic to CO at the coordinating carbon atom; they exhibit broadly similar chemistry (Candlin et al., 1968) and can be attached to rigid, extended R groups (e.g., **8.42**). (Many ligands not mentioned here also have useful properties.)

Ligand supporting structures can maintain substantial strength and stiffness while occupying a reasonably small solid angle. This enables several ligands to be subjected to simultaneous, independent mechanochemical manipulation. The following structures provide one family of examples:

The MM2/CSC stiffness of the following diamondoid support structure (shown in two views):

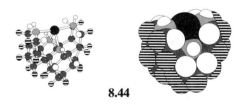

is ~115 N/m for extension and ~20 N/m for bending, both measured for displacements of the carbon atom at the tip relative to the bounding hydrogen atoms at the base.

Alternatively, several ligand moieties can be integral parts of a diamondoid structure, as in the following two views of a bound metal atom with two of six octahedral coordination sites exposed:

The metal-nitrogen bond lengths shown in this structure are appropriate for octahedrally coordinated cobalt.

Single ligands mounted on independently manipulable tips represent one extreme of mobility; multiple ligand moieties in a single rigid structure represent another. An intermediate class incorporates several ligand moieties into a single structure, subjecting them to substantial relative motion by elastic deformation of that structure. This can ensure large interligand stiffnesses, facilitating low-dissipation processes.

c. Mechanically driven processes. Low-stiffness, low-strength bonds are more readily subject to mechanochemical manipulation: both the required force and (for low-dissipation processes) the required stiffness of the surrounding structure are smaller. The stretching frequencies of M—H bonds (Crabtree, 1987) imply stiffnesses between 130 and 225 N/m, ~0.25 to 0.50 times the stiffness of a typical C—H or C—C bond; the stiffness of M—C bonds (which are longer and of lower energy) should likewise be low. Given that bonds as stiff as C—C can be cleaved in a low-dissipation process (Section 8.5.3), a wide range of mechanochemical processes involving transition metals can presumably be carried out in a positive-stiffness, zero-barrier manner (or, with similar effect, in a manner encountering only small regions of negative stiffness, and hence only low barriers).

In solution-phase chemistry, catalysts capable of inserting metals into °alkane C—H bonds, Eq. (8.35), have been unstable, either attacking their ligands or the solvent, and insertion into C—C bonds [Eq. (8.36)] has required a strained reagent (Crabtree, 1987). With an expanded choice of ligands and the elimination of accessible solvent molecules, the first problem should be avoidable. Further, when loads of bond-breaking magnitude can be applied between a transition-metal atom and a potential reagent, intrinsic strain is presumably no longer required in the reagent. Configurations like **8.44** seem well suited for insertion into a sterically exposed bond. Since metal insertion in the above instances has the effect of replacing a strong, stiff bond with two weaker, more compliant bonds, it can facilitate further mechanochemical operations.

Transition-metal complexes with large coordination numbers lend themselves to bond modulation based on manipulation of steric crowding. In octahedral complexes, for example, the metal atom can be anchored by (say) three ligands while two of the remaining ligands are rotated or displaced to modulate overlap repulsion on the sixth. By analogy with processes observed in solution, the introduction of new ligands can be used to expel existing ligands by a combination of steric and electronic effects. Mechanochemical processes can forcibly introduce ligands almost regardless of their chemical affinity for the metal, driving the expulsion of other, relatively tightly bound ligands. Conversely, such processes can remove ligands (having suitable "handles"), even when they are themselves tightly bound. Again, the presence of multiple other ligands to anchor the metal atom facilitates such manipulations.

More subtly (and conventionally), the binding of a ligand in a square or octahedral complex can be strongly affected by the electronic properties of the ligand on the opposite side (the *trans*-effect); changes in this *trans* ligand can alter reaction rates by a factor of ~10^4, suggesting changes in transition-state energy of ~40 maJ (Masters, 1981). Accordingly, mechanical substitution or other alteration of ligands (e.g., double-bond torsions) should be effective in modulating bonding at *trans* sites in mechanochemical reaction cycles.

In summary, although transition metals are of only moderate interest as parts of nanomechanical products, they are of considerable interest as components of tools in mechanochemical systems for building those products. Their comparatively soft interactions, relaxed electronic constraints, and numerous manipulable degrees of freedom suit them for the preparation and recycling of reagent moieties, and (where steric constraints can be met) for direct use as reagents in product synthesis. In light of the broad capabilities of other reagents under mechanochemical conditions, the use of transition-metal reagents is unlikely to expand the range of structures that can be built; it is, however, likely to expand greatly the range of structures that can be built with low dissipation, in a nearly thermodynamically reversible fashion.

8.6. Mechanosynthesis of diamondoid structures

Fundamental physical considerations (strength, stiffness, feature size) favor the widespread use of diamondoid structures in nanomechanical systems. In chemical terms, diamondoid structures comprise a wide range of polycyclic organic molecules consisting of fused, conformationally rigid cages. This section considers the synthesis of such structures by mechanochemical means, based on reagents and processes of sorts described in the preceding section, and using diamond itself as an example of a target for synthesis.

8.6.1. Why examine the synthesis of diamond?

Diamond is an important product in its own right, but here serves chiefly as a test case in exploring the feasibility of more general synthesis capabilities. It is impractical at present to examine in detail the synthesis of numerous large-scale structures. Accordingly, it is important to choose a few appropriately challenging models.

Diamond has several advantages in this regard, as can be seen by a series of comparisons. Synthetic challenges often center around the framework of a molecule, and diamond is pure framework. In general, higher °valence and participation in more rings makes an atom more difficult to bond correctly. At one extreme is hydrogen placement on a surface; at the other is the formation of multiple rings through tetravalent atoms. (Divalent and trivalent atoms such as oxygen and nitrogen are intermediate cases.) Solid silicon and germanium present the same topological challenges as diamond, but atoms lower in the periodic table are more readily subject to mechanochemical manipulation owing to their larger sizes and lower bond strengths and stiffnesses. Thus, a structure built entirely of rings of sp^3 carbon atoms appears to maximize the basic challenges of bond formation, and diamond is such a structure. Further, diamond has the highest atom and covalent-bond density of any well-characterized material at ordinary pressures, maximizing problems of steric congestion. Although diamond is a relatively low-energy structure, lacking significant strain or unusual bonds, existing achievements (Section 8.2.3) suggest that the latter features need not be barriers to synthesis, even in solution-phase processes.

Finally, diamond is a simple and regular example of a diamondoid structure. Accordingly, the description of a small synthetic cycle can suffice to describe the synthesis of an indefinitely large object; this avoids the dilemma of choosing between (1) syntheses too complex to describe in the available space, and (2) syntheses that might in some way be limited to small structures.

8.6.2. Why examine multiple synthesis strategies?

The identification of several distinct ways to synthesize a particular structure suggests that ways can be found to synthesize different but similar structures. Identifying multiple syntheses for diamond provides this sort of evidence

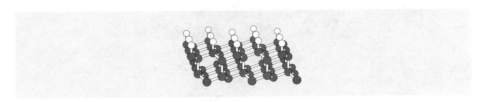

Figure 8.15. A hydrogen-terminated diamond (111) surface.

regarding the synthesis of other diamondoids. Further, diamond itself has several sterically and electronically distinct surfaces on which construction can proceed, and sites on these surfaces can serve as models for the diverse local structures arising in the synthesis of less regular diamondoids. Accordingly, the following sections survey several quite different techniques, not to buttress the case for diamond synthesis (a process already known in the laboratory), but to explore the capabilities of mechanosynthesis for building broadly diamondlike structures by using diamond as an example.

8.6.3. Diamond surfaces

Corners and other exposed sites pose fewer steric problems than sites at steps in the middle of planar surfaces. Section 8.6.4 considers a set of reaction cycles at such sites on low-index diamond surfaces; the present section introduces the surfaces themselves. (In the diagrams here and in the following section, structures are truncated without indicating bonds to missing atoms.)

a.   The (111) surface.   When prepared by standard grinding procedures, a closely packed diamond (111) surface is hydrogen terminated (Pate, 1986), Figure 8.15. When heated sufficiently to remove most of the hydrogen (1200 to 1300 K), this surface reconstructs into a structure with (2 × 1) symmetry (Hamza et al., 1988); the Pandey pi-bonded chain model for this surface, Figure 8.16(b), has received considerable support (Kubiak and Kolasinsky, 1989; Vanderbilt and Louie, 1985). One calculation (Vanderbilt and Louie, 1985) yielded an exoergicity of ~50 maJ per surface atom for this reconstruction, driven by the conversion of radical electrons to bonding electrons (but offset by strain energy).

It is not presently known whether a bare diamond (111) surface, Figure 8.16(a), spontaneously transforms to the (2 × 1) structure at room temperature. Calculations for the similar reconstruction of a silicon (111) surface (Northrup and Cohen, 1982) suggest an energy barrier of ≤5 maJ per surface atom. This energy barrier, however, cannot be equated with an energy barrier for the

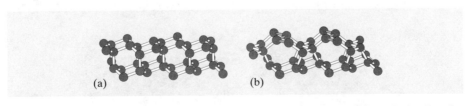

(a)   (b)

Figure 8.16. A bare, unreconstructed diamond (111) surface (a) and the Pandey (2 × 1) reconstruction (b).

Figure 8.17.   An unreconstructed diamond (110) surface.

nucleation of a (2×1) domain on an unreconstructed (1×1) surface, since nucleation requires the simultaneous motion of a number of atoms and stores deformation energy in the domain boundary. Whatever the true situation, since hydrogenation is known to suppress the (2×1) reconstruction, and since donating and abstracting hydrogen atoms from (111) surface sites is straightforward, an unreconstructed (111) surface can be maintained during construction.

b.   The (110) surface.   Termination of the bulk structure leads to a surface like that shown in Figure 8.17. The bonded chains along the surface resemble those formed in the (2×1) reconstruction of a (111) surface, but with the pi bonds subject to greater torsion and pyramidalization. Although this geometry generates no radicals, the pi bonds are quite weak, and hence the *p*-orbital electrons can exhibit radical-like reactivity (Section 8.4.3).

c.   The (100) surface.   Simple truncation of the bulk structure would yield a surface like that shown in Figure 8.18(a), covered with carbene sites. Displacement of surface atoms to form pi bonded pairs yields the stable reconstruction shown in Figure 8.18(b) (Verwoerd, 1981). The resulting surface alkene moieties are strongly pyramidalized and under substantial tension, increasing their reactivity.

d.   Some chemical observations.   The diversity of surface structures possible on a single bulk structure, diamond, shows how a diamondoid structure (like most complex molecules) can be built up through intermediates having widely varying chemical properties. Further, the differing strained-alkene moieties found on (110) and (100) surfaces show that stable diamondoid intermediates can have highly reactive surfaces, facilitating synthetic operations. Finally, the (possible) instability of a bare, radical-dense, unreconstructed (111) surface illustrates how requirements for temporary, stabilizing additions (such as bond-terminating hydrogens) can arise.

8.6.4.   Stepwise synthesis processes

Section 8.6.5 describes synthesis strategies that exploit the regular structure of diamond by laying down reactive molecular strands. Synthesis based on the mechanical placement of small molecular fragments, in contrast, suggests how

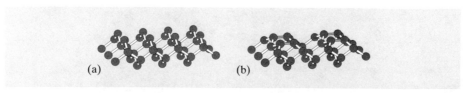

(a)   (b)

Figure 8.18.   A diamond (100) surface, without (a) and with (b) reconstruction.

specific *irregular* structures might be synthesized, and thus provides a point of departure for considering the synthesis of a broad class of diamondoids.

a.   Existing models of diamond synthesis.   Models of synthesis via small molecular fragments have been developed to explain the low-pressure synthesis of diamond under nonequilibrium conditions in a high-temperature hydrocarbon gas, a process of increasing technological importance. Two models have been advanced and subjected to studies using semiempirical quantum mechanics. Both propose mechanisms for the growth of diamond on (111) surfaces, one based on a cationic process involving methyl groups (Tsuda et al., 1986) and one based on the addition of ethyne (Huang et al., 1988). Of these, the ethyne process appears more plausible and more directly relevant to feasible mechanosynthetic processes (regardless of its frequency during diamond growth from the gas phase).

Figure 8.19 (based on illustrations in Huang et al., 1988) shows the addition of two ethyne molecules to a hydrocarbon molecule which, with suitable positional constraints, was used to model a step on a diamond (111) surface. The overall set of calculations used MNDO and consumed ~35 hours of CPU time on a Cray XMP/48. The reaction mechanism originally proposed (Frenklach and Spear, 1988) for the transformations $2 \rightarrow 4$ and $4 \rightarrow 6$ involved multiple steps, later shown to be concerted (steps 3 and 5). The initiation step involves abstraction of a hydrogen atom by atomic hydrogen ($1 \rightarrow 2$; process not shown), and has a significant energy barrier. Steps 3 and 5 were calculated to proceed without energy barriers, suggesting that any energy barriers that actually occur are unlikely to be large.

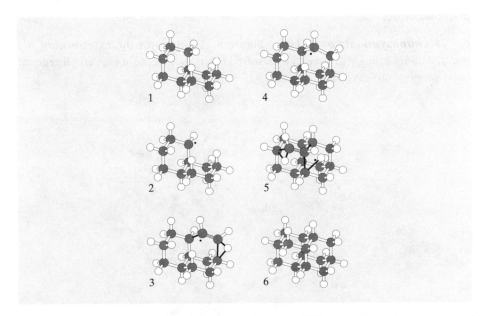

Figure 8.19.   A sequence of reaction steps for the addition of ethyne to a compound serving as a model of a step on a hydrogen-terminated diamond (111) surface. Redrawn from Huang et al. (1988); see Section 8.6.4a. Bonds in the process of breakage and formation are shown in black; radical sites are marked with dots.

In an analogous mechanosynthetic process, ethyne could be replaced by an alkyne moiety bonded to a structure serving as a handle. Step 5 liberates a free hydrogen atom, which is unacceptable in a eutactic environment, but a related approach (Section 8.6.4b) avoids this loss of control.

b.   Mechanosynthesis on (111).   Step 1 of Figure 8.20 illustrates a kink site in a step on a hydrogenated diamond (111) surface; one bond results from a reconstruction, and two hydrogen atoms have been removed to prepare radical sites. In step 2 of the proposed synthetic cycle, an alkyne reagent moiety such as **8.45**

$$-Si \equiv H$$

8.45

is applied to one of the radical sites, resulting in a transition structure analogous to step 3 of Figure 8.19. The chief differences are that insertion occurs into a strained C—C bond rather than an unstrained C—H bond, and that large mechanical forces can (optionally) be applied to drive the insertion process. In step 3, tensile bond cleavage occurs (use of Si or another non-first-row atom reduces the mechanical strength of this bond), and a bond forms to the remaining prepared radical. Deposition of a hydrogen at the newly generated radical site and abstraction of two hydrogens further along the surface step then completes the cycle (not shown), generating a kink site like that in step 1, but with the diamond lattice extended by two atoms. Save for accommodating boundary conditions at the edge of a finite (111) surface, this sequence of operations (like those in the following sections) suffices to build an indefinitely large volume of diamond lattice.

c.   Mechanosynthesis on (110).   Figure 8.21 illustrates the extension of a pi bonded chain along a groove in a diamond (110) surface, using an alkylidenecarbene reagent moiety such as **8.46**.

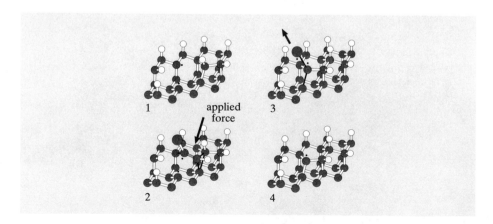

Figure 8.20.   A sequence of reaction steps for the addition of a pair of carbon atoms from an alkynyl moiety to a kink site in a step on a hydrogen-terminated diamond (111) surface (see Section 8.6.4b).

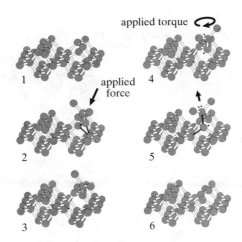

Figure 8.21. A sequence of reaction steps for the addition of a carbon atom from an alkylidenecarbene moiety to a pi-bonded chain on a diamond (110) surface (see Section 8.6.4c).

8.46

Step 1 illustrates the starting state; step 2 illustrates a transitional state in the insertion of the carbene into a strained C—C bond. This process takes advantage of mechanical constraints to prevent the addition of the carbene into the adjacent double bond to form a cyclopropane moiety. Instead, electron density is accepted from the terminal atom of the pi bonded chain into the empty *p*-orbital of the carbene carbon, developing one of the desired bonds, while electron density is donated from the sigma orbital of the carbene carbon to form the other desired bond (shown in step 3). Substantial forces (≥ 4 nN; Section 8.3.3c) can be applied to drive this process, limited chiefly by mechanical instabilities.

Step 4 illustrates the use of torsion to break a pi bond, thereby facilitating tensile bond cleavage, step 5. The final state, step 6, resembles that in step 1 save for the extension of the lattice by one atom. A further cycle (restoring the state of step 1 exactly, save for the addition of two atoms) would be almost identical, except that the equivalent of step 5 involves attack by a newly forming radical on a weak pi bond, rather than its combination with an existing radical.

d. ***Mechanosynthesis on (100).*** Figure 8.22 illustrates a synthetic cycle on diamond (100) in which reactions occur on a series of relatively independent rows of pairs of dimers. In step 1 of Figure 8.22, a strained cycloalkyne reagent moiety such as **8.47**

8.47

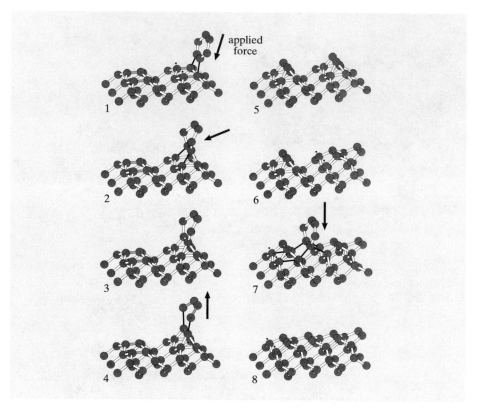

Figure 8.22. A sequence of reaction steps for the addition of pairs of carbon atoms from a strained alkyne moiety to a row of dimers on a diamond (100) surface (see Section 8.6.4d).

is applied. (The division of the supporting structure into blocks is intended to suggest opportunities for modulating bond strength by control of torsional deformations; alternative structures could serve the same role, providing weak bonds for later cleavage.) The reaction in step 1 (promoted by the nearly diradical character of the strained alkyne and by applied mechanical force) is formally a [4 + 2] cycloaddition process, provided that the resulting pair of radicals is regarded as forming a highly elongated pi bond. Step 2 is then formally a (thermally forbidden) [2 + 2] cycloaddition process, but the small energy difference between the bonding and antibonding orbitals in the "pi bond" provides a low-lying orbital of the correct symmetry for bonding, hence the forbiddenness should be weak. Moderate mechanical loads should suffice to overcome the associated energy barrier. Step 4 is an endoergic retro-Diels–Alder reaction yielding a high-energy, highly pyramidalized alkene moiety (which is, however, less pyramidalized than cubene; Section 8.4.3). The energy for this process is supplied by mechanical work.

Step 5 represents the state of the row after bridging dimers have been deposited at all sites. The resulting three-membered rings are analogous to epoxide structures in a model of an oxidized diamond (100) (2×1) half-monolayer surface (Badziag and Verwoerd, 1987).

Steps 6–8 represent a cycle in which dimers are sequentially inserted along the row: In step 6, a dimer has already been added to the right, cleaving the strained rings and generating two radical sites adjacent to a cleft in the surface. Step 7 illustrates the bonds that undergo formation and cleavage as a result of the mechanical insertion of a strained alkyne into the cleft. The nature of the transition state will depend on the spin state of the radical pair and orbital symmetry considerations. The large exoergicity for this process can be seen from a comparison of the bonds lost (two pi bonds in the strained alkene and two strained sigma bonds in the three-membered rings) with the bonds formed (four almost strain-free sigma bonds). The geometry of the cleft permits the application of large loads without mechanical instability, hence large energy barriers to bond formation (in the unloaded state) are acceptable. Afterward, the transition to step 8 (equivalent to step 6, but displaced) is achieved by a retro-Diels–Alder reaction like that in step 4.

8.6.5.   Strand deposition processes

Stepwise synthetic processes like those suggested in Section 8.6.4 need not be applied in regular cycles to build up a regular structure, but could instead be orchestrated to build up diamondoid structures tailored for specific purposes. Where diamond itself is the target, synthesis can take direct advantage of structural regularities.

*a.   **Cumulene strands.***   Cumulene strands, **8.48**, are high-energy, pure-carbon structures that represent promising precursors in the mechanosynthesis of diamond.

$$\left[-C=C= \right]_n$$
8.48

Figure 8.23 represents a pair of similar reaction processes on a dehydrogenated step on a diamond (111) surface and on a groove in a (110) surface: in each, a pi-bonded chain is extended by the formation of additional bonds to a cumulene strand in an exoergic, largely self-aligning process. Substantial forces can be applied through nonbonded interactions (which can also be used to constrain strand motions). The reaction on a hydrogenated (111) surface may

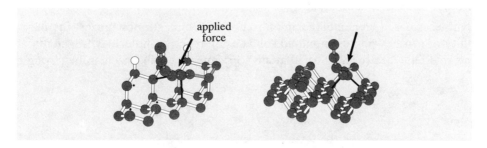

Figure 8.23.   Reaction processes bonding cumulene strands to (left) a dehydrogenated step on a hydrogenated diamond (111) surface and (right) a groove on a diamond (110) surface (see Section 8.6.5a).

require a series of stepwise reactions both to dehydrogenate atoms in the plane to be covered and to hydrogenate new atoms in the plane being constructed; the (110) reaction has no such requirement.

b. *Hexagonal diamond from hexagonal strands.*   The unsaturated, hexagonal, columnar structure **8.49** can be regarded as a tightly rolled tube of graphite; it can made from a °saturated structure by abstraction of all hydrogens. Like a cumulene strand, **8.48**, this is a pure-carbon structure; owing to pyramidalization and torsion of pi systems, it is also relatively high in energy.

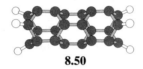

8.49

A semiempirical quantum chemistry study using AM1 on the model structure **8.50** (R. Merkle)

8.50

yielded bond lengths of ~0.1384 nm for the three central, axially aligned bonds, and ~0.1510 nm for the twelve adjacent bonds. These values are close to standard values for pure double and pure single bonds (0.1337 and 0.1541 nm), hence structure **8.49**, representing the hexagonal column as a network of strained alkenes, provides a good description.

Figure 8.24 represents a (100) surface of hexagonal diamond, bounded by similar strained alkenes. Figure 8.25 illustrates a reaction in which a hexagonal column bonds to a groove adjacent to a step on that surface; this can be regarded as proceeding by the attack of a strand radical on a surface alkene, generating a surface radical, which then attacks the next strand alkene, and so forth. Thus, each row of alkenes undergoes a process directly analogous to the free-radical chain polymerization that yields polyethylene, save for the greater reactivity of the participating alkenes and the presence of mechanical forces tending to force each radical addition.

8.6.6. Cluster-based strategies

Syntheses based on cumulene and hexagonal-column strands suggest the feasibility of synthesizing diamondoid solids using reactive molecular fragments of intermediate size (e.g., 10 to 30 atoms). Fragments of this size can be strongly

Figure 8.24.   A (100) surface of hexagonal diamond.

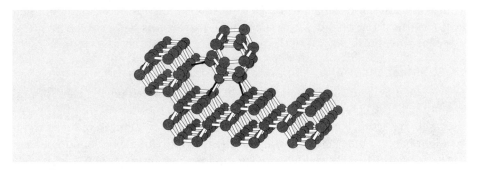

Figure 8.25. A reaction process bonding a tube formed of alkenes to a (100) surface of hexagonal diamond (see Section 8.6.5b).

convex, relaxing steric constraints in their synthesis. Containing tens of atoms, they can embody significant structural complexity and deliver that preformed complexity to a workpiece. By incorporating unsaturated structures, radicals, carbenes, and the like, they can form dense arrays of bonds to a complementary surface. Fullerenes are examples of unsaturated carbon clusters, and C_{60} has been crushed to diamond at room temperature by anisotropic compressive loads of ~20 GPa (Regueiro et al., 1992), suggesting the feasibility of mechanosynthesis of diamond and diamondoid structures from fullerene precursors.

This approach is a form of the familiar chemical strategy of convergent synthesis. Further, the contemplated size range of these fragments is familiar in organic synthesis today; their relatively high reactivity could be achieved (starting with more conventional structures) by a series of mechanochemically guided abstraction reactions in the protection of a eutactic environment.

8.6.7. Toward less diamondlike diamondoids

Members of the broad class of diamondoids can differ from diamond both in patterns of bonding and in elemental composition. The differing bonding patterns created during intermediate stages of the diamond syntheses proposed in this section, together with general experience in chemistry, suggests that nondiamond structures are readily accessible. Deviations from the diamond pattern usually reduce the overall number density of atoms, thereby reducing steric congestion and (all else being equal) facilitating synthesis.

Regarding differences in elemental composition, it is significant that the classes of reagent species exploited in this section (unsaturated hydrocarbons, carbon radicals, and carbenes) have analogues among other chemical elements. For example, most other elements of structural interest (N, O, Si, P, S), all can form double (and sometimes triple) bonds [e.g., C=N, C≡N, C=O, C=S; and the more exotic examples Si=Si and C=Si (Raabe and Michl, 1989), and C=P and C≡P (Corbridge, 1990)]. All these elements can host radical sites. Silicon can form a carbenelike divalent species, silene (Raabe and Michl, 1989). Fluorine, chlorine, and bromine can, like hydrogen, participate in abstraction reactions. The focus on hydrocarbon structures in this chapter has been driven more by limits on time and page space than by limits on chemistry.

The diversity of feasible syntheses for a challenging test-case, diamond, suggests that most reasonably stable diamondoid structures will prove susceptible to mechanosynthesis. Part II of this volume proceeds on this assumption.

8.6.8. Mechanosynthesis of nondiamondoid structures

As we move away from diamondoids, while remaining within the class of reasonably stable structures, synthesis appears to grow easier for reasons of the sort discussed in Section 8.6.1. Within the class of covalent structures, reduced rigidity might hamper mechanosynthesis, but the feasibility of using bound intermediates (using either covalent bonds or nonbonded interactions) can largely compensate. Further, the synthesis of flexible covalent structures is in the mainstream of existing chemical achievement.

Noncovalent solids present different challenges. In general, however, the ability to transfer atoms one at a time or in small clusters, and to perform piezochemical manipulation on a growing surface, can provide broad control. Accordingly, it seems reasonable to assume that most reasonably stable structures—diamondoid or not—will prove susceptible to mechanosynthesis. Aside from considering the use of metallic conductors (Section 11.6.1), the balance of this volume will neglect noncovalent structures.

8.7. Conclusions

The achievements of solution-phase synthesis show that a remarkably wide range of molecular structures can be built, despite the absence of the standard basis for construction, that is, the ability to move and position parts to guide assembly. A comparison reveals several limitations of mechanosynthesis relative to solution-phase synthesis, but displays a more-than-compensating set of strengths. Chief among these strengths are (1) the ability to achieve reactions at specific sites while avoiding them elsewhere by exploiting direct positional control of reagent moieties, (2) as a consequence of this, the ability to apply highly reactive moieties, such as radicals and carbenes, with great specificity, and (3) the ability to accelerate reactions by the application of localized forces of bond-breaking magnitude. Misreaction probabilities are exponentially dependent on relative energies, and with reasonable barrier heights can be $<10^{-15}$; omitted-reaction probabilities are exponentially dependent on repetition and relative well depths, and can likewise be low. It appears that the required conditions for high reliability do not excessively constrain the set of feasible synthetic operations. As a consequence, precise structures having $>10^{12}$ atoms can be made in good yield.

A review of reactive species used in solution-phase chemistry suggests that unsaturated hydrocarbons, carbon radicals, carbenes, and transition-metal complexes (among others) are attractive in a mechanosynthetic context. A consideration of the utility of these species when used in conjunction with positional control and mechanical force indicates the feasibility of a wide range of useful transformations, some of which can be performed with energy dissipation $<kT$.

Finally, the mechanosynthetic production of diamondoid structures has been examined, using several proposed syntheses of diamond as an example. The diversity of feasible reactions (and the availability of analogous reactions involving elements other than C and H) indicates the feasibility of constructing a wide range of diamondoid and other structures. Accordingly, Part II of this volume proceeds on the assumption that a mature mechanosynthetic technology can manufacture most reasonably stable covalent structures.

Some open problems. Mechanosynthesis presents numerous open theoretical problems. The analysis in this chapter indicates the feasibility of a wide range of processes, but it describes only a few in substantial structural and energetic detail. The broader application of *ab initio* molecular orbital methods to potential mechanosynthetic reactions is a high priority. Research is needed to identify and characterize reaction pathways in mechanosynthesis of diamond and diamondoid structures (including the verification, modification, or rejection of the processes suggested in Section 8.6). Similar studies are needed to describe steps in reagent preparation. A major challenge in this work will be the identification and characterization of potential misreaction pathways to determine whether the proposed reaction can be made reliable enough for molecular manufacturing. Semiempirical molecular orbital methods can assist in the exploration of relevant potential energy surfaces. (Note that mechanosynthetic processes can in practice be far less sensitive to errors in PES contours than are solution-phase chemical processes; Section 4.4.)

A long-term objective in this area is to compile a large and well-characterized set of mechanosynthetic reactions suitable for the synthesis of diamondoid structures. This, in turn, can eventually provide the basis for software able to perform retrosynthetic analysis, accepting a component design as an input and (at least in favorable instances) producing a description of a corresponding mechanosynthetic process.

Part II

Components and Systems

9

Nanoscale
Structural Components

9.1. Overview

Nanomachines (like macromachines) are made from components, and their strength, stiffness, shape, and surface properties largely determine what those components can do. Chapter 10 discusses components as moving parts, focusing on sliding and rolling interfaces. The present chapter discusses components (including fixed housings) from a structural perspective, developing approximations and rules for analyzing designs specified in less than atomic detail.

Section 9.2 discusses some general issues in nanomechanical design. Section 9.3 considers components as pieces of material and discusses the relationship between material-based continuum models and atom-based molecular models. Section 9.4 examines component properties, emphasizing surface corrections to bulk quantities. Sections 9.5 and 9.6 discuss the control of shape in designing irregular and symmetric objects. Finally, Section 9.7 adds a discussion of adhesive interfaces for joining components to form larger structures.

9.2. Components in context

Nanomechanical systems differ from systems of biological molecular machinery in their basic architecture: nanomechanical components are supported and constrained by stiff housings, while biological components often can move freely with respect to one another.* It is natural to focus on moving parts, but a typical system has much of its mass in the form of a stiff housing. Gears, bearings, springs, screws, sliding rods, motors—each should be pictured as anchored to or embedded in an extended diamondoid structure tailored to support it in a functional position with respect to other components (Chapter 10 describes conditions under which nonbonded interfaces form good bearings, enabling parts to move in their housings).

---

* E. Tribble has emphasized the importance of such supporting structures for separating degrees of freedom in a system and thereby simplifying design and analysis.

The analogy to macroscale engineering practice is clear: the components of macroscale machines and other systems are commonly mounted on an extended structure, whether this is termed a housing, chassis, casing, airframe, engine block, or framework. (The chief exceptions to this rule are devices that flexibly position an object in space—cranes, robotic arms, desk lamps—and even these constrain the motion of each part with respect to its neighbors.) A housing structure forms a unit, but it need not be monolithic. Macroscale housings can be made from pieces held together by fasteners, adhesives, or welding; nanomechanical housings can likewise be made from pieces held together by fasteners or adhesive interfaces (Section 9.7).

Moving parts typically occupy or sweep out only a modest fraction of the volume enclosed by a housing; the rest, save for small clearances, can be occupied by a solid structure with a diamondlike modulus. Accordingly, housings can often be quite stiff. Note that the thermal positional uncertainty of one region of a homogeneous three-dimensional structure with respect to another approaches a distance-independent constant at large separations. Since the stiffness of a housing can often be much greater than that of the moving parts, a housing can often be treated as rigid on a large scale, making corrections (when necessary) for local compliance.

9.3. Materials and models for nanoscale components

Macromechanical engineering treats components as being made from materials, neglecting the placement of specific atoms. Section 9.3.1 discusses various classes of materials from an engineering perspective, introducing nanomechanical concerns. Section 9.3.2 contrasts viewing a component as a specific molecular structure with viewing it as a piece of material. Section 9.3.3 outlines how the materials perspective can be extended to provide a useful description of many nanoscale mechanical components.

9.3.1. Classes of materials

a. Conventional metallic solids. In common metals (e.g., iron, aluminum, copper), atoms have 8 to 12 nearest neighbors with which they share weakly directional bonds. A simple model of°valence electrons in a metal pictures them as forming a gas in which the ionic cores of the atoms are embedded; in this model, bonding is nondirectional.

Bonds of low directionality allow dislocations to move smoothly, enabling atomic layers to slide over one another with relatively little resistance. This nanoscale behavior permits macroscale ductility, making objects insensitive to cracks and able to undergo large plastic deformations. Brittle materials, in contrast, resist dislocation motion; this makes them stronger, but unable to deform plastically and highly sensitive to microscopic flaws. This sensitivity occurs because cracks under tensile loads concentrate the stress at their tip: ductile materials deform to blunt the tip and reduce the stress; brittle materials break, extending the crack and raising the tip stress, leading to catastrophic failure.

Because conventional manufacturing makes objects with a high density of microscopic flaws, strong-but-brittle materials have proved difficult to use, and weaker-but-ductile metals have dominated macroscale engineering practice.

Molecular manufacturing, however, can limit flaws to rare defects of atomic scale, enabling the wider use of brittle materials. (Other tactics, such as dividing materials into fibers, can also make brittle materials useful; see discussions in the extensive literature on fracture mechanics.) The atomic mobility permitted by metallic bonding sharply limits the usefulness of metals in a nanomechanical context at ordinary temperatures. Although some refractory metals and inter-metallic compounds may in fact be useful, this volume assumes no use of metals as mechanical components.

b. Ionic solids. An ideal ionic solid consists of ions held together almost ex-clusively by electrostatic forces. Because an ion of one charge is strongly at-tracted to its oppositely charged nearest neighbors, but repelled by more distant neighbors, smooth sliding is difficult, and brittle behavior results. Since ionic sol-ids have no great strength to compensate for their brittleness, they find no sig-nificant application in macromechanical engineering and are expected to find no significant application in nanomechanical engineering.

Some solids have mixed covalent and ionic bonding (e.g., alumina, silica) and high strength. These materials (ceramics and glasses) are sometimes regarded as ionic solids; for present purposes, they can be regarded as borderline covalent solids. They find use in macromechanical engineering (limited chiefly by their brittleness), and may find use in nanomechanical engineering (limited chiefly by the availability of covalent materials with better properties).

c. Molecular solids. Weak forces (e.g., van der Waals attraction) can bind inert gas atoms and neutral, covalently bound molecules to form a solid. Unless the molecules are large (Section 9.3.1d), molecular solids are weak and of no use as structural materials.

d. Polymeric solids. In a polymeric solid, weak forces provide effective co-hesion between long molecules, and the covalent bonds within those molecules can dominate the bulk properties of the material. Polymeric solids with well-aligned molecules can be strong and stiff along the axis of alignment. Many polymeric solids are also tough and ductile; they find widespread use in macro-scale engineering.

In nanomechanical engineering, even small-scale mobility of polymer chains must usually be suppressed. This can most directly be accomplished by provid-ing covalent cross links between chains, but in the limiting case, a crosslinked polymer becomes a covalent solid.

e. Covalent solids. In a covalent solid, bonds form a dense, continuous net-work in which most atoms are bonded to three or more neighbors. Examples of covalent solids include diamond and graphite: brittle materials of great strength. Other examples include black phosphorus and germanium, which are of little interest as structural materials. In nanomechanical engineering, interest focuses on the more diamondlike covalent solids.

f. Diamondoid covalent solids. Used narrowly, *diamondoid* refers to mate-rials consisting almost entirely of carbon with sp^3 bonding. As used here, *dia-mondoid* describes strong, stiff covalent solids with a dense, three-dimensional network of bonds. The diamondoid solids of most interest have compositions

Table 9.1. Properties of some strong solids.[a] Many other refractory materials (metal carbides, borides, oxides, nitrides, and silicides) may also be of mechanical (or electrical) interest.

Material	Young's modulus (GPa)	Strength[b] (GPa)	Melting point (K)	Density (kg/m³)
diamond	1050	50	1800[c]	3500
graphite[d]	686	20	3300[c]	2200
H-6 carbon[e]	~1500?	?	?[c]	~3200
SiC	700	21	2570[c]	3200
Si	182	7	1720	2300
Boron	440	13	2570	2300
Al_2O_3	532	15	2345	4000
Si_3N_4	385	14	2200[c]	3100
Tungsten	350	4	3660	19300

[a] Most values in this table are from Kelly (1973).
[b] Strength values are inferred from measurements and may underestimate the values achievable in flaw-free structures.
[c] Material sublimes or decomposes rather than melting.
[d] Modulus and strength of graphite are in the bonded plane.
[e] The stability and properties of H-6 carbon metal are inferred from tight-binding semiempirical calculations (Tamor and Hass, 1990); the ratio of Young's modulus to bulk modulus is here assumed equal to that of diamond.

biased toward multivalent first-row elements that form good covalent bonds (boron, carbon, nitrogen, oxygen), but may make substantial use of the similar second-row elements (silicon, phosphorus, and sulfur), limited use of monovalent elements (hydrogen, fluorine, chlorine), and sparing use of other elements. Among structures fitting this description (which embraces diamond itself) are many that are strong, stiff, and highly stable. Table 9.1 surveys the properties of diamond and several diamondoid materials. Pure-carbon structures falling within this broad definition have been widely discussed (e.g., by Johnston and Hoffmann, 1989; Karfunkel and Dressler, 1992; Merz et al., 1987; Tamor and Hass, 1990).

In the present context, diamondoid structures have a further advantage. With some restrictions on the choice of atoms and bonding patterns, they are reasonably well described by molecular mechanics models (Section 3.3), enabling the computationally economical investigation of mechanical systems specified in atomic detail.

g. Mixed structures. The categories discussed in Sections 9.3.1(a–f), though useful, are not sharply bounded. Many materials have intermediate structures. Ceramics with bonding intermediate between ionic and covalent have been mentioned. Covalent-solid objects of small size can be described as molecules, and combinations of them can be viewed as molecular or ionic solids. Covalent bonding can coexist with metallic conductivity. Molecular objects built on the development pathways from solution chemistry to advanced mechanosynthesis are likely to include extensively crosslinked polymeric structures intermediate between conventional polymers and covalent solids (see Chapter 16).

9.3.2. Materials vs. molecular structures

In macromechanical engineering, components are described as regions occupied by a material (usually isotropic and homogeneous) having certain mechanical properties: strength, density, shear modulus, and Young's modulus. These and other material properties (e.g., fatigue life, thermal coefficient of expansion, damping coefficients, friction coefficients), together with shape, enable a detailed engineering analysis of the component.

From a molecular mechanics perspective, a diamondoid component is a network of bonded atoms. The smallest objects contain few atoms, and must be described in atomic detail, yet at some scale a description in terms of shape and continuum material properties becomes quite useful (especially in preliminary design). To what extent can continuum models be applied to small diamondoid structures?

9.3.3. The bounded continuum approach

In extending continuum models to nanostructures, one must consider both surface effects and the atomic graininess of matter. Surface effects in diamondoid structures can be approximately described by a bounded version of a continuum model, but atomic structure constitutes a fundamental limit to such models. Surface effects on the mechanical properties of metals are more pervasive because of the nonlocal nature of metallic bonding; these effects are not considered here.

a. Surface effects.   A simple continuum model treats the mechanical properties of a material as continuous up to a sharp boundary, but this is physically unrealistic for nanoscale objects. Section 3.5 discusses models of interacting diamondoid surfaces that account for van der Waals attractions and overlap repulsions between their surface atoms, yet treat those atoms as forming a two-dimensional continuum. This approach can be used to convert a standard continuum model into a *bounded continuum model,* treating the boundaries of an object in a physically realistic way.

To apply a continuum model to the interior of an object, however, a sharp boundary must be defined. The natural definition of a surface for mechanical purposes—as the boundary of the region from which other similarly defined objects are excluded—gives misleading results in calculations based on bulk density, modulus, and strength. At ordinary pressures, nonbonded-contact radii are approximately twice as large as covalent radii, hence surface atoms occupy more space than interior atoms. Accordingly, rules for defining surface locations that yield good values for mass (from density), for component stiffness (from modulus), or for component strength (from material strength) define surfaces that fall well within the boundaries of the °excluded volume (see Figure 9.1). Section 9.4 discusses issues of component size and surface effects in more detail.

b. Graininess effects.   The atomic structure of matter constrains the shapes possible for small objects. In macroscale engineering, one assumes that (save for machining geometry and minimum gauge limitations) a part can be made in any desired shape, within some size tolerance and surface roughness. Section 9.5.2 discusses the conditions under which similar assumptions can be applied to nanoscale objects, in effect treating diamondoid structures as materials to be shaped.

Surface roughness can increase friction. Chapter 10 discusses friction, yielding results that can be applied without specifying atomic coordinates; these results are thus applicable in preliminary designs based on analyses at the level of materials, rather than of molecular structures.

c. Anisotropies. The bounded continuum models used in this volume treat materials as isotropic. Although this is a good approximation for many structures, including the most diamondlike diamondoids, it is a poor approximation for others. Continuum models of anisotropic bulk materials (e.g., those developed to describe composite structures) can be useful in describing anisotropic diamondoid components.

9.4. Surface effects on component properties

Density, modulus, size, and shape are ambiguous in nanoscale systems. Size and shape are ambiguous because molecular surfaces are not sharply bounded, and this makes density and modulus ambiguous, despite the clear-cut meanings of mass and stiffness. Nonetheless, a model based on defined surfaces and (isotropic, homogeneous) properties can provide useful design-phase estimates of mass, stiffness, and strength. A conservative application of bounded-continuum estimates can be used to identify design conditions that limit elastic deformation to acceptable bounds, but may provide only a rough description of the residual displacements.

9.4.1. Materials and stiffness

To maximize strength and stiffness, nanoscale structural components are best made from diamondoid solids (Section 9.3.1f). In the systems analyzed in the following chapters, stiffness requirements are usually more constraining than are strength requirements. Stiffness can be directly estimated by computational experiments using molecular mechanics models. Examples of such calculations appear in Section 9.4.3, along with guidelines for estimating the stiffness of diamondoid components.

9.4.2. Assigning sizes

Ordinarily, one regards the region occupied by a component as the region from which other components are excluded. This measure of size varies with compressive loads, and with the shapes and chemical natures of the facing surfaces. A reasonable standard choice is the region occupied by the component atoms of

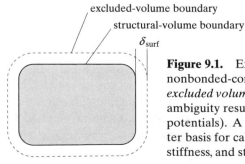

excluded-volume boundary

structural-volume boundary

δ_{surf}

Figure 9.1. Excluded vs. structural volume. The nonbonded-contact radii of surface atoms define the *excluded volume* of a diamondoid component (with some ambiguity resulting from the softness of the repulsive potentials). A smaller *structural volume* provides a better basis for calculations that estimate component mass, stiffness, and strength from bulk material properties.

the structure when each is assigned a summable 0.1 nN radius (Section 3.3.3b); these radii are used in space-filling molecular representations throughout this volume.

Figures 9.2 and 9.3 suggest why this size estimate must be corrected when estimating stiffness. The polycyclic *framework structure* accounts for most of the stiffness, but the excluded volume extends well beyond this framework, because of both the separation between nonbonded atoms and (in some structures) the presence of monovalent surface atoms. To maximize stiffness for a given component size, it is best to avoid monovalent surface atoms, instead favoring structures like the one in Figure 9.3. The size correction δ_{surf} required to describe the structural volume is then approximately the difference between the covalent and nonbonded radii; using the 0.1 nN nonbonded radii, $\delta_{surf} \approx 0.07$ nm for both first- and second-row atoms. A larger correction ($\delta_{surf} = 0.1$ nm is used in Chapter 11) yields a more conservative estimate of stiffness for a component of a given excluded-volume size. (For H-terminated diamond, a comparable correction is $\delta_{surf} \approx 0.14$ nm.)

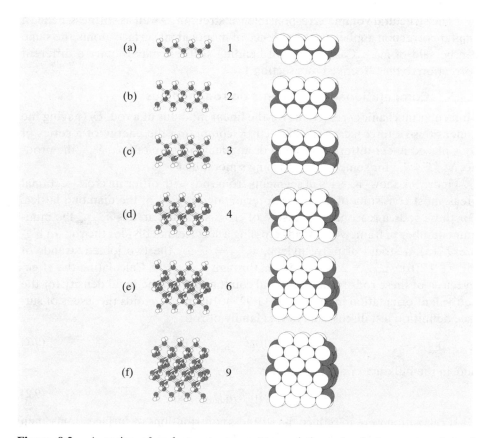

Figure 9.2. A series of rod structures, most consisting of a hydrogen-terminated region of the diamond lattice. Rod (a) is a polyethylene chain; (b)–(f) can be described as additional chains linked side by side (the digits to the right indicate the number of such chains in the corresponding rod). The structure of rod (c) departs from the diamond pattern, adding bonds perpendicular to the rod axis to replace nonbonded H|H contacts. Moduli are graphed in Figure 9.4(c).

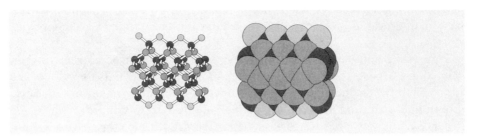

Figure 9.3. A segment of rod resembling Figure 9.2(f), but with N and O termination in place of CH and CH_2. See Section 9.4.3 for discussion and Figure 9.4(a) for modulus.

A further correction can arise from differences in the stiffness contributions between the outermost framework atoms and those in the interior; with fewer constraints on their motions, they may be subject to greater relaxation under stress, reducing their stiffness contributions. This effect can be explored by examining the scaling of stiffness with cross-sectional area in a molecular mechanics model such as MM2.

(The structural volume is responsible for strength as well as stiffness, hence a similar correction applies. In the absence of monovalent surface atoms, the same can be said of mass. Calculations of bending stiffness may require a different correction to the effective cross section.)

9.4.3. Computational experiments on rod modulus

Molecular mechanics predictions for the linear modulus of a rod, E_ℓ (having the dimensions of force), can be obtained by computing the energy of a series of rods placed under differing strains. For an ideal measure of area S_{rod}, the product $S_{rod}E = E_\ell$, for some value of Young's modulus E.

Figure 9.2 shows a series of segments from rods with differing cross-sectional areas, most consisting of a hydrogen-terminated region of the diamond lattice. For these rods, a reasonable measure of cross-sectional area is n_{bonds}, the minimum number of framework bonds crossing a surface that divides the rod. In Figure 9.2(a), a strand of polyethylene, $n_{bonds} = 1$; for the two joined strands of Figure 9.2(b), $n_{bonds} = 2$; and so forth, through $n_{bonds} = 9$. Calculating the effective area of these rods from their bond count and from the bond density for the equivalent orientation in diamond ($\sim 1.925 \times 10^{19}$ m^{-2}) avoids the issues of surface definition just discussed. For this family of rods

$$S_{rod} \approx 5.2 \times 10^{-20} n_{bonds} \ \ (\text{m}^2) \tag{9.1}$$

and, in the bulk-material limit,

$$E_\ell \approx 5.45 \times 10^{-8} n_{bonds} \ \ (\text{N}) \tag{9.2}$$

If relaxation were to reduce the stiffness contributions of surface atoms, then thinner rods would have a lower value of E_ℓ.

Figure 9.4 summarizes the results of computational experiments on the structures in Figures 9.2, 9.3, 9.5, and 9.6. As can be seen, on a per-bond basis, thinner rods have a slightly *greater* value of E_ℓ than thicker rods (presumably owing to contributions from nonbonded interactions involving hydrogen), but rods of all

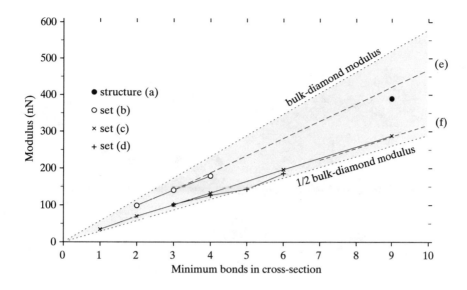

Figure 9.4. Results of computational experiments measuring rod moduli in the MM2/CSC approximation. Structure (a) is the N- and O-terminated rod in Figure 9.3; set (b) consists of the structures in Figure 9.5; set (c), Figure 9.2; set (d), Figure 9.6. The dashed line labeled (e) passes through the origin and a point corresponding to a column of hexagonal diamond structure (with a compact cross-section) having $n_{bonds} = 16$; as can be seen, it almost exactly corresponds to an extrapolation from a similar column with $n_{bonds} = 3$. The dashed line (f) is analogous to (e), but for a structure like family (c) with $n_{bonds} = 18$. The upper dashed line bounding the shaded region corresponds to the modulus of diamond rods oriented along the <111> axis, extrapolating from the properties of bulk diamond; the lower dashed line corresponds to a similar extrapolation with the modulus halved.

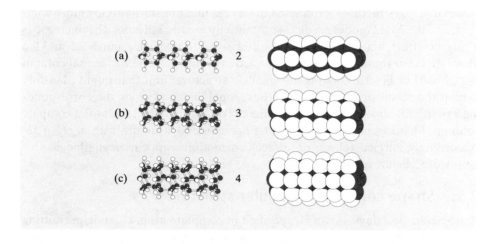

Figure 9.5. A series of rods forming a family of structures suffering increasing strain per bond with increasing diameter; instances with $n_{bond} = 2, 3,$ and 4 are shown. Moduli are graphed in Figure 9.4(b).

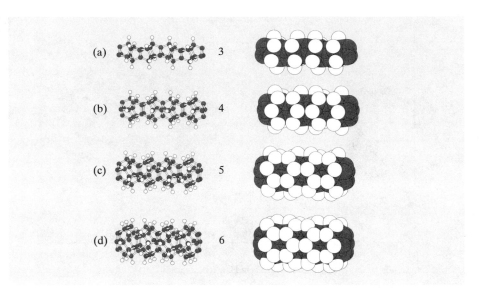

Figure 9.6.   A second series of rods forming a family of structures suffering increasing strain per bond with increasing diameter; instances with $n_{\text{bonds}} = 3$, 4, 5, and 6 are shown. Moduli are graphed in Figure 9.4(d).

sizes have a value substantially lower than that of bulk diamond. Since this trend continues to structures with substantial stiffness contributions from interior atoms (e.g., the $n_{\text{bonds}} = 18$ value indicated in Figure 9.4), it appears that this modulus deficit results from defects in the MM2 model (Section 3.3.2) rather than from surface relaxation. (Since the usual goal is to maximize stiffness, adopting MM2 values usually results in conservative assessments of designs.)

Figure 9.3 illustrates a structure like Figure 9.2(f), but with N and O termination, rather than CH and CH_2. (Chains of sp^3 nitrogen atoms should be stable when each is bonded to carbon in a diamondoid solid of this kind.) For a given value of n_{bonds}, structures terminated in this manner are substantially more compact, and the MM2 model predicts substantially greater stiffness. This increase is a surface effect, but one which it is conservative to neglect. Figures 9.5 and 9.6 illustrate other rods having moduli plotted in Figure 9.4. The hexagonal column ($n_{\text{bonds}} = 3$) of Figure 9.5(b) is the smallest structural unit that might plausibly exhibit the elastic properties of bulk hexagonal diamond along the corresponding axis; it falls almost precisely on a line drawn through the result for a compact column of hydrogen-terminated hexagonal diamond structure with $n_{\text{bonds}} = 16$. Accordingly, surface relaxation effects on modulus appear negligible in this family of structures as well.

9.5.   Shape control in irregular structures

Nanomechanical devices can be specified in complete atomic detail, permitting them to be described (with varying accuracy) by molecular modeling techniques. This is necessary for small structures, and for structures about to be built, but it is not always necessary for large structures during preliminary design and analysis. It will be useful to have some guidelines regarding the boundary between small and large for these purposes.

9.5.1. Control of shape and detail of specification

Nanomechanical engineering differs from macroscale engineering in that the number of possible component shapes can be significantly constrained by atomic sizes and feasible bonding patterns. In sufficiently small components, the set of possible shapes can be exhaustively described. For each type of mechanism, there is some minimum size below which components of the required shapes cannot be constructed (e.g., it is fruitless to attempt to build a gear having 50 teeth in a volume that can contain only 10 atoms; the limiting size for a 50-tooth gear is substantially larger). Near this minimum size, a design must be specified in complete molecular detail to gain any confidence in its validity.

Somewhat above this limit, however, complete atomic specification becomes unnecessary, just as in macroscale engineering and for the same reason: any one of many arrangements of atoms can serve the purpose. On this larger size scale, a specific molecular structure must be chosen before planning a mechanosynthesis, and before performing a fully detailed analysis, but this choice can be made *after* performing a reasonably accurate analysis describing size, shape, strength, stiffness, mass, dynamics, energy dissipation, and performance. Sections 9.5.2 to 9.5.4 discuss how the number of available diamondoid structures varies with volume and structural constraints, drawing conclusions regarding the scale at which molecular detail can be omitted and a bounded continuum model can be adopted. Designs based on these models often are relatively conservative, setting only lower bounds on the performance of nanomechanical systems.

9.5.2. Estimates of the number of diamondoid structures

a. Exponential scaling with volume.   The number of distinct diamondoid structures N_{struct} that can fit in a given volume of reasonable shape is an exponential function of the volume (neglecting surface corrections). This can be seen by considering a hypothetical atom-by-atom construction process: at each surface site in a partially completed structure there is some number of distinct options (with a geometric-mean value N_{opt} which is not necessarily an integer) for how to proceed at the next step, that is, for what kind of atom to bond in what manner. The value of N_{opt} describes the effective number of choices made per atom added, correcting for patterns of choices that in the end produce the same structure. The number of choices defined in this way is proportional to the number of atoms in the final structure (which will, for some typical atomic number density n_v, be proportional to the volume V), and each choice leads to a different structure (by the above definition of *distinct choice*). For a suitable definition of N_{opt}, the number in question is

$$N_{struct} = N_{opt}^{(n_v V)} \qquad (9.3)$$

The proper value of N_{opt} is not obvious, even if the atoms are restricted to a single type. It will depend on the range of permissible local structures, constrained by (for example) the magnitudes and patterns of bond strain permitted in the interior of a solid stable enough for practical use. Knowledge of these constraints would be necessary (yet not sufficient) to enable a simple computation.

b. Estimates from entropy of fusion.   The entropy of fusion of a crystal is a measure of the increase in the available volume in configuration space (Section

4.3.3) that occurs with a transition from a single, regular structure to an ensemble of states which includes many different structures. *If* the potential wells in the liquid state were as narrow as those in the solid state, and *if* each of those potential wells were equally populated and corresponded to a stable amorphous structure (and vice-versa), *then* the entropy of fusion would be a direct measure of the increase in number of wells, and would thus provide a direct measure of the number of available structures. In practice, low-stiffness wells and the thermal population of unstable transition structures tend to make the entropy of fusion an overestimate; the negligible thermal population of high-strain states that would be kinetically stable in corresponding solids tends to make it an underestimate. The absence of a choice of atom types at various sites (in elemental crystals) introduces a strong bias toward underestimation, relative to the systems of interest here.

Measuring the entropy of fusion of carbon (melting point ~3820 K) presents technical difficulties; the values for silicon and germanium crystals, which share the diamond structure, are 4.6×10^{-23} and 4.7×10^{-23} J/K per atom respectively. These values are both about $3.4 k$; if this increase in entropy were solely the result of the increased number of available potential wells per atom, that number would be $N_{\mathrm{opt}} \approx 30$. An estimate of the number of distinct states in argon at liquid density (Stillinger and Weber, 1984) yields a value equivalent to $N_{\mathrm{opt}} \approx 3$ wells per atom; this is compatible with an estimate based on the entropy of fusion.

c. Estimates including multiple elements.   In the set of broadly diamondoid structures built up from C, Si, B, N, P, O, S, H, F, and Cl (with a bias toward the higher-valence and first-row members), N_{opt} is increased as a result of the choice of atom type at each step; a factor of 5 is a plausible estimate of this increase, allowing for constraints and biases in the choices made. Taking the preceding silicon and germanium estimates as a base, this suggests that a reasonable estimate for diamondoid structures is $N_{\mathrm{opt}} \approx 150$, and that calculations based on $N_{\mathrm{opt}} = 30$ are conservative. This conservatism is intended to compensate for various constraints, including structural stability and avoidance of unacceptable electric fields and potentials resulting from patterns of aligned dipoles.

A region with a structural volume of 1 nm^3, $N_{\mathrm{opt}} = 30$, and $n_{\mathrm{v}} = 100$ nm^{-3} has $N_{\mathrm{struct}} \approx 10^{148}$. A similar cubical region with an excluded volume of 1 nm^3 and $\delta_{\mathrm{surf}} = 0.1$ nm has a structural volume of ~0.51 nm^3, and $N_{\mathrm{struct}} \approx 10^{75}$.

9.5.3. Exclusion of structures by geometrical constraints

These enormous numbers imply that objects can be constructed with almost any specified shape, *provided* (1) that the specifications do not describe complex contours on a subnanometer scale, or sharp corners and the like, and (2) that the specifications allow adequate tolerances in the placement of surface atoms, relative to the ideal contours of the shape. Demanding that many surface atoms simultaneously meet tight tolerances can easily result in an overconstrained problem having no physically realizable solution.

For example, one might ask that each of the ~100 atoms on the surface of a 1 nm object be within a distance ε of an ideal surface contour. A randomly chosen structure can be terminated so as to keep only those atoms within a certain boundary, and that boundary can be chosen so that the surface atoms of the terminated structure are close to a preselected ideal surface contour (Section 10.3 discusses a similar picture in more detail). With $n_v = 100$ nm^{-3}, the positions of these surface atoms typically vary over a range of ~0.2 nm about that contour. If we demand that each surface atom be within $\varepsilon = 0.01$ nm of this contour, then on the order of $(0.01/0.2)^{100} \approx 10^{-130}$ of a set of randomly chosen structures will meet this condition. Given $N_{struct} \approx 10^{75}$, the probability that some structure satisfies this condition is on the order of 10^{-55}. If, however, one asks that 50 atoms forming particular surfaces be accurately placed within $\varepsilon = 0.02$ nm, then on the order of $(0.02/0.2)^{50} \approx 10^{-50}$ of a set of randomly chosen structures will meet this condition, and the probability that a suitable structure does *not* exist is on the order of 10^{-25}. (Brute-force search procedures would be inadequate to find acceptable structures, but various options exist for developing more sophisticated search procedures to solve problems of this general kind; doing so presents an interesting and important challenge for computer-aided design of molecules.)

Tight constraints on object shape arise chiefly in the working interfaces of moving parts, for example, where surfaces must slide smoothly. Chapter 10 develops several models for such interfaces; some are based on the choice of regular structures of high symmetry (to which the statistical analysis in this section does not apply); others describe the constraints on an irregular structure imposed by the requirement that it slide smoothly over a regular surface. Other working interfaces bind molecules; these are considered in the next section. Nonworking surfaces are exposed to free space and have few constraints.

9.5.4. Exclusion of structures by molecular binding requirements

Surfaces with °binding sites (receptors) for specific molecules are useful, for example in acquiring and processing molecules from solution (Section 13.2). The principles of selective binding are familiar from molecular biology, and chiefly involve detailed surface complementarity between the molecule and its receptor: complementary surface shapes to provide strong van der Waals attraction, complementary patterns of charge and dipole orientation to provide electrostatic attraction, and so forth (in aqueous solutions, so-called hydrophobic forces can play a large role). These requirements for complementarity impose constraints on surface structure, again reducing the size of the set of acceptable structures by many orders of magnitude.

Mammalian immune systems provide an example in which structures capable of strong, selective binding are chosen from a set of structures of calculable size. Mammals can produce highly specific antibodies able to bind any one of a wide range of small molecules (when a molecule of a particular type is presented as an immunologically stimulating *hapten*); this includes molecules not found in nature. These antibodies are developed by a process of variation and selection chiefly within a space of possibilities defined by differing sequences of the 20 genetically encoded °amino acid residues within the antibody hypervariable domains (variations at other sites are relatively rare). These domains contain ~38 amino acid residues; hence most antibodies are selected from a set of structures containing ~$20^{38} \approx 3 \times 10^{49}$ members. Equation (9.3) with $N_{opt} = 30$ suggests that comparable diversity can be provided by a set of structures each containing

<35 atoms, although the geometric and structural constraints of receptors often require larger structures. (Note that use of rigid structures for *most* machine components does not preclude the use of flexible structures, e.g., to enable a flap on a binding site to fold over a ligand.)

9.5.5. Kaehler brackets

T. Kaehler has proposed a systematic approach to designing irregular interfaces to meet particular geometrical constraints (Kaehler, 1990; Kaehler, 1992). First, one defines an anchoring surface [e.g., a hydrogenated diamond (111) surface]. Second, one defines a large family of brackets extending from this surface. (These can resemble the hexagonal column of Figure 9.5(b), but adding various modifications: substitution of N, B, P, or SiH for CH, replacement of a C—C linkage or a CH|HC contact with C—CH$_2$—C, C—O—C, or C—S—C, etc., all subject to stiffness and stability constraints; see Figure 9.7. The size of the resulting family of structures increases exponentially with bracket length and soon reaches millions.) Third, one determines the geometry of each bracket and notes the positions and orientations of its terminal bonds. Finally, one compiles an index based on this data, organized to enable one to find the brackets with terminal bonds most nearly matching a desired position and orientation.

How might an indexed set of Kaehler brackets be used in design? The brackets define a relatively comprehensive set of slim diamondoid structures, indexed according to the geometrical relationship between their ends. These can be used

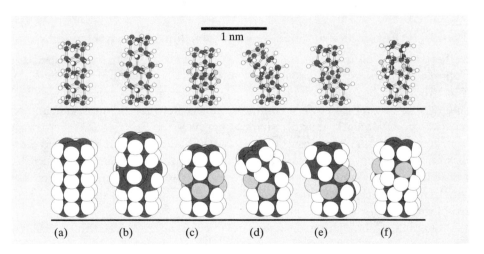

Figure 9.7.   Six short Kaehler brackets (~0.9 nm between the top and bottom six-membered rings), illustrating a portion of the available range of structural variations and geometries. Bracket (a) is a simple column (a hydrogen-terminated fragment of hexagonal diamond; all column ends truncated). Bracket (b) is lengthened by substituting Si for C at six sites. Bracket (c) is shortened by substituting C—O—C for a CH|HC contact at six sites. Bracket (d) is bent by one substitution of the sort in bracket (a) and three of the sort in (c). Bracket (e) has its top offset with respect to the bottom because of four substitutions like those in (c), the substitution of Si for C at two sites closer to the ends, and the substitution of F for H at two sites. Bracket (f) has its top rotated with respect to the bottom because of three substitutions as in (c) and three substitutions of HC—CH for C. If brackets are anchored to a diamond structure with a choice of locations, the range of end displacements need only span a diamond lattice cell. (MM2/CSC)

to link structures that must be held in specific relative orientations; several such bridges can form a stiff connection. Further, the design-phase substitution of one Kaehler bracket for another bracket of similar end geometry is like moving a small robotic arm by a small distance (and perhaps changing its stiffness). One can imagine algorithms that in effect push and pull a flexible structure into a desired shape (e.g., to fine-tune a receptor design), choosing brackets of the shape necessary to apply the needed forces. Discrepancies between predicted and actual geometries could be rectified by further substitutions among brackets of similar end geometry, but perhaps quite dissimilar structure.

9.6.   Components of high rotational symmetry

In building rotating machinery, components having high-order rotational symmetry are often desirable. Chapter 10 discusses applications of such components in bearings and gears. (Perturbations from an asymmetric environment always break perfect symmetry, but perfection is unnecessary.) Rotationally symmetric components divide roughly into two classes: *strained-shell* structures made (at least conceptually) by bending a straight slab into a hoop, and *special-case* structures that are more intrinsically cylindrical. Members of an intermediate class of *curved-shell* structures resemble strained-shell structures, but with regular arrays of dislocation-like features that reduce the strains. These terms correspond to rough categories, not to precise distinctions.

9.6.1.   Strained-shell structures

Figure 9.8 illustrates a relatively small strained shell having two distinct layers of cyclic structure. The thickness-to-radius ratio (t/r) of strained shells is constrained by the net hoop stress and the permissible bond tensile strain at the outer surface. For diamondoid structures with negligible net hoop stress, permissible bond strains may be as large as 0.035 nm ($\sim 1.23 r_0 \approx 0.187$ nm); assuming a Morse potential (Section 3.3.3a), a C—C bond with smaller strain has positive stiffness, tending to stabilize its length. In the presence of a strain gradient perpendicular to the axis of strain (as in a strained shell), even a layer somewhat beyond the positive-stiffness limit can be stabilized by restoring forces imposed by the less-strained layers closer to the shell axis (the layers are coupled by shear stiffness). *Note that this situation differs from the single-bond, pure-tensile-*

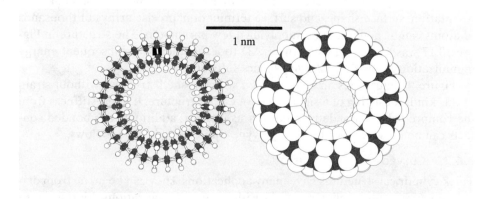

Figure 9.8.   A strained shell structure with a relatively thin wall and small diameter.

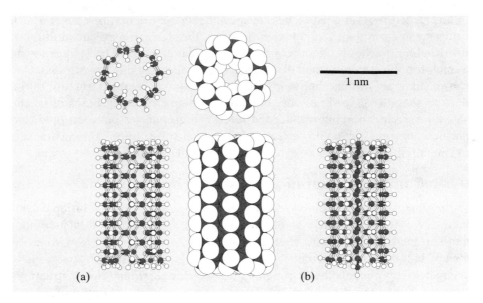

Figure 9.9.   A strained shell structure with a thin wall and small diameter (a); structure (b) adds a nonbonded polyyne chain along the axis.

load model used to derive bond failure rates in Section 6.4.4a. In general, the stability analysis for highly strained shells is complex, because of the possibility of concerted bond rupture and rearrangement; the designs in this volume assume considerably smaller bond strains.

If the material of the shell in Figure 9.8 had linear elastic properties, a 0.035 nm strain would occur with $t/r \approx 0.46$ (using a mid-thickness reference for r). The worst-case nonlinearity would prohibit bond compression entirely, resulting in a limit of $t/r \leq {\sim}0.23$. Both these estimates neglect the favorable effects of angle-bending relaxation. The appropriate measure of thickness for these calculations is smaller than that of the structural volume discussed in Section 9.4.2; atomic centers at the edge of the framework structure provide a conservative marker for the surface.

R. Merkle (1991; 1992a) has developed computer-aided design (CAD) software able to generate a wide range of strained-shell structures based on the diamond lattice, providing flexible control of radius, thickness, crystallographic orientation, surface shape, and surface termination; precise arrays of thousands of atoms can be generated by supplying a few parameters. The structure in Figure 10.17 was generated using this CAD tool together with subsequent energy minimization using molecular mechanics software.

Figure 9.9 illustrates an example of a strained-shell structure without strain of this kind, consisting of a single layer of cyclic structure. It gains stiffness from the compressed nonbonded contacts in its interior; additional nonbonded contacts can be provided by placing a polyyne chain along its axis (below).

9.6.2. Curved-shell structures

Thick cylindrical structures have many applications. They can be made from diamondlike structures, relieving strain with arrays of dislocations. Merkle notes that the Lomer dislocation is stable and has useful geometrical properties for

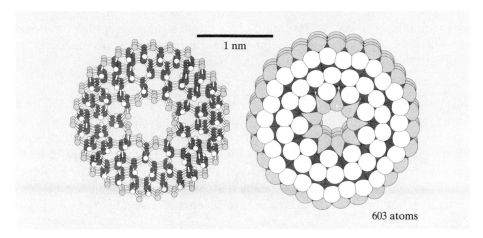

Figure 9.10. A curved shell structure containing an array of Lomer dislocations (structure courtesy of R. Merkle, Xerox PARC, and L. Balasubramaniam, MIT).

this purpose. Figure 9.10 illustrates a cylindrical structure incorporating several Lomer dislocations; preliminary work by R. Merkle and L. Balasubramaniam has produced software that aids in the design of tubular structures of this kind. Merkle suggests that computer-aided design tools able to automatically generate diamond lattices modified by (potentially dense) arrays of dislocations offer a promising approach to the generation of diamondoid structures of considerable diversity.

9.6.3. Special-case structures

Some cylindrical structures are not members of scalable families. The rods illustrated in Figures 9.5 and 9.6 are examples of small-radius structures having significant rotational symmetry. The cylindrical structure of Figure 9.11 provides

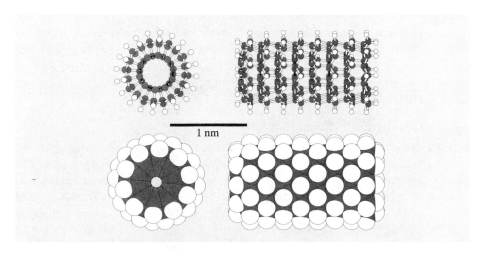

Figure 9.11. A cylindrical structure with sp^3 surface structure and an sp^2 inner structure (the ten-membered conjugated rings will exhibit °aromatic stabilization).

an example of intermediate size; the outer ring of the bearing in Figure 1.1 provides a still larger example. The set of diamondoid structures of high rotational symmetry, small length, and diameter ≤ 1 nm is large, but may be small enough to permit the development of an exhaustive catalogue.

9.7. Adhesive interfaces

Large structures can be built directly from small reactive moieties and clusters, but there are substantial advantages to building indirectly via intermediate-scale blocks (Sections 8.6.6 and 14.2). These blocks can be joined by *adhesive interfaces* that can exploit any of the attractive forces responsible for the cohesion of solids. Designing a system to be assembled in this way imposes some constraints on its structure, but in many instances these need not be severe.

9.7.1. Van der Waals attraction and interlocking structures

The simplest adhesive interface consists of two surfaces of complementary shape, with no special provision for bonding. If the interface has suitable corrugations, the interface will resist shear. Van der Waals attraction provides tensile strength and (together with balancing overlap repulsions) interfacial stiffness. Figure 3.12 indicates that a simple contact between complementary diamondoid surfaces can have a tensile strength ~ 1 GPa and a stiffness per unit area of >30 N/m·nm^2. The compliance of an interface with this stiffness is equal to that of a ~ 30 nm thick slab of diamond.

Introducing a simple van der Waals interface thus degrades the strength of a diamondoid structure by about two orders of magnitude, and substantially degrades the stiffness of systems built on a 100 nm scale. Where strength is unimportant and stiffness is not at a premium, such interfaces may be of considerable use. Convoluted, interlocking interfaces (e.g., with knobs that snap into holes) can greatly increase strength and stiffness by making greater use of overlap repulsions.

9.7.2. Ionic and hydrogen bonding

Complementary patterns of charge or °hydrogen bonding capacity can increase interfacial strength and stiffness. The strength and stiffness of interfaces with extensive ionic bonding can be estimated from the properties of sodium chloride. Theoretical calculations of the breaking stress yield estimates around 2.7 GPa (Kelly, 1973), and modulus data indicates an interlayer stiffness per unit area of ~ 170 N/m·nm^2.

9.7.3. Covalent interfacial bonding

The strongest and stiffest adhesive interfaces result from the formation of dense arrays of strong bonds; various reactive diamondoid surfaces can be designed to join in this way. Diamond (111) surfaces terminated with complementary patterns of B and N atoms can join by forming an array of °dipolar bonds. Arrays of dienes and dienophiles can join through Diels–Alder reactions to form an array of C—C bonds. Hexagonal diamond surfaces like that illustrated in Figure 8.24 can apparently bond to form continuous hexagonal diamond (bonding two planar surfaces is electronically analogous to bonding a hexagonal column to a surface); a pair of diamond (110) surfaces can apparently bond to form continuous cubic diamond. In these instances, adhesive surfaces can unite blocks of the strongest, stiffest materials known to experimental science without leaving a distinct interface, and hence without loss of strength or stiffness.

A potential difficulty in forming strong interfaces is avoiding the sudden release of unacceptable amounts of mechanical energy: the energy of a diamond (111) surface is ~ 5.4 J/m^2, which if suddenly thermalized would be enough to raise the temperature of a 1 nm thick slab by $\sim 10^3$ K. Adhesion, however, can be viewed as the inverse of crack formation. The gradual application of inhomogeneous stress distributions can cause controlled crack growth; the relaxation of similar stress distributions can enable controlled joining. Alternatively, interposed structures could force joining to proceed at a controlled rate, or surface structures might be chosen such that adhesion is slowed by steps with substantial thermal activation barriers. Section 14.2.2a notes the utility of alignment pegs in large interfaces.

9.8. Conclusions

Nanoscale structural components can serve either as housings or as moving parts. The best materials for nanoscale components are diamondoid in structure: these materials have unusual strength and stiffness, and lend themselves to design using molecular mechanics methods. Computational experiments indicate that bulk-material stiffnesses for diamond (and diamondoid solids) can be used to describe components of subnanometer dimensions, provided that the modulus is applied to a cross-sectional area that is descriptive of the structural framework (here termed the *structural volume*), omitting an outer layer of the *excluded volume* resulting from the relatively long range of overlap forces, and from any monovalent surface atoms. At a ~ 1 nm size scale, the number of stable diamondoid structures becomes enormous, and essentially any shape can be built, provided that tight (<0.02 nm) tolerances are not applied too widely. (A discussion of the compatibility of irregular structures with smooth sliding motion is deferred to the next chapter.) Components of high rotational symmetry can be constructed in several ways; Chapter 10 describes the use of such structures in building bearings and gears.

Some open problems. Open problems in the field of nanoscale structural components divide into problems of design and problems of analysis. Regarding the latter, further research using molecular mechanics studies to improve bounded continuum models and explore their range of validity would be valuable. A challenging objective is to develop software that merges classical finite-element methods with bounded continuum models and molecular mechanics methods; this would enable a designer to control the level of detail in a structural description, keeping atomic detail where it is significant and omitting it elsewhere to improve computational efficiency.

In design, it would be useful to specify and characterize many small, regular structures useful as shafts, gears, and so forth; means for indexing and recovering designs are of comparable importance. Still more useful would be the development of software for designing irregular structures that meet particular boundary conditions. The Kaehler bracket approach is promising for a range of applications, as is Merkle's suggestion of dense dislocation networks; another might be a simulated annealing process (using a potential function developed for the purpose) in which atom number and atom types are permitted to change.

In another area entirely, studies of the joining of diamond surfaces could resolve questions regarding the practicality of using the strongest adhesive interfaces in component assembly. Further, characterization of the properties of B—N dipolar bonds (and other dipolar and coordination bonds) could yield descriptions of desirable adhesive interfaces, and might yield improved designs for interfaces suitable for use in nanomechanical gears (Section 10.7).

Mobile Interfaces
and Moving Parts

10.1. Overview

Moving parts—gears, bearings, and so forth—typically have sliding or rolling interfaces. Mobile interfaces and strong, stiff nanostructures like those discussed in Chapter 9 can be combined to build a wide range of nanomechanical systems, including mechanical computers (Chapter 12) and molecular manufacturing systems (Chapter 14).

Stable molecular liquids show that interfaces between molecules can be both stable and mobile at ordinary temperatures. Familiar examples include saturated hydrocarbons, fluorocarbons, °ethers, and °amines; diamondoid structures having analogous surface moieties appear in the examples of this chapter. A more diverse range of surface moieties are stable when incorporated into diamondoid structures operating in well-ordered environments.

Two mutually inert surfaces in contact can be characterized by their potential energy of interaction. Models like those described in Section 3.5 can estimate the potential energy as a function of separation for two relatively smooth surfaces, but in gears and bearings, the potential energy function associated with sliding motions is of central importance. The two cases of greatest interest are those in which the potential energy is nearly flat along the direction of motion, permitting smooth sliding, and those in which it has large corrugations, entirely blocking sliding. Smooth interfaces can be used in bearings; corrugated interfaces can be used in gears.*

The following sections explore mobile interfaces and their applications in gears, bearings, and related devices. Section 10.2 begins by characterizing properties of the spatial Fourier transforms of interatomic nonbonded potentials, which prove useful in analyzing the properties of sliding interface bearings. Section 10.3 considers the problem of sliding motion between irregular covalent

---

* The two conditions compatible with zero energy dissipation are zero sliding resistance and zero slip. Interfaces with smooth potentials can approximate zero resistance at low sliding speeds; interfaces with strongly corrugated potentials can exhibit zero slip (though the interface deforms and other energy dissipation mechanisms intervene).

objects and regular covalent surfaces, developing a Monte Carlo model that predicts the fraction of irregular structures expected to exhibit smooth sliding motion with respect to such surfaces. Section 10.4 develops the theory of symmetrical sleeve bearings, presenting results from analytical models of idealized bearings, and characterizing two specific designs using molecular mechanics methods. Section 10.5 generalizes from these results to several other systems incorporating sliding-interface bearings, including nut-and-screw systems, rods sliding in sleeves, and constant-force springs. Section 10.6 briefly describes bearings that exploit single atoms as axles. Section 10.7 moves from sliding interfaces to nonsliding interfaces, examining analytical models of gears, and using these as a basis for examining the properties of roller bearings and systems resembling chain drives. Section 10.8 examines some general issues arising in the construction of extended systems that have nearly flat potentials. Finally, Section 10.9 briefly surveys devices that use surfaces intermediate between freely sliding and nonsliding: dampers, detents, and clutches.

10.2. Spatial Fourier transforms of nonbonded potentials

In the design of nanomechanical systems, smooth sliding motions can be most simply achieved when one or both surfaces at an interface have a periodic or nearly periodic structure of high spatial frequency, that is, when the surface has a series of closely spaced features that repeat at regular intervals, with each feature found in essentially the same local environment. In one-dimensional sliding motion (by convention, along the x axis), only periodicity along the x axis is significant; the corresponding spatial frequency is k rad/m, or f_x cycles/m ($= k/2\pi$).

The variations in the potential $\mathcal{V}(x)$ associated with sliding of a component over a surface can in the standard molecular mechanics approximations be decomposed into a sum of the pairwise nonbonded potentials between the atoms in the object and those in the surface, together with terms representing variations in the internal strain energy of the object and the surface. For stiff components under small interfacial loads, the soft, nonbonded interactions dominate the variations in $\mathcal{V}(x)$, and structural relaxation causes only small deviations from straight-line motion of the interfacial atoms. Under these conditions, variations in the total potential $\mathcal{V}(x)$ are accurately approximated by a sum of the pairwise nonbonded potentials between atoms in straight-line relative motion. Figure 10.1 plots a set of such interaction potentials for pairs of sp^2 carbon atoms moving on paths with differing closest-approach distances d, based the MM2 exp–6 potential, Eq. (3.8).

10.2.1. Barrier heights and sums of sinusoids

Barrier heights $\Delta\mathcal{V}_{barrier}$ for sliding of components over periodic surfaces can be described in terms of a sum of contributions associated with integral multiples of the spatial frequency f_x. This sum can be divided into contributions each resulting from an atom in the sliding object interacting with a row of evenly spaced atoms in the periodic surface. The energy of an atom with respect to a row consists of a sum

$$\mathcal{V}(x) = \sum_{i=-\infty}^{\infty} \mathcal{V}_{vdw}\left(\sqrt{d^2 + (x + i/f_x)^2}\right) \tag{10.1}$$

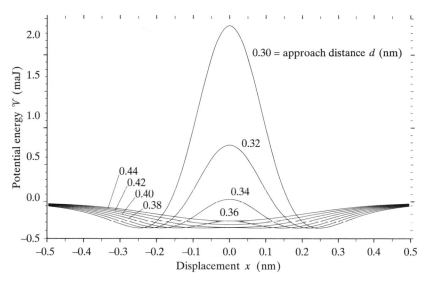

Figure 10.1. Nonbonded potentials, Eq. (3.8), as a function of x-axis displacement for two sp^2 carbon atoms moving in straight lines of differing closest-approach distance d.

where $\mathcal{V}_{\text{vdw}}(r)$ is the nonbonded energy as a function of distance for the two atom types, Eq. (3.8). Graphically, such a sum can be represented as an evenly spaced set of sample points, and the difference between two sums separated by one half cycle can be represented by an evenly spaced set with weights alternating between $+1$ and -1 (Figure 10.2); in the typical cases of interest, the maximum value of this difference equals the barrier height for sliding of a single atom with respect to a single row.

Figures 10.3 and 10.4 plot amplitude spectral densities,* $|\mathrm{H}(f_x)|$, derived from the spatial Fourier transforms of a set of interaction potentials like those in Figure 10.1. At large spatial frequencies f_x, the associated amplitudes drop steadily on a logarithmic scale as f_x increases. Exceptions to the "typical cases of interest" in the previous paragraph can arise when $f_x \approx f_{x,0}$ (see Figure 10.3), but these exceptional cases have unusually low barrier heights. The designs discussed here do not exploit this phenomenon, as it relies on a delicate cancellation rather than on robust properties of intermolecular potentials.

The sampling shown in Figure 10.2 can be described as a sequence of positive and negative delta functions. The Fourier transform for such a sequence yields the relationship

$$\Delta\mathcal{V}_{\text{barrier}} \propto \left|\mathrm{H}(f_x) - \mathrm{H}(3f_x) + \mathrm{H}(5f_x) - \mathrm{H}(7f_x)...\right| \qquad (10.2)$$

---

* The concept of spectral density stems from the analysis of waves carrying power. The *power spectral density* of a radio transmission, for example, measures its power per unit frequency-range (substantial in a narrow band of radio wavelengths, negligible at x-ray wavelengths), and can be derived from the time-domain shape of the emitted electromagnetic wave by Fourier transform methods. These methods are equally applicable to functions in the spatial domain, but the resulting "power spectral density" has a purely nominal relationship to power. An amplitude spectral density is proportional to the square root of a power spectral density. Because only ratios of values are considered here, units for these quantities are omitted.

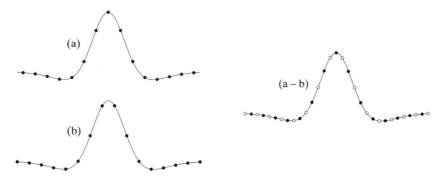

Figure 10.2. A nonbonded potential (as in Figure 10.1) sampled at regular intervals (a), and sampled with a shift in phase (b). The difference in the sum of the sample energies as the phase shifts is the energy of a sum of samples with positive and negative weights (a – b); this observation can be used to relate spatial Fourier transforms of pairwise potentials to energy barriers in sliding motion.

From the results plotted in Figures 10.3 and 10.4, it can be seen that with reasonable values of d (e.g., ≥ 0.2 nm) and with a moderately high spatial frequency (e.g., $f_x \geq 3.0$ cycles/nm), the first term in Eq. (10.2) dominates the rest by multiple orders of magnitude because contributions from spatial frequencies ≥ 9.0 cycles/nm are quite small. For a diamond (111) surface in the high-f_x direction, $f_x \approx 4.0$ cycles/nm; for graphite, $f_x \approx 4.1$ cycles/nm. As a consequence, the potential of a component sliding over such a surface can be accurately approximated as a sum of sinusoidal contributions from interactions between the atoms of the component and each of the component rows of the surface, all of spatial frequency f_x, but of varying amplitudes and phases.

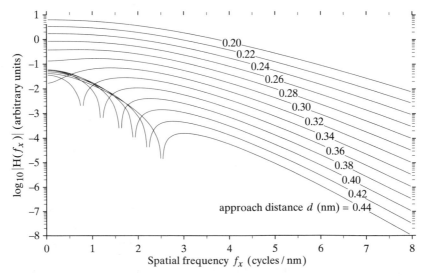

Figure 10.3. Amplitude spectral densities derived from spatial Fourier transforms of nonbonded $C|C$ (sp^2) potentials like those in Figure 10.1, for a range of closest-approach distances d. For relatively large values of d, the Fourier transform changes sign, resulting in a zero at some spatial frequency $f_{x,0}$.

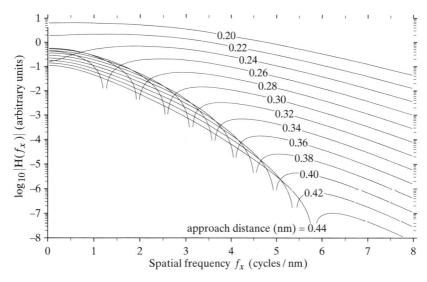

Figure 10.4. Amplitude spectral densities as in Figure 10.3, but for nonbonded H|H potentials. Note that these graphs would be identical (within a constant energy factor) if all distances were measured in units scaled to the equilibrium nonbonded separation.

10.3. Sliding of irregular objects over regular surfaces

Section 9.5 estimates the enormous number of possible nanometer-scale diamondoid structures, and observes that constraints on surface structure can drastically reduce the size of this set. In particular, requiring that surface structures be regular imposes a requirement that interior structures be regular (to a depth dependent on the tolerance for residual irregularities); this reduces the set of possible structures to a minute fraction of that available without this constraint. Regular structures can make excellent bearings (as shown in Section 10.4), and this is their chief value in the present context. This section examines the bearing performance that can be achieved in the far larger set of irregular structures.

10.3.1. Motivation: a random-walk model of barrier heights

Consider an atom sliding over a regular surface with spatial frequency f_x at a height h (note that $h \leq$ the minimum value of d with respect to any of the rows of the surface). As shown in Section 10.2, the potential energy of such an atom can be represented as a sum of sinusoids, each characterized by some amplitude $\Delta \mathcal{V}/2$ and phase ϕ. For a surface of high spatial frequency, and for values of h corresponding to modest loads, the sum is dominated by a single sinusoid with a spatial frequency f_x.

Consider an irregular object sliding over a surface with N such atoms in contact. In the approximation just described, the interaction energy of each is dominated by a sinusoid of the same f_x, but with differing values of $\Delta \mathcal{V}$ and ϕ. In the vector representation (Figure 10.5), the summing of these sinusoids can be visualized as a walk over a plane. For a set of irregular objects with randomly distributed values of ϕ and bounded values of $\Delta \mathcal{V}$, the resulting random walk has familiar properties: in the limit of many steps, the probability density for the end

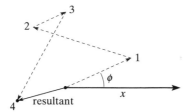

Figure 10.5. Sinusoidal energy terms (defined by magnitude and phase) represented as vectors, illustrating the magnitude and phase of the sum as the result of a random walk over the plane.

points is Gaussian, and the mean value of the radius is

$$\overline{\Delta \mathcal{V}}_{\text{sum}} = \overline{\Delta \mathcal{V}}_{\text{barrier}} \propto \sqrt{N} \tag{10.3}$$

hence the area over which the end points are scattered varies as N. For a set of irregular structures in which there are n_{opt} choices for the properties of the N atoms, the number of possible structures increases as n_{opt}^N, and thus the density of endpoints in the plane (Figure 10.5) near the origin varies as

$$\rho_{\text{area}} \propto n_{\text{opt}}^N / N \tag{10.4}$$

and the mean distance from the origin to the closest point (i.e., the value of the smallest barrier for any member of the family of structures) is

$$\overline{\Delta \mathcal{V}}_{\text{barrier}} \propto \sqrt{N / n_{\text{opt}}^N} \tag{10.5}$$

Thus, although the expected barrier height for any given irregular structure increases with increasing N, the minimum expected barrier height for a family of structures decreases.

10.3.2. A Monte Carlo analysis of barrier heights

The scaling principles discussed in Section 10.3.1 do not translate directly into an accurate statistical model; such a model would require a treatment taking account of the nature of the available choices of interacting atoms and their interactions with the regular surface, which in turn affects the value(s) of n_{opt} and the distribution of values of $\Delta \mathcal{V}$. Further, as values of $\Delta \mathcal{V}_{\text{barrier}}$ become small, sinusoidal terms of higher spatial frequency become important, and the end points must be treated as being scattered in a space of higher dimension. The complexity of these interacting physical effects and design choices suggests the use of a Monte Carlo model to estimate the distribution of $\Delta \mathcal{V}_{\text{barrier}}$ for model systems of interest.

a. Approximations and assumptions. To reduce the computational burden while retaining the essential physical and statistical features of the problem, a set of approximations was adopted:

- Use of the MM2 exp–6 potential to represent nonbonded interaction energies. The exp–6 potential is realistic over a wide range of separations (Section 3.3.3b), and the neglected electrostatic terms (Section 3.3.2d) would have little effect on $\Delta \mathcal{V}_{\text{barrier}}$.
- Use of a straight-line translation model, which effectively treats structures as infinitely stiff. For x-axis sliding, neglect of y- and z-axis relaxation is

conservative in this context, while neglect of x-axis relaxation is the reverse; at modest loads, the neglected effects are minor.

- Lack of rotational relaxation of the sliding object. An irregular object pressed against a surface tends to rotate to distribute load over several contact points; neglect of this artificially increases the disparity in contact loads, which tends to increase barrier heights.

With these approximations, the interaction energy of an object is just the sum of the energies of a set of atoms, each interacting with the surface independently. Atomic interaction energies can be precomputed and then combined to represent different structures.

Families of structures can be modeled using a further set of assumptions and approximations:

- Each family is characterized by a framework structure having a set of N sp^3 surface sites within a certain range of distances (h_{min} to h_{max}) from the regular surface.

- To generate the members of a family, each surface site can either be occupied by an N atom with a lone pair, or by a C atom bonded to H, F, or Cl, or can be deleted (locally modifying the framework); each site thus has 5 states, giving each structural family 5^N possible members.

- To model irregular structures, sites are assumed to be randomly distributed with respect to the unit cells of the regular surface, and bond orientations are assumed to be randomly distributed within a cone directed toward the regular surface with an angle of 40° (a wider angle is expected to be more favorable).

- Site positions are treated as fixed, with lone pair, H, F, and Cl positions determined by the site coordinates and bond orientation, together with standard bond lengths (Section 3.3.2a).

- To ensure that each structure examined is in firm contact with the regular surface, site-states are characterized by their z-axis stiffnesses relative to the regular surface. Members of a structural family containing a site with a stiffness less than some threshold $k_{s,thresh}$ are discarded, as are those with less than three nondeleted sites. These exclusions ensure that the z-axis stiffness for each structure is $\geq 3k_{s,thresh}$, and usually reduce the number of retained members of a structural family to far fewer than 5^N.

b.   Computational procedure. Based on the preceding assumptions, the computation proceeds by generating a set of sites characterized by locations and bond orientations (200 sites were used), constructing a set of site states for each site, filtering this set for stiffness $\geq k_{s,thresh}$, and computing the energy of each remaining site state at a series of displacements (16) spanning one cycle of motion over the regular surface. A structural family is defined by randomly choosing N sites from the final set, and a list of the acceptable members of that family is generated by forming all possible combinations consisting of one state from each site and discarding those with more than $N-3$ deletion states. The energy of each structure as a function of displacement is computed by summing the corresponding precomputed values for each of its constituent site states, and the barrier height for sliding motion of the structure is taken as the difference between the maximum and minimum energies in the resulting sum. Finally,

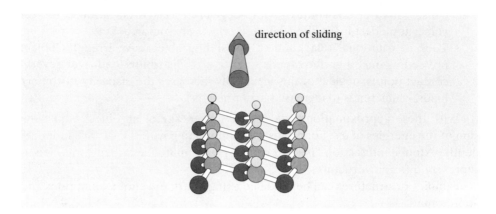

Figure 10.6.   View across a nitrogen-terminated diamond (111) surface with lone pairs illustrated, indicating the direction of sliding assumed in the calculations summarized in Figures 10.7 and 10.8 (this direction is taken as the x axis).

$\Delta \mathcal{V}_{barrier}$ for the structural family is taken as the minimum of the energy differences found for any member of that family.

*c.   **Results.**   The results of a set of calculations based on the preceding model for sliding motion of irregular structures over a strip of nitrogen-substituted diamond (111) surface (as illustrated in Figure 10.6) are summarized in Figure 10.7, taking the initial sampling bounds for site generation as $h_{min} = 0.2$ nm and $h_{max} = 0.5$ nm. The statistical distribution of values of $\Delta \mathcal{V}_{barrier}$ is presented as a

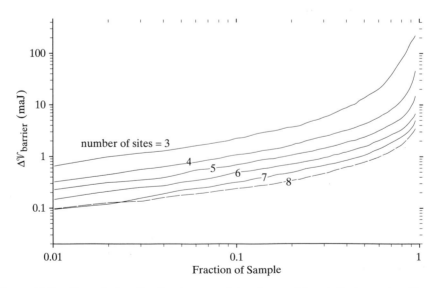

Figure 10.7.   Cumulative distributions resulting from a Monte Carlo study of barrier heights encountered by irregular structures sliding over a regular surface, based on the model described in Section 10.3.2. Each curve is the result of 1000 trials, where each trial selects the best member of a particular family of structures (as described in the text); note that sampling statistics result in substantial scatter, particularly toward the left tail of the distributions. The surface modeled is nitrogen-substituted diamond (111), Figure 10.6.

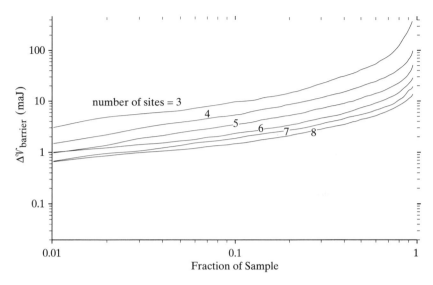

Figure 10.8. Cumulative distributions like those in Figure 10.7, but with minimum values of z-axis stiffness 10 times higher. The author thanks L. Zubkoff for his help with the computations used to acquire the data plotted in Figures 10.7 and 10.8.

series of cumulative distributions for samples of randomly generated structural families with $N = 3, 4, 5, 6, 7$, and 8. As can be seen, for $N \geq 5$, about 10% of randomly selected structural families have a member yielding $\Delta \mathcal{V}_{barrier} \leq 1$ maJ ($<0.25 kT_{300}$). Barriers this low will be surmounted on most encounters, and hence will for most purposes fail to act as barriers: the result is a bearing surface without static friction. Further, if one assumes that the potential is characterized by a sinusoid with the period of the lattice ($\sim$0.28 nm), then the peak negative stiffness during sliding along the x axis is $\sim$−0.25 N/m; this low magnitude enables a bearing interface to be coupled to relatively distant structures with sufficient stiffness that the *net* stiffness is positive (Section 10.8), enabling smooth motion even at low temperatures.

Figure 10.8 presents the results of a similar calculation with $h_{min} = 0.19$ nm, $h_{max} = 0.49$, and $k_{s,thresh} = 10$ N/m. Values of $\Delta \mathcal{V}_{barrier}$ are higher owing to the combined effects of higher energies and a shift toward higher spatial frequencies in the interatomic potentials (Figures 10.3 and 10.4).

10.3.3. Implications for constraints on structure

Assume that a nanomechanical component has been designed to meet some set of functional constraints along one surface (the *functionally constrained surface*), and that some other surface of the component (the *sliding surface*) must slide smoothly over a regular surface. One would like the design of the sliding surface to place few additional constraints on the structure of the functionally constrained surface. One can define a set of *compatible-framework structures* that consists of diamondoid frameworks that satisfy the constraints of the functionally constrained surface and extend for some indefinite distance toward (and beyond) the desired location of the sliding surface. If the two surfaces are separated by a distance on the order of 1 nm, then the set of compatible-framework structures (now considering only variations in the region that falls short of the desired sliding surface) contains a large combinatorial number of elements, typically $>10^{10}$. Candidate families of sliding surface structures can be generated (in a design sense, not a fabrication sense) by truncating these compatible-framework structures so as to generate a set of sp^3 sites falling between $h_{\min}$ and $h_{\max}$; members of each family can then be generated by modifying these sites in the manner suggested by the model of Section 10.3.2a.*

The results of the model described in Section 10.3.2 suggest that (for modest loads applied to regular surfaces of high spatial frequency) a substantial fraction of these truncated compatible-framework structures permits low values of $\Delta \mathscr{V}_{\mathrm{barrier}}$, provided that the number of available sites between $h_{\min} = 0.2$ nm and $h_{\max} = 0.5$ nm is >5. If all atoms in the slab could serve as sp^3 sites, then for structures with an atomic number density $n_v = 100$ nm^{-3}, a cross-sectional bearing area $S_{\mathrm{bear}} = 0.25$ nm^2 would often suffice. If the areal density of available sites equaled that of atoms in a diamond (111) surface, then a somewhat larger area, $S_{\mathrm{bear}} = 0.28$ nm^2, would be necessary. For design work, values of $S_{\mathrm{bear}} \geq 0.5$ nm^2 should prove ample. Larger areas (or the selection of structures from a set of many compatible frameworks) can permit both greater z-axis stiffness and lower $\Delta \mathscr{V}_{\mathrm{barrier}}$.

In summary, it is safe to assume that any component with a surface in contact with a strip of regular, high-spatial-frequency structure can be made to slide smoothly with respect to that strip, provided that loads are modest, that the contact surface is of a reasonable shape (with a length of at least $2\pi/k$ in the x direction) with $S_{\mathrm{bear}} \geq 0.5$ nm^2, and that the contact surface is only loosely constrained in structure. The latter condition is usually satisfied a few atomic layers from a tightly constrained surface. This conclusion generalizes to irregular structures that slide along regular grooves and ridges, and to irregular structures sliding along regular curved surfaces in rotary bearings.

10.3.4. Energy dissipation models

a. Phonon scattering.   Interaction between a small sliding contact and ambient phonons can be modeled as scattering from a moving harmonic oscillator (Section 7.3.4). With surface interaction stiffnesses of several N/m per atom, a total stiffness on the order of 30 N/m will not be atypical. This corresponds to the example in Section 7.3.4, which yielded an estimated energy dissipation of $\sim 3 \times 10^{-16}$ W at a sliding speed of 1 m/s, and $\sim 3 \times 10^{-20}$ W at 1 cm/s.

---

* This discussion implicitly assumes that modification of surface sites does not make compatible-framework structures incompatible. Strains introduced by modifying the sliding surface will have little effect on the constrained surface if the separation is adequate (~ 1 nm), and if the modifications do not cause large net tensile or compressive stresses. These conditions will often be met. Where they are not, the conversion of compatible to incompatible structures by distortions induced by surface modification will, in typical design problems, be balanced statistically by the conversion of incompatible structures to compatible structures by similar distortions. In circumstances where this is significant, however, the task of designing a sliding surface cannot be treated separately from the task of designing a compatible framework.

b. Acoustic radiation. A small sliding contact exerts a time-varying force with an amplitude roughly proportional to the amplitude of the energy variation $(= \Delta \mathcal{V}_{barrier}/2)$. From Eq. (3.19), the approximate force amplitude is

$$F_{max} \approx 1.7 \times 10^{10} \, \Delta \mathcal{V}_{barrier} \qquad (10.6)$$

(F_{max} in N, $\Delta \mathcal{V}_{barrier}$ in J). Assuming sinusoidal variations in force, Eqs. (10.6) and (7.8) yield

$$P_{rad} \approx 2.9 \times 10^{20} \, \Delta \mathcal{V}_{barrier}^2 \, v^2 k^2 \, \sqrt{\rho} \, \frac{1}{8\pi M^{3/2}} \qquad (10.7)$$

Assuming $\Delta \mathcal{V}_{barrier} = 1$ maJ, a surface with a stiffness and density like those of diamond, and $k \approx 2.2 \times 10^{10}$ m^{-1}, Eq. (10.7) predicts a radiated acoustic power of $\sim 3 \times 10^{-19}$ W at 1 m/s, and $\sim 3 \times 10^{-23}$ W at 1 cm/s. These losses are small compared to those resulting from phonon scattering; in geometries in which the energy variations reflect variations in pressure (for example, sliding in a tube or a slot with balanced forces on either side), then a relationship like Eq. (7.21) applies and losses are still lower.

c. Thermoelastic damping. Eq. (7.50) can be applied to estimate losses resulting from thermoelastic damping. Using diamond material parameters for the surface (save for K_T, which is taken as 10 W/m·K), and treating the alternating forces in the system as being applied to square nanometer areas and cubic nanometer volumes, the estimated energy loss per cycle (at a sliding speed of 1 m/s) is $\sim 3 \times 10^{-30}$ J, or $\sim 1 \times 10^{-20}$ W. In addition, the total (nonalternating) force varies with time from the perspective of a site on the surface. Assuming that the force is adequate to ensure a nonbonded stiffness of ~ 30 N/m (~ 1 nN) and that the length scale ℓ of the loaded region is ~ 1 nm, the pressure is ~ 1 GPa; at a sliding speed $v = 1$ m/s the estimated energy loss per cycle ($t_{cycle} \approx \ell/v$) is $\sim 10^{-27}$ J, or $\sim 10^{-18}$ W. Thermoelastic-damping losses thus are small compared to phonon-scattering losses.

10.3.5. Static friction

Where $\Delta \mathcal{V}_{barrier} \ll kT$, static friction is effectively zero. At low temperatures, however, and neglecting tunneling, the static friction of a sinusoidal potential can be identified with the maximum value of $\partial \mathcal{V}/\partial x$, where x measures the displacement of the surface. Where the sinusoid has a period d_a (as is the case whenever amplitudes are significant),

$$F_{frict} = \left(\frac{\partial \mathcal{V}}{\partial x} \right)_{max} = 4\pi \frac{\Delta \mathcal{V}_{barrier}}{d_a} \qquad (10.8)$$

For a typical value of d_a (~ 0.25 nm), $F_{frict}/\Delta \mathcal{V}_{barrier} \approx 0.05$ nN/maJ.

10.3.6. Coupled sites

An extended object (such as a rod) can contact a periodic surface at regions spread over a considerable distance. For smooth motion, the positive stiffness between different regions within the extended object must exceed in magnitude the negative stiffnesses resulting from the interaction of these regions with the periodic surface (i.e., as they cross energy barriers). This condition must hold on all length scales.

Consider a contact region for which families of compatible structures yield N vectors (in the space considered in Section 10.3.1) falling within a disk of radius $\Delta\mathcal{V}_1$. Within the approximations used here, the distribution of these vectors is (for small values of $\Delta\mathcal{V}_1$) essentially random. Now consider a second such contact region, with an independent set of families of structures. Given that the density of vectors is approximately constant within a region of radius $1.5\Delta\mathcal{V}_1$, each vector in the first region will have (on the average) $N/4$ vectors in the second region that are within a radius $\Delta\mathcal{V}_1/2$ of $-\mathcal{V}$. Choosing a pair of structures for which this holds would, in the rigid coupling approximation, yield a system with $\Delta\mathcal{V}_{barrier} \leq \Delta\mathcal{V}_1/2$; the negative stiffness of the interaction between the two sites is bounded by

$$|k_s| \leq 3\Delta\mathcal{V}_1 (\pi/d_a)^2 \qquad (10.9)$$

For $d_a = 0.25$ nm and $\Delta\mathcal{V}_1 = 1$ maJ, $|k_s| \leq 0.5$ N/m. So long as the stiffness coupling the two regions is large compared to this small value, the two regions can be treated as rigidly coupled for purposes of the present analysis. For comparison, the stretching stiffness of a diamond rod with a length of 10 nm and a structural cross sectional area of 1 nm^2 is ~100 N/m; relatively long, slim, low-modulus rods will accordingly have stiffnesses >0.5 N/m.

The mean number of candidate structures with this property is $N^2/4$. If we require

$$N^2/4 \geq N \qquad (10.10)$$

then the same reasoning can be applied between *pairs* of regions of the sort just described, with $\Delta\mathcal{V}_2 = \Delta\mathcal{V}_1/2$ playing the role of $\Delta\mathcal{V}_1$. By induction, sets of regions can be constructed on all length scales, with $\Delta\mathcal{V}_{barrier}$ varying inversely with size. Note that, in a rod, stiffness varies as ℓ^{-1}, but for a system with these scaling properties, the magnitude of the negative stiffness between regions likewise varies as ℓ^{-1}. A threshold value for this argument to proceed (neglecting statistical considerations) is $N \geq 4$; larger values ensure reliability in the face of statistical fluctuations, and can ensure that the magnitude of the negative stiffness between regions varies as ℓ^{-b}, b > 1.

10.4. Symmetrical sleeve bearings

The results of the previous section indicate that irregular sleeve bearings with small energy barriers are feasible provided (1) that either the shaft or the sleeve has a rotational symmetry such that a rotation corresponding to a small tangential displacement is a symmetry operation, and (2) that the other component is sufficiently weakly constrained in its structure that a design can be selected from a large set of possible structures. Sleeve bearings, however, lend themselves to analysis and design that exploit symmetry in both components. This section extends a preliminary study (Drexler, 1987b) that indicated the promise of this

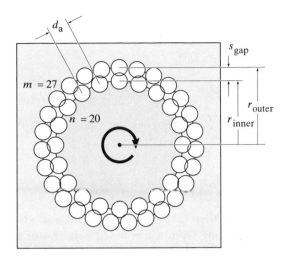

Figure 10.9. Coplanar ring model for a symmetrical sleeve bearing.

class of structures. The resulting analyses can in several instances be extended directly to sliding-interface bearings with noncylindrical geometries.

10.4.1. Models of symmetrical sleeve bearings

For calculations involving bearing stiffness, interfacial stiffness, and dynamic friction, a sleeve bearing can often be approximated as a cylindrical interface with a certain stiffness per unit area k_a for displacements perpendicular to the surface, and a distinct stiffness per unit area $k_{a,para}$ (which can be low) for displacements of the surface parallel to the axis of the bearing.

Where static friction is concerned, sleeve bearing models must take account of atomic detail. Both the outer surface of a shaft and the inner surface of a sleeve can be decomposed into rings of atoms, each having the rotational symmetry of the corresponding component. In the no-relaxation approximation, the potential energy of the system as a function of the angular displacement of the shaft with respect to the cylinder can be treated as a sum of the pairwise interaction of each inner ring with each outer ring. These pairwise potentials are well approximated by sinusoids of a single frequency, which will in the worst case add in phase, and in the best case will substantially cancel. A single ring-ring interaction thus captures the essential characteristics of a shaft-sleeve interaction, save for the omission of (potentially favorable) cancellations. For concreteness, coplanar rings are used as a model in the following section, with parameters illustrated in Figure 10.9.

10.4.2. Spatial frequencies and symmetry operations

Consider an inner ring with n-fold rotational symmetry and an outer ring of m-fold symmetry. If $n = m$, then the inner ring must be displaced by an angle $\theta_{sym} = 2\pi/n$ to restore the initial geometry and potential energy, and the spatial frequency corresponding to this symmetry operation is approximately that of the interatomic spacing in the inner ring, d_a ($r_{inner} = d_a n/2\pi$). In a better approximation, it can be taken as the interatomic spacing of the inner ring projected to the mean radius

$$r_{eff} = \frac{r_{inner} + r_{outer}}{2} = r_{inner} + \frac{s_{gap}}{2} \tag{10.11}$$

yielding an effective spatial frequency for rotational displacements of the inner ring

$$f_\theta = \frac{1}{r_{eff}\theta_{sym}} = \left[d_a \left(1 + \frac{s_{gap}}{2r_{inner}} \right) \right]^{-1}, \quad n = m \tag{10.12}$$

When $n \neq m$, a smaller relative displacement can yield a geometry that is identical to one resulting from a *combined* rotation of the two rings. Since a combined rotation leaves the energy unchanged, this smaller rotation is a symmetry operation for the potential energy function (note that this still holds in the presence of relaxation). Numerical experiments indicate that the required angle is related to the least common multiple of n and m,

$$\theta_{sym} = 2\pi/\text{lcm}(n,m) \tag{10.13}$$

yielding an effective spatial frequency

$$f_\theta = \left[\frac{nd_a}{\text{lcm}(n,m)} \left(1 + \frac{s_{gap}}{2r_{inner}} \right) \right]^{-1} \tag{10.14}$$

Merkle (1992b) has constructed a mathematical proof of Eq. (10.13) based on results in number theory.

10.4.3. Properties of unloaded bearings

Figure 10.10 presents a logarithmic plot of calculated barrier heights for coplanar-ring bearing models for $1 \leq n, m \leq 15$ in the concentric (unloaded) case. Figure 10.11 shows the closely corresponding pattern of spatial frequencies calculated from a relationship based on Eq. (10.14); this correspondence results from the smooth fall-off of amplitude spectral densities $|H(f_x)|$ with increasing f_x shown in Section 10.2; calculations using curved trajectories and f_θ would differ little. Figure 10.12 shows the effect of variations in s_{gap} for two examples with fixed n and m. (Note that H interactions, small values of s_{gap}, and relatively large values of d_a have been used in these calculations because each choice is significantly adverse; larger atoms, larger gaps,* and smaller interatomic spacings all are feasible and reduce $\Delta\mathcal{V}_{barrier}$.)

Unloaded bearings for which $\text{lcm}(n,m) \gg n$ and d_a is reasonably small usually have minute values of $\Delta\mathcal{V}_{barrier}$; the best values of n and m are relatively prime. Small loads, however, destroy the symmetries upon which this result depends.

Anisotropies in the potential for displacements perpendicular to the axis (ideally characterized by a single stiffness k_s) give an indication of how rapidly

---

* A bearing interface can support a net tensile load yet retain a positive stiffness.

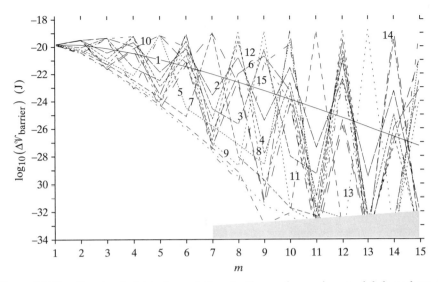

Figure 10.10. Barrier heights for rotation in the coplanar-ring model, based on the MM2 exp–6 potential, Eq. (3.8), for the H|H interaction, using parameters from Table 3.1. All rings were constructed with $d_a = 0.3$ nm and $s_{gap} = 0.2$ nm. (In the shaded region of this and similar graphs, barriers are negligible and roundoff errors dominate the calculation.)

barrier heights increase with load. Where k_s is nearly isotropic, small loads perpendicular to the axis store nearly equal energies, independent of the angle of rotation; where k_s varies greatly, so will differences in stored energy. Where $n - 2 \leq m \leq n + 2$, problems of commensurability and anisotropic stiffness tend to be severe. For relatively isotropic ring systems with the parameters used in Figure 10.10, $k_s \approx 10n$ N/m. Sleeves with multiple rings have correspondingly greater stiffnesses.

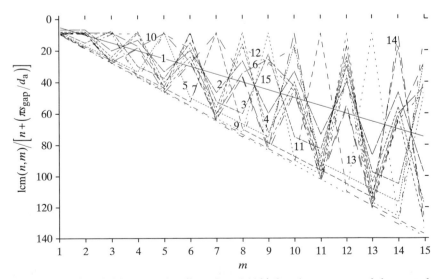

Figure 10.11. Spatial frequencies (in units of $1/d_a$) for ring systems of the sort shown in Figure 10.9.

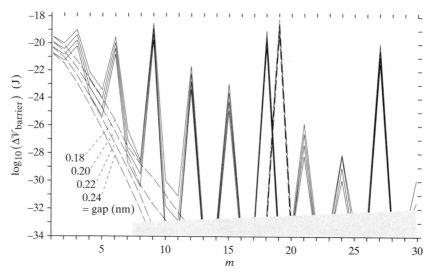

Figure 10.12. Barrier heights as in Figure 10.10, for sleeve bearings with $n = 9$ and 19, $m = 1$ to 30, and values of s_{gap} varying from 0.18 to 0.24 nm. Lines for $n = 9$ are solid, for $n = 19$, dashed.

10.4.4. Properties of loaded bearings

The effects of load perpendicular to the bearing axis can be modeled by displacing the center of the inner ring by a distance Δx perpendicular to its axis and examining the potential as a function of angular displacement of the inner ring about its new, offset axis. Figures 10.13 and 10.14 show the results of such an investigation for a series of ring systems with $n = 9$ and 19 respectively. As can

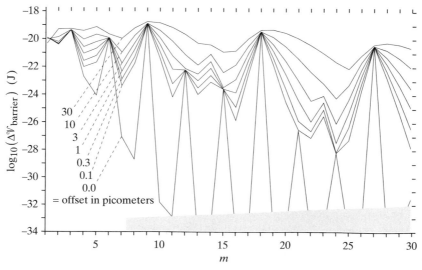

Figure 10.13. Barrier heights as in Figure 10.10, for sleeve bearings with $n = 9$, $s_{gap} = 0.2$ nm, and varying values of transverse offset.

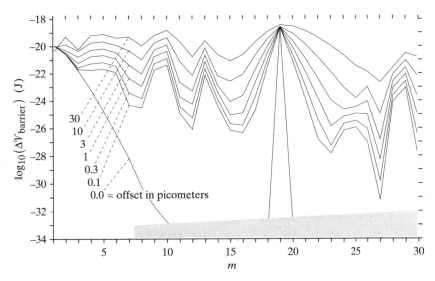

Figure 10.14. Barrier heights as in Figure 10.13, for sleeve bearings with $n = 19$.

be seen, even small displacements destroy the delicate cancellations required for extremely low $\Delta\mathcal{V}_{barrier}$, but for suitably chosen systems with n and $m \geq 25$, values of $\Delta\mathcal{V}_{barrier}$ can be negligible compared to kT_{300}, even at substantial displacements. This continues to hold for bearings having multiple interacting rings, so long as their number is modest or (as can often be arranged) they contribute sinusoidal potentials that add out of phase and approximately cancel.

For $n = 9, m = 14$, and the parameters of Figure 10.10, a displacement of $\Delta x = 0.01$ nm corresponds to a mean restoring force of ~0.96 nN, and an energy barrier of ~1 maJ. The restoring force, however, fluctuates by ~0.1 nN, and since k_s is ~100, the associated fluctuations in stored energy from this source ~$(\Delta F^2/2k_s)$ would be ~0.05 maJ for shafts subject to no other source of stiffness. For $n = 19$, $m = 27$, the corresponding force and stiffness are ~2 nN, 210 N/m, with fluctuations in force <0.0001 nN and fluctuations in stored energy from this source ~10^{-29} J. These quantities have been neglected in the calculations described above.

Load along the axis can be supported by interlocking circumferential corrugations, and do not disrupt the rotational symmetry of the bearing. Effects of load on sliding-interface drag are considered in Section 10.4.6.

10.4.5. Bearing stiffness in the transverse-continuum approximation

Section 3.5.2 develops a model of surfaces that averages overlap repulsion and van der Waals attraction over displacements transverse to the interface. This model can be used to estimate the energy, force, and stiffness per unit area as a function of separation for a pair of nonpolar, nonreactive, diamondoid surfaces (so long as those surfaces are smooth, regular, and out of register with one another), and provides a good approximation for a broad class of sleeve bearings.

A simple sleeve bearing can be characterized by a cylindrical interface of radius r_{eff}, length ℓ, and interfacial stiffness per unit area k_a. As indicated by Figure 3.12, k_a can range from large positive values to moderate negative values, depending on the separation s_{gap} (these models assume $r_{eff} \gg s_{gap}$). Where $k_a \approx 0$, the tensile stress across the interface can be ~ 1 GPa, but this is only a few percent of the tensile strength of diamond.

So long as $k_a > 0$, an unloaded shaft remains centered in the sleeve; small displacements perpendicular to the bearing axis are characterized by a positive stiffness $k_{s,bear}$ (larger displacements result in larger stiffnesses, owing to the nonlinearity of overlap repulsions). The contribution of a patch of interface to $k_{s,bear}$ varies with the angle θ between the normal and the direction of displacement, yielding an expression for $k_{s,bear}$ in terms of the interfacial area and k_a:

$$k_{s,bear} = \int_0^{2\pi} k_a \ell r_{eff} \sin^2(\theta) \, d\theta = \pi k_a \ell r_{eff} \tag{10.15}$$

10.4.6.   Mechanisms of energy dissipation

Energy dissipation in sleeve bearings has several sources. These include acoustic radiation (Section 7.2), shear-reflection drag (Section 7.3.6), band-stiffness scattering (Section 7.3.5c), band-flutter scattering (Section 7.3.5d), and thermoelastic damping (Section 7.4.1). Aside from acoustic radiation, all of the following models estimate dissipation at the cylindrical interface between the shaft and the sleeve using relationships developed for the limiting case of a flat interface between indefinitely extended solids. Sample calculations are presented for bearings of large stiffness (1000 N/m) and moderate size ($r_{eff} = \ell = 2$ nm).

a.   Acoustic radiation.   For unloaded symmetrical sleeve bearings of high spatial frequency, oscillating forces are zero and oscillating torques and pressures are negligible. Significant radiation can be expected only from loaded bearings, where oscillating torques are on the order of $\pi \Delta \mathcal{V}_{barrier} r_{eff}/d_a$ N·m, and the radiated torsional acoustic power, Eq. (7.15), is accordingly on the order of

$$\begin{aligned} P_{rad} &\approx \left(\frac{\pi \Delta \mathcal{V}_{barrier} r_{eff}}{d_a} \right)^2 \left(\frac{2\pi v_{inter}}{d_a} \right)^4 \frac{\rho^{3/2}}{48\pi G^{5/2}} \\ &\approx \left(\Delta \mathcal{V}_{barrier} r_{eff} \right)^2 v_{inter}^4 \frac{102 \rho^{3/2}}{d_a^6 G^{5/2}} \end{aligned} \tag{10.16}$$

With diamondlike material properties, $\Delta \mathcal{V}_{barrier} = 1$ maJ, $r_{eff} = 1$ nm, the interfacial sliding speed $v_{inter} = 1$ m/s, and $d_a = 0.25$ nm, the power resulting from torsional acoustic radiation $P_{rad} \approx 10^{-25}$ W.

Oscillating forces in loaded bearings depend on the nature of the load, the frequency of the oscillation, and the mass of the rotor. From Eq. (7.8),

$$P_{rad} = F_{max}^2 \left(\frac{2\pi v_{inter}}{d_a} \right)^2 \sqrt{\rho} \, \frac{1}{8\pi M^{3/2}} \tag{10.17}$$

In typical instances, $F_{max} \ll 0.1$ nN, and hence with the previous assumptions $P_{rad} \ll 10^{-17}$ W at $v_{inter} = 1$ m/s; for the example with $n = 19$, $m = 27$ (Section

10.4.4) $F_{max} < 0.0001$ nN, and $P_{rad} \approx 10^{-23}$ W. With good designs, losses from acoustic radiation decrease as bearing size increases, since n and m can then have large and favorable values.

The fluctuating forces in a loaded bearing can excite transverse vibrations of the bearing with respect to the sleeve. To avoid this, it is necessary to avoid more than transient operation near the condition

$$\frac{\omega_{bearing} r_{inner}}{d_a} = \frac{v}{d_a} = \frac{1}{2\pi}\sqrt{\frac{k_{s,trans}}{m_{rotor}}} \qquad (10.18)$$

For rotors with transverse stiffnesses ~ 1000 N/m and dimensions of a few nanometers, the critical interfacial sliding speed $v \approx 100$ m/s, and the critical frequency $\omega_{bearing} > 10^{12}$ rad/s.

b. Shear-reflection drag. Several of the following energy dissipation mechanisms depend on the thermally averaged phonon transmission coefficient of the bearing interface, T_{trans}. Using the approximations of Eqs. (7.41) and (7.42) and assuming materials with the modulus, atom number density, and Debye temperature of diamond, the transmission coefficient at 300 K is

$$T_{trans} \approx \frac{z}{1+3z}, \quad \text{where} \quad z = 2.4 \times 10^{-37} k_a^{1.7}, \quad \text{and} \qquad (10.19)$$

$$T_{trans} \approx 2.4 \times 10^{-37} k_a^{1.7}, \quad \text{if} \quad k_a \leq 300 \text{ N/m} \cdot \text{nm}^2 \qquad (10.20)$$

As discussed in Section 7.3.5f, these approximations should overestimate the transmission coefficient for interfaces that are curved or that have mismatched acoustic speeds, and hence should prove conservative in the present context.

Applying the approximations of Section 7.3.6 and again assuming the material properties of diamond at 300 K yields an estimate of the drag power as a function of the sliding speed v, area S, and stiffness k_a of the interface:

$$P_{drag} < 1.8 \times 10^{-33} k_a^{1.7} v^2 S \qquad (10.21)$$

or, applying Eq. (10.15) with the same restriction,

$$P_{drag} < 1.6 \times 10^{-33} k_{s,bear}^{1.7} (\ell r_{eff})^{-0.7} v^2 \qquad (10.22)$$

For a bearing with $r_{eff} = \ell = 2$ nm and $k_{s,bear} = 10^3$ N/m, $P_{drag} \approx 3 \times 10^{-16}$ W at $v = 1$ m/s, and $\approx 3 \times 10^{-20}$ W at 1 cm/s.

Sleeve bearings can be used both to support loads along the shaft axis and to provide stiffness in resisting axial displacements; this can be accomplished by ensuring that the shaft and sleeve have interlocking circumferential corrugations (e.g., as shown in Figure 10.17). An interface of this sort exhibits a stiffness per unit area for transverse displacements (in the axial direction) of $k_{a,trans}$. This coupling mode permits phonons of a different polarization to cross the interface, providing an independent mechanism for energy dissipation characterized by expressions like those above, but with $k_{a,trans}$ substituted for k_a (and in a more thorough analysis, a different modulus, etc.). Where the axial stiffness equals $k_{s,bear}$, the increment in energy dissipation is $\sim 0.5^{1.7} P_{drag} \approx 0.3 P_{drag}$.

c. Band-stiffness scattering. Following the procedure in the previous section, expressions analogous to Eqs. (10.21) and (10.22) for power dissipation from band-stiffness scattering, Eq. (7.35), are

$$P_{drag} < 3.0 \times 10^{-33} \, k_a^{1.7} R^2 \frac{\Delta k_a}{k_a} v^2 S \tag{10.23}$$

and

$$P_{drag} < 2.7 \times 10^{-33} \, k_{s,bear}^{1.7} (\ell r_{eff})^{-0.7} R^2 \frac{\Delta k_a}{k_a} v^2 \tag{10.24}$$

In the coplanar ring model, R (Section 7.2.6c) equals $|m/(m-n)|$; if the interatomic spacings in the inner and outer rings are equal, then $R \approx 10$ when $s_{gap} \approx 0.2$ nm, and $r_{inner} \leq 2$ nm. Regardless of bearing radius, differences in surface structure or strain on opposite sides of the interface can be used to ensure that $R \leq 10$.

The parameter $\Delta k_a/k_a$ can be estimated from variations in the stiffness of nonbonded interactions between rows of equally spaced atoms as a function of their offset from alignment. Like many differential quantities, it is strongly dependent on the spatial frequencies involved. For first-row atoms (taking carbon as a model), $\Delta k_a/k_a \approx 0.3$ to 0.4 (at a stiffness-per-atom of 1 and 10 N/m, respectively) where $d_a = 0.25$ nm, and ~ 0.001 to 0.003 where $d_a = 0.125$ nm. This value of d_a cannot be physically achieved in coplanar rings, but it correctly models a ring sandwiched between two other equidistant rings having $d_a = 0.25$ nm and a rotational offset of 0.125 nm.

With the parameters used in the previous section, and $R = 10$ and $\Delta k_a/k_a = 0.4$, $P_{drag} \approx 2 \times 10^{-14}$ W at $v = 1$ m/s; with $\Delta k_a/k_a = 0.003$, $P_{drag} \approx 1.5 \times 10^{-16}$ W.

d. Band-flutter scattering. Again following the procedure in Section 10.4.6b, the expression for power dissipation from band-flutter scattering, Eq. (7.37), becomes

$$P_{drag} < 1.2 \times 10^{-31} \, k_{s,bear}^{1.7} (\ell r_{eff})^{-0.7} R^2 (A/d_a)^2 v^2 \tag{10.25}$$

The amplitude of the interfacial deformation, A, can be roughly estimated from Δk_a, R, d_a, and a characteristic elastic modulus M. From Eq. (3.18) and the associated discussion, it can be seen that Δp in the interface $\leq 3 \times 10^{-11} \Delta k_a$. A pressure distribution varying sinusoidally across a surface with a spatial frequency k produces displacements that decrease with depth on a length scale $w = 1/k \approx R d_a/2\pi$, or ~ 0.4 nm for systems with $d_a = 0.25$ nm and parameters like those described in Section 10.4.6b. The amplitude $A \approx w \Delta p/M$; with $k_a = 8 \times 10^{19} (\text{N/m·m}^2)$, $\Delta k_a/k_a = 0.4$, and the previous values, $P_{drag} < 5 \times 10^{-18}$ W at $v = 1$ m/s; this is negligible in comparison to band-stiffness scattering.

e. Thermoelastic damping. Using diamond material parameters and the effective thickness w from the previous section together with $\tau_{therm} = 10^{-12}$ s yields the following specialization of Eq. (7.54):

$$P_{drag} \approx 4.3 \times 10^{-27} 2 \pi r_{eff} \ell w (\Delta p v/d_a)^2 \tag{10.26}$$

Applying the assumptions of Section 7.4.1 with the parameters assumed in Section 10.4.6c and the estimate of ℓ and Δp from the previous section yields $P_{\text{drag}} \approx 6 \times 10^{-16}\,\text{W}$ ($\Delta k_a/k_a = 0.4$) and $\sim 8 \times 10^{-20}\,\text{W}$ ($\Delta k_a/k_a = 0.003$), for $v = 1\,\text{m/s}$. This too is negligible in comparison to band-stiffness scattering.

f. Summary. For well-designed bearings on a nanometer scale at 300 K, acoustic radiation losses typically are negligible compared with losses resulting from phonon interactions. The latter loss mechanisms all scale as v^2. Dropping the apparently negligible contributions from mechanisms other than shear-reflection and band-stiffness scattering interactions, and multiplying drag by 1.3 to approximately account for the effects of a transverse interfacial stiffness giving an axial bearing stiffness equaling $k_{\text{s,bear}}$, yields an estimated bound on the total bearing drag:

$$P_{\text{drag}} < \left(2.0 \times 10^{-33} + 3.5 \times 10^{-33} \frac{\Delta k_a}{k_a} R^2 \right) \frac{k_{\text{s,bear}}^{1.7}}{(\ell r_{\text{eff}})^{0.7}} v^2 \qquad (10.27)$$

With $k_{\text{s,bear}} = 1000\,\text{N/m}$, $r_{\text{eff}} = \ell = 2\,\text{nm}$, and $R = 10$, $P_{\text{drag}} \approx 2.7 \times 10^{-14} v^2\,\text{W}$ ($\Delta k_a/k_a = 0.4$), dominated by band-stiffness scattering, or $\approx 5.8 \times 10^{-16} v^2\,\text{W}$ ($\Delta k_a/k_a = 0.003$), with a substantial contribution from shear-reflection drag. Using the higher value and $v = 1\,\text{m/s}$, a bearing of this sort is estimated to dissipate $< 0.06 k T_{300}$ per rotation. This energy dissipation (which can be reduced with changes in design) is consistent with high system efficiency, if a rotation of the bearing is associated with some process that does many times $k T_{300}$ of useful work (e.g., in a motor, Section 11.7, or a mechanosynthetic system, Section 13.3). It is high, however, if one instead considers power dissipation per unit volume ($\sim 10^{11}\,\text{W/m}^3$).

(Note that the approximations made in deriving the drag expressions in this section are intended to provide only a gross upper bound on the magnitude of the drag. Constants in these expressions are sometimes written to two significant digits, but usually lack the accuracy that this notation might imply.)

10.4.7. Sleeve bearings in molecular detail

Sleeve bearings can usefully be examined at two levels of molecular detail: interfacial structure and overall structure. The design of relatively large bearings can exploit strained cylindrical shells (Section 9.6.1). Bearings of this sort can be viewed as forming families with specific unstrained structures and crystallographic orientations (relative to the interface and the bearing axis), and with specific surface terminations at the sliding interface. Within such a family, the bearing radius r_{eff} and the spacing between the surfaces s_{gap} are determined within broad limits by the choice of the inner and outer circumferences (in lattice units). For these bearings, specification of the interfacial structure is primary, since the overall structure is simple and repetitive.

For smaller bearings, in contrast, strained cylindrical shells become a poor model. The structure of the shaft then becomes a special case rather than a member of a parameterized family, and the overall structure must be considered as a unit. Examples drawn from both classes are given in the following sections.

a. Bearing interfaces for strained-shell structures.   Figures 10.15 and 10.16 illustrate several pairs of terminated diamond surfaces, each forming an interface suitable for use in a symmetrical sleeve bearing. These interfaces have differing properties with respect to axial stiffness and drag.

Each of the interfaces shown exhibits substantial axial stiffness at suitably small values of s_{gap}. Along the axis, the surface atomic rows on opposite sides of each interface have identical spacings, hence the sinusoidal potentials for sliding

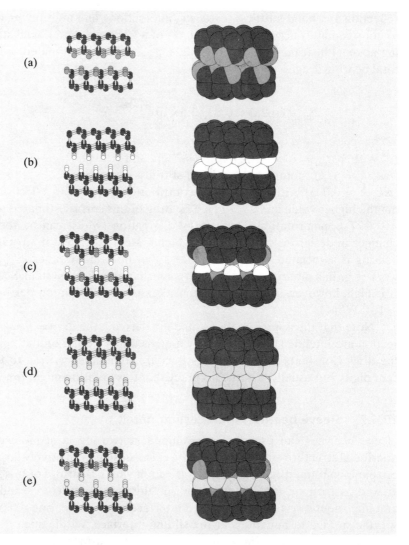

Figure 10.15.   Several sliding interfaces based on diamond (111) surfaces, with nitrogen termination (a), hydrogen termination (b), alternating, interlocking rows with nitrogen and hydrogen termination (c), fluorine termination (d), and alternating, interlocking rows of nitrogen and fluorine termination (e). Note that pairs of surfaces from (a), (b), and (d) could be combined, as could a pair from (c) and (e). In each diagram the direction of sliding is nearly perpendicular to the page. Different terminations offer different functional relationships among spacing, shear stiffness, and compressive stiffness.

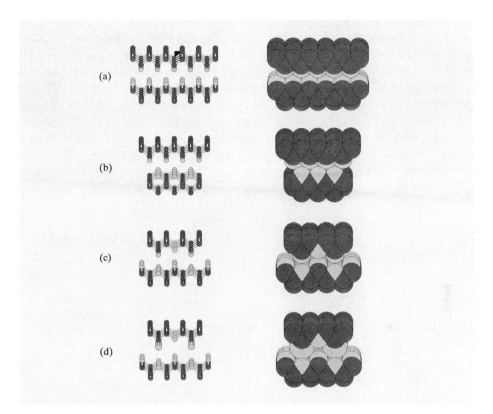

Figure 10.16. Several sliding interfaces based on diamond (100) surfaces, with oxygen termination and aligned rows (a); oxygen termination and crossed rows (b); oxygen termination with alternating deleted rows, providing an interlocking surface (c); and surfaces like (c), but with the exposed terminating rows on one surface consisting of sulfur rather than oxygen (d). In each diagram the direction of sliding is nearly perpendicular to the page. Different terminations offer different functional relationships among spacing, shear stiffness, and compressive stiffness.

in this direction add in phase over the entire interface. Interfaces 10.15(c), 10.15(e), 10.16(c), and 10.16(d) include interlocking grooves, increasing the energy barrier for axial sliding and (in most instances) the stiffness as well. In these structures, the axial stiffness can equal or exceed the transverse-displacement stiffness $k_{s,bear}$. An interface with mismatched spacings in both dimensions [e.g., combining 10.15(b) and 10.16(a), can permit both rotation and axial sliding (see Section 10.5.2].

Interfaces 10.15(c), 10.15(e), and 10.16(a–d) exhibit values of $d_a \approx 0.25$ nm, corresponding to the high-drag cases in Sections 10.4.6c and 10.4.6d. Interfaces 10.15(b) and 10.15(d), however, will (in the absence of significant axial loads) exhibit $d_{a,eff} \approx 0.125$ nm: an atom on one surface interacts equally with two staggered rows on the other, halving the effective spacing for most purposes. This value of $d_{a,eff}$ corresponds to the low-drag cases in Sections 10.4.6c and 10.4.6d. A nitrogen-terminated (111) surface, Figure 10.15(a), has $d_{a,eff} \approx 0.125$ nm where first-layer interactions are concerned, but the second atomic layer introduces significant interactions with $d_a \approx 0.25$ nm.

b.   Interfacial stability.   Each of the interfaces in Figures 10.15 and 10.16 (and that shown in Figures 10.17 and 10.18) appears stable enough for practical use under the baseline conditions assumed in this volume (i.e., no extreme temperatures, no UV exposure, no extraneous reactive molecules, no extreme mechanical loads); further, the symmetrical and low-polarity interfaces should guarantee the absence of contact charging (Section 6.4.7). The low-valence atoms used to terminate each surface form strong bonds to carbon and (usually) weaker bonds to one another. A reaction between one surface and the other typically must form a bond between two low-valence atoms at the expense of cleaving two bonds to carbon. Since this would be a strongly endoergic process, the energy barrier is large (>500 maJ) and the rate of occurrence negligible

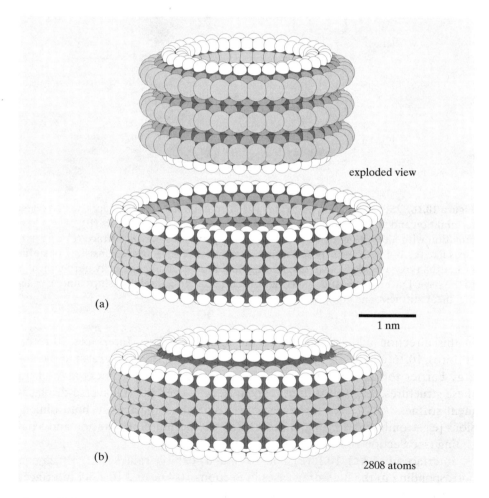

exploded view

(a)

1 nm

(b)

2808 atoms

Figure 10.17.   A 2808-atom strained-shell sleeve bearing with an interlocking-groove interface derived by modifying a diamond (100) surface; (b) shows the shaft within the sleeve, (a) shows an exploded view. This design was developed in collaboration with R. Merkle at the Xerox Palo Alto Research Center, using an automated structure-generation package (Merkle, 1991). It was then energy minimized and analyzed using the Polygraf molecular modeling software (Polygen Molecular Simulations, Inc., Waltham, MA).

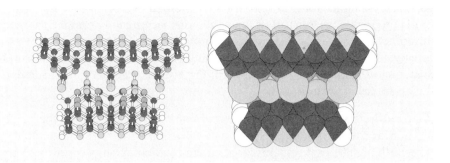

Figure 10.18. A section through the interface of the bearing shown in Figure 10.17. The view roughly parallels the planar diagrams shown in Figure 10.16, differing in the presence of curvature and in the use of a different (100)-based surface modification for the interface structure. The use of sulfur bridges on the outer shaft surface rather than oxygen both reduces strain (via longer bonds) and increases interfacial stiffness (via larger steric radii).

($< 10^{-39}$ s^{-1}). These remarks apply with equal force to a wide variety of sliding and rolling interfaces with similar termination by low-valence atoms.

Graphitic interfaces are also desirable, but the potential reactivity of their unsaturated tetravalent atoms demands attention. Experiments show that graphite transforms to a transparent, nondiamond phase at room temperature under pressures of ~18 GPa (Utsumi and Yagi, 1991); this pressure corresponds to a compressive load of ~0.5 nN per atom (with an associated stiffness on the order of 20 N/m). The transformation is nucleated at specific sites in a crystal, and can be observed to spread over a period of ~1 hour. Small areas of graphitic bearing interface can presumably be made that lack suitable nucleation sites, hence interfacial pressures of ~18 GPa should be consistent with chemical stability.

c. A specific strained-shell structure. Figure 10.17 illustrates a strained-shell structure containing 2808 atoms, with a shaft of 34-fold rotational symmetry and a sleeve of 46-fold rotational symmetry; lcm$(n,m)/n = 23$. The dimensions of the interface are somewhat ill defined, but approximate values are $r_{eff} \approx 3.5$ nm and $\ell \approx 1.0$ nm; the external radius at the illustrated termination surface is ~4.8 nm. This bearing was designed (perhaps over designed) for large axial stiffness, achieved by combining interlocking ridges with large nonbonded contact forces (Figure 10.18 shows a section through the interface). A molecular mechanics model (see Figure 10.17 caption) predicts that the maximally strained C—C bonds of the outer surface of the sleeve have lengths of ~0.166 nm, well within the estimated positive-stiffness limit of 0.187 nm (Section 3.3.3a). The closest nonbonded contacts across the interface are ~0.284 (for S|O) and ~0.270 (for S|N); the corresponding forces are ~0.56 and ~1.1 nN; the corresponding stiffnesses are ~20 and ~40 N/m. The compliance of the interface stemming from nonbonded forces is ~10^{-21} m^2·m/N (from MM2/CSC), comparable to the shear compliance of a ~0.5 nm thickness of diamond. Bond angle bending adds significant compliance to the interface, but the total compliance does not exceed the shear compliance of a sheet of diamond a few nanometers thick. For the bearing as a whole, the axial stiffness should exceed 2000 N/m.

Energy minimization of this structure yielded no rotation of the shaft with respect to the sleeve, regardless of angular displacement. Examination of energy differences for minimized structures as a function of angular displacement indicates values of $\Delta \mathscr{V}_{barrier}$ <0.03 maJ. Owing to high interfacial stiffnesses, however, the estimated drag in a bearing of this sort is larger than that calculated in the examples of Section 10.4.6.

d. Small sleeve-bearing structures. In small sleeve bearings, the pursuit of high-order rotational symmetry dramatically limits the set of acceptable structures, although this set remains large enough to make enumeration challenging. A segment of any of the structures shown in Figures 9.5 and 9.6 could serve as a shaft, given a suitable sleeve. Thin strained shells, backed up by layers with regular dislocation-like structures, could serve as shafts or sleeves, as could less-regular, special-case structures.

e. A specific small sleeve-bearing structure. An example combining a special shaft with a special sleeve is shown in Figure 10.19. This structure makes use of chains of sp^3 nitrogen atoms to form ridges (having high stiffness and spatial frequency) on both the shaft and the sleeve. These features require attention because N—N bonds are known to be weak, and *isolated* chains of this sort are apparently unstable (a search of the chemical literature failed to identify a well-characterized example). In the present instance, however, these chains are not isolated. Each N atom is also bonded to carbon and subject to the familiar constraints of a diamondoid structure: momentary thermal cleavage of an N—N bond is resisted (and usually reversed) by elastic restoring forces from the surrounding structure. The most accessible failure mode in this system is appar-

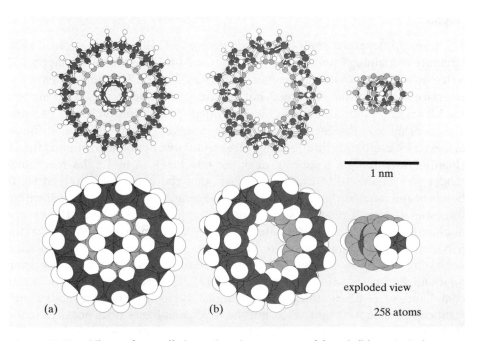

1 nm

(a)   (b)   exploded view

258 atoms

Figure 10.19. Views of a small sleeve-bearing structure: (a) end, (b) exploded.

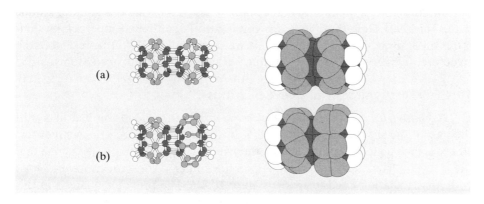

(a)

(b)

Figure 10.20. The shaft of the bearing shown in Figure 10.19 (a), showing a hypothetical failure mode (b).

ently the transformation of a nitrogen chain in the shaft into a series of N=N dimers (Figure 10.20), but estimates of bond energies and strain energies suggest that this transformation is only moderately exoergic (~70 maJ?) on a per-bond-cleaved basis. Moreover, the formation of a single pi-bonded dimer by cleavage of two adjacent sigma bonds is strongly endoergic (sacrificing two sigma bonds to create, initially, one strongly twisted pi bond), and the simultaneous cleavage of six sigma bonds should be energetically prohibitive. Accordingly, these structures appear sufficiently stable for use, although none of the following design and analysis relies on this.

The stiffness of the bearing interface can be computed from the change in nonbonded interaction energies as a function of relative displacement of the shaft and sleeve, in the absence of structural relaxation. Calculations yield an axial stiffness of ~360 N/m and an isotropic transverse stiffness of ~470 N/m. Rotational energy barriers (computed with structural relaxation) were found to be <0.001 maJ. All computations used the MM2/CSC model extended with parameters to accommodate N—N—N bond angles. The parameter θ_0 for N—N—N angle bending was set at 114.5° to fit AM1 semiempirical computations on H_2NNHNH_2 (performed in collaboration with R. Merkle); k_θ was set at 0.740 aJ/rad^2, equaling the MM2 value for N—N—C angle bending. Bearing stiffnesses (but not barriers) are sensitive to the choice of θ_0. Supplemental torsional parameters are of less significance; values were chosen to match analogous MM2 values.

The total strain energy in this structure is large (~12 aJ), but only ~530 maJ of this is in the form of bond stretching, and this energy is well distributed over many bonds; ~71% of the strain energy is in the form of bond angle-bending, much of this owing to the presence of 22 cyclobutane rings within the structure. The closest nonbonded distance between shaft and sleeve is ~0.26 nm (N|N). For a relaxed model of a shaft popping into (or out of) a sleeve, with the nitrogen chains approximately coplanar, the total energy is increased by ~1.7 aJ, and the closest N|N distance is ~0.236 nm: bond lengths remain reasonable, the N—N bonds are under stabilizing, compressive loads, and the estimated peak forces in achieving this configuration are small compared to bond tensile strengths. Assembly of this bearing from separate components thus appears feasible.

This structure is the first nanomechanical sleeve bearing designed in atomic detail (in 1990). Note that the bearing in Figure 1.1, although superficially similar, is an entirely different structure. It was designed (in 1992) to resemble the structure of Figure 10.19 while omitting questionable sp^3 nitrogen atoms.

10.4.8. Less symmetrical sleeve bearings

a. Asymmetries to compensate for load. Transverse loads on bearings shift the axis of the shaft with respect to that of the sleeve; the use of a symmetrical sleeve is then no longer motivated by the symmetry of the situation. Assume that the sleeve is fixed, and that the shaft rotates under a constant transverse load with an orientation fixed in space (Figure 10.21). The previous analysis in this chapter assumes that this asymmetric load is supported by the topmost atomic layers, chiefly through differences in overlap repulsion caused by shaft displacement. Alternatively, however, the load on the shaft can (in many instances) be supported by differences in van der Waals attraction resulting from an asymmetric structure having regions of different Hamaker constant $\mathcal{A}$ (e.g., differing atom number densities) immediately behind the surface layer. This approach can reduce $\Delta\mathcal{V}_{barrier}$, and can increase load-bearing capacity without increasing interfacial stiffness.

In small loaded bearings, $\Delta\mathcal{V}_{barrier}$ can be large (Figure 10.13). Where the shaft is of high symmetry and the sleeve has >10 interfacial atoms, the calculations of Section 10.3 suggest that an asymmetric placement of sleeve atoms can be chosen to ensure that $\Delta\mathcal{V}_{barrier} < 1$ maJ. In many instances, a slightly perturbed version of a symmetrical structure can accomplish this.

b. Asymmetries to simplify construction. Bearings having interfaces with shallow grooves or relatively large s_{gap} often can be assembled from separate shafts and sleeves in the obvious manner, by pressing the shaft into the sleeve. Large loads often can be applied, and the final energy minimum with respect to axial displacement can be deep and stiff.

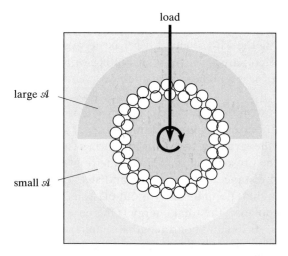

Figure 10.21. Schematic diagram of a loaded bearing, with compensating asymmetric van der Waals attractions.

Structures with more strongly interlocking grooves cannot be assembled in this fashion. If they are to have full symmetry, the sleeve must be synthesized *in situ* around the shaft (e.g., building out from polymeric bands that become the ridges on the sleeve), or must have a final assembly step that closes an adhesive interface. Sleeves made from two C-shaped segments with an atomic-scale discontinuity at the two seams have lower symmetry but can nonetheless be designed to exhibit low $\Delta \mathcal{V}_{\text{barrier}}$.

10.5. Further applications of sliding-interface bearings

The preceding results regarding irregular objects and symmetrical sleeve bearings shed light on a wide range of other sliding-interface systems. Among these are nuts turning on screws, rods sliding in sleeves, and a class of constant-force springs. Energy dissipation analyses are omitted, but follow the principles discussed in Section 10.4.6.

10.5.1. Nuts and screws

The thread structure of a nut-and-screw combination can formally be generated by dividing a grooved, strained-shell sleeve bearing parallel to the axis along one side, shifting one cut surface in the axial direction by an integral number of groove spacings, and reconnecting. The result is locally similar to the original bearing, save for the introduction of a helical pitch in the grooves and ridges, which accordingly must come to an end at some point (in any straight, finite structure).

What is the static friction of such a structure, assuming that the helical atomic rows of the inner and outer surfaces have nonmatching spacings? (Note that these spacings need not be commensurate; in good designs, the spacings can lack small common multiples.) Figure 10.5 approximates the potential of an atom with respect to a row as a single sinusoidal potential with some amplitude and phase, represented as a vector magnitude and angle in a plane. In this representation, the potential of a *series* of atoms in one surface of a uniform, nonmatching interface takes the form shown in Figure 10.22: all vectors are of the same magnitude, and each is rotated with respect to the last by a fixed angle. Where that angle is not zero, the resultant vector always lies on a circle passing through the origin. Its magnitude oscillates between fixed bounds and periodically assumes a small value; $\Delta \mathcal{V}_{\text{barrier}}$ and the static friction accordingly do the same. A bound on the magnitude of the barrier is

$$\Delta \mathcal{V}_{\text{barrier,max}} = \frac{\Delta \mathcal{V}_{\text{barrier, single}}}{2} \sqrt{1 + \tan^2 \left(\frac{\pi - \phi}{2} \right)} \qquad (10.28)$$

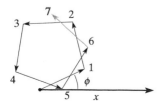

Figure 10.22. Vector representation of interatomic potentials (see Figure 10.5) for a row of regularly spaced atoms in one surface sliding over regularly spaced atoms in another. Each atom experiences an approximately sinusoidal potential, and the phase difference between the sinusoid experienced by one atom and that experienced by its successor in the row corresponds to a constant difference in the angle between their vectors.

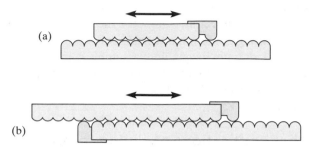

Figure 10.23. Linear representation of the sliding of two finite but otherwise regular surfaces over one another. In (a), the range of motion of one surface places it within the width of the other surface at all times; this corresponds (for example) to a nut turning on the middle of a long screw. In (b), the range of motion of the surfaces enables each to extend beyond one limit of the other at all times; this corresponds (for example) to a screw partially inserted into a deep threaded hole. In (a), the irregularities corresponding to both ends of the overlap region move together over a surface of a single spatial frequency; tuned irregularities at one end can keep $\Delta\mathcal{V}_{\text{barrier}}$ low. In (b), the irregularities move in opposite directions over surfaces of differing spatial frequency, and tuned irregularities are usually needed at both ends (that is, on both sides of the interface) if the two sinusoidal components of the potential are both to be cancelled.

where ϕ is the phase angle between succeeding atoms in the surface under consideration.

This result indicates that the static friction of a nut-and-screw system (under low loads) depends chiefly on the end conditions. With the right choice of interface length, $\Delta\mathcal{V}_{\text{barrier}}$ is low because the resultant vector is of small magnitude. For other choices of interface length, $\Delta\mathcal{V}_{\text{barrier}}$ can be made low by the methods discussed in Section 10.3, that is, tuned structural irregularities can be introduced at one end of a nut in such a way that the amplitude and phase of their contribution to the potential cancels the residual contribution from the regular portion of the nut. As shown in Figure 10.23, where the overlap of the nut and screw is variable, minimizing the sinusoidal component of the potential can require tuned irregularities at two sites.

Nut-and-screw systems with high potential barriers can be used as components of adjustable-length struts (see the related discussion in Section 10.9.2). Systems with low potential barriers can be used in jacks or power screws.

10.5.2. Rods in sleeves

The analysis of rod-in-sleeve systems is entirely analogous to that for nut-and-screw systems, save that the helical grooves take the degenerate form of straight lines. Again, end conditions determine the magnitude of the static friction, and again, choice of length or tuning of irregular structures can yield low values of $\Delta\mathcal{V}_{\text{barrier}}$. As noted in Section 10.4.7a, cylindrical interfaces can be designed to permit simultaneous axial sliding and rotation in any proportion.

10.5.3. Constant force springs

In the variable-overlap case mentioned in Section 10.5.1, suppression of sinusoidal potentials does not leave a flat potential, because the energy of the system is in general a function of the overlap. Where local interactions are dominant, the potential energy is proportional to the overlap, with a positive or negative constant of proportionality depending on the net energy of the surface-surface interaction. A sliding rod in a sleeve of this sort, tuned for smooth motion, acts as a constant-force spring over its available range of motion.* Since the characteristic attractive interaction energies of surfaces (Section 3.5.2) are on the order of 100 maJ/nm^2, the force with which a rod can be made to retract into a sleeve is on the order of $0.6r$ nN, where r is in nanometers (linear scaling fails for very slim rods). Since repulsive interaction energies can be much larger, forces for a constant-force spring operating in this mode can be much larger, and considerable energy can be stored in elastic deformation of the rod and sleeve.

10.6. Atomic-axle bearings

10.6.1. Bonded bearings

Sigma bonds permit rotation, in the absence of mechanical interference between the bonded moieties. Barriers for rotation vary. Two model systems are cubylcubane **10.1** and phenylcubane **10.2**

10.1   **10.2**

with MM2/CSC values of $\Delta \mathcal{V}_{barrier}$ equaling 11.5 and 0.3 maJ respectively; note that the latter has a structure with 3-fold rotational symmetry interacting with one with 2-fold symmetry. The stiffness for shearing displacements of sigma bearings of this sort is about twice the transverse-displacement stiffness of a sigma bond (Section 3.3.2b), ~60 N/m for structures like cubylcubane. A rotor supported by two such bearings has a transverse displacement stiffness of ~120 N/m, and (with a proper choice of relative phases) small values of $\Delta \mathcal{V}_{barrier}$. Figure 10.32 shows a planetary gear system that includes nine sigma-bond bearings.

Using a $-C{\equiv}C-$ unit in place of a sigma bond can drop $\Delta \mathcal{V}_{barrier}$ to near zero (Knox, 1971), at the expense of increasing the volume and greatly decreasing resistance to shearing displacements. Organometallic bearings based on the ferrocene structure can have fairly low barriers [~7 maJ (Huheey, 1978)] and should have good stiffness.

10.6.2. Atomic-point bearings

A rotor with protruding atoms on opposite sites can be captured between a pair of surfaces with matching hollows. Values of $\Delta \mathcal{V}_{barrier}$ can be low if the interacting surfaces are well designed. Under substantial compressive loads, stiffness

---

* Conventional springs, in which force is approximately proportional to displacement, can be implemented in many ways. Most stable diamondoid structures (as well as many rod and cage structures of lower compliance and less diamondlike structure) are perfectly elastic, save for energy dissipation mechanisms that depend on the speed of deformation.

can be moderately high. A macroscale analogue of a bearing of this sort would have sliding interfaces, but the reduction of the contact region to a single atom (on one side of the interface), together with the placement of that atom on the axis of rotation, makes atomic-point bearings qualitatively different.

10.7. Gears, rollers, belts, and cams

10.7.1. Spur gears

Spur gears find extensive use in machinery, chiefly for transmitting power between shafts of differing angular frequency. Spur gears achieve zero slip (so long as the teeth remain meshed), and ideally exhibit minimal energy barriers during rotation, minimal dynamic friction, and maximal stiffness in resisting interfacial shear.

Nanomechanical gears can exploit various physical effects to implement "teeth." These include complementary patterns of charge, of hydrogen bonds, or of °dipolar bonds. The most straightforward effect to analyze, however, is overlap repulsion between surfaces with complementary shapes (the effect exploited by macroscale gears).

In conventional gearing, teeth are carefully shaped (e.g., in involute curves) to permit rolling motion of one tooth across another. Because a tooth on one gear meshes between two teeth on the opposite gear, but can only roll across one of them, a clearance (backlash) must be provided. Accordingly, on reversal of torque, teeth lift from one surface before contacting another. These complexities of geometry are both impossible and unnecessary in nanomechanical gears. Since sliding of one tooth over another is just another variation on a sliding-interface bearing, there is no need to construct involute surfaces. Since the steric interaction between atoms is soft, and since sliding contacts between teeth are acceptable, no backlash need be provided (and would not be a well-defined quantity if it were).

Small nanomechanical gears of this class use single atoms or rows of atoms as steric teeth; any of the grooved interfaces in Figures 10.15 and 10.16 could serve this role in a strained-shell structure. As with bearings, small special-case structures are also feasible. And again, models based on rigid, circular arrays of atoms reduce computational costs while preserving the essential physics.

a.   A relaxation-free model of meshing gear teeth.   Figure 10.24 illustrates a model of steric gears as rigid, coplanar rings of atoms. The calculations described in this section, like those in the discussion of bearings, assume the MM2 exp–6 nonbonded potential, Eq. (3.8), with H atoms separated by $d_a =$ 0.3 nm. Again, the general nature of the results is insensitive to the choice of potential, atom type, and interatomic spacing.

Figure 10.25 plots barrier heights as a function of the number of teeth n (here, equal for both gears) and the separation s_{gap} (see Figure 10.24). The lower family of curves shows barriers for corotation at a uniform angular velocity without slippage. The upper family of curves shows barriers for slipping of one ring with respect to another, based on a search for minimum-energy pathways at a range of rotational angles. As can be seen, with small values of s_{gap} (<0.12 nm)

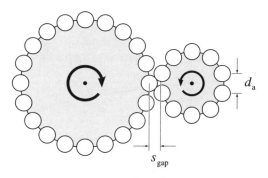

Figure 10.24. Coplanar ring model for a steric gear.

and moderate numbers of teeth (>20), energy barriers to slippage are large (>500 maJ) and energy barriers to corotation are small (<0.01 maJ).

Under interfacial shear loads (required, for example, to transmit power), the symmetry of the system is degraded, and corotational barriers are larger. The effects of load can be modeled by a constant angular offset between the rings relative to their minimum-energy, symmetrically meshed geometry. Figure 10.26 shows corotational barrier heights for two values of n at several values of the offset (measured by displacement at the ring circumference). For n as small as 10 and offsets as large as 0.01 nm, the corotational barrier remains <1 maJ; at small values of s_{gap}, this offset is adequate to transmit a shear force of ~1 nN per fully meshed tooth. At a gear rim speed $v = 1$ m/s, this corresponds to a transmitted power of 1 nW.

A prominent feature of Figures 10.25 and 10.26 is the presence of sharp dips in the barrier height at locations that depend both on n and s_{gap}. Sharp dips in barrier height are a robust feature of a broad class of nanomechanical systems,

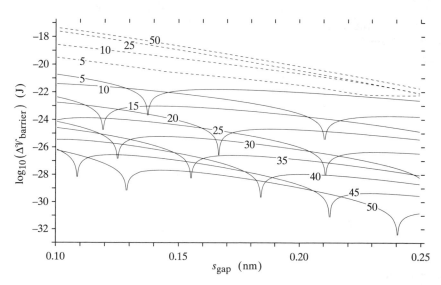

Figure 10.25. Barriers for gear corotation (solid lines) and slippage (dashed lines) in the rigid coplanar ring model (Section 10.7.1a), for a range of between-ring separations s_{gap} and several tooth-numbers n. Energies increase as matching rings are added.

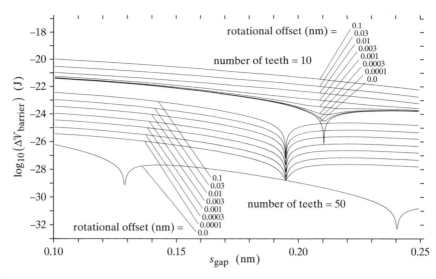

Figure 10.26. Barriers for gear corotation as in Figure 10.25, for $n = 10$ and 50, modeling various torsional loads with varying values of between-ring rotational offset.

for reasons illustrated by the diagrams in Figure 10.27. For a single tooth, the energy is low (and nearly constant) before and passing through the meshing zone, and is high (and nearly constant) as it passes directly between the two gears (as in the potential energy curves shown in Figure 10.27). The potential energy of a gear (in the present model) is simply the sum of the nonbonded interaction energies of its teeth, occupying evenly spaced points along a suitable potential curve. When a single tooth, or a gap between a pair of teeth, is precisely between the two gears, the slope of the potential energy function is zero by symmetry [see (a) and (b) in Figure 10.27]. The potential is periodic (again by symmetry); in the sinusoidal approximation, the slope of the potential is maximal at position (c), halfway between (a) and (b). The slope of the bearing potential energy function at position (c) can be represented as the sum of the slopes at the points in diagram (d), folding points from the left side to the right, and giving them negative weights in the sum.

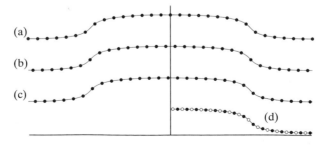

Figure 10.27. Schematic illustration of the potential energies of gear teeth as a function of rotational angles (shoulders exaggerated). Diagrams (a) and (b) represent symmetrical positions, and (c) an intermediate position. Diagram (d) represents (c) with the left-hand points folded over and reversed in sign, to illustrate certain properties of the slope of the potential function with respect to gear rotation angle.

The sign of this slope depends on the positions and weights of the points on the shoulder of the potential curve. Where this shoulder is well away from the center (as it was not in Figure 10.2), changes in the spacing of the points, or in the position of the shoulder (e.g., by changing n or s_{gap}), can readily change the sign of the slope. For a clear example, consider an interpoint spacing that places only one point in a region of high slope: by varying the parameters, points of either positive or negative weight can be placed in that position.

Finally, where state (c) is of zero slope, it becomes an extreme of the potential energy function, the spatial frequency of barrier crossing is doubled, and the barrier heights are greatly reduced (much as barrier heights are reduced by higher spatial frequencies in bearings). In moving along a curve like those in Figures 10.25 and 10.26, the energy difference between states like (a) and (b) in Figure 10.27 changes sign. A result of this sort can be expected in a wide range of circumstances.

b.   Energy dissipation in gear contacts.   The energy dissipation mechanisms in gears parallel those for sliding contacts on a surface (within the approximations already adopted). Acoustic radiation losses are small for most well-designed gears. Relative to the figures used in Section 10.3.4, typical stiffnesses and compressive loads are anticipated to be ~10 times larger (~300 N/m, ~10 nN), resulting in phonon scattering cross sections and thermoclastic effects ~100 times larger. The corresponding power levels (at a 1 m/s interfacial rolling speed) are ~3×10^{-14} and ~10^{-16} W for phonon scattering and thermoelastic drag, respectively. For a gear system operating at a shear force of 1 nN, these losses amount to ~0.003% of the transmitted power.

c.   Integrated bearings and gears.   Gears of the sort discussed in this section require substantial compressive loads at the interface to ensure good meshing and stiff contact between teeth. In macroscale gearing this would be achieved (were it necessary) by means of loads transmitted through bearing surfaces on adjacent shafts. In nanomechanical gearing, however, the distinction between a gear interface and a bearing interface is chiefly a matter of relative interatomic spacings and curvatures of the opposing surfaces. The computational examples of bearings and gears in this chapter have used the same spacings and atom types on the convex surface of each interface, differing only in the nature of the opposed surfaces. Thus, a single surface can serve both roles, as in Figure 10.28,

gear interface

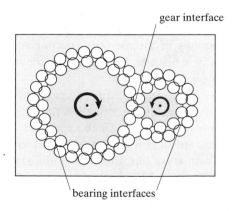

bearing interfaces

Figure 10.28.   Schematic diagram of two gears supported by bearing surfaces.

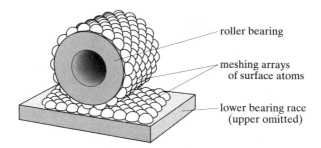

Figure 10.29. Schematic illustration of a roller between two race surfaces (one omitted), showing gearlike meshing of atomic rows.

and the required compressive loads can be transmitted to the housing structure by the gear surfaces themselves. In a system of this sort, symmetry constraints do not guarantee low values of $\Delta\mathcal{V}_{barrier}$, but tuning of the interface can readily be applied.

10.7.2. Helical gears

Gears can be constructed from concentric strained shells with opposing torsional shear deformations locked into each shell by bonding them together at the ends. The resulting structures can subject surface atomic rows to substantial helical deformations. Matching gear surfaces of this sort can be made to mesh more smoothly than spur gears of the same radius; in effect, n is larger.

10.7.3. Rack-and-pinion gears and roller bearings

The principles of spur and helical gears are equally applicable when one of the toothed surfaces is flat (producing a rack-and-pinion gear system) or concave (as in planetary gearing). A gear without a shaft can serve as a roller bearing between two flat surfaces (Figure 10.29), or between two concentric cylindrical surfaces. The barrier to slippage in this instance need not be high enough to transmit large shear forces, but if it is sufficiently high to prevent thermally activated transitions, then a series of roller bearings can be made to keep a uniform spacing around the raceway without requiring a cage.

10.7.4. Bevel gears

Where the axes of two coplanar shafts intersect at an angle, power can be transmitted from one to another by means of bevel gears (essentially, rolling cones). In nanomechanical systems, the soft interactions of atomic teeth permit two noncoplanar rings of teeth to mesh essentially as well as coplanar rings; bevel gears with single-atom teeth thus present no special problem.

Larger nanomechanical bevel gears, however, cannot be directly patterned on macroscale gears, owing to the impossibility of making atomic rows that converge smoothly (in spacing and steric radii) toward the tips of the cones. One alternative is to make bevel gears from conical surfaces with matching patterns of teeth and holes, without attempting to form rows like those in traditional bevel gears. The modified (111) surfaces in Figure 10.30, developed into cones,

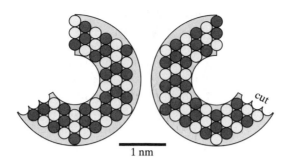

Figure 10.30. Surfaces for a matched pair of small bevel gears; light and dark circles represent protruding atoms (e.g., F) and hollows (e.g., N termination sites) on a modified (111) surface.

have this property; with a modest degree of strain (~6%), they can form cones having a 45° half angle, suitable for shafts meeting at 90°. In small bevel gears based on this approach, the structure immediately beneath the surface must depart from that of a simple strained lattice, and the degree of achievable regularity is at present uncertain. Where all radii of curvature exceed 1 nm, however, irregularities can be buried more deeply and the structural choices for those irregularities become ample to ensure solutions permitting low values of $\Delta \mathcal{V}_{\text{barrier}}$.

10.7.5. Worm gears

Worm gears combine characteristics of nuts and screws with those of gears to yield large gear ratios in a small package. A simple implementation combines a driven screw tangent to (and meshing with) the rim of a helical gear, creating an interface which has short-term fluctuations in potential energy driven by the sliding of the screw with respect to the gear surface, with superimposed long-term fluctuations driven by the meshing and unmeshing of helical teeth with the thread of the screw.

Where static friction is concerned, the low-frequency components are of relatively low importance, owing to the high gear ratio. Further, the smoothness properties of the low-frequency component resemble those of a spur (or more accurately, rack and pinion) gear system with a comparable number of teeth; these properties can be favorable even in the absence of mechanical advantage. The high-frequency components of the fluctuations in potential energy can be made to cancel (to a good approximation) given suitable choices of gear geometry and compressive load, but the analysis is complicated by the superimposed rotation of the helical gear.

10.7.6. Belt-and-roller systems

In engineering practice, tension members stretched over rotors commonly are used to transmit either materials or power (or both). Examples include conveyor belts moving over rollers, drive belts moving over pulleys, and chain drives moving over sprockets.

In nanomechanical structures, surfaces commonly have periodicity like chains and sprockets; accordingly, it is natural to design devices in such a way that belts and rollers mesh. The meshing and unmeshing of a belt and roller then locally resembles the meshing and unmeshing of a rack and pinion, and values of $\Delta\mathcal{V}_{\text{barrier}}$ can have parallel behavior. A significant difference is that the larger region of contact (where the belt wraps around the roller) decreases the required contact pressure and areal stiffness for a given interfacial shear load, and this, in turn, decreases thermoelastic damping and phonon scattering. The discussion in following chapters assumes that the dominant energy dissipation mechanism is shear in the bearing interface of the roller.

The meshing of bumps on rollers and belts can excite resonant transverse vibrational modes in the belts. For example, in a belt with a length of 20 nm and a ratio of tension to linear mass density equal to that of diamond under a tensile stress of 10 GPa, the transverse wave speed is $\sim 1.7 \times 10^3$ m/s, and the frequency of the lowest normal mode is $\sim 1.3 \times 10^{11}$ rad/s. If bumps on the rollers have a spacing $d_a = 0.25$ nm, then the lowest resonant belt speed is ~ 5 m/s. Operation of equipment between (rather than below) resonant modes is common practice in macroscale engineering, but requires attention to start-up dynamics as resonant conditions are traversed; nanomechanical engineering is no different in this regard.

Belt-and-roller devices play a central role in the mechanochemical systems discussed in Section 13.3, providing for molecular conveyance and reactive encounters. It might seem that mechanical conveyance must consume more energy than spontaneous diffusion in the solution phase, but motions in eutactic systems can in fact be more efficient than comparable motions in disordered systems. The entropic free-energy cost of diffusive transport down a concentration gradient is exactly the same as the free-energy cost of the work required to force the molecules to move through the liquid at the same mean velocity against the resistance of viscous drag. Belts in vacuum move molecules more efficiently.

10.7.7. Cams

In a cam mechanism, a contoured surface moves beneath a follower, and its bumps and hollows cause the follower to rise and fall. This provides a way of deriving complex follower motions from a simple motion of the cam surface, and is thus of broad usefulness in mechanical design.

A contoured surface can be constructed (conceptually, as distinct from mechanosynthetically) as a thin slab of diamond distorted by a buried array of dislocations. The resulting surface can be reasonably smooth and regular, save for the irregularities necessitated by the contour itself.

A follower can either slide or roll over the cam surface. Owing to the irregularity associated with the varying slope (and elastic strain) of the cam surface, the principles discussed in Sections 10.3 through 10.5 do not guarantee that a follower can be designed to slide with low energy barriers. A follower surface that is left free to pivot, thereby keeping it flat to the cam surface, could in some instances reduce the effect of these irregularities. A structure with a region of sliding contact that is wide but short (in the sliding direction) should permit the design of cam surfaces with irregularities that sum to a smooth potential; this appears attractive where the cam surface contour must change slope substantially over small distances. Finally, many cam surfaces based on distorted crystals are compatible with a rolling follower having smoothly meshing teeth and low energy barriers.

10.7.8.   Planetary gear systems

Planetary gears can be used to convert shaft power from one angular frequency to another in a stiff colinear mechanism. Figure 10.31 provides a schematic illustration; Figure 10.32 shows the first such mechanism that has been designed in atomic detail (several earlier designs were unstable: energy minimization extruded the gears and produced only wreckage). This design and the following analysis results from a collaboration with R. Merkle.

If the planets are to be arranged symmetrically, the structure must meet the condition

$$n_{\text{outer}} = i\,n_{\text{planets}} - n_{\text{inner}} \qquad (10.29)$$

where n_{outer} is the number of teeth in the outer ring gear, n_{inner} is the number of teeth in inner sun gear, n_{planets} is the number of planet gears, and i is an integer (note that the number of teeth on a planet gear does not appear in this relationship). A quantity that correlates with the smoothness of the potential energy as a function of rotation is the number of distinct orientations m of the planet gears with respect to the radius vector

$$m = n_{\text{planets}} / \gcd\left(n_{\text{planets}}, n_{\text{inner}}\right) \qquad (10.30)$$

Where the number of teeth per planet is small, a large value of m is desirable. In the mechanism illustrated in Figure 10.32, $n_{\text{inner}} = 16$, $n_{\text{planets}} = 9$, and $n_{\text{outer}} = 29$; m has its maximal value, $m = n_{\text{planets}} = 9$. The planet gears have 6

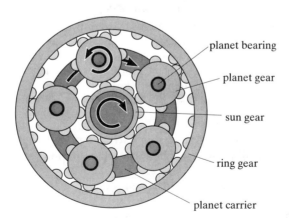

Figure 10.31.   Schematic diagram of a planetary gear (end view). Fast rotation of the sun gear drives slower rotation of the planet carrier, if the ring gear is held fixed. In general, either the sun gear, the ring gear, or the planet carrier can be held fixed, imposing a constraint on the relative motion of the other two components.

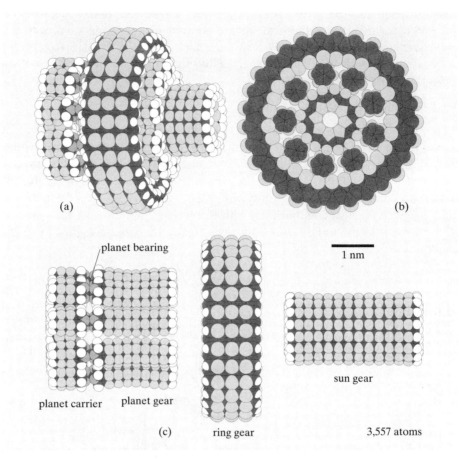

Figure 10.32.   A planetary gear, (a) end view, (b) side view, (c) exploded view. The ring gear structure is a strained Si shell with S termination; the sun gear is a special-case structure related to an O-terminated (100) diamond surface; the planet gears resemble multiple hexaasterane structures with O (rather than CH_2) bridges between the parallel rings; view (c) retains elastic deformations occurring in (a), hence gears are bowed. The planet carrier is adapted from the structure shown in Figure 9.10, and linked to the planet gears using C—C bonded bearings (Section 10.6.1). These bonded linkages are visible in (b). This design results from a collaboration with R. Merkle at the Xerox Palo Alto Research Center; energy minimization was performed using the Polygraf molecular modeling software (Polygen Molecular Simulations, Inc., Waltham, MA).

teeth, hence a motion of the system that carries a planet gear through $1/(6\cdot9) = 1/54$ of a full rotation is a symmetry operation for the potential energy function.

10.8.   Barriers in extended systems

In a typical nanomechanical subsystem, a series of components is mechanically coupled, moving as a nearly rigid unit with respect to some motion coordinate q. Examples include rotating shafts supported by multiple bearings, sliding rods moving over multiple surface regions, and sets of shafts and rods linked by gearing. In each case, where the local negative stiffnesses of energy barriers in the

components are small compared to the positive stiffnesses of the structures linking those components, the energy barriers for the system as a whole are not
those resulting from the potentials of the components taken individually, but
those resulting from the sum of the component potentials with respect to q. In
systems of this sort, the overall barriers are bounded by the sum of the component barriers, but can be much lower if nearly sinusoidal contributions from different components have the same period and are made to cancel.

More generally, a coupled subsystem of this sort can commonly be extended
by adding a *tuning component* that undergoes simple linear or circular motion in
an environment with no function save that of adjusting the overall subsystem
potential. Using levers or gears, the ratio of physical displacements in the tuning
component to those found elsewhere in the subsystem can be made ≥ 1, and
therefore the characteristic frequencies (with respect to q) of interactions
between moving and stationary surfaces in the tuning component can be made
greater than or equal to those in the subsystem as a whole. (Note that the high
spatial frequencies resulting from special symmetries are associated with low
barriers, and hence seldom motivate the introduction of a tuning component.)

10.8.1. Sliding of irregular objects over irregular surfaces

One class of tuning component could be used to smooth the potential associated
with the sliding of an irregular object over an irregular surface, for example, a
rod with irregular features sliding in a sleeve with irregular features. There is no
obvious procedure for choosing object irregularities so as to result in a nearly
constant potential at all displacements with respect to a given irregular surface;
in general, there is some fluctuating potential $\mathcal{V}(q)$. Given that $\mathcal{V}(q)$ does not
result in excessive negative stiffnesses, however, a tuning component can be
introduced (Figure 10.33) in which one or more atoms on the moving part interact with an irregular surface shaped so as to provide a compensating potential
$\mathcal{V}'(q) \approx -\mathcal{V}(q)$.

This design task is feasible if each strip of the fixed surface of the tuning
component interacts with only one moving atom. Each such moving atom can
interact strongly with several stationary atoms at each point, and $\mathcal{V}'(q)$ can be
generated as a sum of the interactions of an indefinitely large number of moving atoms. Multiple strips of interacting atoms in a system of this sort suffice to
provide (1) an indefinitely large energy range for $\mathcal{V}'(q)$, (2) multiple, independent contributions to $\mathcal{V}'(q)$ in each atomic-scale range of q, permitting fine control of its magnitude, and (3) control of $\mathcal{V}'(q)$ with a spatial resolution
comparable to that of the features in $\mathcal{V}(q)$. This provides sufficient freedom to
design systems with $\mathcal{V}'(q) \approx -\mathcal{V}(q)$.

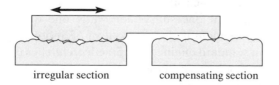

irregular section   compensating section

Figure 10.33.   Linear representation of irregular surfaces sliding over one another.

10.9. Dampers, detents, clutches, and ratchets

The previous sections of this chapter focus on components that can move with smooth or flat potential energy functions (implying low static friction), and with low energy dissipation. Macroscale engineering practice demonstrates the utility of components with properties quite different from these, which are briefly discussed here.

10.9.1. Dampers

Dissipation of energy (e.g., vibrational energy) is often desirable in a dynamical system, and can be accomplished in various ways in nanoscale systems. Stiff, interlocking siding interfaces with large values of R and $\Delta k_a/k_a$ (Section 7.3.5) can exhibit relatively large phonon drag (although the models developed here should not be used to estimate this drag, since they are expected to be conservative when drag is to be minimized, not maximized). Alternatively, interfaces with short, flexible protrusions (e.g., $-C \equiv C-H$) can be designed such that the protrusions hop from one potential well to another as they slide over a facing surface; this can provide substantial damping together with a threshold shear strength for the interface. Energy release per transition must be limited to avoid damage from vibrational stresses. Other linear and nonlinear damping devices can be developed based on the energy dissipation mechanisms described in Chapter 7.

10.9.2. Detents

Violations of the design principles for bearings result in devices with relatively deep potential wells along some motion coordinate. These devices can serve as detents, snap fasteners, and the like. One class of devices resembles a sleeve bearing with a moderately large value of $\mathrm{lcm}(n,m)$, but with flexible protrusions bearing the interfacial load. For a proper choice of (positive) bending stiffness and (negative) interaction stiffness between a protrusion and the opposite surface, the protrusions will approach flexural instability as they move from one potential well to the next, and the single-protrusion potential at that point can have arbitrarily high spatial frequency components, relative to the overall rotational coordinate of the device. The result can be a system with $\mathrm{lcm}(n,m)$ equally spaced potential wells, where $\mathrm{lcm}(n,m) \gg n$ or m.

10.9.3. Clutches

The depth of the potential wells encountered as one surface moves past another depends on their relative positions and the magnitude of the interfacial load. Where two complementary, gearlike surfaces can be pressed together or separated under the control of one subsystem, while one of them moves in an orthogonal degree of freedom under the control of another subsystem, the result is a clutch. As in standard engineering practice, clutches can be used to couple and decouple mechanisms and power sources. The design of such devices is constrained by the potential for excessive stresses or energy dissipation during coupling.

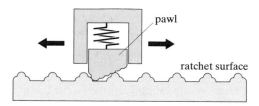

Figure 10.34. Schematic diagram of a ratchet-and-pawl mechanism able to move more easily to the right than to the left under the influence of applied force. Thermal excitation does not constitute an applied force, and the asymmetry of the interface does not bias the direction of thermal hopping.

10.9.4. Ratchets and reversibility

Ratchet mechanisms typically are based on an asymmetric, spring-loaded contact that can resist a large applied force in one direction, yet permit relatively free motion in the other (Figure 10.34). These can readily be implemented on a nanoscale, given suitable teeth and a force (whether from a spring or other source) large enough to make thermally activated hopping rare.

It may be tempting to think that a nanoscale ratchet mechanism would exhibit biased motion purely as a result of thermal excitation, but it cannot. A mechanically-driven motion in one direction may be blocked, but the energy barriers are the same in both directions. In terms of the PES, asymmetries of shape do not destroy the symmetry of detailed balancing (Section 4.3.3a): at equilibrium, transitions in one direction always occur at the same rate as those in the other. A ratchet with biased thermal motion could steadily extract useful work from a heat reservoir at a single temperature, in violation of the second law of thermodynamics.* Thermally excited ratchets are discussed in (Feynman et al., 1963).

10.10. Perspective: nanomachines and macromachines

Macroscale machines are familiar; their properties shape our intuitive sense of what machines are and how they work. Accordingly, it is useful to compare and contrast them with nanomachines, both to encourage free use of sound analogies and to warn of crucial differences.

10.10.1. Similarities between nanomachines and macromachines

The similarities between nanomachines and macromachines are pervasive and fundamental. At the analytical level, systems of both kinds can be described by applying classical mechanics to objects that occupy space, exclude other objects from that space, and resist deformation. At the design level, systems of both kinds must apply forces, guide motions, limit friction, and so forth.

As this chapter illustrates, nanoscale components can serve functions familiar in the macroscale world. Because functions at the system level can usually

---

* If this could be done by a mere molecular ratchet, the complex machinery of photosynthesis and digestion might never have evolved.

be implemented in many different ways at the component level, the parallels between macro and nanoscale systems can be even stronger than those between their components. Accordingly, many of the lessons of macroscale mechanical engineering can be applied directly. When nanomechanical designs are drawn at a scale and resolution that omits atomic detail, they can be almost indistinguishable (save for dimensioning labels) from designs for macromachines.

10.10.2. Differences between nanomachines and macromachines

Macroscale machines can be made of many different materials and parts, but many consist chiefly of steel and a lubricant; these suffice to build gears, bearings, drive shafts, cams, and so forth. Lubricated steel systems are accordingly taken as the basis for comparison in this section.

a. Soft surfaces, different frictional behavior. On a macroscopic scale, steel surfaces are hard: objects have well-defined boundaries, and any attempt to force a geometric overlap between two steel objects instead results in contact forces and elastic or plastic deformation. On a nanometer scale, diamondoid surfaces are soft: if one defines object boundaries by (for example) assigning radii to the surface atoms, the repulsion forces will increase smoothly as the objects approach and overlap. This nonlinear surface compliance is distinct from (and additive with) the compliance from elastic deformations of the objects, and has a characteristic length scale of ~0.03 nm. This length is small enough to make the concepts of "shape" and "surface" useful, yet large enough to bar the direct and uncritical application of macroscopic concepts.

Because of these surface properties, nanoscale components need not meet precise dimensional specifications to achieve a good fit. Centimeter-scale steel components commonly are designed to tolerances less than 0.001 of their linear dimensions. For a nanometer-scale component, a fractional deviation of ≤ 0.01 from a hypothetical ideal is insignificant, and a deviation of 0.03 or more is often acceptable. This tolerance for deviations from ideal geometries helps compensate for the discrete nature of nanomechanical structures, and for the consequent unavailability of a continuum of sizes and shapes.

As a result of these surface properties, nanomechanical components are better viewed as moving smoothly in a force field than as sliding subject to friction. Static friction (in the conventional sense) does not exist in well-designed eutactic bearings: its closest analogues—energy barriers and force gradients—can be *lower* for two coupled objects sliding over a surface than for either object considered separately (Section 10.8), even though their contact forces are additive. This is alien to the familiar tribology of disordered interfaces. These interfacial properties make distinct lubricating substances needless and harmful (an oil molecule is not a lubricant on this scale: it is an object).

b. No wear, no contaminants. In macromachines with sliding or rolling interfaces, cumulative wear commonly limits device lifetimes. The strong, precise surfaces of nanomachines, in contrast, experience no change in a typical operational cycle, and hence no wear. Within the single-point failure model (Section 6.7.1a), the first step in a wear process is regarded as fatal, and hence *cumulative* wear plays no role in determining device lifetimes.

In macromachines, wear particles often contaminate lubricants, accelerating wear. In eutactic nanomachines, contaminants of all kinds are rigorously excluded and the equivalent of wear particles cannot appear until the device has failed. Accordingly, contaminants are not found in devices operating within the guidelines assumed here. (The design of devices that tolerate contamination is relatively complex; Section 13.2 touches on this subject in discussing the purification of feedstocks for molecular manufacturing.)

c.   No fatigue, tolerance of high strains.   In a typical load cycle, a well-designed nanomechanism undergoes no structural change. As with wear, the single-point failure assumption implies that devices are not subject to cumulative damage before failure, hence cumulative fatigue plays no role in determining device lifetimes (Section 6.7.1a).

Conventional macroscale structures contain numerous defects (e.g., dislocations) of varying mobility. Even a slight load usually moves some of these defects (or biases their thermally activated diffusion), causing slight plastic deformation. Eutactic nanomechanisms need not have defects of significant mobility, and hence can entirely escape plastic deformation and resulting fatigue. Conventional steel mechanisms commonly operate for years with cycle frequencies of 1 to 100 Hz, with stresses chosen to avoid fatigue failure. Diamondoid nanomechanisms can operate for years at $>10^9$ Hz, executing many more cycles.

Few steel objects can tolerate deformations as large as 0.01 without substantial plastic flow. Some diamondoid structures, in contrast, can tolerate deformations >0.1 without plastic flow or fatigue. Accordingly, nanomechanical devices relying on the repeated, large-amplitude elastic deformation of strong, stiff components are practical, though their macroscale analogues would promptly fail. Exploiting large elastic strains can further relax constraints on component dimensions, where parts can stretch to fit. Further, components locked together by mechanical interferences (e.g., the bearing in Figure 10.17) can often be assembled via strained intermediate geometries. Where elastic deformation is concerned, typical diamondoid structures act more like an extremely stiff elastomer than like a ductile metal.

d.   Insensitivity to parts count.   The cost and reliability of conventional macromachines depends strongly on the number of parts they contain. To make a part, surfaces must be shaped; to use a part, it must be assembled to other parts. These steps add costs and potential defects. In operation, most moving parts have interfaces subject to wear, decreasing reliability. Good design practice in macroscale engineering aims to minimize parts count.

The reliability and fabrication cost of nanomachines will be almost independent of the number of mechanical parts they contain. In molecular manufacturing, the number of assembly operations is roughly proportional to the number of atoms in the product, and hence roughly proportional to the mass. Costs will be insensitive to the number of nonbonded interfaces in the product, and hence insensitive to the number of separate mechanical parts. If the design follows the principles and analytical assumptions outlined in Section 6.7.1b, component lifetimes are determined by radiation damage, and hence (approximately) by mass; this again is independent of the number of separate parts. (Component interiors

are somewhat less sensitive to damage than component surfaces, but the resulting incentive to reduce the number of parts—here, merely to reduce the total
area of interfaces—is weak compared to that in conventional engineering.)

10.11.   Bounded continuum models revisited

As discussed in Section 9.5, diamondoid components can approximate any
desired shape, so long as strong precision and symmetry requirements are not
imposed, and so long as features either have a specified structure or dimensions
≥ 1 nm. Further, the mechanical properties of these components can approximate those of diamond, save for a surface correction to the effective component
size (to allow for the difference between nonbonded and covalent radii), and
degradation of stiffness by a factor no worse than 0.5, to account for less dense
arrays of bonds, the use of second-row elements, and the like. Finally, so long as
one of the conditions for smoothly sliding or rolling interfaces holds (Sections
10.3 to 10.5), smooth motion may be assumed. These generalizations permit the
use of bounded continuum models in the design of a certain range of structures.

Within a bounded continuum model, as in the bulk-material models used in
mechanical engineering, design work can proceed without reference to the positions of individual atoms. In nanomechanical engineering, the production of
devices requires atomically detailed specifications, but—if one is willing to
accept a performance penalty resulting from conservative assumptions regarding size, stiffness, and so forth—this specification process can in many instances
be postponed to a phase relatively late in the design process. Models of this sort
are used for much of the analysis in the following chapters.

10.12.   Conclusions

Both sliding and meshing interfaces can be constructed in various nanoscale
geometries, as can interfaces with intermediate properties and specially tailored
potential energy functions along a sliding coordinate. As a consequence (and
together with more elementary devices, such as springs), it is feasible to construct nanomechanical rotary bearings, sliding shafts, drive shafts, screws and
nuts, power screws, snaps, brakes, dampers, worm gears, constant-force springs,
roller bearings, levers, cams, toggles, cranks, clamps, hinges, harmonic drives,
bevel gears, spur gears, planetary gears, detents, ratchets, escapements, indexing
mechanisms, chains and sprockets, differential transmissions, Clemens cou-
plings, flywheels, clutches, Stewart platforms, robotic positioning mechanisms,
and suitably adapted working models of the Jacquard loom, Babbage's Difference and Analytical Engines, and so forth.

The analysis presented in this chapter does *not* establish that all classes of
sliding-interface bearing are feasible in the nanometer size range. For example,
a smoothly sliding ball-and-socket joint requires a potential energy function
that is smooth with respect to three rotational degrees of freedom; neither the
symmetry properties exploited in Sections 10.3 and 10.4, nor the tuning
approaches discussed in Section 10.8 guarantee that this can be accomplished in
a stiff, nanoscale device. (Note that most properties of a ball-and-socket joint
can be provided by a Hooke joint.)

The conclusions of Sections 10.3 and 10.8 support the extension of bounded continuum models to the design of moving parts and their interfaces. As discussed in Section 9.5, diamondoid components with dimensions ≥ 1 nm can be of almost any desired shape. These components can be assumed to have smoothly sliding interfaces so long as one of the conditions established in this chapter holds. These conditions include either (1) translational or rotational symmetry in one component and adequate size and design flexibility in the other (Section 10.3), (2) suitable symmetry properties in both components (Sections 10.4 and 10.5), or (3) the ability to couple an interface with a given irregular potential to another with a designed and compensating irregular potential (Section 10.8). Energy dissipation can be estimated using the methods of Chapter 7. With this extension, bounded continuum models can provide a basis for the analysis of conservatively designed nanomechanical systems specified in less than atomic detail.

Some open problems. The most immediate open problems associated with mobile nanomechanical components are chiefly those of design. Devices of several classes have been described and abstractly characterized in this chapter, but examples specified in atomic detail are sparse. It would be useful to have a large set of examples spanning many different classes, and to characterize those examples in detail (e.g., considering all plausible instabilities, measuring stiffness in all important degrees of freedom, and so forth), including studies of the sensitivity of device properties to variation in the assumed potential energy function. (The particular structures used in this chapter to illustrate desirable geometries and intercomponent potential energy functions may prove defective in other respects, but alternatives are abundant.) Software like Merkle's diamond-tube generating package can greatly help in this work, and should be developed and extended. Beyond this, the design of devices and their integration into mechanical systems using existing chemistry-oriented molecular modeling software can better define requirements for nanomechanical CAD systems, and can better define the objectives of mechanosynthesis. Eventually, one would like to have a large library of specifications for well-characterized components, procedures for designing systems that use those components, and procedures for specifying mechanosynthetic sequences to construct such systems. Although there are many easy and rewarding steps toward this objective, reaching it will require an enormous amount of work, both in nanomechanical design and in software development.

11

Intermediate Subsystems

11.1. Overview

Between the simple components described in Chapters 9 and 10 and the complex systems described in Chapters 12–14 are subsystems of intermediate complexity. These subsystems find use in more complex systems, but are worthy of a separate description.

Section 11.2 describes mechanical measurement devices; these are useful in signaling (Section 16.3.2), in mechanical logic systems (Chapter 12), and in °fault-tolerant molecular manufacturing systems (Section 13.3.6b). Section 11.3 describes stiff, high gear-ratio mechanisms suitable for use in molecular manipulation systems (Section 13.4.1). Section 11.4 surveys the principles of fluid flow in small volumes, then describes a set of components and subsystems for handling fluid flow: seals, valves, and pumps, including vacuum pumps. Section 11.5 describes systems using branched pipes for high-capacity convective cooling. Section 11.6 surveys the principles of electrical current flow in small conductors, then describes electromechanical devices including modulated tunneling junctions and electrostatic actuators. Section 11.7 describes direct-current (DC) electrostatic motors and generators.

11.2. Mechanical measurement devices

Mechanical measurement systems can latch an *indicator* into one of two or more distinct positions, making the position chosen depend on (or correlate with) some feature or property being measured. Two limiting cases of mechanical measurement (among others) are measurement of force and of shape. With suitable transducers, measurements of magnetic field, electric field, and pressure can be transformed into measurements of force. Measurements of conditions imposed from the outside (e.g., pressure resulting from acoustic waves, Section 13.3.6b) can be used as a means of receiving signals. Measurements of shape (or position) can discriminate among the internal states of a system, for example, in order to detect defective products in a manufacturing system. They can likewise be used to discriminate among external states, for example, in characterizing molecular structures in a biological context. Chapter 13 describes the use of shape measurements to detect errors in manufacturing.

This section does not attempt to describe measurement devices of optimal size, reliability, and so forth. It merely illustrates some basic approaches, and some conditions under which they enable good reliability. Of the wide variety of methods that might be used, only processes based on direct mechanical measurement of indicator displacements are discussed.

11.2.1. Well partitioning and indicator latching

Measurement systems of the sort considered here latch an indicator as shown in Figure 11.1. Latching is usually necessary if the measurement must represent the state of a time-varying measured system at a particular moment, or if the position of the indicator is subject to substantial thermal motion. The following will consider a latching mechanism that makes a binary distinction, but generalizations to more locations are straightforward.

An indicator moves in a potential well, driven by thermal excitation. Section 10.8.1 shows that the potential energy of an object sliding over a surface can be made nearly flat, or can be given more complex contours subject to resolution limits stemming from the finite scale lengths of interatomic potentials. A well can be partitioned by introducing an energy barrier; concretely, this can be accomplished by inserting a physical barrier that obstructs the motion of an atomically wide feature on the indicator. The resulting latched position of the indicator can then determine subsequent mechanical behaviors, most generally, by providing an input to a computational system (Chapter 12).

In the logic rod systems described in Section 12.3, each rod can be regarded as making a measurement of the position of the rod (or register) in the preceding stage, yet latching is unnecessary. All changes in state are sequenced by a clocking mechanism, the state of the measured system is stable for the duration of the measurement, and thermal displacements are small. Latching is, however, required in registers (Section 12.4), which must display the results of a measurement after the measured system has been reset.

At high speeds ($\sim 10^{-9}$ s per operation) the energy cost of well partitioning is likely to be dominated by nonisothermal compression losses (Section 7.5). When the wells are later merged, further losses can occur if the input signal has changed (well merging losses are treated in Section 7.6.2). If the input has not

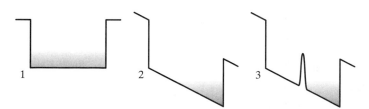

Figure 11.1. Well partitioning in a measurement system. Each curve represents the potential energy of the indicator as a function of its sliding displacement; the shading represents the probability density over that range of motion. In (1), the potential is flat, and the PDF is uniform. In (2) an applied force has skewed the potential and the PDF. In (3), a barrier has been introduced, trapping the indicator on the right hand side with high probability as a direct consequence of the presence of the applied force.

changed, then well merging is simply the reverse of well partitioning, and can be nearly thermodynamically reversible (save for further nonisothermal losses).

11.2.2. Force discrimination

A force can push an indicator toward one end of a potential well; a bias force can push it toward the other. In a system intended to distinguish between two forces differing by ΔF, a typical bias force might have a magnitude of $\Delta F/2$. If the length of the potential well is ℓ, then the probability of an erroneous measurement is

$$P_{err} \approx \frac{\exp(-\ell\Delta F/4kT) - \exp(-\ell\Delta F/2kT)}{1 - \exp(-\ell\Delta F/2kT)} \tag{11.1}$$
$$\approx \exp(-\ell\Delta F/4kT), \quad \text{if } \exp(-\ell\Delta F/4kT) \ll 1$$

where the chief approximations are (1) that under an applied force of $-\Delta F/2$ the well is square and uniform, and (2) that the partitioning process divides the well in the middle before disturbing the initial probability density function. (The no-disturbance assumption overestimates the error rate; the effect of the square-ness assumption depends on the definition of ℓ.) For $\ell = 1$ nm, $\Delta F = 1$ nN, $P_{err} \approx 6 \times 10^{-27}$ at 300 K.

More generally, the magnitude of the bias force can be chosen to minimize the cost of errors, given differing costs for reading "high" when the sensed force is low vs. reading "low" when the sensed force is high. A well that is deeper at the ends (creating subwells) can reduce error rates, so long as the intervening barrier does not substantially impede equilibration of the subwells. Multivalue measurements are best made by partitioning a more nearly parabolic well.

11.2.3. Shape and position discrimination

The position of an indicator structure can be coupled to the position of a probe resting against a surface. If the probe and indicator are parts of a lever, small probe displacements can cause large indicator displacements; levers can thus be used to avoid scale-length problems in measuring probe displacements that are small compared to an atomic diameter.

A typical small difference in surface position is ~0.1 nm; a typical probe load is ~2 nN; a typical stiffness associated with contact of the probe tip is ~50 N/m (allowing for some compliance in the probed surface itself). A rough estimate of the error rate in partitioning an indicator well that corresponds to a probe displacement distance of 0.1 nm is ~10^{-5}. For many purposes, error rates of this magnitude are unacceptable.

11.2.4. Reliability through iterated measurements

For weak forces and small displacements, and for small differences in shape and position, thermal excitation can render individual measurements unreliable. Iterated measurement processes can be devised, however, that correspond to incrementing a counter by one with each positive measurement, while decrementing it by one with each negative measurement. A series of these operations, followed by a measurement of the state of the counter, provides a measurement procedure that can reduce overall error rates to any desired level.

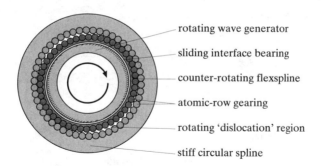

- rotating wave generator
- sliding interface bearing
- counter-rotating flexspline
- atomic-row gearing
- rotating 'dislocation' region
- stiff circular spline

Figure 11.2. Cross sectional diagram of a harmonic drive. One revolution of the wave generator in the indicated direction causes the flexspline to turn in the opposite direction with respect to the stiff circular spline, moving it by a two-tooth increment.

11.3. Stiff, high gear-ratio mechanisms

Reliable positioning of parts in the presence of thermal excitation requires drive mechanisms of high stiffness. If these are to be powered by mechanisms of lower stiffness, then they must have high gear ratios. Both harmonic drive and toroidal worm drive subsystems can meet these requirements.

11.3.1. Harmonic drives

Some macroscale robots use harmonic drives as stiff, high gear-ratio mechanisms. Figure 11.2 illustrates the structure of a harmonic drive: the basic kinematics are those of a pair of dislocations moving along an interface bent into a cylinder, with the dislocation motions driven by the rotation of a wave generator, which in turn deforms a flexspline. Each rotation of the wave generator turns the flexspline by a two-tooth increment relative to the surrounding gear surface.

Harmonic drives are well suited to nanomechanical systems. Diamondoid materials have large elastic limits, enabling them to accommodate the required deformation of the flexspline. Any of several kinds of gear teeth (Section 10.7.1) can be used, and those that can operate without a substantial compressive load across the interface (e.g., by exploiting a bonding interaction) can impose lower loads on the stiff circular spline.

11.3.2. Toroidal worm drives

Another class of stiff, high gear-ratio mechanism is the *toroidal worm drive*. Since the toroidal worm drive as described here exploits unique properties of nanomechanical structures, it may be novel; since it has so few moving parts, however, it may well be described in some form in the existing literature.

A toroidal worm drive is built around a triply-threaded torus (illustrated in Figures 11.3 and 11.4). The torus would, if cut and allowed to relax, take the form of a straight rod. The inverting rotational motion shown in Figure 11.2 can occur essentially without energy barriers, since a small displacement in this mode is (in a good structure) a symmetry operation for the potential energy function. A structure of this sort is more practical here than in conventional engineering,

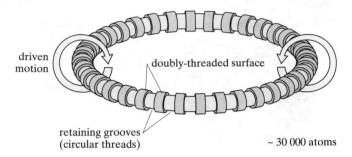

driven
motion

doubly-threaded surface

retaining grooves
(circular threads)

~ 30 000 atoms

Figure 11.3. Overall geometry of the triply-threaded torus of a toroidal worm drive, showing one set of threads and the inverting rotation (resembling that of a smoke ring) imposed on the torus. With a suitable choice of major and minor radii, the substantial strains occurring during this motion fall well within the elastic limit of diamondlike structures; cyclic strains will not cause fatigue (Section 6.7.1a). A similar design with conventional materials on a macroscopic scale would present serious difficulties.

owing both to the availability of strong, stiff materials having high elastic limits (strains >0.2), and to the feasibility of constructing rings that both relax to a straight rod if cut, *and* have no seams resulting from bending and joining. Figure 11.5 illustrates and discusses the structure of a triply-threaded torus in more detail. Figure 11.6 shows the components of a toroidal worm drive in cross section; its caption summarizes the kinematics of the device. The structure and dimensions of the torus must, of course, be chosen to avoid excessive strains; J. Soreff has suggested discontinuous designs with similar kinematics that avoid this constraint.

The energy barriers to the inverting rotation of a torus in an intersegment groove can be analyzed using the symmetry and spatial-frequency principles of Sections 10.4 and 10.5. Each segment of the torus is an object of low symmetry, and turns in an environment of low symmetry. Good design practice can readily avoid large, abrupt energy barriers by avoiding bad steric contacts in the sliding interfaces, but the resulting potential energy function will almost inevitably

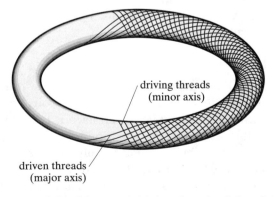

driving threads
(minor axis)

driven threads
(major axis)

Figure 11.4. Geometry of the driven and driving threads of the triply-threaded torus of a toroidal worm drive. The angles chosen for these threads are purely illustrative; in practice, angles can be chosen to provide the desired mechanical advantage for the drive system. The retaining grooves (not shown; see Figure 11.3) cut across this pattern.

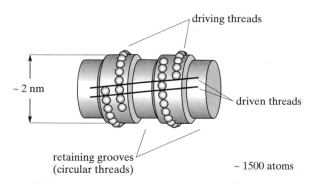

Figure 11.5. A more detailed (but still schematic) illustration of two segments of a triply-threaded torus. The driving threads are formed by rows of atoms alternating with grooves; interfacial structures might resemble Figure 10.18. The driven threads are shallower, formed by the alignment of rows of atoms and shallow interatomic gaps in a direction roughly perpendicular to that of the driving threads. These features can be formed from a helical variant of one of the structures discussed in Section 10.4.7a; the angle of the driven threads can be determined by a suitable choice of the locked-in torsional strain of the torus. The final set of threads, forming the retaining grooves, fall at an angle to the natural lattice orientation of the structures just described, and hence have (at best) low-order symmetry with respect to the inverting rotation.

exhibit substantial residual fluctuations, which are periodic in the inverting-rotation coordinate. A series of segments, however, forms a coupled system that can readily meet the conditions described in Section 10.3.6 for treating the energy barriers using the rigid-coupling approximation. With a favorable choice of structural parameters, each segment has a different phase angle with respect to the inverting-rotation coordinate, giving the overall potential energy function

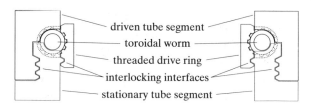

Figure 11.6. A cross-sectional diagram of a toroidal worm drive. Power (and positional control) are provided via the threaded drive ring, the inner surface of which can be geared to a drive gear in the arm's core structure (see Figure 13.11). The outer surface of the drive ring bears a helically threaded band consisting of ridges like those on the surfaces in Figure 10.18; this surface interlocks with the driven threads of the triply-threaded torus. The outer surface of the drive also bears a circularly threaded band forming an interlocking bearing surface with the lower, stationary tube ("stationary" only in the present frame of reference). Rotation of the threaded drive ring about its axis thus drives the inverting rotational motion of the triply-threaded torus within the groove formed by the junction of the stationary and driven tube segments (which in turn are held together by an additional interlocking bearing surface). Within this groove, the stationary tube bears ridges which interlock with the retaining rings of the torus, and the driven tube bears ridges which interlock with the driving threads of the torus. Thus, as the torus undergoes its inverting rotation, the driven tube segment is forced to turn.

a period of $2\pi/N_{seg}$ in that coordinate, where N_{seg} is the number of segments in the torus. If the minor diameter is 2 nm and $N_{seg} = 50$, the spatial frequency of the potential energy function (measuring displacements at the sliding interface) is ~ 8 nm^{-1}. This results in strong mutual cancellation of the residual fluctuations in potential energy associated with the individual segments, yielding a potential energy function with negligible barriers for motion of the torus as a whole. (These parameters are appropriate for the manipulator joints described in Section 13.4.1.)

11.4. Fluids, seals, and pumps

Nanomechanical systems often must be in contact with a fluid medium. Typical operating environments contain fluids, whether gases or liquids; control of the internal environment then requires that external fluids be excluded by some combination of walls, seals, and pumps. Similar devices are likewise required for fluid-based convective cooling systems (Section 11.5). Further, hydraulic devices may be useful in nanomechanical systems, although this possibility is not further explored here. The following sections survey the nature of fluid flow in nanoscale systems, then consider walls, seals, and pumps.

11.4.1. Fluid micromechanics

a. Scale lengths in gases and liquids. Classical continuum models of fluid flow fail when the dimensions of tubes (or other structural features) are comparable to the characteristic molecular length scales of the fluid. In a gas, the largest characteristic length scale is the mean free path for a molecule between collisions. In a liquid, the characteristic length scale is the molecular size.

The *Knudsen number* measures the ratio of the mean free path to the feature size r in the flow system; where it is ≥ 100, the process is termed *free molecular flow;* where it is ~ 1, the process is termed *transition flow*. In macroscale devices, transition flow occurs at low pressures ($\sim 10^{-5}$ atmospheres). In nanoscale devices ($r \leq 100$ nm), transition flow occurs in air at atmospheric pressure. Free molecular flow can be modeled as a process of repeated scattering from surfaces.

The behavior of liquids in small spaces is more complex and is a subject of active research (much of the following paragraph is based on Israelachvili, 1992). Liquids have short-range order, and that order is strongly perturbed near surfaces. For example, two smooth surfaces brought into close proximity in a liquid (≤ 6 to 10 molecular diameters) experience forces that oscillate strongly as a function of separation: gaps that accommodate an integral number of molecular layers are preferred; slightly smaller separations result in repulsive forces, slightly larger separations result in attractive forces. These oscillations are smaller with rough surfaces and in liquids that mix molecules of varying size. With smooth mica surfaces in water, force oscillations virtually disappear at separations ≥ 2 nm. (In water, *hydrophilic* forces can cause short-range surface-surface repulsion and *hydrophobic* forces can cause short-range attraction, both with ranges of a few nanometers.) In liquid films a few molecular layers thick, viscosity increases and finite shear strength can develop; at a thickness of 7 to 10 layers or more, the bulk fluid viscosity gives a good description of shear stresses (Israelachvili, 1992).

b. Shear rates. Nanoscale fluid flows can exhibit extreme rates of shear: for example, a 1 m/s shear speed imposed on a 1 nm gap corresponds to a shear rate of $10^9 \, \text{s}^{-1}$. For a slab of fluid a meter thick to undergo shear at this rate, one surface would have to exceed the speed of light. In typical liquids (composed of small molecules with only local order), viscosity is little affected by shear rate so long as the velocity differential across a molecular diameter is small compared to the mean thermal velocity. A shear rate of $10^9 \, \text{s}^{-1}$ easily meets this criterion.

c. Formulas for drag and pressure drop. With the above caveats regarding phenomena on short length-scales, fluid flow in small structures can usually be described by the classical continuum equations. The Reynolds number

$$R = r\rho v / \eta \tag{11.2}$$

(where η is the viscosity of the fluid, v is a characteristic flow speed, ρ is the fluid density, and r is a characteristic dimension) indicates whether flow will be laminar or turbulent around (or within) an object of a given shape. For nanoscale objects in fluids of ordinary viscosities and velocities, R is low, and the flow laminar.

In this regime, Stokes's law relates the radius r of a sphere, the viscosity η of a fluid, and the speed of the sphere v to the drag force F

$$F = 6\pi r\eta v \tag{11.3}$$

This expression is routinely used to describe the motion of objects as small as protein molecules (Creighton, 1984). For $\eta = 10^{-3} \, \text{N·s/m}^2$ (= 1 centipoise, the viscosity of water at 20 C), $r = 100$ nm, $F = 1$ nN, $v \approx 0.5$ m/s, and the drag power is ~0.5 nW. The Reynolds number (based on the radius r) is 0.05, well within the range for laminar flow.

The Hagen-Poiseuille law relates the volumetric flow rate $\Delta V/\Delta t$ through a long tube to its length ℓ, its radius r, and the pressure difference Δp between its ends

$$\Delta V/\Delta t = \pi r^4 \Delta p / 8\ell\eta \tag{11.4}$$

This expression (like Stokes's law) assumes incompressible laminar flow. The incompressibility assumption is significantly violated in gases where the pressure differential is substantial compared to the total pressure. The laminar flow assumption is seldom violated save in relatively large tubes with relatively high flow speeds. A tube with $r = 100$ nm, $\ell = 1000$ nm, $\Delta p = 1$ MPa, transports a fluid of $\eta = 10^{-3} \, \text{N·s/m}^2$ at ~$4 \times 10^{-14} \, \text{m}^3\text{/s}$, at a mean fluid speed of 1.25 m/s. The Reynolds number is again low (flow is laminar up to a diameter-based R of ~2000).

An approximation for the pressure gradient along a tube containing a fluid in turbulent flow is the Darcy-Weisbach formula

$$\Delta p/\Delta \ell = f\rho v^2 / 4r \tag{11.5}$$

where v is the mean velocity of the fluid, and f is a *friction factor* that depends on the Reynolds number R of the flow and the roughness of the wall. The parameter f can be evaluated by methods described in Tapley and Poston (1990); a high value (for a rough pipe at low R) is 0.1, a low value (for a smooth pipe at $R > 10^7$) is 0.008.

11.4.2. Walls and seals

A nonbonded interface can serve as a seal, permitting relative motion of the sur-faces but hindering the flow of fluid between them. The bearing interfaces explicitly considered in Section 10.4 can also serve as sealing interfaces, pro-vided that no deep, unfilled groove crosses the seal band.

a. Molecular mechanics analysis of sealing interfaces.   A pair of surfaces in nonbonded contact has an equilibrium separation. An interposed molecule will introduce additional nonbonded repulsions that locally increase this separation. The energy of interposition includes the strain energy of the facing structures, any strain energy in the interposed molecule (none, if it is monatomic), and the energy of the nonbonded repulsions. The energy is greater if the interface is compressed, but can be large even without compression.

Computational experiments using molecular mechanics (MM2/CSC with added parameters for H_2 and He) indicate the magnitude of the energy required to interpose a molecule between a pair of N-terminated (111) diamond surfaces, modeled as a pair of clusters. With an equilibrium separation of ~0.30 nm be-tween the facing planes, the energy of a He atom at one of the local minima of the relaxed structure is estimated to be ~170 maJ, and the restoring force confin-ing it to the plane of symmetry has $k_s \approx 150$ N/m. For H_2, an analogous energy is ~470 maJ, and $k_s \approx 380$ N/m. The energy of a molecule passing through the tran-sition state for insertion into the interface is presumably higher than that of the observed minima.

Once interposed between two surfaces, a molecule diffuses from site to site and may exit from the far side of a seal. It is conservative to assume that all mol-ecules that enter the seal gap immediately exit on the far side, thus treating the gap as an aperture of zero length and internal resistance. Using the He parame-ters just discussed, at 300 K and $k_s = 150$ N/m, the effective width of the gap [Eq. (6.5)] is ~0.013 nm; the Boltzmann factor corresponding to an interposition energy of 170 maJ decreases the number density of molecules within that effec-tive width by ~1.5×10^{-18}. Accordingly, the interface at worst behaves like a gap with a width of ~2×10^{-29} m. Assuming an atmospheric He number density of ~10^{20} m^{-3}, the leakage rate is ~10^{-15} atom/nm·s. At this rate, the mean waiting time for a seal of 1000 nm length to pass a single He atom is >10^4 years. Leakage for H_2 and larger species appears negligible.

b. Sealing against ionic species.   Positive ions such as Li^+ and Be^{2+} have smaller radii than helium atoms, and from a purely steric perspective would more easily pass through a seal. (A bare proton, H^+, is a chemically nonexistent species; protons rapidly bind to other molecules.) These ions will rarely be found in appreciable numbers outside a solvating liquid environment, owing to the large electrostatic (Born) energy of a free ion in the gas phase or in a nonpolar fluid.

The heat of solvation of a monovalent ion in water is ~800 maJ, relative to the gas phase. Any substantial desolvation of an ion in moving from water to a site within a bearing interface will increase the energy of the ion by a substantial fraction of 800 maJ, in part because the rigid seal surface cannot rearrange to provide solventlike coordination of polar groups [in water, Li^+ and Be^{2+} are coordinated to some 4 to 6 water molecules (Israelachvili, 1992)]. Further, a sealing interface can be surrounded by a guard band that provides an adverse electrostatic environment, resembling a nonpolar, low dielectric constant solvent. In this circumstance, the interposition energy can easily exceed 200 maJ, and the leakage rate for ionic species from an aqueous (or other highly polar) medium is negligible.

c. Applications of seals. Since tight seals can also serve as good bearings, eutactic nanomechanical systems can be placed in intimate contact with fluids. Internal actuators can drive sliding and rotating shafts that extend into liquid media, hydraulic mechanisms with pistons and cylinders are feasible, many standard macroscale valve designs can be implemented, and so forth.

d. Permeability of walls. The relative gas tightness of nonbonded interfaces suggests that continuous walls of diamondoid structure can be effectively impermeable at ordinary temperatures. Diffusion of gases through metal and glass barriers is accelerated by the presence of defects: conventional metals have high densities of dislocations and grain boundaries; silicate glasses are relatively open and irregular structures that permit substantial diffusion of helium. Some metals dissolve hydrogen and permit it to diffuse relatively freely.

Defect-free layers of diamond and graphite differ from these materials. It is difficult to see how a molecule can pass through these materials without undergoing a reactive transformation. The species of chief concern is hydrogen, but calculations indicate that the energy required to move atomic hydrogen from free space into a minimum-energy site in diamond is thermally prohibitive [≥ 800 maJ (Chu and Estreicher, 1990)]. If the initial species is H_2, the required energy is far higher.

In ordinary environments, it seems likely that walls can be constructed that transmit gas molecules at negligible rates. If this assumption proves wrong, then double-walled structures with an intervening space containing either getter materials or a pumped vacuum can provide the same result.

11.4.3. Pumps and vacuum systems

a. Positive displacement pumps. Seals enable the construction of valves, pistons and cylinders, and so forth (Section 11.4.2), therefore positive-displacement pumps can be made. Use of diamond and diamondoid structural materials can enable operation at pressures ranging from high vacuum to >10 GPa, assuming that allowance is made for elastic deformation of the structures at high pressures. The dominance of viscous forces in fluid flow at small scales favors the use of positive displacement pumps over centrifugal pumps.

b. Vacuum pumps. Vacuum has advantages as an operational medium for nanomechanical systems: unlike typical solvents, it is unreactive, causes no viscous drag, and cannot jam a mechanism. Accordingly, nanomechanical vacuum pumps are of interest.

In macroscale systems, the performance of vacuum pumps frequently is limited by the vapor pressure of their lubricants. Sliding interface seals and bearings need no separate lubricants, and hence present no such problem. Positive displacement pumps can accordingly serve as high-vacuum pumps.

The high strength to density ratio of graphitic and diamondoid materials lends itself to the construction of turbomolecular pumps (Figure 11.7): their

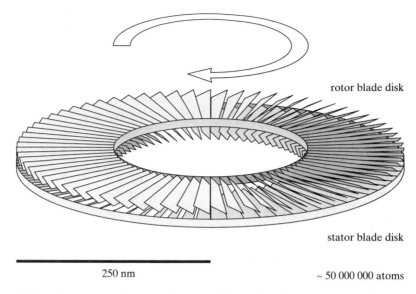

Figure 11.7. Schematic diagram of a pair of blade disks in a turbomolecular pump; the illustrated rotor direction corresponds to downward pumping. (Axle, bearings, casing not shown; disks not of minimal size.)

effectiveness depends on the ratio of the blade speed to the characteristic thermal speed of the lightest gas molecule to be pumped. With diamondoid materials, this speed can easily exceed that of hydrogen, providing a compression ratio of ≥ 10 per blade row (Chu and Hua, 1982). Blade thickness can be <1 nm, and the pump length per blade row can be <10 nm, permitting compression ratios $>10^{10}$ in a 100 nm length. Turbomolecular pumps are designed to operate under free-molecule flow conditions; with blade row spacings <10 nm, free molecule flow occurs at pressures up to atmospheric.

Conventional systems that cycle from dirty conditions at atmospheric pressure to high vacuum are designed to permit bakeout, the use of high temperatures to speed outgassing. Nanomechanical systems built in accord with the room-temperature design rules assumed in this volume would not tolerate the required temperatures. Note, however, that the systems and system-construction processes described in later chapters avoid this requirement: they either build in a clean vacuum environment (Chapters 13 and 14), or can produce such an environment gradually, by diluting a fixed burden of contaminants over an increasing volume of mechanisms (Chapter 16).

11.5. Convective cooling systems

Because operating nanomechanical systems dissipate energy, they must be cooled if they are to operate for an extended time. Owing to the short distance between the interior of a submicron structure and its environment, conductive cooling usually suffices for small, isolated devices. Larger systems, such as macroscopic arrays of nanomechanical computers, require some form of convective cooling. Although mechanical convection based on solid strips moving with the aid of bearings may prove superior, cooling systems based on forced convection of liquids are examined here. It will be assumed that heat is generated at a uniform rate throughout a macroscopic volume that must be cooled by fluid flowing from an input port to an output port.

11.5.1.   Murray's Law and fractal plumbing

For a given volumetric flow rate, large diameter tubes minimize resistance and hence pump energy. A single large tube linking the input port to the output port would, however, be an ineffective cooling system owing to poor thermal contact. Thin tubes improve thermal contact but long, thin tubes would cause excessive resistance.

The appropriate compromise uses large tubes to carry fluid over most of the distance, smaller tubes to carry fluid over most of the remaining distance, and so forth, yielding a branching structure resembling a biological circulatory system that links large arteries and veins via smaller vessels ultimately linked by capillary beds. Murray's Law states that, for a system of tubes containing a fluid in laminar flow, the minimum volume for a given pressure drop occurs when the radii of the tubes at a branch point satisfy the relationship

$$r_0^3 = r_1^3 + r_2^3 + r_3^3 + \ldots + r_n^3 \tag{11.6}$$

where r_0 is the radius of the incoming tube, and r_1, r_2, etc., are the radii of the outgoing tubes. This condition equalizes shear stresses in the tubes, and is approximately obeyed in many biological circulatory systems (LaBarbera, 1990). Murray's Law is obeyed by an approximately fractal structure of the sort illustrated in Figure 11.8, in which tube lengths are approximately proportional to their radii, giving each level of the hierarchical structure an approximately equal volume and fluid residence time. In a macroscale system, the larger tubes may operate in turbulent flow; these tubes have different optimal dimensions.

11.5.2.   Coolant design

To increase heat capacity and minimize temperature variations in a system, it is desirable to use a coolant that absorbs large quantities of heat in a narrow temperature range. This can be accomplished by means of a phase transition or by the expansion of a slightly supercritical gas. The latter process resembles vaporization in that thermal energy is converted to intermolecular potential energy as molecules separate, but avoids problems of bubble formation.

Suspended particles can add their heat of fusion to the heat capacity of a fluid. An attractive class of coolants combines encapsulated submicron ice particles (having surface structures that prevent aggregation) with a low-viscosity, low-melting point carrier, such as a light hydrocarbon. An expression (Hiemenz, 1986) derived from the Einstein relationship relates the viscosity of a fluid suspension η, the volume fraction of spherical particles f_{vol}, and the viscosity without the particles η_0

$$\eta / \eta_0 = (1 - f_{vol})^{-2.5} \tag{11.7}$$

and is reasonably accurate for $f_{vol} \leq 0.4$. Packaged ice particles with a volume fraction of 0.4 increase fluid viscosity by a factor of ~3.6; if pentane is used as a carrier, the viscosity of the resulting suspension is $\sim 8 \times 10^{-4}$ N·s/m^2. On melting, ice particles at this volume fraction absorb $\sim 1.2 \times 10^8$ J/m^3, based on coolant volume. This use of distinct heat-carrying bodies parallels the use of red blood cells as oxygen carriers in the circulatory system.

11.5.3. Cooling capacity in a macroscopic volume

A preliminary design exercise has been performed based on the coolant suspension described in Section 11.5.2 and a 4-level system of tubing of the sort illustrated in Figure 11.8, but with a higher branching ratio. This calculation (which takes account of turbulent flow in the largest tubes) indicates that $\sim 10^5$ W of thermal power can be extracted at ~ 273 K (the melting point of ice) from a cube a centimeter on a side (or, more generally, $\sim 10^5$ W/cm^3 from a slab not more than 1 cm thick). The thinnest tubes are < 1 μ in diameter and pass within <5 μ of each point in the volume that is not itself inside a tube. The volume fraction of the tubing system is 0.25, and the overall pressure drop is 20 MPa. A more nearly optimal design should yield better results.

The desirability of minimizing communications delays between processors favors the construction of macroscale systems full of high-speed, closely-packed computational devices; this encourages the design of high-capacity cooling systems. The large manufacturing systems described in Chapter 14 must be convectively cooled, but the required capacity is orders of magnitude smaller than that just described. The microscopic systems described elsewhere can (in small numbers) be cooled adequately by conduction to a surrounding environment.

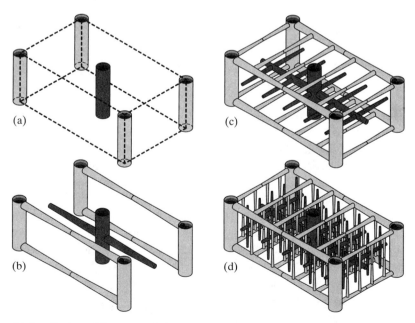

Figure 11.8.   Series of four panels illustrating hypothetical stages in the construction of a nearly fractal system of cooling tubes. After this pattern has been extended to a sufficiently fine scale, it would be broken by introducing connections between the light (inflow) and dark (outflow) tubes.

11.6. Electromechanical devices

Except in discussing certain damage and error mechanisms, all systems considered so far have been assumed to be in their electronic °ground states, with their properties determined entirely by the arrangement of their atoms. In insulators with relatively large band gaps (e.g., diamond) this is a common and stable condition.

This section considers several classes of electromechanical devices. In these, mechanical motion is coupled to the motion of electric charge, allowing mechanical motion to control charge motion, and charge motion to control mechanical motion. In true molecular electronic devices (e.g., Aviram, 1988; Hopfield et al., 1988), electronic degrees of freedom would be central and mechanical displacements would play a secondary role; these will not be considered here. The flow of current in nanoscale structures, even in electromechanical systems, is a complex topic. The following makes only rough estimates of conductivity in the pursuit of estimates of device size and performance.

11.6.1. Conducting paths

a. Conductivity.   As the diameter of a wire becomes comparable to the electron mean free path, surface scattering can increase resistance above the value suggested by scaling laws and bulk resistivity. This can be avoided by suppressing surface scattering. In particular, structures can be chosen that have matching lattice spacings across the interface between metal insulator, measuring the spacings along the axis of current flow (matching usually requires strain). Electrons with wave vectors that enable propagation without reflection from metallic lattice planes will then experience no degradation of their longitudinal momentum as a result of diffractive scattering from the metal-insulator interface; in the longitudinal direction, specular reflection will dominate.

Doped graphite and organic polymeric materials can exhibit conductivities comparable to or greater than that of copper (Kivelson, 1988). Organic polymeric materials such as doped polyacetylene are termed quasi-one-dimensional conductors and presumably show little degradation of conductivity through surface scattering.

By engineering semiconductor structures with suitable band gaps and doping patterns, conducting filaments can be constructed that are narrow enough that electron wave functions having a transverse node are substantially higher in energy than those lacking a transverse node. These filaments, termed quantum wires, strongly suppress the small-angle scattering processes that dominate the degradation of electron momentum in ordinary conductors; they can accordingly exhibit unusually high conductivities (Sakaki, 1980; Timp et al., 1990).

Experimental and theoretical work on conducting polymers and other low-dimensional conducting structures suggests that eutactic electronic structures can be built that provide conducting paths of substantially lower resistance than would be estimated from their dimensions and bulk metallic resistivities. Indeed, it is conceivable that the ability to build a far wider range of metastable structures will lead to the discovery of superconductors with critical temperatures substantially higher than those now known. The balance of this work, however, assumes conductors of bulk metallic conductivity; as just noted, this can be achieved by tailoring the metal-insulator interface. Conductors with diameters <1 nm will not be considered. For bulk aluminum, the room-temperature resistivity equals $\sim 2.7 \times 10^{-8}\ \Omega\cdot m$.

b. Current density.   In integrated circuits, high current densities can cause *electromigration,* in which metal is redistributed from one region of a conductor to another, eventually breaking circuit continuity. This phenomenon limits acceptable current densities.

Electromigration results from biased diffusion of metal atoms and vacancies, chiefly along dislocations and grain boundaries. It is strongly suppressed in fully-dense, single-crystal metallic structures where the energy required to form a vacancy is large (e.g., >150 maJ in Cu). The required energy is still larger if the displaced atom cannot occupy a normal lattice site (in a slightly expanded crystal), but is instead forced to occupy a high-energy interstitial site. Accordingly, a single-crystal metal wire surrounded by a strongly bonded, lattice-matched sheath should be stable at far greater current densities than a conventional wire of the same material. The structural stability of (some) conducting polymers should likewise minimize electromigration and similar phenomena. Despite these considerations, the following sections assume current densities no higher than those used today; a typical limit for aluminum conductors in integrated circuits is $\sim 3 \times 10^9\ A/m^2$ (Mead and Conway, 1980).

11.6.2. Insulating layers and tunneling contacts

As a result of electron tunneling and field emission, current can flow between conductors separated by nominally insulating layers. To keep the tunneling conductance of an insulator within acceptable bounds requires some minimum thickness of material (which increases with increasing voltage); in addition, conductance from field emission (tunneling into vacuum or the interior of an insulator, as distinct from tunneling into another conductor) places limits on the maximum electric field permissible at a negatively charged surface. At small voltages (≤ 1 V), an insulating gap of ~ 3 nm can limit tunneling current densities to magnitudes that can usually be neglected in a nanomechanical context (e.g., $\leq 10^{-18}\ A/nm^2$); metals exposed to vacuum can sustain fields of ~ 2 V/nm with a negligible current density from field emission (Farrall, 1980).

At small separations, tunneling currents can provide a mechanism for electrical contact without the complications of a conventional mechanical contact. Experiment shows that tunneling currents crossing a vacuum gap in an STM apparatus can exceed 300 nA at 20 mV (Gimzewski, 1988), implying a resistance near zero voltage of $< 10^5\ \Omega$; resistance falls with increasing field strength (for comparison, the resistance attributed by classical continuum models to an interatomic spacing in bulk aluminum is $\sim 100\ \Omega$). In the STM, most of the current is transmitted through a single atom at the tip, hence devices with parallel surfaces of several square nanometers should exhibit resistances of $\sim 10^4\ \Omega$. The design of good tunneling contacts will be complex, requiring attention to surface stability, stability of pairs of surfaces at small separations, and factors (such as work functions) affecting the gap resistance; surfaces of negative electron affinity may be useful [e.g., hydrogenated diamond (111) surfaces (Himpsel et al., 1979) with suitable doping of the underlying material].

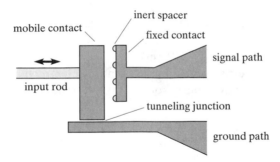

Figure 11.9. Schematic diagram of a modulated tunneling junction for interfacing mechanical inputs to electrical outputs (moving parts and electrical components only).

11.6.3. Modulated tunneling junctions

The transduction of mechanical signals to electrical signals can be performed by mechanically modulated tunneling junctions of the sort diagrammed in Figure 11.9. Bonding forces tend to snap metallic surfaces together (Israelachvili, 1992), but this can be resisted by small-area compressive contacts between non-bonding protrusions. Together with other means of tailoring the potential energy as a function of displacement, these can enable the device to operate without requiring substantial stiffness in the driving mechanism.

Section 11.6.2 concludes that tunneling junction resistances as low as $10^4 \, \Omega$ should be feasible with junction areas of a few square nanometers. Separation of the surfaces to a distance of a few nanometers increases the resistance of a typical junction by many orders of magnitude (roughly a factor of 10 per 0.1 nm) and permits substantial voltages to be supported. Accordingly, a modulated junction can serve as a mechanically controlled electrical switch.

11.6.4. Electrostatic actuators

Electrostatic actuators (Figure 11.10) can produce substantial mechanical forces. In this mechanism, one plate of a small capacitor is fixed and connected to a signal source; the other plate is grounded via a tunneling junction, leaving it free to

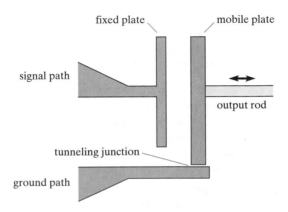

Figure 11.10. Schematic diagram of an electrostatic actuator for interfacing electrical inputs to mechanical outputs (moving parts and electrical components only).

move within a small range of displacements. In a concrete example, an actuator with a stroke length of 1 nm (moving between plate separations of 3 and 4 nm) and a potential ranging from 0 to 5 V can apply a force ranging from zero to 1 nN with a plate area of ~150 nm$^2 \approx (12 \text{ nm})^2$.

Although this is perhaps the simplest kind of electrostatic actuator, many others are possible. For example, a solenoidlike system can draw a high-dielectric constant plate into a gap with a force depending on the field; this can produce a large displacement. A long structure with multiple actuator mechanisms along its length can produce a large force with a small displacement. Finally, the range of actuators that can be built by molecular manufacturing will likely have no sharp boundary separating electrostatic from piezoelectric devices. In a mechanism of a given size, piezoelectric structures can typically provide larger forces but smaller displacements than electrostatic actuators of the sort just described. The use of a direct piezoelectric drive mechanism for manipulators of the sort described in Section 13.4.1 would increase their similarity to existing proximal probes.

11.6.5. Electrostatic motors

Motors based on electrostatic forces (like those based on electromagnets) can be built to run on alternating or direct current, and can have a wide range of mechanical properties. Alternating-current (AC) motors can synchronize a mechanical system to an electrical input, but a substantial part of their complexity is in the power supply. Direct-current (DC) motors are discussed in the following section.

11.7. DC motors and generators

Electrostatic DC motors provide a straightforward means of supplying power to nanomechanical systems. In one implementation (Figure 11.11), electric charge is placed on the rim of a rotor as the rim passes within a dee electrode, and is then transported across a gap to the interior of the opposite dee electrode, where it is removed and replaced by a charge of the opposite sign. If a voltage of the proper sign is applied across the dees, the charges in transit between them experience a force that applies a torque to the rotor, delivering power. This resembles a Van de Graff generator operating in reverse.

11.7.1. Charge carriers and charge density

The rim of a rotor can be made of insulating material with embedded conductive electrodes (Figure 11.11) that serve as charge carriers. If the rim electrodes are separated by 3 nm and no large voltages are imposed between them, then interelectrode charge transfer can be neglected. Neglecting the beneficial effects of insulator polarization, the surface charge density of the rim electrodes corresponding to a (modest) field of 0.2 V/nm is ~0.0018 C/m^2. Assuming an electrode diameter of 3 nm and length of 20 nm, the charge per electrode is ~3.3 × 10^{-19} C, or ~2 electronic charge units; the corresponding charge per unit rotor circumference is ~5.5 × 10^{-11} C/m. A vacuum gap of several nanometers can insulate the rim electrodes from a nearby dee electrode surface.

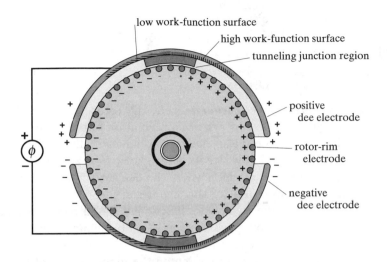

Figure 11.11. Schematic diagram of an electrostatic motor (not to scale). The dee electrodes and rotor-rim electrodes are conducting, the rotor structural material is insulating. Rotation and torque are in the direction shown.

11.7.2. Electrode charging mechanism

a. Work functions and contact potentials. The work function ϕ_w of a material is the energy required to remove an electron from its interior to a nearby exterior region. The work function of a material often depends strongly on its surface condition, and extreme values of ϕ_w can presumably be achieved by carefully tailoring surface structures. The range of work functions observed on clean surfaces of elemental metals of reasonably high melting-point places a lower bound on breadth of the range to be expected in a broader class of stable structures. Among the lower values (Lide, 1990) is $\phi_w = 2.7$ eV (for samarium, melting point ~1350 K); among the higher values is $\phi_w = 5.7$ eV (for platinum, melting point ~2050 K).

Placing two metals with differing work functions in electrical contact equalizes their electron energies and results in a potential difference (the *contact potential*) between the region of space immediately outside one surface and the region immediately outside the other. This potential difference implies an electric field and a corresponding surface charge density. Contact potentials provide a convenient mechanism for charging the rim electrodes.

b. Charging via contact potentials. Consider a rotor-rim electrode in electrical contact with the interior of a dee electrode, in an environment where its surface and the surrounding dee electrode surface share a common work function $\phi_{w,middle} = 4.2$ eV. The surface charge density is accordingly zero. Now, while maintaining electrical contact through a tunneling junction, allow the rotor-rim electrode to move into a region in which the inner surface of the dee electrode has a work function $\phi_{w,high} = 5.7$ eV. The magnitude of the contact potential is accordingly 1.5 V; with a capacitance of ~2.1×10^{-19} F, the charge on the rotor-rim electrode will have a magnitude of 2 electronic units. Modeling the rim electrodes as well-separated cylinders parallel to a plane at a distance of 3 nm from their surfaces yields a capacitance per cylinder of ~6.3×10^{-19} F; modeling them as a continuous strip at the distance of their centers yields a capacitance per rim-electrode length of $\sim 2.4 \times 10^{-19}$ F. Accordingly, the assumed charge could be induced in a system with a substantially smaller value of $\phi_{w,high} - \phi_{w,middle}$. (Since $\phi_{w,high} - \phi_{w,middle}$ can equal $\phi_{w,middle} - \phi_{w,low}$, charge of the opposite sign can be induced in a similar fashion.)

c. The rim electrode charge/discharge cycle.

A rotor-rim electrode can be made to go through the following cycle in a revolution of the rotor: Starting in the middle of the positive dee, it is in electrical contact with the dee via a tunneling junction in an environment where its induced charge is zero. Moving clockwise (Figure 11.11), it enters a region where the dee surface has a higher work function, inducing a charge of $+2e$. This change in mean equilibrium charge occurs gradually, because the mean work function of the surrounding electrode is made to vary gradually from position to position by changing the fraction of area occupied by surfaces of different kinds.* Once this charge is induced, the tunneling contact (made to a narrow ridge on the inner surface of the dee) is gradually broken in surroundings that keep the equilibrium charge on the rim electrode constant. After traveling a further distance, the positively charged rim electrode exits the positively charged dee and enters the negatively charged dee, dropping through a potential difference of ϕ_{motor} as it does so. Once beyond the fringing fields at the entrance to the negative dee, the rim electrode encounters an electrostatic environment that is the same as it was before exiting the positive dee (in accord with basic electrostatic principles, the charge and field resulting from the dee's overall potential are chiefly distributed across its outer surface), and the approach to the middle of the negative dee is a nearly precise reversal of the sequence of environments and charge-transfers that occurred in the positive dee, ending with no net charge on the rim electrode. Continuing through the negative dee, exiting, entering the positive dee, and returning to the initial position proceeds in the same manner as the first half of the cycle, except that low surrounding work functions induce a negative charge on the rim electrode.

This cycle never permits electrical contact between conductors in electrical disequilibrium, and is directly analogous to mechanochemical cycles that avoid well merging losses (Section 7.6.5). By connecting conductors only at equilibrium, current flow is postponed until the contact resistance is low, enabling the charge-transfer process to be nearly thermodynamically reversible. (Note that reversing the direction of rotor motion would in fact reverse the directions of current flow and of work, making the system function as a DC generator.)

---

* Under some conditions, Coulomb blockade effects (Barner and Ruggiero, 1987; Kuzmin et al., 1989) can become important. In particular, if the tunnel resistance is large compared to the quantum unit of resistance $\pi\hbar/2e^2 \approx 6.5 \times 10^3 \, \Omega$, then electrons will be sufficiently well localized that a rim electrode can be viewed as having a discrete number of electrons. If, in addition, the change in electrostatic energy caused by adding one electronic charge ($e^2/2C$, where C is the electrode capacitance) is large compared to kT, then the mean charge will vary in a staircase fashion, rather than smoothly. Under these conditions, a more uniform rate of charging can be achieved if the potential induced by work-function effects changes swiftly in the range where the mean equilibrium charge approximates an integral number, and more slowly in the range around half-integral values.

11.7.3. Motor power and power density

Neglecting resistive and frictional losses, DC electrostatic motors of this kind deliver a shaft power equal to the product of the current and the voltage ϕ_{motor}. The current, in turn, is twice the product of the rim speed and the charge per unit length. With diamondoid structures to support centrifugal loads and an evacuated volume to avoid fluid drag, a rim speed of ~1000 m/s is readily achievable. With a charge per unit length of ~5.5×10^{-11} C/m (Section 11.7.1), the current is ~110 nA (including contributions from both sides of the rotor). If $\phi_{\text{motor}} = 10$ V, the delivered power is ~1.1 μW.

Allowing for the geometrical constraints of minimum gap sizes for insulation, and so forth, a motor radius of 50 nm is generous (to permit direct application of sample calculations in a later section, a radius of 195 nm is assumed in the following). A motor thickness of 25 nm is consistent with the rim electrode dimensions assumed in Section 11.7.1, yielding a motor volume of ~2×10^{-22} m^3. The power density is large compared to that of macroscale motors: >10^{15} W/m^3. For comparison, Earth intercepts ~10^{17} W of solar radiation. (Cooling constraints presumably preclude the steady-state operation of a cubic meter of these devices at this power density.)

11.7.4. Energy dissipation and efficiency

a. Resistance in conductors. Since electrostatic motors do not exploit current-generated fields, they need no long, coiled conductors and have only small internal resistive losses. (Bearing friction could be regarded as a resistive loss associated with mechanical charge transport by the rotor, but is considered separately below.) Resistive losses are dominated by conducting paths outside the motor proper, making the following estimate somewhat arbitrary. At a current density of 1 nA/nm^2, the cross sectional area of the leads for a 110 nA motor is 110 nm^2. If each lead has a length equal to the motor diameter, their series resistance is ~200 Ω, assuming a resistivity equal to the bulk value for aluminum (~2.7×10^{-8} Ω·m). This yields an estimated resistive power dissipation in the leads of ~2.4 pW. The current flow responsible for distributing charge along a rim electrode contributes ~0.1 pW, calculated on a similar basis.

b. Resistance in tunneling contacts. Since the length of the rim contained in a dee is ~500 nm, >80 rim electrodes can be in contact in a given dee at any given time. Assuming that the mean charge on a rim electrode increases steadily during its transit of a dee, the tunneling contacts in each dee can be treated as resistances in parallel. If a tunneling contact to a rim electrode has a resistance of <10^4 Ω (Section 11.6.2), then the total resistance of the tunneling contacts through both dees is <250 Ω, with a corresponding power dissipation of <3 pW. As noted in Section 11.7.4d, however, larger gaps and resistive losses may prove desirable in order to reduce contact drag.

c. Bearing drag. A rotor remains below its lowest critical angular velocity so long as the condition

$$\sqrt{k_{\text{s,t}}/m_{\text{rotor}}} > \omega_{\text{rotor}} \tag{11.8}$$

is met (where $k_{s,t}$ is the stiffness of the bearing structure in resisting displacement of the shaft in a direction perpendicular to its axis, and m_{rotor} and ω_{rotor} are the rotor mass and angular velocity). With a rotor radius of 195 nm (chosen for compatibility with Section 12.7.5) and rim speed of 1000 m/s, $\omega_{rotor} \approx 5 \times 10^9$ rad/s and $m_{rotor} = 8 \times 10^{-18}$ kg, requiring a bearing with $k_{s,t} \geq \sim 200$ N/m, a condition which is easily met (see Sections 10.4.5 and 10.4.6). Applying the combined drag model (Section 10.4.6f) to an axle bearing with a stiffness of 500 N/m (ample for rotor stability), $r_{eff} = 10$ nm, $\ell = 20$ nm, $v = 50$ m/s, $R = 10$, and the pessimistic value $\Delta k_a / k_a = 0.4$, the estimated bearing drag power is ~ 1.3 pW.

A more serious instability results from the negative stiffnesses associated with the attractive forces in tunneling contacts (Chen, 1991; Dürig et al., 1988). In a sliding-contact design, these may require bearing stiffnesses larger by two orders of magnitude, resulting in a bearing drag power on the order of several nanowatts.

d.   Contact drag.   A sliding tunneling contact between a rim electrode and the tunneling contact structures inside a dee can be expected to result in drag associated with electron transfer. Transfer of a single electron across the gap would (on the average, in a classical system) be expected to dissipate the kinetic energy corresponding to its mass and the relative speed of the rotor: $\sim 5 \times 10^{-25}$ J, or $\sim 3 \times 10^{-7}$ of the energy delivered by an electron in dropping through a 10 V potential. Efficient, near-equilibrium electrode charging requires many electron exchanges for the net transfer of a single electron, multiplying the energy loss. As suggested by J. Soreff, however, the use of low-dimensional conductors lacking substantial conductivity parallel to the direction of motion should strongly suppress energy losses from electron transfer (and from associated scattering mechanisms).

The attractive forces associated with sliding tunneling contacts also cause losses through thermoelastic effects and phonon scattering. These remain to be evaluated and may prove to be the dominant loss mechanism in this design. If they are sufficiently large, alternative designs with rolling contacts would be preferable. This shift in design could also reduce the requisite bearing stiffnesses and resulting bearing drag.

e.   Summary.   Omitting contact drag, the energy dissipation appears to be dominated by the bearing drag (several nanowatts). If contact drag does not greatly change this picture, then motor efficiencies $>.99$ should be achievable at an extremely high power density; efficiencies can be higher at lower power densities and could likely be increased by replacing sliding with rolling electrical contacts. None of the analyses in later chapters require high motor efficiencies.

11.7.5.   Motor start-up

If a motor is initially disengaged from any load, start-up can occur spontaneously. The mean thermal speed of the rotor is ~ 0.01 m/s, with little damping. Rotation in reverse is effectively prohibited when the dees are charged; forward rotation is permitted and will begin to carry charge across the gap, producing torque and faster rotation, in $\sim 10^{-5}$ s.

Forcible motor start-up can be ensured if the rim electrodes in the gap between the dees have the proper charge, producing a strong torque. This can be accomplished by providing extensions from each dee that provide a weak tunneling current to the rim electrodes in the gap (on the side where they have the same charge in normal operation). If this rate is low, it has little effect on an operating motor, yet can ensure charging of electrodes in the gaps of a stationary motor.

11.7.6. Speed regulation

Precise speed regulation calls for a mechanism based on a frequency standard, such as a mechanical oscillator or an external AC source. For a system operating asynchronously with respect to other systems in its environment, approximate speed regulation can often suffice. A spring-loaded variant of the Watt governor can serve this purpose, opening and closing a tunneling contact instead of a steam valve. If designed for bistability, it will oscillate between having a closed contact and having an open contact, dissipating little energy in either state.

11.8. Conclusions

This chapter surveys several classes of subsystems of intermediate size and complexity. These include mechanical measurement systems, stiff mechanisms of high gear ratio, components for handling fluids, convective cooling systems, electrostatic switches and actuators, and DC electric motors of high efficiency and power density. A conclusion emerging from the chapter as a whole is that functional analogues of many macroscale systems can be constructed on a scale of tens to hundreds of nanometers; their detailed structures, operating principles, and performance parameters, however, often diverge widely from those of the corresponding macroscale systems.

Some open problems.   As in Chapter 10, the task of expanding the range of well-characterized, atomically detailed models for devices defines a large set of open problems. Many subsystems (measurement devices, stiff drive mechanisms, seals and pumps) fall within the size range accessible with existing molecular modeling systems. Detailed studies of various combinations of seals and infiltrating species, using various modeling techniques, would be of considerable interest. The design of more nearly optimal convective cooling systems is of interest, and progress can be made using standard relationships for the pressure drop in pipes. Electromechanical devices present a set of open problems revolving around electron transport in well-designed, stable, eutactic structures on a nanometer scale, and in tunneling junctions in particular. Further study is needed to characterize drag mechanisms in sliding tunneling junctions. The performance estimates for electromechanical devices in this chapter are crude at best, and the junction structures remain to be specified.

12

Nanomechanical Computational Systems

12.1. Overview

This chapter examines a representative set of components and subsystems for nanomechanical computers, chiefly applying the bounded continuum model (Sections 9.3.3, 9.4, and 9.5). The range of useful components and subsystems is, however, larger than that considered here. The following sections describe systems capable of digital signal transmission, fan-out, switching, data storage, and interfacing with existing electronics; this suffices to demonstrate the feasible scale, speed, and efficiency of nanomechanical technologies for computation. A brief discussion surveys other components, including carry chains, random-access memory (RAM), and tape memory systems. The discussion of logic rods parallels that in Drexler (1988b) but applies a wider range of analytical tools to a different set of physical structures.

Within the bounded continuum model, the design of nanomechanical systems resembles that of macromechanical systems. In neither case are structures specified in atomic detail, and in both, structural properties are described in terms of parameters such as strength, density, and modulus. The special characteristics addressed in the bounded continuum model include surface corrections to strength, density, and modulus; constraints on feature size and shape; models for static friction derived from analyses based on intermolecular potentials; and phonon-interaction based models for dynamic friction. Further, many nanomechanical designs are significantly constrained by statistical mechanics and the resulting trade-offs among structural stiffnesses, positional tolerances, and error rates.

12.2. Digital signal transmission with mechanical rods

12.2.1. Electronic analogies

In conventional microelectronics, digital signals are represented by the voltages of conducting paths: a high voltage within a certain range can be taken as a 1, and a low voltage within a certain range can be taken as a 0. Propagation of voltages through circuits requires the flow of current, with associated delays and energy losses.

Analogously, digital signals in nanomechanical computers can be represented by displacements of solid rods; for example, a large displacement falling within a certain range can be taken as a 1, and a small displacement falling within a different range can be taken as 0. Propagation of displacements along rods requires motion, with associated delays and energy losses.

The parallels between existing microelectronics and proposed nanomechanical systems are not exact. The time constants in microelectronics are dominated by resistance, not inductance; time constants in the nanomechanical systems are dominated by inertia (analogous to inductance), not by drag (analogous to resistance). Elastic deformation of rods plays a role resembling (but different from) that of parasitic capacitance in conductors.

12.2.2. Signal propagation speed

The speed of signal propagation in rods is limited to the speed of sound, in diamond ~17 km/s ($~6 \times 10^{-5}c$). Energy dissipation caused by the excitation of longitudinal vibrational modes in a rod is small provided that the characteristic motion times are long compared to the acoustic transit time (Section 12.3.4b). Adhering to this constraint lowers the effective signal propagation speed, but delays can still be in the range familiar in microelectronic practice: at an effective propagation speed of only 1.7 km/s, for example, the delay over a 100 nm distance is ~60 ps, and over a 1 μ distance is ~0.6 ns. As will be seen, these distances arc comparable to the diameter of a °CPU. Note that the exemplar parameters developed in the following sections assume a material of lower modulus (500 GPa, vs. ~1 TPa) and acoustic speed (~12 km/s).

Signals can be transmitted over greater distances at higher speeds by acoustic pulses, at the cost of either dissipating the acoustic energy or requiring accurate frequency control in the drive system to permit its recovery. (Further discussion of acoustic signal propagation is deferred to Section 12.5.4, after gates, drive systems, etc., have been introduced.)

12.3. Gates and logic rods

12.3.1. Electronic analogies

In conventional microelectronics, logic °gates are built using transistors. In °CMOS systems, a transistor makes the current-carrying ability of a path dependent on the voltage applied to a gate, which can be constructed either to transmit current at a high gate voltage and block it at low, or to block current at a high gate voltage and permit it at low.

Analogously, logic gates in nanomechanical computers can be built using *interlocks*. These can resemble CMOS transistors, in that interlocks can make the mobility of a rod dependent on the displacement applied to a gate knob, either permitting motion when the gate knob is at a large displacement and blocking it at low displacements, or *vice versa*.

Again, the parallels are not exact. In particular, CMOS transistor gates have a large capacitance relative to a comparable length of simple conducting path, causing substantial propagation delays; interlock gates, in contrast, only slightly delay signal propagation along a logic rod. Accordingly, fan-out has less effect on speed in rod logic than in CMOS logic.

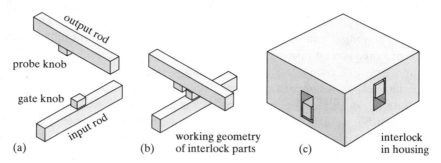

Figure 12.1.   Components of an interlock: an input rod with a gate knob and an output rod with a probe knob, shown separated (a), in their working positions (b), and constrained by a housing structure (c).

12.3.2.   Components and general kinematics

Figure 12.1 schematically illustrates the components of an interlock, including two rods, two knobs, and a housing structure. Figure 12.2 uses a more abstract graphical notation to represent the structure and kinematics of a small logic rod system (some components are included only for compatibility with descriptions of rods having greater fan-out). In the following, the term *logic rod* will refer to a particular rod under consideration, and the otherwise-similar rods that interact with it will be termed *input* and *output* rods.

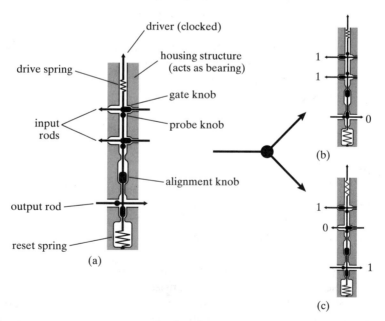

Figure 12.2.   Schematic diagram of a short logic rod with two inputs and one output. Diagram (a) shows the initial state, with both input rods in the blocking position. Diagram (b) shows a state in which both input rods have been moved to the unblocking position, enabling the driver to move the vertical rod and switch the output gate; diagram (c) shows a state in which one of the input rods has remained in the blocking position, forcing the drive spring to accommodate the motion of the driver and leaving the vertical rod unmoved.

a. Drivers and drive springs. The rod logic systems considered here are clocked, with a distinct clock signal for each level of gates in a combinational logic system (this approach helps minimize energy dissipation). Each rod is accordingly attached to a *driver*, a source of periodic, nonsinusoidal displacements; the implementation considered here achieves this motion using a follower sliding on a sinusoidally oscillating cam surface (Section 2.7.1). (The cam surface, in turn, is part of a thick *drive rod*, which in turn is part of a power distribution and clocking system ultimately driven by crankshafts coupled to a motor-flywheel system; Section 12.7.5.) Multiple rods can be attached to a single driver mechanism.

Driver displacements are coupled to a single rod via a *drive spring*. This can be a constant force spring that retains a fixed length until the force exceeds a threshold. If the rod is not blocked, as in Figure 12.2(b), displacements of the driver are transmitted through the drive spring and cause comparable displacements of the rod. If the rod is blocked, as in Figure 12.2(c), displacements of the driver chiefly cause stretching of the drive spring. The driver and drive spring thus form a *drive system* that periodically tensions and de-tensions the rod, without forcing motion.

b. The housing structure. A stiff housing structure surrounds the moving parts of a rod logic system, constraining their motions (within small excursions) to simple linear displacements. The surface of this housing structure serves as a bearing for the sliding motions of rods, and the rod-housing interaction can be analyzed as an extended system along the lines developed in Chapter 10.

c. Gate knobs, probe knobs, and interlocks. Each logic rod bears a series of protrusions termed *knobs*. A *gate knob* and a *probe knob* in a suitable housing form an interlock, as shown in Figure 12.1. A gate knob on a rod in its 0 position can be positioned to either block or fail to block the corresponding probe knob. Displacement of the gate-knob rod from 0 to 1 can thus either unblock or block a probe knob, depending on the how the interlock has been constructed. (A knob-free crossover structure can permit motion in either state.)

d. Input rods. The mobility of a logic rod can be determined by the state of an indefinitely large number of *input rods,* each bearing a gate knob and interacting with a probe knob on the logic rod. When none of the input rods blocks its corresponding probe knob, the logic rod is free to move when the drive system next applies tension. Figure 12.2 shows a gate with two input rods, both of which must be displaced to unblock the vertical logic rod; it implements the NAND (not-and) operation.

e. Alignment knobs. To fix the displacements of the two logic states requires an alignment mechanism. The *alignment knob* of a logic rod slides within a certain range, bounded by *alignment stops,* so that a net force in one direction places the rod in one positional state, and a net force in the other direction places it in the other positional state. If rods were rigid, the alignment knob could be located at any point; given a finite rod compliance per unit length, a location immediately adjacent to the gate knob (output) segment of the rod minimizes thermal displacements. Placing it between the probe and gate knob segments isolates the gate knob segment from fluctuations in tension caused by the drive system, and thus ensures greater dimensional stability and better gate knob alignment relative to the output rods.

(A small advantage in alignment stiffness could be gained at the cost of a less regular structure by placing the alignment knob somewhat inside the gate knob segment. A greater advantage can be gained by placing a second alignment knob at the far end of the gate knob segment.)

f. Output rods. The displacement of a logic rod determines further steps in a computation by blocking or unblocking an indefinitely large number of *output rods,* each bearing a probe knob and interacting with a gate knob on the logic rod. An interlock in the output segment of one rod is in the input segment of the next. The state of the interlocks of an output segment switch when the input segment is unblocked, the drive system applies tension, and the seating knob shifts from the 0 to 1 position. In the NAND gate of Figure 12.2, the single output switches from unblocked to blocked if the rod is mobile.

g. Reset springs. When the drive system de-tensions a logic rod, a restoring force returns the rod to its resting state. This can be provided by a constant-force spring (drawn in Figure 12.2 as a large, low-stiffness spring). With this choice, the tension (and hence the strain) in the gate knob segment remains fixed throughout the cycle, allowing the gate knobs to be properly aligned with the interlocks in both the 0 and 1 logic states. Note that the tension in the probe-knob segment varies, but that the resulting fluctuations in length do not affect the reliability of the logic operations.

12.3.3. A bounded continuum model

To explore how system performance parameters such as gate density, speed, error rate, and energy dissipation vary with device geometry and other parameters, a bounded continuum model can be applied. For components of sufficient size, and for a suitable choice of material and interface parameters, it can provide a realistic description. (For smaller components, it can give a preliminary indication of the performance to be expected, provided that a structure can be found having the appropriate geometry and properties.)

a. Geometric assumptions and parameters. For simplicity, it will be assumed that probe knob segments and gate knob segments are perpendicular, and that rods and knobs can be approximated as rectangular solids (Figure 12.3). Probe knobs can be spaced at regular intervals d_{knob}. The position of each gate knob depends on its logical function, but the spacing between knobs of the same kind

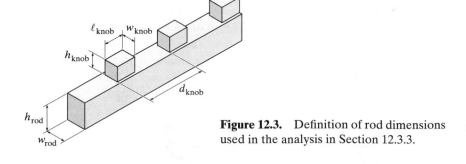

Figure 12.3.   Definition of rod dimensions used in the analysis in Section 12.3.3.

(blocking on 1 or blocking on 0) is here assumed to be the same as the spacing of probe knobs, d_{knob}. Accordingly, each interlock in a grid of intersecting probe segments and gate segments (as in a programmable logic array, Section 12.5) occupies a square region with sides of length d_{knob}. Gate knobs and probe knobs are assumed to have the same dimensions.

The geometry of a rod can in some regions be described in terms of hypothetical sliding-contact surfaces (with no overlap and no gap between objects). This description implies a choice of atomic radii, which can (for uniformity) also be applied where surfaces do not make a sliding contact. For atoms in the carbon row of the periodic table, typical radii of this kind are ~0.14 to 0.17 nm. Rod dimensions can then be described by a set of widths w_{rod} and w_{knob} and heights h_{rod} and h_{knob} defined as shown in Figure 12.3. The total height of the moving parts in an interlock is then $h_{knob} + 2h_{rod}$.

The function of a rod logic system demands that the knob length ℓ_{rod} meet the condition

$$\ell_{knob} \leq d_{knob} - w_{knob} \qquad (12.1)$$

A simple and attractive set of choices that meets this condition is

$$w_{knob} = w_{rod} = \ell_{knob} = d_{knob}/2 \qquad (12.2)$$

With these choices, the thickness of the housing structure also equals w_{rod}, and the minimum feature dimensions are uniform throughout. For the exemplar calculations in the following sections, it will be assumed that $w_{rod} = 1$ nm, and that $h_{knob} = 0.5$ nm.

The overall dimensions of a rod can be characterized by the number of input rods n_{in} and output rods n_{out}. The length of the input segment is $\ell_{in} = d_{knob}n_{in}$; the length of the output segment is $\ell_{out} = d_{knob}n_{out}$; and (with a correction for the length of the alignment knob mechanism) the total length of the rod is

$$\ell_{rod} = d_{knob}(n_{in} + n_{out} + 1) \qquad (12.3)$$

This length will be used in estimating several dynamical quantities; for estimating system dimensions, however, allowances must also be made for the drive system and reset spring, and their associated structures. The exemplar calculations will assume $n_{in} = n_{out} = 16$, implying $\ell_{rod} = 66$ nm.

b. Interactions and applied forces. The dynamics and error rates of a logic rod can be described in terms of the potential energy function of the rod, including interactions with its environment (the housing and crossing rods) and the forces applied to it by the drive and reset springs. From the description in Section 12.3.2, it can be seen that the environment of a specific *mobile* logic rod is the same in every cycle, with the exception of crossovers that permit mobility in either of two states. If the interaction energy of the crossing rod with the logic rod is made nearly equal in both states (e.g., with differences in van der Waals attraction compensated by differences in steric and electrostatic repulsion), then the potential energy function of the logic rod in its mobile state is nearly invariant and can be adjusted to near constancy within the permitted range of motion (as discussed in Section 10.8).

The force with which the alignment knob is pressed against a limit stop affects the PDF for thermal displacements that carry the knob away from the stop, and hence affects the error rate. Symmetry considerations suggest that the alignment force applied to one stop in the tensioned state and that applied to the other in the de-tensioned state should both be of the same magnitude, F_{al}. Accordingly, the (constant) force applied by the reset spring equals F_{al}, and the peak force applied by the drive spring equals $2F_{al}$. The exemplar calculations will assume $F_{al} = 1$ nN.

c. *Stiffness and mass.* In the bounded continuum model, the stiffness of knobs and rods is estimated by applying a bulk-phase modulus of elasticity to component dimensions modified by a surface correction. Assuming that rods are of diamondoid structure, with surface termination chiefly using di- and trivalent (rather than monovalent) atoms, it appears reasonable to compute stiffnesses and strengths on the basis of effective dimensions that discard a surface layer $\delta_{surf} = 0.1$ nm thick; this allows for the difference between steric and covalent radii, along with a further margin for the effects of surface relaxation. The effective cross sectional area of the rod is then

$$S_{eff} = (w_{rod} - 2\delta_{surf})(h_{rod} - 2\delta_{surf}) \tag{12.4}$$

or 0.64 nm^2 in the exemplar case.

A conservative value of Young's modulus, $E = 5 \times 10^{11}$ N/m^2, is used here; this is about 1/2 the modulus of diamond and only moderately greater than the modulus of silicon carbide, silicon nitride, or alumina. In strong covalent solids, the shear modulus G commonly is $\sim 0.5 E$, as it is in diamond.

The bending stiffness of the rod and the transverse constraint forces from the housing are large enough that the effect of the transverse vibrational modes of the rod on longitudinal stiffness and positional variance (Section 5.6) can be neglected. The stretching stiffness of a segment of rod of length ℓ is then simply

$$k_s = S_{eff} E / \ell \tag{12.5}$$

(neglecting the stiffening effect of the knobs). For $\ell = \ell_{rod}$, $k_{s,rod} = 4.85$ N/m.

In the bounded continuum model, masses are estimated by combining a density with adjusted component dimensions. In general, a slightly different value of δ_{surf} is appropriate for mass calculations, but for simplicity $\delta_{surf} = 0.1$ nm will be used throughout the exemplar calculations. With these assumptions,

$$m_{rod} \approx \rho \ell_{rod} (w_{rod} - 2\delta_{surf})[(h_{rod} - 2\delta_{surf}) + h_{knob}(\ell_{knob} - 2\delta_{surf})/d_{knob}] \tag{12.6}$$

A density $\rho = 3500$ kg/m^3 (approximately that of diamond, somewhat higher than that of silicon carbide or silicon nitride) will yield conservative estimates of device performance. With the exemplar parameters, $m_{rod} = 1.9 \times 10^{-22}$ kg.

12.3.4. Dynamics and energy dissipation in mobile rods

The small drag forces in the exemplar system can be neglected in calculating the magnitude of dynamical quantities, then reintroduced to estimate energy dissipation. Since the elastic deformation of a mobile rod is small relative to its overall displacement, accelerations and kinetic energies are well approximated by a rigid-body analysis. Again, corrections are introduced later, both in estimating energy dissipation and in computing requirements for the drive system. This section discusses mobile rods; Section 12.3.5 considers blocked rods.

a. Dynamics in the rigid-body approximation.   The use of a cam surface in the drive mechanism (Sections 10.7.7 and 12.7.1) enables flexible control of the displacements $\Delta x(t)$ imposed on the driver, and hence of the forces applied by the drive spring. In particular, these can be chosen such that a mobile rod executes a smooth motion that (in the rigid-body approximation) can be approximated by

$$\Delta x(t) = \begin{cases} 0, \ t < 0 \\ \dfrac{d_{\text{knob}}}{4}\left[1 - \cos(\pi t/t_{\text{switch}})\right], \ 0 \leq t \leq t_{\text{switch}} \\ d_{\text{knob}}/2, \ t > t_{\text{switch}} \end{cases} \qquad (12.7)$$

This motion switches the logic state of the rod in a time t_{switch}. (A more accurate model would include a finite rate of onset of acceleration.)

The rod mass

$$m_{\text{rod}} = \rho \, \ell_{\text{rod}} \left(w_{\text{rod}} - 2\delta_{\text{surf}}\right)\left(h_{\text{rod}} + h_{\text{knob}}/2 - 2\delta_{\text{surf}}\right) \qquad (12.8)$$

can be combined with the peak rigid-body acceleration to yield an estimate of the peak drive force for rod acceleration

$$F_{\text{accel}} \approx m_{\text{rod}} \left(\pi/t_{\text{switch}}\right)^2 d_{\text{knob}}/4 \qquad (12.9)$$

The peak speed in this model is

$$v_{\text{max}} \approx \pi d_{\text{knob}}/4t_{\text{switch}} \qquad (12.10)$$

With the exemplar parameters, if $t_{\text{switch}} = 0.1$ ns, and $m_{\text{rod}} = 1.94 \times 10^{-22}$ kg, then $F_{\text{accel}} \approx 0.096$ nN, and $v_{\text{max}} = 15.7$ m/s. (The balance of this analysis assumes $F_{\text{accel}} \leq F_{\text{al}}$, as is true for the exemplar parameters.)

b. An estimate of vibrational excitation.   A detailed examination of the vibrational dynamics of a logic rod (e.g., taking account of the effects of alignment knob contacts and drive system force profiles on each vibrational mode) is beyond the scope of the present work. The chief interest in vibrational excitation in the present context is its role as a mechanism of energy dissipation. The energy of excitation can be estimated from a harmonic-oscillator model of the lowest vibrational mode of the rod (coupling of energy from the drive system to higher-frequency modes is much lower). The energy of vibrations excited in the housing structure should be small by comparison.

A harmonic oscillator with mass m, stiffness k_s, and natural frequency ω will, if suddenly subject to a constant longitudinal acceleration a, acquire a vibrational energy

$$\Delta \mathcal{E} = (ma)^2/2k_s = k_s a^2/2\omega^4 \qquad (12.11)$$

Substituting the peak acceleration derived from Eq. (12.9) yields

$$\Delta \mathcal{E} = \frac{k_s d_{\text{knob}}^2}{32}\left(\pi/\omega t_{\text{switch}}\right)^4 \qquad (12.12)$$

For $\lambda \gg d_{knob}$, the speed of a longitudinal wave along a rod is

$$v_s \approx \sqrt{E/\rho(1+h_{knob}/2h_{rod})} \tag{12.13}$$

which includes an approximate correction for knob mass (for the exemplar system, $v_s \approx 11$ km/s). The angular frequency of the fundamental mode of the rod (in which the far end, lacking restoring forces, is essentially free) is

$$\omega_0 \approx \pi v_s/2\ell_{rod} \tag{12.14}$$

To estimate the vibrational energy introduced into a rod, ω_0 can be substituted for ω in Eq. (12.11), and $\ell_{rod}/2$ for ℓ in Eq. (5.22); the resulting value of k_s can (conservatively) be substituted for k_s in Eq. (12.11). Vibrational excitations induced by deceleration can add in phase with, or cancel, those induced by acceleration. Averaging over these cases introduces a factor of two in the total vibrational energy per displacement, and yields the estimate

$$\Delta\mathscr{E}_{vib} \approx \frac{2S_{eff}d_{knob}^2\rho^2\ell_{rod}^3}{Et_{switch}^4}(1+h_{knob}/2h_{rod})^2 \tag{12.15}$$

As can be seen from this expression, energy dissipation via the excitation of irrecoverable vibrational energy in logic rods is strongly sensitive to design choices. In particular, choosing sufficiently small values of ℓ_{rod} or sufficiently large values of t_{switch} can reduce $\Delta\mathscr{E}_{vib}$ to negligible values. In the exemplar system, $\Delta\mathscr{E}_{vib} \approx 0.56$ maJ.

A more detailed analysis might choose a force profile to be applied by the drive spring that delivers the alignment knob to the alignment stop with approximately zero mean relative velocity, and then tensions the input with a load of $2F_{al}$. The smaller displacement (with the exemplar parameters, $\sim d_{knob}/8$) and higher characteristic frequency ($\sim 4\omega_0$) of this tensioning motion allow it to be fast, yet induce little vibrational excitation.

c. An estimate of sliding-interface drag. The sample calculations of Section 10.4.6 indicate that band-stiffness scattering is the dominant energy dissipation mechanism in systems where $\Delta k_a/k_a = 0.4$ and $R = 10$. Since the sliding interface between a logic rod and a housing is analogous to the sliding interface in other bearing systems, Eq. (10.23), applied with these parameters, should yield a conservative estimate of the drag in the logic rod system.

The contact area can be taken as

$$S = \ell_{rod}(w_{rod} + 2h_{rod}) \tag{12.16}$$

and Eq. (10.23) with the preceding values of $\Delta k_a/k_a$ and R yields the estimate

$$\Delta\mathscr{E}_{drag} \approx 3.3\times10^{-32}\ell_{rod}(w_{rod}+2h_{rod})\frac{d_{knob}^2 k_a^{1.7}}{t_{switch}} \tag{12.17}$$

(the use of mean square rather than peak speed introduces a factor of 1/2). Assuming an interfacial stiffness $k_a = 10$ N/m·nm^2 and the exemplar parameters, this yields an energy dissipation of 0.052 maJ per displacement. Since this energy loss is small compared to the kinetic energy of motion (~ 23 maJ), damping can (as has been assumed) be neglected in computing velocities, accelerations, and so forth.

d. Sliding-interface drag in the cam surface. If the maximum slope of the cam surface in the driver mechanism (Section 12.7.1) is 1/2, then the sliding speed of the cam surface is

$$v_{cam} \approx 2v_{max}\left(1 + 4F_{al}\ell_{in}/d_{knob}S_{eff}E\right) \qquad (12.18)$$

where the expression in parentheses provides an upper-bound correction for the effect of the elastic deformation of the input segment. Applying the exemplar parameters, $v_{cam} \approx 38$ m/s. In the contemplated system context, multiple logic rods share a drive-system cam. Reasonable values for the sliding interface in the cam are ~4 nm^2 per logic rod, with an interfacial stiffness $k_a = 10$ N/m·nm^2. The energy dissipation per rod per 0.1 ns period is then ~0.012 maJ. Since the cam surface continues to move when the rod is stationary, this energy must be multiplied by the reciprocal of the fraction of time spent in rod motion, ~5, yielding ~0.06 maJ per switching event.

e. An estimate of thermoelastic losses. In the transition to the tensioned state, the tension in the probe segment increases by $2F_{al}$, causing thermoelastic losses. A bound on the energy dissipation in this process can be derived from Eq. (7.49)

$$\Delta\mathscr{E}_{therm} < 8.2 \times 10^{-5}\beta^2\frac{(2F_{al})^2}{S_{eff}}\ell_{in} \qquad (12.19)$$

in which a volumetric heat capacity $C_{vol} = 1.7 \times 10^6$ J/K·m^2 and temperature $T = 300$ K have been assumed. The value of the volumetric thermal coefficient of expansion β varies widely among materials, and is occasionally negative. The values found in Table 7.1 suggest that $\beta = 5 \times 10^{-6}$ will prove conservative for many diamondoid structures chosen with some attention to this parameter. With this assumption, the exemplar parameters yield $\Delta\mathscr{E}_{therm} < 0.41$ maJ.

f. Other energy dissipation mechanisms in mobile logic rods. Treating the knobs as phonon-scattering centers of large mass leads to estimated scattering cross sections ~10^{-23} m^2, and to comparatively negligible power dissipation. Thermoelastic effects caused by the motion of knobs with respect to the housing are likewise small. Compression of the alignment knob against its limit stop results in minimal nonisothermal losses because the knob is in good thermal contact with the rest of the rod, which provides a substantial heat sink.

g. The resetting process. A logic rod is reset after its output rods have been reset and before its input rods are reset. Thus, the resetting process occurs in the same environment as the switching process. Further, the motion of the driver approximates the time reversal of the switching motion, since it is driven by the same follower tracking the same cam surface, but on its return stroke. Accordingly, the *overall* dynamics of resetting approximate the time reversal of the switching motion, in those systems for which vibrational excitation of the rod is small. Energy losses for a switching cycle, including resetting, are therefore approximately twice those that result from the switching motion itself.

h. Nonthermal vibrations as source of noise. The motion of components generates vibrational excitations in the structure as a whole, and these are not immediately thermalized. The magnitude and effects of these vibrations depend on the design, and nonthermal vibrations can render sufficiently bad designs unworkable. Good design practices will include the use of stiff housing structures, vibration cancellation between different components, the avoidance of structures that resonate at the major frequencies of excitation, and provision for effective damping at those frequencies.

12.3.5. Dynamics and energy dissipation in blocked rods

If one or more probe knobs is blocked by an input gate knob, no large displacement of the logic rod occurs, and sliding interface losses are minimal. As the driver displacement increases, the tension climbs to $\sim 2F_{al}$, the drive spring begins to extend, and the tension becomes constant. The lowest longitudinal modal frequencies of a blocked input segment are $\geq \sim 4\omega_0$, reducing vibrational energy losses. The tensioned length is $\leq \ell_{in}$; hence the thermoelastic losses are comparable to or less than those for a mobile rod.

12.3.6. Fluctuations in stored energy

The flow of energy between the drive system and rod is large compared to the energy dissipated. Logic-state-dependent fluctuations in the energy stored in the rods (at a given point in the clock cycle) determine certain requirements for the drive system (Section 12.6), and hence must be estimated.

The greatest mechanical difference in the state of a rod occurs between a cycle in which it is mobile and one in which it is blocked near the drive spring. In a mobile cycle, the energy stored by tensioning is dominated by the energy in the reset spring ($\Delta\mathscr{E}_{reset}$) and the strain energy of the input segment ($\Delta\mathscr{E}_{input}$). In a blocked cycle, it is dominated by the energy stored in the drive spring ($\Delta\mathscr{E}_{drive}$). The energy difference is thus

$$\Delta\mathscr{E}_{state} \approx \Delta\mathscr{E}_{drive} - \left(\Delta\mathscr{E}_{reset} + \Delta\mathscr{E}_{input}\right)$$

$$\approx 2F_{al}\left(\frac{d_{knob}}{2} + 2F_{al}\frac{\ell_{in}}{ES_{eff}}\right) - \left(F_{al}\frac{d_{knob}}{2} + \frac{(2F_{al})^2}{2}\frac{\ell_{in}}{ES_{eff}}\right)$$

$$\approx F_{al}\left(\frac{d_{knob}}{2} + 2F_{al}\frac{\ell_{in}}{ES_{eff}}\right) \tag{12.20}$$

With the parameters of the exemplar system, $\Delta\mathscr{E}_{state} \approx 1.2$ aJ.

12.3.7. Thermal excitation and error rates

Thermal displacement of gate knobs can cause errors in switching. In principle, these errors could occur either because a gate knob blocks a probe knob when it should pass, or because it lets a probe knob pass when it should be blocked. Erroneous blockage, however, is transient relative to the switching time because (given low damping, Section 12.3.4c) the maximum delay before the obstructing knob moves from the path is $\sim 2\ell_{out}/v_s$, or ~ 0.006 ns for the exemplar system. Accordingly, the probe knob will begin to pass the gate knob early in the switching period, even if a thermal fluctuation places the gate knob in a blocking position for a brief time.

Erroneous probe knob passage, in contrast, does not undergo fast, spontaneous reversal: it causes a faulty output. The resulting error rate can be estimated from a model of positional uncertainty in gate knob position.

a. A conservative bound on acceptable gate knob displacement. Gate knobs and probe knobs are here regarded as blocks with facing surfaces of equal extent, w_{knob}. In the bounded continuum model, corners are regarded as occupied by atoms, and hence are rounded. If the contact between gate knob and probe knob occurs solely through a glancing contact of corner atoms, the probe knob will exert a force tending to push the gate knob aside. For a wide range of geometries and stiffness and force parameter, this tendency can be resisted by gate-knob restoring forces.

A simple, conservative model neglects these restoring forces, and treats as an error condition any contact in which the line of travel of the center of the corner atom of a probe knob falls beyond the center of the corner atom of the corresponding gate knob. In terms of the effective steric radii for the atoms, r_{eff}, the effective width of the knobs is $w_{\text{eff}} = \ell_{\text{eff}} = w_{\text{knob}} - 2r_{\text{eff}}$. The acceptable gate knob displacement is then $\Delta x_{\text{thresh}} = w_{\text{eff}}$ (assuming that the knobs are aligned at $\Delta x = 0$). For the exemplar system, assuming $r_{\text{eff}} = 0.15$ nm, $\Delta x_{\text{thresh}} = w_{\text{eff}} = 0.7$ nm.

b. Probability densities and error rates.   Transverse elastic displacements of the probe knob with respect to the housing are characterized by a stiffness $k_{\text{s,p}}$, and the parallel, longitudinal displacements of the gate knob with respect to the alignment knob are characterized by a stiffness $k_{\text{s,g}}$. In the size and temperature range considered here, quantum mechanical effects on positional uncertainty are negligible (Section 5.3), hence the variance in relative positions resulting from these elastic displacements is

$$\sigma_{\text{el}}^2 = kT\left(1/k_{\text{s,p}} + 1/k_{\text{s,g}}\right) \tag{12.21}$$

and the associated PDF for the elastic contribution to the relative displacement is

$$f_{\Delta x_{\text{el}}}(\Delta x_{\text{el}}) = \frac{1}{\sqrt{2\pi}\sigma_{\text{el}}}\exp\left[-\tfrac{1}{2}\left(\Delta x_{\text{el}}/\sigma_{\text{el}}\right)^2\right] \tag{12.22}$$

In the approximation of hard-surface interactions between the alignment knob and its limit stops, the PDF for the corresponding contribution to the relative displacement is the exponential

$$f_{\Delta x}(\Delta x) = \frac{F_{\text{al}}}{kT}\exp(-F_{\text{al}}\Delta x/kT) \tag{12.23}$$

The probability of a relative displacement greater than Δx_{thresh} is then

$$P_{\text{err-disp}} = \int_0^\infty f_{\Delta x_{\text{al}}}(\Delta x_{\text{al}}) \int_{\Delta x_{\text{thresh}}-\Delta x_{\text{al}}}^\infty f_{\Delta x_{\text{el}}}(\Delta x_{\text{el}})\,d(\Delta x_{\text{el}})\,d(\Delta x_{\text{al}}) \tag{12.24}$$

$$\approx \exp\left[\frac{1}{2}\left(\frac{\sigma_{\text{el}}F_{\text{al}}}{kT}\right)^2 - \frac{\Delta x_{\text{thresh}}F_{\text{al}}}{kT}\right]; \quad \frac{\sigma_{\text{el}}F_{\text{al}}}{kT} < \frac{\Delta x_{\text{thresh}}}{\sigma_{\text{el}}} \tag{12.25}$$

Applying the exemplar parameters, $k_{\text{s,p}} = 10$ N/m; assuming (conservatively) $k_{\text{s,p}} = 40$ N/m, $\sigma_{\text{el}} \approx 0.023$ nm. The condition for Eq. (12.25) then holds, implying $P_{\text{err-disp}} \approx 1.3 \times 10^{-67}$.

A worst-case analysis of the resulting error rate assumes that a properly positioned rod is not latched in place by impinging probe knobs, and hence can subsequently move, allow knobs to pass, and permit an error. The characteristic frequency of rod vibration is $<10^3$ times the frequency of the CPU, permitting $<10^3$ opportunities for an error per logic-rod motion, and hence an error rate $<10^{-64}$. Error rates of this order are negligible by almost any standard.

12.3.8.   Summary observations based on the exemplar calculations

The sample calculations in the preceding sections describe the characteristics and performance of a nonoptimized logic rod design with 16 inputs and 16 outputs. One cycle of this rod represents one-half cycle for 32 interlocks, or a total of 16 complete switching events.

a.   Volume per interlock.   Allowing a 0.5 nm thickness of material to partition one layer of interlocks from the next, the volume per interlock in a tightly packed array is 12 nm^3. Including the length of the alignment knob and adding 6 nm at one end for the drive system and 2 nm at the other for the reset spring (and adjusting the effective width of a logic rod assembly to account for similar additions to the length of the crossing rods) yields an estimated volume ~16 nm^3 per interlock.

b.   Energy per switching cycle.   For a mobile rod with a switching time of 0.1 ns, the estimated energy loss in a switching cycle (including reset) is ~2 maJ. This amounts to ~0.013 maJ per interlock per switching cycle, or ~$0.031\,kT_{300}$. This low dissipation is possible because the rod displacements in combinational rod logic systems in effect store each input and intermediate result, rendering all computations logically reversible. Note that in the limit of slow motion all of the identified energy dissipation mechanisms in combinational rod logic systems approach zero. It can be seen by inspection that the energy barriers responsible for ensuring reliable operation (Section 12.3.7) are essentially unrelated to the energy dissipated in switching (Section 12.3.4), hence there is no need to dissipate energy to ensure reliability. This is consistent with results in the extensive literature on thermodynamically reversible computation (e.g., Landauer, 1961; Toffoli, 1981; Bennett, 1982; Fredkin and Toffoli, 1982; Feynman, 1985), but inconsistent with statements in some older textbooks (e.g., Mead and Conway, 1980). Several reversible computation schemes, both electronic (based on transistors) and mechanical (based on buckling structural elements), have recently been described by Merkle (1992c, 1992d).

c.   Comparison to transistors.   Like transistors, interlocks are switching devices with the essential properties of a "good computer device" as defined by Keyes (1985): they support fan-out and fan-in, make reliable binary decisions, switch rapidly, pack densely, work with interconnections on their own scale, restore signals to a reference level at each step, and display high gain (in the form of a large change in output rod displacement when the gate displacement passes a threshold). In meeting these conditions, they are superior to the numerous failed device technologies described by Keyes (1989), although they suffer from the practical disadvantage of lacking an extant manufacturing technology.

In speed, interlocks are inferior to present experimental transistors, having switching times of ~100 ps rather than ~10 ps. Transistors continue to improve, while the exemplar system may be within a modest factor of real physical limits. Accordingly, there is every reason to expect that the fastest devices in the future will exploit electronic rather than mechanical phenomena; electronic devices will, of course, benefit from the precise control possible with molecular manufacturing.

In density, interlocks are greatly superior to present transistors. The area occupied by an exemplar interlock (~4 nm^2) is orders of magnitude smaller than the area occupied by a modern transistor (~10^6 nm^2). Further, there is no obstacle to stacking interlocks with a vertical spacing of ~3 nm, while planar semiconductor processing, together with packaging, results in vertical transistor spacings >10^6 nm. The overall improvement in volumetric packing density is thus >10^{11}.

In energy dissipation, interlocks are greatly superior to present experimental transistors. These have a switching energy of ~100 aJ, >10^6 times the calculated switching energy of the exemplar interlock. Relative to present commercial CMOS systems, the advantage in energy dissipation is ~10^9. To a first approximation, this advantage in switching energy translates into a ~10^9 increase in the quantity of computation that can be performed in a box of a certain size within given cooling limitations. (Systems like those discussed in Section 11.5 can substantially expand cooling capacity as well.)

12.4. Registers

Logic rods suffice to implement combinational logic systems of arbitrary depth and complexity. Without other components, however, the resulting state is destroyed when the rods are reset. To implement systems capable of iterative computation requires devices able to record the output of one computational cycle and present it as an input for a later cycle. Registers serve this function.

12.4.1. Kinematics of an efficient class of register

Figures 12.4 and 12.5 illustrate the structure and kinematics of an efficient class of register that abstractly parallels previous proposals based on a particle in a system with time-varying potential wells (e.g., Landauer, 1961; Bennett, 1982). The present description is based on a bounded continuum model; accordingly, the rod widths in Figure 12.4 are taken to be ~1 nm. The four components are as follows:

1. The *input rod* to the register is part of the output segment of a logic rod; it sets the state of the register.
2. The *state rod* stores the state of the register; it functions as an indicator in a measurement system (Section 11.2.1).
3. The *spring rod* provides a clocked bias force, helping to set the state of the register without determining the result.
4. The *latching rod* provides a clocked latching force, blocking movement of the state rod after the input rod is reset.

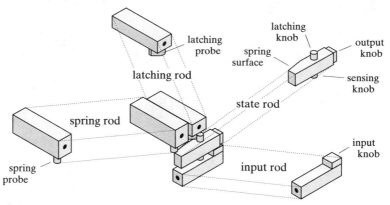

Figure 12.4. An exploded diagram of the moving parts of a thermodynamically efficient class of register, with molecular detail suppressed in accord with the bounded continuum approximation and the housing structure omitted for clarity. Rod widths are ~1 nm; sensing and latching knobs are assumed to be single protruding atoms (to minimize van der Waals interaction energies in sensing, and to minimize knob length in latching). The rod ends marked with dots represent truncated surfaces of longer structures. See Section 12.4 for a description.

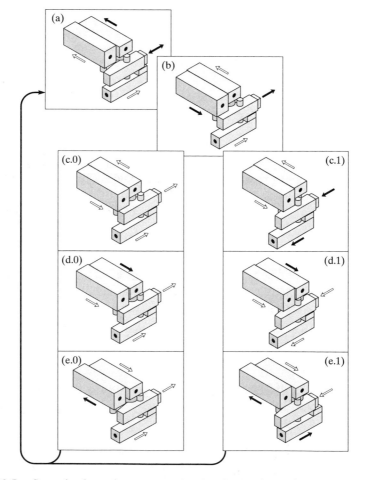

Figure 12.5. Steps in the write-erase cycle of a thermodynamically efficient register, illustrated as in Figure 12.4. A housing structure (not shown) constrains rods to move along their axes. Black arrows show motion; white arrows show displacement.

356

The interaction of these components is diagrammed in Figure 12.5 and described in the following sections.

a. **Compression.** In frame (a) of Figure 12.5, both the spring rod and the latching rod are retracted. This leaves the state rod subject to thermal motion, traversing a range that carries the output knob through both its blocking and nonblocking positions (relative to a probe knob on a logic rod to which the state of the register cell is an input). In the transition to frame (b), the spring rod advances, pressing a small, compliant knob against the spring surface of the state rod, which is thereby forced toward the right, compressing its range of motion into the nonblocking 0 position.

b. **Input.** In the transition from (b) to (c.1), the input rod moves to its 1 state, and the contact between the input knob and the sensing knob forces the state rod to the left, into its 1 position. In doing so, the input rod does $\sim 30 kT_{300} \approx$ 120 maJ of work against the force applied by the spring knob; assuming a displacement of 0.5 nm, the required force is ~ 0.24 nN.

In the transition from (b) to (c.0), the input rod remains in its 0 state, and no displacements occur. The state rod remains in its 0 position. (In either sequence, the state rod can now be read as an input to the next level of logic.)

c. **Latching.** In the transition from (c.0) to (d.0), the latching rod advances, pressing the latching knob further to the right and stabilizing the state rod in its 0 position. In the transition from (c.1) to (d.1), the latching probe presses the latching knob further to the left, stabilizing the state rod in its 1 position.

d. **Input reset.** In the transition from (d.0) to (e.0), the spring rod retracts, leaving the state rod latched in place. In the transition from (d.1) to (e.1), both the spring rod and the input rod retract, again leaving the state rod latched in place. These retractions need not be synchronized. In both instances, the resetting of the input rod (along with others controlled by the same clocking phase) can initiate a chain of resetting operations that returns a preregister combinational logic system to its initial state. Meanwhile, the latched state rod continues to serve as an input gate knob for a logic rod in the first stage of a postregister combinational logic system.

e. **Expansion.** After the postregister combinational logic system has completed its cycle and been reset, the latching rod can be retracted, returning the state rod to the condition represented by frame (a). The reexpansion of the range of motion of the state rod discards the stored information and increases entropy by $\ln(2)k$.

12.4.2. Device size and packing

State rods can have the same widths and separations as the exemplar logic rod (1 nm; see Section 12.3.3a), making the registers geometrically compatible with densely packed arrays of logic rods. With the geometry shown in Figure 12.4, the length of the assembly is ~ 4 nm, and its height is ~ 5 nm (including an allowance for housing structure), yielding a total volume of ~ 40 nm^3. This volume should prove conservative.

12.4.3. Energy dissipation estimates

Unlike logic rods in a purely combinational system, these registers undergo a logically and therefore thermodynamically irreversible cycle. Their energy dissipation mechanisms can be divided into frictional mechanisms (which in a good design can approach zero loss per cycle as speed approaches zero) and fundamental mechanisms (which impose a fixed loss per cycle, regardless of speed).

a. Vibrational, thermoelastic, and drag losses.   Owing to small size and high modal frequencies, losses resulting from the excitation of vibrations can be made negligible compared to those in logic rods. Thermoelastic losses depend on the square of the stress, which is substantial only in the spring rod, and then only if one spring rod serves many register cells. With a cross-sectional area of ~1 nm^2, a spring rod serving 16 register cells has a mean-square tensile stress of ~4 GPa. This yields an estimated thermoelastic loss of ~0.04 maJ per cell per cycle.

The speeds of sliding motions are uniformly equal to or less than those in the associated system of logic rods. Since the sliding interfaces are similar, the energy losses per unit area per cycle are similar. The sliding interface area of an interlock with the exemplar parameter is ~12 nm^2; for the register system, it is roughly twice that. Accordingly, drag losses can be estimated at ~0.01 maJ.

b. Nonisothermal compression and expansion losses.   Losses from nonisothermal compression and expansion can be estimated using the model of Section 7.5.1. With the stated dimensions and a density of 3500 kg/m^3 of (surface-corrected) volume, the mass of the state rod is ~6×10^{-24} kg, giving it a root mean square thermal velocity of ~25 m/s at 300 K.

During the compression phase, the extension of the spring rod reduces the range of motion of the state rod from ~0.5 nm to a small region (e.g., an effective width of ~0.16 nm, assuming that the stiffness of the final potential well is ~1 N/m). If the compression operation is assumed to take 0.1 ns, with an approximately linear motion over a distance of $0.5 - 0.16 = 0.34$ nm, then the speed is 3.4 m/s. Assuming that the input knob and the spring probe are built for reasonably efficient energy exchange with state knob thermal motions, Eq. (7.62) with an accommodation coefficient $\alpha = 0.5$ should provide a reasonable estimate of the energy loss. With the above parameters, the result is ~1.4 maJ.

During the expansion phase, the withdrawal of the latching rod allows expansion of the range of motion of the state rod from a small region to a 0.25 nm long region before free expansion occurs (assuming, for the moment, that barrier crossing is switched on instantaneously) with an estimated energy loss of ~0.14 maJ.

Nonisothermal compression and expansion results in the largest speed-dependent, nonfundamental losses identified here. These can be reduced by reducing speed (undesirable but direct and effective), reducing compression and expansion lengths (for which there is only modest scope), reducing state rod mass (substantially, in a good design), and improving the coupling of state rod motions to thermal modes (which will involve trade-offs between drag and nonisothermal compression losses).

c. Free expansion losses.   As the latching probe withdraws, the barrier dividing the potential wells corresponding to 0 and 1 disappears; in good designs, these potential wells have approximately equivalent free energies throughout the merging process. As discussed in Section 7.6.3, the merger of symmetrical potential wells leads to an increase in entropy of $\ln(2)k$, and hence to a loss of free energy of $\ln(2)kT$, or ~ 2.87 maJ at 300 K. This loss is irreducible.

d. Summary of losses. Losses dependent on design and speed of operation are estimated to total ~ 1.6 maJ. Thermodynamically irreducible losses resulting from register erasure are ~ 2.87 maJ. Total losses are thus ~ 4.5 maJ per register cell per cycle.

12.4.4. Fluctuations in stored energy

The energy stored in a register cell during the input phase depends on the logic state: in writing a 1, the input rod does work against the spring probe, causing a net transfer of energy (on spring-probe withdrawal) from the input-rod drive system to the spring-rod drive system. This energy transfer is on the order of 120 maJ, with the precise value depending on design details and the required reliability of the register.

12.5. Combinational logic and finite-state machines

Combinational logic systems compute sets of Boolean functions (mapping from bit vectors to bit vectors), and commonly are built using programmed logic arrays (a.k.a. programmable logic arrays, programmable array logic, PALs, and PLAs). PLAs can be combined with registers to build finite state machines and much of the control circuitry for a CPU (Mead and Conway, 1980). The following discussion of PLAs illustrates how logic rods and registers can be combined to build clocked computational systems.

12.5.1. Finite-state machine structure and kinematics

The PLA-based finite state machine illustrated in Figure 12.6 is nearly a device-for-device translation of a design described in Mead and Conway (1980). The chief differences are (1) that the use of devices abstractly resembling CMOS rather than °NMOS transistors halves the number of device columns in the AND-plane, and (2) that state information transmitted from the output register to the input register passes through a simple PLA structure, providing a delay during which the rods in the main PLA are reset.

Computation proceeds as follows: First, the vertical rods to the left, set 1 of PLA (a), are tensioned; their mobility is determined by the bit vector latched into the register to the lower left, register (ba). After a switching time, the horizontal rods, set 2 of (a), are tensioned; their mobility is determined by the state of set 1. After another switching time, the vertical rods to the right, set 3 of (a), are tensioned; they move as determined by set 2. Finally, the state of set 3 is latched into the register to the lower right, (ab), by tensioning the latching rod (L2). A similar process then starts with this register and propagates through the lower PLA (b) while the series of motions in the upper PLA is reversed.

12.5.2. Finite-state machine timing and alternatives

a. Two-register systems. The upper panel of Figure 12.7 presents a timing diagram for a two-stage finite state machine. Note that the time for a cycle (including two register-to-register transfers with intervening combinational logic)

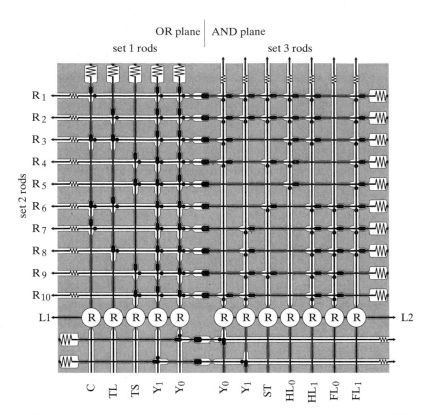

Figure 12.6. A PLA-based finite state machine translated from an NMOS design (Mead and Conway, 1980) to nanomechanical logic, shown in the notation of Figure 12.1, but omitting several drive and spring systems owing to the constraints of a flat representation. Rod labels correspond to those in the source. PLA (a) is above the L rods; a smaller PLA, (b), is below. Rows of circled Rs represent cells in two registers.

is $10t_{switch}$, or 1 ns for the parameters used in the exemplar system. In this timing sequence, both the register compression processes (driven by the spring rod) and the latching processes (driven by the latching rod) take t_{switch}, as assumed in Section 12.4.3. Motion of the first set of rods in the postregister PLA, however, does not immediately follow the displacement of the last set of rods in the preregister PLA (as it could, based on purely local constraints); it is instead delayed by $2t_{switch}$ in order to satisfy global timing constraints involving the requirement that a PLA and its input register be reset when the other PLA output is ready to be written. In the overall cycle, $4t_{switch}$ out of the cycle time of $10t_{switch}$ is consumed in these delays.

b. Four-register systems. These delays can be eliminated by moving to a different architecture. The lower panel of Figure 12.7 shows a timing diagram for a four-stage finite state machine, in which a cycle involves four register-to-register transfers with intervening combinational logic. If PLA (a) and (c) are the same as (a) in the above example, and if (b) and (d) are the same as (b) in the two-register example, then (aside from an alternation in which register contains the most current bit vector) the systems are equivalent, with one full four-stage cycle corresponding to two cycles of the two-stage system. If (a) $\neq$ (c) and

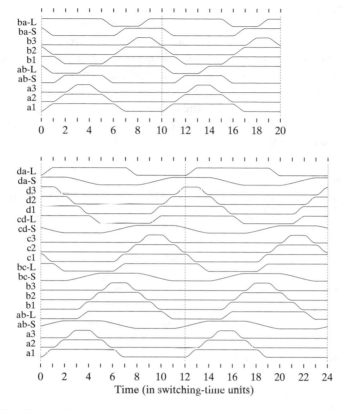

Figure 12.7. Timing diagrams for two PLA-based systems, with time measured in units of the logic rod switching delay t_{switch}, and driver displacement represented on an arbitrary scale (tensioned is high, de-tensioned is low). Upper panel: with two PLAs, (a) and (b), two registers, (ab) and (ba). Lines (a1), (a2), (a3) represent the driver displacements for the sets of rods in PLA (a); lines (ab-S) and (ab-L) graph the displacements of the spring rod and the latching rod drivers (respectively); other labels are analogous. Lower panel: with four PLAs and registers, otherwise analogous to the upper panel.

(b) $\neq$ (d), then the four-stage system is equivalent to a more complex system of two-stage register-to-register PLAs.

The four-stage architecture performs register-to-register combinational logic operations at 5/3 the frequency of the two-stage architecture, using twice the volume and device count to do so. The register compression process, moreover, can be allotted a time period of $3t_{\text{switch}}$ to complete; the energy dissipation model for this process predicts that this change will divide energy dissipation by a factor of 3. The estimates in Sections 12.3.4 and 12.4.3 suggest that register compression is the largest nonirreducible energy loss mechanism. Accordingly, in systems where mass and volume are inexpensive, and where speed and low power dissipation are at a premium, it appears that a four-stage architecture is superior to a two-stage architecture.

c. Estimated instruction execution rate in a CPU. The time required for register-to-register transfer through a PLA (t_{PLA}) can be used to estimate the clock rate in a CPU. A clock cycle in a well-designed RISC (reduced instruction set computing) machine requires $2t_{PLA}$ to $3t_{PLA}$, with the latter being a conservative value for the present purposes (Knight, 1991). In a RISC machine operating with a typical mix of instructions, the rate of instruction execution can equal or (in a superscalar architecture) exceed the clock rate (Hennessy and Patterson, 1990). With a four-register architecture based on the exemplar parameters of previous sections, $4t_{PLA} = 1.2$ ns; hence a conservatively designed rod logic technology should enable the implementation of RISC machines capable of >1000 MIPS. This is competitive with the speed of current microelectronics, but is presumably inferior to the speed of future electronic systems.

12.5.3. Fan-in, fan-out, and geometric issues

The exemplar logic rod described in Section 12.3 is straight and has a fan-in and fan-out of 16. Real systems will seldom require a rod with exactly these properties.

a. Changes in fan-in.   Reducing fan-in shortens the input segment and increases its stiffness, improving feasible switching speed and reducing the total volume and energy dissipation per rod. The overhead of the drive mechanism, alignment knob, and reset spring are spread over a smaller number of devices, however, increasing the volume and energy dissipation per device.

Fan-in can be increased by lengthening the input segment, but at the cost of decreasing the feasible switching speed. Alternatively, multiple input segments can be mounted in parallel (e.g., yoked together at the drive end and at the alignment knob), yielding a system with the same logical properties as a series connection, but with higher vibrational frequencies and hence a higher feasible switching speed. (Indeed, this strategy might be desirable for $n_{in} \leq 16$.)

b. Changes in fan-out.   Reductions in fan-out parallel reductions in fan-in where stiffness, switching speed, energy dissipation, and volume are concerned. The resulting increases in stiffness also reduce error rates resulting from thermal noise; in optimized designs (where error rates are significant but acceptable), this shift will typically permit the use of either slimmer rods or lower values of F_{al}.

Increasing fan-out increases both switching speed and error rates, but errors can be kept in check by proportional increases in S_{eff} (and hence stiffness). Again, yoking a set of shorter output segments together in parallel (e.g., at the alignment knob and the reset spring) results in a system with the same logical properties but higher frequencies and switching speeds. This strategy reduces errors from thermal excitation and reduces the required magnitude of F_{al}, on a per-rod basis. (And again, this strategy might be desirable for $n_{in} \leq 16$.)

c. Flexible rod geometries.   Interlocks can be linked by flexible rods passing through curved housings. Flexible rods can be made from structures that are one atom thick in the direction of curvature, such as polyyne chains or strips having a graphitic structure. These can have a radius of curvature ~1 nm, and the potential energy functions for sliding motions in the housings can be smooth (see Sections 10.5 and 10.8). Curved linkages enable broad geometric flexibility in the design of rod logic systems, permitting (for example) the use of PLAs stacked face to face that share access to a register.

12.5.4. Signal propagation with acoustic transmission lines

In the exemplar system, the state of an output gate is switched in 0.1 ns over a distance of 64 nm, yielding an effective signal propagation speed of only 640 m/s. In larger logic systems, it is desirable to transmit signals at the full acoustic speed, ~17 km/s in diamond. This can be accomplished using systems that launch acoustic pulses at one end of a rod and probe displacements at the other.

a. Kinematics and positional uncertainty. For concreteness, consider a rod 5 μ in length, with $S_{eff} = 3$ nm^2 and $E = 10^{12}$ N/m^2. The one-way signal propagation time along this rod is ~0.3 ns. As with the exemplar logic rods, a drive system and drive spring apply forces to one end, and an input segment with one or more interlocks then either blocks or fails to block displacement of the rod. If the rod is in a mobile state, the force applied by the drive spring accelerates the rod. After 0.1 ns, a force of 1.8 nN ($\approx 2F_{al}$ in the exemplar system) will have displaced the end of the rod by 1 nm and propagated a region of tensile stress 1.7 μ along the rod. Ceasing to apply tension for a further 0.1 ns, then applying a reverse force for 0.1 ns restores the rod end to its initial position, while launching a wave which, as it passes, displaces the rod by 1 nm for 0.1 ns. At the far end, the displacement can be probed by a conventional interlock controlling a short, briefly displaced logic rod that writes into a register and is immediately reset.

The positional uncertainty at the receiving end of the rod is adequately constrained by positional control at the sending end. The energy associated with a 1 nm stretching deformation of the rod is ~600 maJ, or ~$150kT_{300}$. Nonthermal vibrational modes are a concern, but clocked damping can be introduced to remove energy from these modes between signal pulses, without dissipating energy from the signal pulses themselves.

b. Energy recovery. The energy of a wave of this sort is ~3.6 aJ. The round-trip signal propagation time is ~0.6 ns; hence a pulse can be launched and recovered in a single clock period. In systems with stable clocking at the correct speed, the energy delivered by the reflected pulse can be recovered by a suitably structured drive system. Since an outgoing pulse consisting of a rarefaction followed by a compression is reflected from a free end as a compression followed by a rarefaction, the return pulse resembles a time-reversed version of the outgoing pulse, and a simple reversal of drive shaft motion produces a sequence of cam displacements that couples the energy or the return pulse into the drive system with reasonable efficiency. Energy dissipation is larger than that in the exemplar rods, but the need to propagate signals over such a distance should be reasonably rare; a sphere of 5 μ radius can contain >10^{10} interlocks.

12.6. Survey of other devices and subsystems

Although registers and PLA-based combinational logic in principle suffice to implement general-purpose computational systems, devices of several further classes are of considerable practical importance. The following sections briefly survey gates suitable for non-PLA combinational logic, carry chains for efficient implementation of arithmetic logic units (ALUs), and both random-access and tape-based memory systems; it closes with a discussion of devices for interfacing to macroscale systems.

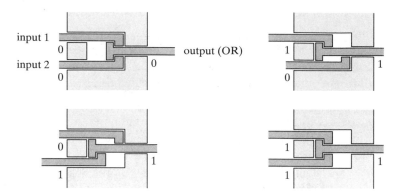

input 1
0
input 2   output (OR)
0

Figure 12.8. Schematic diagram of an OR gate operating within a single rod displacement cycle.

12.6.1. Gates for non-PLA combinational logic

A PLA system requires three successive rod displacement cycles to compute a set of Boolean functions. The functions AND, OR, and NOT can, however, be computed in a single displacement cycle by what is here termed *direct* logic.

The direct logical OR of a set of inputs can be computed by a linkage in which any input rod can displace the output rod without displacing any other input rod. This can be done by a mechanism like that in Figure 12.8. The direct logical AND of a pair of inputs can be computed by a mechanism like that in Figure 12.9 (note that the inputs need not be synchronous). If 1 and 0 states are defined by the direction of rod displacement, then a direct logical NOT can be computed by a lever having input and output rods attached on opposite sides of the fulcrum.

Direct gates add length and compliance, thereby increasing the minimal switching time relative to a connection not passing through a gate. This effect is small, however: several layers of direct combinational logic can be performed within the 0.1 ns switching time of a rod with the exemplar parameters. Regular recourse to clocked, interlock-based logic can provide the power and logic-level restoration required for complex computations.

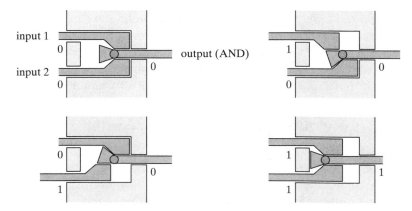

input 1
0
input 2   output (AND)
0

Figure 12.9. Schematic diagram of an AND gate operating within a single rod displacement cycle.

12.6.2. Carry chains

Carry chains play an essential role in ALUs. A familiar example is the propagation of a carry bit when 00000001_{binary} is added to 01111111_{binary} to yield 10000000_{binary} (note that the left-most bit of the output can in general depend on the state of any bit in either input). In a carry chain, one cell represents each bit in a binary number, and a series of operations propagates carry information along the series of cells. In a fast carry chain, propagation does *not* require a ripple of switching operations moving from the low bit to the high. For example, a transistor-based Manchester carry chain (see Mead and Conway, 1980) forms a set of conducting paths through the cells by switching transistors to a conductive state where a carry signal should propagate to the next cell, and switching transistors to a nonconductive state where a carry signal should be blocked. This occurs in one step; a subsequent step applies carry-determining signals that propagate through the cells without switching delays.

An analogous scheme can be used in a nanomechanical ALU. Special-purpose rod segments (sliding in parallel grooves set at a shallow angle to the axis of the carry chain) can be coupled or decoupled by pins that are displaced by input rods. These specialized rod segments can then be driven by a set of carry-determining input rods; the carry signals then propagate through sets of precoupled segments. For a 32-bit word size, the length of a carry chain is about the same as the length of the exemplar rods (Section 12.3.3a). The compliance of the junctions between segments decreases the mean stiffness of the rod, reducing the low-frequency acoustic speed by about 1/2. Carry signal propagation in a 32-bit system is accordingly estimated to take about twice the switching time for an exemplar rod, or ~0.2 ns.

12.6.3. Random-access memory

Rod logic systems can be used to transmit and decode requests to a RAM and to carry the information read or written. As with registers and carry chains, however, the RAM itself requires specialized mechanisms.

To read (or write) one of many different words from a strip of RAM using a single set of read/write rods, an address signal must be able to engage the correct row of bit cells with the rods while other similar rows are left disengaged. One design approach that achieves this has features in common with register cells (Section 12.4), moving entire rows of cells as a unit to engage and disengage them. The size, speed, and energy dissipation of a RAM cell should be similar to that of a register cell.

12.6.4. Mass storage systems

RAM permits fast access and good energy efficiency, but at the cost of considerable bulk per bit stored. Denser information storage can be provided by a tape memory system based on polymer chains with side groups of two distinct kinds.

For example, a partially fluorinated polyethylene molecule can store two bits per carbon atom by representing a 1 with an F atom and a 0 with an H atom. A chain of this sort can be read by a measurement system (Section 11.2) and written by a mechanochemical device able to perform abstraction and deposition reactions. The design of the reading mechanism can be simplified if one surface of the molecule has a uniform structure (all H); the design of the writing system can be simplified if each carbon on the other surface has exactly one F (writing can then proceed exclusively by removing an H and then depositing an H on the proper face of the resulting pi radical). With these restrictions, the storage density for the tape itself is one bit per two carbon atoms, or ~15 bits/nm^3. Alloting twice the tape volume for other functions (to accommodate reels, packing inefficiency, drive systems, read-write mechanisms, and so forth) decreases the mean storage density to ~5 bits/nm^3, or 5×10^{21} bits/cm^3.

A volume of 10^4 nm^3 appears ample for a read-write mechanism. If 10 times this volume is allocated to the corresponding tape (and reels, etc.), then the tape-unit capacity is ~5×10^5 bits. At 10^9 bits/s, reading the entire tape takes 500 μs. At a seek speed of 10 m/s, traversing the length of the tape takes ~15 μs. Accordingly, the expected access times of molecular tape storage systems are substantially less than those of macroscale disk storage systems.

12.6.5.   Interfaces to macroscale systems

Demonstrating the feasibility of interfacing nanomechanical computational devices to conventional microelectronics shows how nanomechanical systems can be interfaced to systems that are in turn interfaced to the macroscale world. This can be accomplished using electromechanical devices (Section 11.6).

a.   Input.   Electrostatic actuators of the sort described in Section 11.6.4 can be used to transmit data from a 5 V microelectronic system to a rod logic system; doubling the plate area (and hence the force) enables the actuator to replace the input segment of an exemplar logic rod. Electrostatic forces of the stated magnitude can move a plate with a mass equaling that of a 1 nm thick slab of diamond by 1 nm in ~0.03 ns. Data input rates can accordingly be consistent with the rod logic speeds calculated in Section 12.5.

b.   Output.   Modulated tunneling junctions like those described in Section 11.6.3 can convert logic rod displacements into electrical signals that can be detected by microelectronic systems. At a junction potential of 1 mV and resistance of 10^4 Ω (Section 11.6.2), the current is 1 μA and the resistive power loss is 10 nW. Existing transistors can switch with ~10^{-15} C of charge, requiring ~10^{-9} s of current flow at this rate; accordingly, data can be output at a clock rate like that estimated in Section 11.5.2. The energy dissipated in the junction per bit communicated is ~10 aJ. Sliding-contact switches (e.g., based on conductors terminated by graphitic planes) might reduce resistance and reduce power requirements.

c.   System context.   In typical systems containing many nanocomputers, nanoscale systems will communicate chiefly with other nanoscale systems, and little of the communications traffic will be with macroscale systems. Accordingly, the energy dissipated per bit in communicating with the macroscale world is relatively unimportant. Finally, since molecular manufacturing techniques can produce large devices, communication between nanoscale and macroscale systems need not proceed through conventional microelectronics (even if this were still in use), but could instead be mediated by a chain of amplifiers better matched to the problem.

12.7. CPU-scale systems: clocking and power supply

In electronic digital logic, clocking systems with two, four, or eight phases are not uncommon. What were termed two- and four-stage architectures in Section 12.5.2 might seem similar, but are in fact substantially different: each stage includes three levels of logic and a register with two control rods; this requires a total of twenty distinctly clocked inputs in a four-stage system. Further, the clocking signals in electronic digital logic are used to modulate power distributed by a DC system, while rod logic systems use the clocking signals as a power supply. The architecture of the clocking and power supply system is accordingly quite different.

To estimate clocking and power-distribution parameters appropriate for CPU-scale systems requires a model describing the size and contents of a CPU. The following will assume that a CPU-scale system contains 10^6 interlocks, 10^5 logic rods, and 10^4 register cells, which (together with interconnects, power supply mechanisms, wasted space, etc.) occupy a cube 400 nm on a side. (Assuming a mean density of 2000 kg/m^3 implies a mass of 1.6×10^{-17} kg and hence a half-life against radiation damage in Earth ambient background radiation of ~100 years; see Figure 6.13.)

12.7.1. Clocking based on oscillating drive rods

Consider the timing diagram for a two-stage system (shown in the upper panel of Figure 12.7). The tensioning and de-tensioning of each set of rods can be characterized by the duration of the tensioned and de-tensioned intervals and by the time marking the middle of the tensioned interval; this time can be taken to define the phase of the clock for that set of rods. Inspection of this diagram shows four phases: one for the logic rods of PLA (a) and the spring rod of register (ab), another for the latching rod of register (ab), and another two phases for PLA (b) and register (ba).

As illustrated in Figure 12.10, a sinusoidally oscillating cam surface on a drive rod can generate clocked drive-system impulses having intervals determined by the position of the follower with respect to the mean position of the ramp on the cam surface. The drive system for a set of rods could use several follower-and-ramp sets on the same drive rod; these can move in synchrony. A single drive rod (or system of drive rods sharing single phase) can power several drive systems having different intervals of tensioning. Further, since a ramp can slope in either direction, sets of rods differing in phase by 2π can likewise be driven by a set of drive rods sharing a single phase. Accordingly, the ten different patterns of clocked impulses required for the two-stage system in Figure 12.7 can be generated by drive rods having only two distinct phases; the four-stage system similarly requires only four driver phases.

12.7.2. A CPU-scale drive system architecture

A drive system requires a mechanism for linking drive rods so that different sets have the correct relative phases, a mechanism for buffering energy to compensate for fluctuations in the energy stored in the logic system in different logic states, and a source of motive power to compensate for energy losses. The source

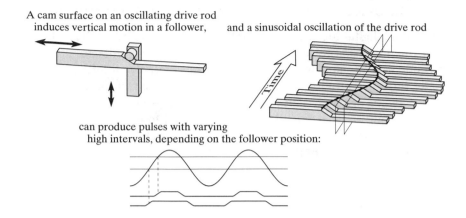

A cam surface on an oscillating drive rod induces vertical motion in a follower, and a sinusoidal oscillation of the drive rod

can produce pulses with varying high intervals, depending on the follower position:

Figure 12.10. Diagrams and text illustrating the generation of clocked drive-system impulses of varying intervals from the sinusoidal motion of a cam surface (only one ramp and one follower are shown).

of motive power can be a DC electrostatic motor (Section 11.7). The mechanism for buffering energy can be a flywheel rotating at the frequency of the system clock. The linking mechanism can be a crankshaft able to convert a single rotary motion into an indefinitely large number of linear, approximately sinusoidal motions of differing relative phases. (Note that subsidiary crankshaft mechanisms can be used at remote locations to couple drive subsystems of differing phase, permitting energy transfer between them and increasing the resistance to local desynchronization. These subsidiary crankshafts also provide one of several mechanisms for transmitting drive power around corners.)

The dynamics of this clocking system differ from the dynamics of clocking systems in integrated circuits, which are dominated by damping, leading to the diffusive spread of a clock signal along the conducting paths. The mechanical system, in contrast, is dominated by inertia, leading to resonant behavior in which all parts of the system could (in the absence of load) share a single phase. In real systems, however, loads will cause clock skew.

12.7.3. Energy flows and clock skew

During a clock cycle, a drive rod executes an oscillation, forcing movement in a set of logic, spring, or latching rods. *If* the energy stored in these rods as a function of time were invariant from cycle to cycle, and *if* there were no energy dissipation, then the potential energy of the drive rod as a function of position could be adjusted to allow it to move as a harmonic oscillator, executing sinusoidal oscillations with no external energy input or constraining force. In practice, variations in the energy stored in rods from cycle to cycle cause fluctuating forces and displacements in the drive system, and energy must on the average flow into the drive rod from an external source to compensate for energy losses.

a. Energy fluctuations. Fluctuations in stored energy are >100 times greater than fluctuations in dissipated energy (per clock cycle), and hence dominate the cycle-to-cycle fluctuations in drive-system dynamics. Further, fluctuations in the energy stored in logic rods are ~10 times greater than comparable fluctuations in register cells. An energy-based analysis focusing on logic rod states can provide an estimate of the magnitude of the drive system stiffness and inertia required to ensure that these fluctuations do not cause excessive disturbances in local clock phase.

An examination of the timing diagrams in Figure 12.7 shows that no more than 1/3 of the 6 sets of logic rods in the two-stage system are tensioned simultaneously, as are no more than 5/12 of the sets in the four-stage system. Accordingly, the estimated 1.2 aJ $\Delta \mathscr{E}_{\text{state}}$ derived for logic rods with the exemplar system parameters (Section 12.3), together with the assumed 10^5 rods for a CPU-scale system, yields an estimated maximum variation in stored energy (relative to the mid-range value) of $\Delta \mathscr{E}_{\text{max}} = 1/2 \times 5/12 \times 10^5 \times 1.2 \text{ aJ} = 2.5 \times 10^{-14} \text{ J}$.

(Note, however, that a logic system could be built in which every logic rod set to a 1 state is mirrored by a rod set to a 0 state, thus cancelling the major contributions to $\Delta \mathscr{E}_{\text{max}}$; this approach would approximately double system volume and energy dissipation. This scheme has electronic parallels.)

b.   Acceptable clock skew and required drive-system stiffness.   In a system with the exemplar rod parameters, switching causes a 1 nm displacement in the gate knobs, and a comparable displacement in the probe knobs. An interlock switching from a nonblocking to a blocking state can produce an error if substantial probe knob displacements precede substantial gate knob displacements; significant added energy dissipation and an increased probability of error can occur if the motion periods overlap more than slightly.

A system that undergoes a premature probe-knob displacement of ≤ 0.05 nm in the worst-case logic state will suffer negligible degradation of reliability or energy dissipation. With the exemplar parameters, this displacement corresponds to a clock skew of ~0.014 ns. At a drive-rod speed of ~38 m/s, (Section 12.3.4d) this skew corresponds to a displacement of ~0.53 nm relative to the optimal drive-rod position (a phase-angle error $\Delta \phi \approx 0.073$ rad). If the energy occurring in a maximal fluctuation, $\Delta \mathscr{E}_{\text{max}} = 2.5 \times 10^{-14}$ J, were delivered in a single period $t_{\text{switch}} = 0.1$ ns over a stroke length 0.1 ns × 38 m/s = 3.8 nm, the mean (fluctuating-component) force applied to the drive system during this period would be ~6.6×10^{-6} N.

To transmit this force while limiting the elastic displacement to ~0.53 nm requires a system with a stiffness of ~1.2×10^4 N/m. If this is to be achieved using drive mechanism components with a typical length of 200 nm (~1/2 the system diameter) and a modulus $E = 10^{12}$ N/m^2, the required total cross sectional area of the rods is ~2500 nm^2, and their volume is ~0.0078 that of the reference CPU-scale system. Accordingly, a drive system stiff enough to limit clock skew to an acceptable value need occupy only a small fraction of the total system volume.

Clock signal distribution will likely have to differ in larger-scale synchronous systems. Control of local phase using optical or electrical signals offers one approach. As M. Miller observes, propagation of acoustic pulses at known speeds over known distances (e.g., with propagation times that are an integral number of clock periods) can also serve as a basis for large-scale synchronization. At some scale, designers of computer systems typically abandon synchronization.

12.7.4. Power requirements

The total power dissipation for the model CPU-scale system can be estimated from the component energy dissipations and the clock rate. Assuming a 1.2 ns clock with 0.03 maJ per switching operation per interlock (Section 12.3.4), 4.4 maJ per register-cell storage cycle (Section 12.4.3), and operation of every device on every clock cycle yields an estimated power dissipation of 74 aJ per clock cycle, or ~60 nW. This is less than 1/100 the magnitude of the power flows corresponding to fluctuations in stored energy.

12.7.5. Power supply and energy buffering

a. Flywheels for energy buffering. A cylindrical flywheel of radius $r = 195$ nm and height $h = 20$ nm can be placed within a 400 nm CPU-scale package. Its kinetic energy is

$$\mathscr{E}_f = \frac{1}{2} I_f \omega_f^2 = \pi^2 r^4 \rho h \big/ t_{clock}^2 \qquad (12.26)$$

With $\rho = 3500$ kg/m^3, $t_{clock} = 1.2$ ns, and the above geometric parameters, $\mathscr{E}_f \approx 6.9 \times 10^{-13}$ J. Since $\Delta\mathscr{E}_f/\mathscr{E}_f \approx 2\Delta\omega_f/\omega_f$ (when $\Delta\mathscr{E}_f/\mathscr{E}_f \ll 1$), $\Delta\mathscr{E}_{max} = 2.5 \times 10^{-14}$ J implies $\Delta\omega_f/\omega_f \approx 0.018$.

Note that the rim velocity, ~1000 m/s, is well below the bursting speed for a diamond hoop, and that the presence of a significant vacuum gap avoids frictional drag.

b. Integration of motor and flywheel. The rotor of a motor with the exemplar parameters described in Section 11.7.3 can serve as a flywheel like that just described. The motor can deliver more power (~1.1 μW) than is needed to operate a CPU-scale system (~0.06 μW).

c. Start-up. Save for acoustic transmission lines, all of the computational mechanisms can work at low frequencies and dissipate less energy per operation as they do so. Static friction can be made effectively nil, with barriers $\ll kT_{300}$. Accordingly, a motor torque that can power a system at full speed is more than adequate to start it. Acoustic transmission lines operating with energy recovery are more dissipative just below their design frequency, but can be locked during start-up.

12.8. Cooling and computational capacity

In nanomechanical systems of sufficiently large scale, the ~10^{12} W/m^3 power dissipation density of ~1 GHz nanomechanical logic systems described here exceeds any possible means of cooling. On a sufficiently small scale, however, cooling poses no problem. For example, the equilibrium ΔT of an isolated 100 nW system of 400 nm dimensions is <0.01 K in a medium with a thermal conductivity of 10 W/m·K.

On an intermediate scale, convective cooling is appropriate. Section 11.5 outlines a cooling technology based on branched channels, inhomogeneous phase-changing coolants, and high pressures capable of removing ~10^5 W of thermal energy from a 1 cm^3 volume at ~273 K. This is sufficient to cool ~10^{12} CPU-scale systems with an aggregate instruction-execution rate of ~10^{21} per second. A more modest 10 W system can deliver ~10^{11} MIPS.

Present techniques in software engineering could be used to apply parallel systems of this scale to a wide range of problems (e.g., fluid dynamics and many other physical simulations), but many other problems are less simple and uniform. To exploit growing computational resources, software engineering has developed techniques (e.g., object-oriented programming) that reduce the need for detailed, centralized planning. An analysis suggests that using market mechanisms to organize complex, decentralized processes has fundamental advantages in computation, particularly in large-scale parallel systems (Miller and Drexler, 1988a; 1988b; Drexler and Miller, 1988). This approach can be viewed as a further step in the direction taken by modern software engineering.

12.9. Conclusions

Nanomechanical computational systems can be implemented using logic systems based on sliding rods having switching times of ~0.1 ns, with energy dissipation $\ll kT_{300}$ per gate. Register cells can be constructed that approach the theoretical minimum energy dissipation of $\ln(2)kT$. Logic rods and registers can be joined to build register-to-register combinational logic systems that achieve four register-to-register transfers in ~1.2 ns; this performance (together with comparisons to existing microelectronic practice) suggests that nanomechanical RISC machines can achieve clock speeds of ~1 GHz, executing instructions at ~1000 MIPS. Fast carry chains, RAM, tape, and I/O systems all appear feasible.

A CPU-scale system containing 10^6 transistor-like interlocks (constructed with the parameters used in previous example calculations) can fit within a 400 nm cube. Compatible systems for clocking, power supply, and cooling have been described and analyzed. The power consumption for a 1 GHz, CPU-scale system is estimated to be ~60 nW, performing $>10^{16}$ instructions per second per watt.

Some open problems. The design calculations in this chapter are based on bounded continuum models; an open and accessible problem is the design and modeling of nanomechanical logic components (gates, registers, carry chains, RAM cells) in atomic detail. With good designs, substantially improved performance should be possible. At a higher level, the design of general-purpose computers that minimize irreversible operations (e.g., register erasure) is of substantial interest, as is the design of systems that reduce nonessential dissipation from logically reversible processes. The ongoing work in molecular and quantum electronics is of considerable interest. The anticipated fabrication capabilities of molecular manufacturing, however, may encourage consideration of a wider range of physical structures. For all but the highest-speed (as distinct from highest-throughput) computers, R. Merkle emphasizes that good switching technologies must eventually enable the use of nearly reversible operations to limit power dissipation; this may deserve greater attention in molecular and quantum electronics research.

Molecular Sorting, Processing, and Assembly

13.1. Overview

Nanomechanical mechanisms and processes have natural applications to molecular manufacturing. Previous chapters show that nanometer-scale structural components can have diverse shapes and good stiffness (Chapter 9); that nanometer-scale gears, bearings, chain drives, cams, and the like are feasible (Chapter 10); that mechanical and electromechanical components (Chapter 11) can be combined to build systems of power-driven machinery capable of executing complex, programmed patterns of motion (Chapter 12); and that mechanical manipulation of reactive moieties can be used to direct a broad range of synthetic operations with reliable positional control (Chapter 8). The present chapter considers how these and related capabilities can be combined to acquire, order, and process feedstock molecules, first preparing reagent moieties and then applying them to workpieces to build up complex structures.

This process begins with small molecules in a liquid and ends with complex structures in an evacuated region. Liquid solutions are a natural starting point for molecular manufacturing: they are dense, thoroughly disordered, and can carry a wide variety of molecules. A vacuum environment is a natural destination: the use of vacuum avoids both viscous drag and mechanical interference from stray molecules, while minimizing unwanted chemical reactions. These characteristics improve the performance of molecular manufacturing systems and simplify their design and analysis. (For many applications, of course, the manufacturing system must ultimately deliver its product to a nonvacuum external environment.) To accomplish these steps, the required processes include:

- Acquiring and ordering molecules from solution
- Transforming streams of molecules into streams of reagent moieties
- Assembling reagent moieties to form complex molecular objects

Analyzing these processes requires attention to:

- Assembly processes
- Materials transport
- Cycle times
- Energy requirements
- Error rates
- Error sensitivities

This chapter addresses these issues in several contexts. Section 13.2 describes the sorting and ordering of molecules to produce nearly deterministic streams of molecules of single type starting from a disordered external solution. Section 13.3 describes the transformation of streams of input molecules into streams of bound reagent moieties suitable for use in molecular manufacturing processes, adopting a molecular mill approach. Section 13.4 describes programmable molecular manipulators for combining reagent moieties (and larger reagent structures) to build complex products. The next chapter builds on these results to describe systems capable of assembling macroscale products.

13.2. Sorting and ordering molecules

Nearly deterministic molecular manufacturing processes require well-ordered inputs, derived ultimately from a disordered feedstock. In designing a system, the contents of this feedstock can, within broad limits, be chosen for convenience. For concreteness, this section considers liquid feedstock solutions that can be prepared today: moderately pure mixtures consisting of small, relatively stable molecules of selected kinds in chosen proportions. (A typical small molecule, for present purposes, is one containing 20 or fewer atoms each of which has a degree of surface exposure equal to or greater than that of the central carbon atom in propane.) An input port to a manufacturing system can be presented with a feedstock consisting predominantly of a single chemical species (with trace contaminants) or with a mixture of species each present in substantial concentrations (again with trace contaminants).

The major species in a feedstock mixture can be chosen to serve as convenient fuels or raw materials; the trace contaminants must be assumed to be diverse. It is this diversity that poses the major challenge in analysis, forcing consideration of a set of molecules with a wide range of properties, and (in the early stages of processing) too large to be exhaustively described.

In constructing nanomechanical systems, only relatively exotic devices require isotopic refinement (e.g., for the construction of small, high-speed rotors with a precisely balanced mass distribution). Neither devices of this kind nor isotopic refinement are considered here.

This section first describes a class of mechanisms capable of selectively binding and transporting chemical species from a feedstock solution. It then discusses the combination of such mechanisms to build purification systems able to deliver streams of molecules that are (with high reliability) of a single type. Finally, it discusses the conditions that must be met in order to capture and orient single molecules from such a stream in a reliable manner, providing a eutactic input stream for subsequent processing.

13.2.1. Modulated receptors for selective transport

a. Basic concepts. Molecular biology provides numerous examples of systems that selectively bind molecules from solution (e.g., antibodies, enzymes, and various signal-molecule °receptors) and of systems that transport bound molecules against concentration gradients (e.g., active-transport systems in cell membranes). Nanomechanical systems can perform similar operations. The analysis of purification operations is fundamentally different from the analysis of processes in a eutactic environment, in that the population of impurity molecules is not specified. Accordingly, portions of this analysis are cast in terms of general properties of molecular shapes, although practical systems are likely to gain great advantages from the exploitation of other properties of particular molecules of interest.

Figure 13.1 illustrates a class of mechanisms capable of using mechanical power to transport molecules of a particular type from regions of lower to higher free energy, and from regions of lower to higher purity (see caption for description). Analogous mechanisms include devices with differing geometries (e.g., cam surfaces to the side of the rotor, rather than in the hub) and differing mechanisms for modulating the binding interaction (e.g., elastic deformation of the receptor, alteration of local charge distributions). A slightly more elaborate class of mechanism, motivated by specificity concerns, is discussed in Section 13.2.1d. The general capabilities of such devices can be analyzed using a combination of principles from statistical mechanics and experimental observations from molecular biology.

The capabilities of general structures can equal or exceed those of antibodies, about which much is known from experimentation. As noted in Section 9.5.4, antibodies can be developed that will bind any one of a wide range of small molecules with small °dissociation constants [typical values of K_d are 10^{-5} to 10^{-11} nm^{-3} (Alberts et al., 1989)] and strong specificity (relatively poor binding of all but very similar molecules). The probability that a receptor will be found occupied by a molecule of the desired type is related to the concentration of a bindable molecule (ligand), c_{ligand}, and K_d by the expression

$$P_{\text{occupied}} = \frac{c_{\text{ligand}}}{K_d} P_{\text{unoccupied}} \qquad (13.1)$$

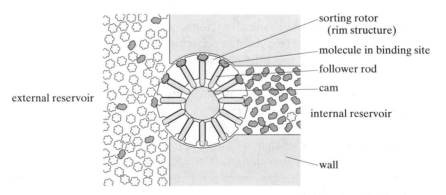

Figure 13.1. A sorting rotor based on modulated receptors. In this approach (illustrated schematically), a cam surface modulates the position of a set of radial rods. In the binding position (mapping the illustration onto a 12-hour clock dial, 10:00), the rods form the bottom of a site adapted to bind molecules of the desired type. Between 10:00 and 2:00, the receptors undergo transport to the interior, driven by shaft power (coupling not shown). Between 2:00 and 4:00, the molecules are forcibly ejected by the rods, which are thrust outward by the cam surface. Between 4:00 and 8:00, the sites, now blocked and incapable of transporting molecules, undergo transport to the exterior. Between 8:00 and 10:00, the rods retract, regenerating an active receptor. Section 13.2.1c discusses receptor properties; Section 13.2.1e discusses energy dissipation.

Values of K_d depend on the interaction energies of the ligand and receptor, the ligand and solvent, and the solvent and receptor. Among differing solvent systems in which the interactions of the solvent with the ligands and the receptor are relatively weak and nonspecific, K_d will be relatively solvent independent.

The number of atoms required to equal the basic structural diversity of antibodies was estimated on combinatoric grounds to be <35 (Section 9.5.4). On geometric grounds, however, a receptor that encloses a molecule of substantial size (e.g., ~0.9 nm in diameter) with a layer of equal thickness must have an excluded-volume diameter of ~2.7 nm, and contain ~10^3 atoms. (The diameter of an active, Fv fragment of an antibody provides a highly conservative upper bound of ~4 nm on receptor size.) Allowing 2.7 nm of circumference per receptor, and 12 receptors per rotor yields a rotor diameter of ~10 nm. With a rotor thickness of 2.7 nm, the total number of atoms in the rotor system is ~2×10^4.

Modulated receptor mechanisms can be applied either to concentrate a molecular species from a relatively dilute feedstock solution, or (in a multistage cascade, Section 13.2.2) to purify a stream of molecules by excluding impurities. For use in concentration, receptors having K_d considerably less than the actual solution concentration are of value, since these can (given reasonable specificity) deliver a product stream dominated by the desired species in a single operation. Since diffusion-controlled rate constants for binding of small molecules are typically ~10^9 s^{-1} nm^{-3} (Creighton, 1984), and since several tens of molecular species can be simultaneously present in solution at concentrations ≥ 0.1 nm^{-3}, ~100 encounters occur in a 10^{-6} second exposure time, ensuring equilibration. With $K_d \leq 10^{-3}$ nm^{-3}, the probability of receptor occupancy is $\geq \sim .99$.

Delivering the contents of these receptors to a compartment containing a high concentration of product molecules reduces entropy, requiring a work of concentration amounting to ~20 maJ per molecule at 300 K. This can readily be provided by coupling to a mechanical drive mechanism (details of such drive mechanisms are not considered here; they can be built up from gears, bearings, springs, cams, and motors of the sorts described in Chapters 10 and 11). With the application of greater drive forces, a modulated receptor mechanism can easily pump against pressures of several gigapascal (i.e., several nanonewtons per square nanometer), at an energy cost on the order of 100 maJ per molecule. Where concentrations and pressures differ little from one side to the other, as in the later stages of purification cascades (Section 13.2.2), the input work required can be quite small.

b. Diamondoid vs. protein structures for receptors.

Diamondoid structures have advantages over proteins in their feasible affinities and specificities. In general, stiffer structures permit greater specificity. At one extreme, a sufficiently soft receptor can exhibit solventlike specificity, that is, very little. (Receptors containing flexible, oligomeric chains can provide mechanically modulated, broad-spectrum affinity.) Proteins can exhibit substantial stiffness, enabling strong specificity (Creighton, 1984). Diamondoid structures can exhibit stiffness greater by several orders of magnitude, permitting significant improvements in specificity, particularly where exclusion of a particular species depends on overlap forces resulting from a bad geometric fit.

Diamondoid structures have a greater atom number density, leading to stronger van der Waals attractions and hence to greater binding energies, which can contribute to greater affinities. Further, the greater design-stage flexibility of general structures will often permit better matching of ligand and receptor, with respect to both geometry and electrostatics.

c.   General considerations in designing for specificity.   Receptor affinity (that is, $1/K_d$) and specificity (roughly, differences in affinity for different molecular species) are related but distinct properties. In a practical system, what matters is the ability to discriminate between the desired molecule and the competitors *actually present* at that location in the process, not specificity for the desired molecule with respect to all possible competitors. Accordingly, the optimal receptor structure may differ at different stages in a cascade.

A high binding affinity for the preferred ligand can be a poor design choice if it leads to slow dissociation kinetics for a competitor, even if the differential in affinities remains large. Under these circumstances, the occupancy of receptors in transit through the barrier is controlled by competitive binding kinetics, rather than by equilibria; this typically results in poor discrimination. This problem does not arise in systems with $K_d > {\sim}10^{-3}$ nm^{-3} and having ${\sim}10^{-6}$ s receptor exposure times.

If two similar molecules can compete for the same receptor, then (neglecting differences in solvation) the ratios of their dissociation constants are directly related to the differences in their free energies of binding in vacuum and (neglecting pV effects and differences in entropy) are approximately related to the differences in their potential energies of binding in vacuum

$$K_{d,1}/K_{d,2} \approx \exp[(\Delta \mathcal{F}_1 - \Delta \mathcal{F}_2)/kT] \approx \exp[(\Delta \mathcal{V}_1 - \Delta \mathcal{V}_2)/kT] \qquad (13.2)$$

The potential energy differences in Eq. (13.2) can readily be estimated by energy minimization in a molecular mechanics model; evaluation of the free energy differences in Eq. (13.2) requires calculations that either explore a region of the PES (Mitchell and McCammon, 1991; Straatsma and McCammon, 1991) or perform an integration over a region (see Section 4.3.2).

Consider a receptor that, unlike those in Figure 13.1, entirely surrounds its ligand (to permit the entrance and exit of ligands, the receptor must have at least one moving part). Assume that the receptor conforms closely to the contours of the preferred ligand, with the interfacial atoms placed near their minimum-energy distances, or somewhat closer. Alternative ligands can now be classified into three categories based on their size and shape: (1) those that resemble the preferred ligand, but occupy a strictly smaller volume in some conformation (e.g., having a similar structure, but with H substituted for CH_3); (2) those that resemble the preferred ligand, but occupy a strictly larger volume in all conformations (e.g., by having the reverse substitution); and (3) those of all other shapes. Ligands in category 1 typically bind more weakly, owing to loss of van der Waals attraction energy; those in category 2 bind more weakly (or are effectively excluded) owing to overlap repulsion; those in category 3 suffer both difficulties simultaneously, since they can neither fill the receptor nor avoid overlap.

Discrimination against ligands in category 1 increases with the Hamaker constant (Section 3.5.1a) of the receptor structure; this constant is larger for typical diamondoid structures than for proteins. Even for proteins, however, the discrimination against ligands that lack a CH_3 group (putting H in its place) can be ~24 maJ (Creighton, 1984), altering binding affinity by a factor of ~0.003 at 300 K. (To model binding of molecules in ordinary solvents, this energy difference has been reduced by subtracting contributions from hydrophobic interactions.)

Discrimination against ligands in categories 2 and 3 can be maximized by using a stiff receptor structure that surrounds the preferred ligand at less than the minimum-energy distance. The use of such *tight-receptor structures* sacrifices °binding energy in exchange for a particular kind of specificity, and may *reduce* discrimination against ligands in category 1. One of the smaller increments in steric bulk occurs between structures in which CH replaces N, shifting regions of similar overlap energy outward by ~0.09 nm over a small region. If the initial separation between an N atom in a ligand and an atom of the receptor is chosen to be ~0.3 nm, computational experiments in the MM2 approximation indicate that the adverse overlap energy experienced by the competitor can be ~45 maJ, sufficient to alter affinities by a factor ~5×10^4 at 300 K, according to Eq. (13.2). (Note that in the present application, preferred ligand molecules can be selected with ease of purification as a criterion, hence structures presenting unusual difficulties can be avoided during the design stage.)

This example—discriminating between two species differing by the substitution of CH for N—also illustrates the utility of nongeometric mechanisms for molecular discrimination. Only small differences in shape arise from replacing a lone pair with a bond to hydrogen, but the resulting structures differ greatly in their electronic properties. In this instance, the nitrogen lone pair (but not the CH) can participate in the formation of a dipolar bond or a hydrogen bond. Geometrically, such bonds permit a closer contact between the N and the interface of the receptor. Energetically, they permit stronger binding. In general, different species with similar shapes have dissimilar electronic properties, permitting nongeometric mechanisms for discrimination.

d. A class of tight-receptor mechanism. A tight-receptor mechanism can be implemented as shown in Figure 13.2. In this paired-rotor approach, all competing molecular species are in fast equilibrium with the fully exposed receptors on the left, but only the preferred species (and strictly smaller ligands) can pass through the fully enclosed position between the two rotors without encountering a large barrier caused by overlap repulsion. According to the estimate made in Section 13.2.1c, barriers >30 maJ should be easily implemented. Tuning of the rotor potential energy function can effectively eliminate barriers when passing the preferred ligand (the common case for a purification process).

The paired-rotor, tight-receptor mechanism requires a compliant drive mechanism to operate properly. With an excessively stiff drive mechanism, even bulky competing ligands would routinely be forced through the gap (which must permit their passage in order to avoid failure by mechanical obstruction, regardless of drive stiffness). With a sufficiently compliant drive mechanism, however, the rotor moves (on short time scales) as if in free rotational diffusion. The rotational relaxation time for a rotor this size is $<10^{-7}$ s, assuming a mean viscosity

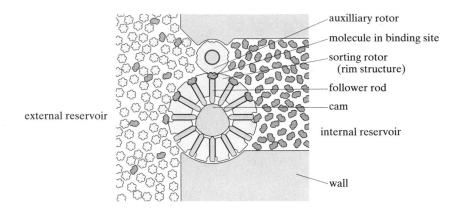

Figure 13.2.   A sorting rotor as in Figure 13.1, but with an auxiliary rotor that forces bound molecules to pass through a totally enclosed state, effectively excluding molecular species that are unable to fit within a volume of defined size and shape (for discussion, see Section 13.2.1d). The alignment of the primary and auxiliary rotors can be ensured by means of a geared interface (not shown).

of the surrounding medium like that of water (Creighton, 1984). Accordingly, in a system transferring receptors at mean intervals of 10^{-6} s, a bulky competing ligand momentarily blocked by a barrier has ample time to be carried backward by several receptor diameters and escape into the surrounding solution. A nonlinear compliance can provide a small driving force (e.g., $\sim 10^{-12}$ N, measured at the rotor rim) with negligible stiffness over a certain angular range of motion (e.g., several radians), while providing a strong, stiff constraint that prevents larger excursions of the rotor relative to an underlying drive mechanism of ordinary stiffness. This appears to meet the relevant constraints.

Mechanisms of this sort can strongly discriminate against ligands unable to fit within the space occupied by the preferred ligand. This capability is exploited in the analysis of staged cascades for purification of input streams.

e.   Energy dissipation.   Properly designed sorting rotor systems can approach thermodynamic reversibility in the limit of slow motion when handling fluids of nearly pure composition (otherwise, entropies of mixture can be significant). Aside from speed-dependent (chiefly viscous) drag mechanisms, free energy is dissipated when potential wells in disequilibrium are merged (Section 7.6.2). Well merging occurs in a system like that of Figure 13.1 or 13.2 whenever a receptor is exposed to a molecular reservoir; to avoid energy dissipation during this process, the wells must, in each configuration as they are exposed, have a probability of occupancy that is precisely the probability of occupancy that they would have after a long, equilibrating period of exposure in that configuration. This condition is automatically achieved for well-blocked receptors in the outbound transportation process: their probability of occupancy is zero when exposed to the interior reservoir, and remains zero when they are exposed to the exterior reservoir. This condition can also be achieved for receptors during the inbound transportation process between two reservoirs containing known concentrations: a receptor with a particular configuration has a certain probability of occupancy as it leaves the external reservoir, and would (if unchanged) have a higher probability of occupancy on exposure to the internal, higher-concentration reservoir. A downward modulation of the affinity of the receptor during the inbound transportation process, however, can make these probabilities correspond. This can be achieved by adjusting the shape of the cam surface.

Viscous drag can be estimated from rotor dimensions, given some choice of effective viscosity for the surrounding solution environments.* With the dimensions described in Section 13.2.1a, the wetted area per rotor is ~50 nm^2, and the characteristic length for shear (which is subject to considerable design-stage control) is ~2.7 nm. At a rim speed of 0.0027 m/s (10^6 receptor-sites per second), the drag power in a fluid of viscosity $\eta = 10^{-3}$ N·s/m^2 is ~10^{-16} W, or ~0.1 maJ per receptor transferred. Sliding-interface drag in associated bearings can be made negligible in comparison.

The energy dissipation per operation can be reduced somewhat by using thick rotors with several rows of receptors, thereby increasing the characteristic scale length of fluid shear and reducing fluid structure effects. Dissipation can be reduced to a greater extent by increasing the number of rotors (or the number of rows of receptors per rotor) and slowing the rotor speeds in proportion. If a single-row rotor has ~2×10^4 atoms (Section 13.2.1a), and a rotor with its housing and prorated share of the drive system contains ~10^5 atoms, then at 10^6 operations per second, it processes its own mass in <0.1 s. A considerably lower frequency would remain compatible with high productivity.

f. Damage mechanisms and lifetime. Modulated receptors for concentrating and purifying molecules differ from other mechanisms considered in this volume in that they are, by the nature of their task, exposed to an ill-defined chemical environment. Their susceptibility to damage caused by chemical reactions varies with structure and environment, and can most easily be estimated by analogy to other ill-defined chemical systems.

The surface reactivities of diamondoid structures (if designed with low surface strain, etc.) resemble the reactivities of smaller organic molecules. The stability and lifetime of solvent molecules in relatively pure, unreactive solutions can thus provide a model for the stability and lifetime of small patches of surface structure, including receptors, exposed to a reasonably pure, unreactive feedstock solution. This suggests reasonably long operational lifetimes.

Damage resulting from trace quantities of highly reactive contaminants can be minimized by flowing feedstock solutions past surfaces bearing bound moieties resembling those used on critical rotor surfaces, but selected for higher reactivity. Sacrificial moieties of this sort can combine with and neutralize many reactive species, including free radicals.

---

* At this size scale, liquid structure effects can be important. Designs having small liquid-filled gaps with adverse geometrical and surface properties could produce solidlike ordering among small feedstock molecules (Section 11.4.1a). At ordinary temperatures, however, different choices of geometry and surface structure should avoid this problem, but bulk-phase viscosity values should be taken as no more than a rough guide to liquid behavior near surfaces.

Proteins in living systems provide a model for molecular machines in a relatively complex, chemically aggressive environment. Metabolic enzymes can have lifetimes of several days (Creighton, 1984), despite the relative fragility of protein structures. Lifetimes of diamondoid sorting rotors of greater stability in a more benign environment should be longer, but even a one-day lifetime is sufficient for a device to process $\sim 10^6$ times its own mass before requiring replacement. (Issues of redundancy and system reliability are addressed in more detail in Chapter 14.)

13.2.2. Cascades of modulated receptors

To achieve low error rates in a molecular manufacturing process requires an input stream with a low fraction of contaminant molecules. For example, if a device is to perform 10^6 operations per second for several decades without an error, the fraction of contaminant molecules must be held to $\leq 10^{-15}$. This degree of purity can readily be achieved (for a wide range of reasonably stable feedstock molecules) by using a cascade of modulated receptors.

Multi-stage cascade systems are ubiquitous in chemical engineering practice, with implementations including leaching systems, distillation columns, extraction columns, and the like. The analysis of cascades is in general complex, and is described in books such as (McCabe and Smith, 1976). The following section presents an analysis of a simple case that illustrates the essential properties of cascades for the present application.

a. A model of staged cascade systems. Cascades (see Figure 13.3) combine multiple separation stages in a counterflow system. For simplicity, consider a system with N reservoirs of fluid containing predominantly molecules of a single type (the desired product), but contaminated by small concentrations c_0, c_1, $c_2,... c_{N-1}$ of impurity molecules (measuring concentrations in molecule fractions). Each reservoir n (save the last) is linked to reservoir $n + 1$ by an inbound flow (from the source to the destination) carrying f_{in} molecules per second, and each save the first is linked to reservoir $n - 1$ by an outbound flow carrying f_{out} molecules per second. (This is a simple instance of a flow pattern that permits mass balance in each reservoir.) Each flow is the product of a selective transport mechanism characterized by a differential affinity for impurity and product molecules; the inbound stream from compartment n thus has an impurity concentration $c_n \alpha_{in}$, and the outbound stream has an impurity concentration $c_n \alpha_{out}$. The condition for mass balance for impurities in reservoir n is then

$$c_{n-1}\alpha_{in}f_{in} + c_{n+1}\alpha_{out}f_{out} - (c_n\alpha_{in}f_{in} + c_n\alpha_{out}f_{out}) = 0 \qquad (13.3)$$

The assumption of stage-independent values for f_{in}, f_{out}, α_{in}, and α_{out}, together with a constant value of c_0, leads to a steady-state solution in which c_n declines exponentially with n. Convenient variables are then

$$f_{out}/f_{in} = f_{rat}; \quad \alpha_{out}/\alpha_{in} = \alpha_{rat}; \quad c_n/c_{n-1} = c_{n+1}/c_n = R \qquad (13.4)$$

where the value of R, which defines the rate of the exponential decline, is determined by

$$R = (f_{rat}\alpha_{rat})^{-1} \qquad (13.5)$$

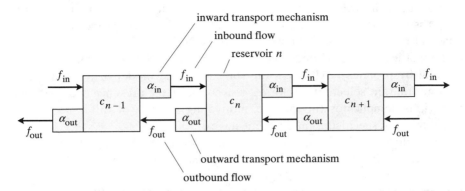

Figure 13.3. Schematic diagram of a portion of a staged-cascade process (Section 13.2.2a).

The final impurity concentration is then

$$c_N = c_0 R^N \tag{13.6}$$

which for R ≪ 1 can be extremely small for modest values of *N*. Although real systems will never correspond precisely to this model, it can be used to estimate the relationships among receptor properties, system size, relative flow rates, and product purity. Figure 13.4 illustrates a modulated-receptor implementation of a cascade system.

b. Purification using staged cascades of modulated receptors. For concreteness, consider a system that must deliver a stream of product molecules with a contaminant fraction $\leq 10^{-15}$, starting with a feedstock mixture containing

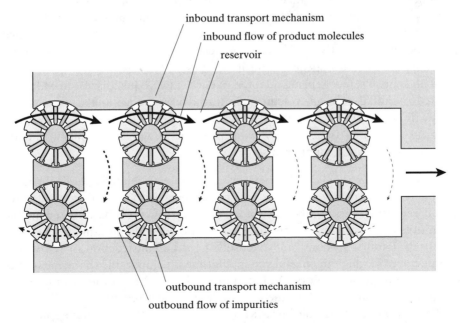

Figure 13.4. Schematic diagram of a staged-cascade process based on sorting rotors like those of Figure 13.1. (Note the inversion of the cams in the lower rotors.)

product molecules at a fractional concentration of 10^{-2} (~ 0.1 nm^{-3}). For generality, assume that only the universally available geometric properties of molecules are exploited to achieve receptor specificity, neglecting the use of electrostatics, hydrogen bonds, and so forth (all of which play prominent roles in binding by proteins). Although the staged-cascade model of Section 13.2.2a assumes that the selectivity factor α_{rat} is a constant, the presence of a mixture of impurities complicates this picture, motivating the use of different receptor structures and specificities at different stages.

Receptors for the inward transport mechanism of the first stage are best designed for high product affinity in the presence of a dominant background of quite different molecules. Analogies with antibodies (Section 13.2.1a) suggest that an inward-transport rotor can deliver a product stream with impurity fractions of $\sim 10^{-4}$ to 10^{-9}, depending on affinities, specificities, and the concentrations of the effectively competing ligands. The value of α_{rat} for the transport stage depends on the specificity of the outbound mechanism. If this is completely unselective, and if $f_{\text{rat}} = 0.1$ (corresponding to a modest outbound flow rate), then the value of R for this stage can range from $\sim 10^{-3}$ to 10^{-8}, yielding impurity fractions of similar numerical magnitudes.

Receptors for the subsequent purification stages should usually be designed with lower product affinities (Section 13.2.1c), to ensure equilibrium rather than kinetic control of transport selectivity. With tight-receptor mechanisms (Section 13.2.1d), the concentrations of competing ligands having excess bulk is reduced rapidly, with feasible values of $R \geq \sim 5 \times 10^3$, given $f_{\text{rat}} = 0.1$ (Section 13.2.1c). Ligands substantially smaller than the product molecule are effectively discriminated against by loss of van der Waals attraction energy.

After several stages of this sort, the chief impurities will typically be molecules that resemble the product, save for lower steric bulk in some region. The number of molecular types that can stand in this relationship to an already-small product molecule are few. Low values of R can be ensured in this circumstance by use of several selective, tight-receptor mechanisms in the outbound direction, each adapted to bind a specific impurity species.

With transport mechanisms adapted to each stage as just discussed, values of $\alpha_{\text{rat}} \geq 10^4$ should be readily achievable throughout, hence analyses using $R = 10^{-3}$ will yield conservative results. Accordingly, producing a stream with an impurity fraction $\leq 10^{-15}$, starting with a feedstock with an impurity fraction $\sim .99$, should require a number of stages $N \leq 5$.

c. Overall performance.   To construct a system with relatively large throughput, many sorting rotors can be used in parallel, each with multiple rows of modulated receptors; the number of rotors in a set can then be adapted to the magnitude of the stream of molecules to be handled, and operating rates can be equalized. Parallel systems lend themselves to the construction of redundant architectures in which the useful system lifetime is comparable to the mean lifetime of the individual rotors. As noted in Section 13.2.1e, rotors with multiple receptor rows can also reduce the viscous drag loss per receptor.

The rotors handling the inward flow streams are the dominant source of energy dissipation, assuming low values of f_{rat} as above. For $N = 5$ and capacities of $\sim 10^6$ molecules per second per row, losses from this mechanism amount to

~0.5 maJ per molecule delivered, using the parameters assumed in Section 13.2.1e. This is small compared to the useful work performed in molecular concentration (~20 maJ per molecule, Section 13.2.1a); losses from entropy of mixing are greater than this, amounting to ~10% of the useful work (assuming $f_{rat} =$ 0.1). With the estimated per-rotor masses (etc.) assumed in Section 13.2.1a, a purification system can deliver a quantity of product equaling its own mass in <0.5 s.

d. Handling harder cases. Separation on the basis of shape will fail (or, more properly, require many steps) if shapes are nearly identical. Among small molecules (defined at the beginning of Section 13.2) that consist of the elements of greatest interest (Figure 1.3), differences in atomic valence, bond length, and nonbonded radius guarantee substantial differences in shape. Among molecules containing transition or lanthanide metals, however, these differences can become quite subtle. In these instances, alternative strategies relying more heavily on chemical and electronic properties (e.g., differential complexation of ionic species, as in chromatography) may be desirable; laboratory experience demonstrates that these elements can be separated in this way. In manufacturing systems consuming the mix of elements expected to be typical, separation of these more difficult elements will be a quantitatively small portion of the overall processing task, having little impact on such system parameters as mass, energy consumption, and productivity.

13.2.3. Ordered input streams

Sorting processes of the sort described in Section 13.2.2 can keep a reservoir filled with molecules that are reliably of a single type. To provide deterministic inputs to a molecular manufacturing process, these molecules must be transformed into a stream of bound moieties (e.g., in a molecular mill of the sort described in Section 13.3) such that each site in the stream is reliability occupied. This could be accomplished by combining an *un*reliable mechanism for ordering and reactively binding molecules with a mechanism for probing and sorting the resulting structures; this section will instead describe approaches capable of achieving reliability directly.

a. Steps in ordering and input. One approach to providing ordered inputs comprises the following steps:

1. Expose a receptor to a pure liquid under conditions that ensure reliable occupancy of that site by a bound molecule.
2. Expose the bound molecule to a reagent moiety (on a separate transport mechanism) under conditions that ensure the reliable formation of a specific covalent bond.
3. Transfer the molecule to the other transport system as a covalently bound moiety, emptying the receptor for reuse.

This approach to molecular ordering is illustrated in Figure 13.5, with transfer to a reagent belt mechanism like those discussed in Section 13.3.1. In a mechanism of this sort, step 3 is a natural consequence of the pattern of motion, leaving steps 1 and 2 as the primary design problems.

b. Ensuring reliable receptor occupancy. A receptor will be occupied reliably ($P_{empty} \leq 10^{-15}$) if the free energy of binding has a magnitude greater than ~143 maJ. The loss of entropy that occurs when a molecule is bound from solution is comparable to the loss of entropy in freezing to a crystal; for small organic molecules, a typical value is 6×10^{-23} J/K, hence the required potential energy of binding is $\Delta \mathcal{V}_{binding} \approx -161$ maJ.

Two sources of driving force can be applied to increase $\Delta \mathcal{V}_{binding}$: attractive intermolecular forces to reduce the energy of the bound molecule, and hydrostatic pressure to increase the Gibbs free energy of the unbound molecules. In general, feasible binding energies increase in magnitude with increasing molecular size, polarizability, and dipole moments. Feasible hydrostatic pressures (i.e., those that do not cause the liquid to solidify) tend to decrease with the same variables. The magnitudes of feasible binding energies can be evaluated using molecular mechanics methods; the magnitudes of feasible hydrostatic pressures are indicated by data on liquids at high pressures.

Figure 13.6 shows a model receptor for a small nonpolar molecule, ethyne; molecular mechanics calculations indicate $\Delta \mathcal{V}_{binding} = -97$ maJ (this example is chosen to illustrate receptor principles with a simple, symmetrical structure, not necessarily to recommend ethyne as a feedstock). A larger model (including more atomic layers) would increase van der Waals attractions, increasing the magnitude of the binding energy. Achieving $\Delta \mathcal{G}_{binding} = -161$ maJ requires an increment of 64 maJ; this can be supplied by hydrostatic pressure.

Many fluids of small organic molecules remain liquid at pressures of ~1 to 2 GPa at 300 K (e.g., methanol, >3 GPa; ethanol, >2 GPa; acetone and carbon

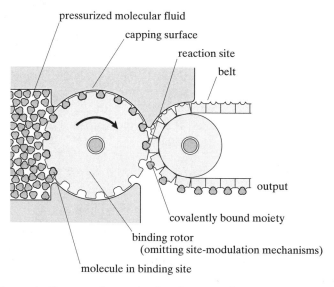

Figure 13.5. Schematic diagram of a mechanism for removing molecules from a liquid phase and covalently binding them to a moving belt (of the sort described in Section 13.3). This diagram omits mechanisms for modulating the properties of the receptor; these would be necessary in order to approach thermodynamic reversibility (exposure of an empty receptor as shown would be inherently dissipative). Generalized mechanisms for modulating the receptor can also relax constraints on reaction geometry and conditions (for example) by forcing compression and motion of the confined molecule.

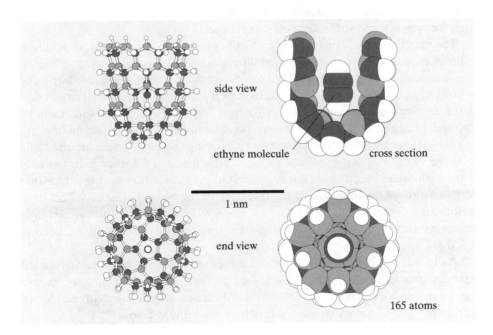

side view

ethyne molecule   cross section

1 nm

end view

165 atoms

Figure 13.6.   MM2 model of an ethyne receptor.

disulfide, >1 GPa), and in this pressure range, typical organic liquids have a volume per molecule ~0.75 that at atmospheric pressure (Gray, 1972). The volume per molecule of ethyne (low-pressure liquid density = 621 kg/m^3) is accordingly ~5 × 10^{-29} m^3/molecule in the pressure range of interest. Supplying an energy increment of 64 maJ per molecule thus requires a pressure of ~1.3 GPa, at which liquids consisting of small, nonpolar organic molecules like ethyne are still far from solidification.* Larger molecules require less pressure for a given energy, and require less energy to make up for deficits in the interaction energy with the receptor. Conversely, smaller molecules require more pressure and a larger energy increment, but tolerate greater pressures without solidification. The range of feedstock molecules that can be ordered in this manner is large.

c.   *Bonding to a molecule in a receptor.*   For suitably chosen feedstock molecules and receptor designs, it can be assumed that the molecule is either well oriented, or that patterns of symmetry and reactivity make molecular orientation unimportant. The properties of the receptor itself need not constrain the reaction environment: the bound molecule is confined by moving under the capping surface (Figure 13.5), and can then be subjected to direct, forcible mechanical

---

* The *chemical* stability of liquid ethyne itself at this pressure (in small volumes of pure substance, bounded by inert walls) is an open question; polymerization is exoergic, but the low-pressure activation energy is large. Sample calculations in Chapter 14 assume the use of acetone as a carbon source. Various strategies can enable the use of reactive molecules like ethyne as feedstocks; for example, they could be made to form bound complexes in an early, low-pressure stage, permitting later purification and orientation stages to handle a less reactive structure.

manipulation. Such processes can move it to a new location characterized by high free energy and stiff intermolecular contacts.

The reaction that transfers the molecule must take the form of an addition (unless means are provided for handling displaced groups). For unsaturated molecules, radical additions or Diels–Alder cycloadditions are candidates. For saturated molecules, additions to transition metals [e.g., Eq. (8.34)] are candidates. To overcome activation energy barriers, large piezochemical forces can be applied. Strategies for forming strong bonds with little energy dissipation are available (Section 8.5.6). The principles underlying these processes are the same as those in the mechanosynthetic processes discussed in Chapter 8. In working with small molecules, however, it is difficult to provide stiff restoring forces that resist motion of the molecule in the direction of the attacking moiety. This limitation may, for example, preclude low energy dissipation in radical coupling reactions involving small molecules (see the related discussion in Section 8.5.3b).

As is true throughout the design of materials input subsystems, any candidate feedstock molecule that presents an intractable problem in some regard can be discarded in favor of an alternative. Feedstock molecules can be chosen to provide the necessary elemental inputs in convenient forms.

13.3. Transformation and assembly with molecular mills*

Chapter 8 assumes that accurately controlled mechanical motions can be imposed as a boundary condition in molecular systems, implying control of positions, trajectories, and forces during chemical transformations. This section describes mechanisms that can provide such control using systems of belts and rollers; Section 13.4 describes more flexible (and less efficient) mechanisms based on programmable manipulators.

13.3.1. Reactive encounters using belt and roller systems

Systems of belts and rollers can be used to implement mill-style molecular processing systems. Belts moving over rollers and along guides find widespread use in macroscale manufacturing: they transmit power and transport materials and parts. They can serve this function in nanoscale systems as well. With the help of auxiliary mechanisms—such as moving parts in °reagent devices, rollers and cam surfaces to drive those parts, devices for transfer to and from other transport systems—molecular mills built in this fashion can perform a wide range of mechanochemical operations within a larger manufacturing system.

a. Simple encounters.   The simplest reactive encounters involve direct contact and separation, a motion which can be provided by a device like that shown in Figure 13.7(a). An encounter of this sort can transfer a group through an abstraction reaction (Section 8.5.4) or through a more complex chemical mechanism (e.g., those illustrated in Figures 8.12 and 8.14). Among the mechanical parameters subject to design control are the peak compressive load (which can be >10 nN, Section 8.3.3c), the stiffness of reagent moieties relative to one

---

* **mill** *n.*: Any of various machines for making something by some action done again and again. (Webster's New World Dictionary)

another along the line of motion ($\sim$90 N/m with sufficiently stiff bearings; see Section 8.5.3d), and the relative stiffness of the moietal supporting structures perpendicular to this line (with pairs of reagent devices having sterically complementary interfaces placed under compressive load, this can easily exceed 50 N/m). An encounter of this sort can be performed with a belt-belt contact, a belt-roller contact, or a roller-roller contact (only belt-belt contacts are illustrated in Figure 13.7).

The reaction time depends on belt speed, moietal trajectories (determined by the belt and wheel geometries), and the range of positions regarded as comprising the encounter state. Assuming that the trajectories are characterized by simple circular motions of radii r_1 and r_2 at a speed v with a positional tolerance of $\pm\delta$ (assumed small), the reaction time is

$$t_{\text{trans}} \approx \frac{2}{v}\left(r_1^{-1}+r_2^{-1}\right)^{-1}\cos^{-1}\left(1-2\delta\left(r_1^{-1}+r_2^{-1}\right)\right) \approx \frac{4}{v}\sqrt{\delta\left(r_1^{-1}+r_2^{-1}\right)^{-1}} \quad (13.7)$$

For $v = 0.005$ m/s, $r_1 = r_2 = 10$ nm, and $\delta = 0.01$ nm, $t_{\text{trans}} \approx 1.8 \times 10^{-7}$ s, moderately greater than the typical value assumed in Chapter 8. Fast, simple-encounter reactions could be performed with higher speeds and smaller mechanisms. Reaction times can be indefinitely prolonged by placing belts in contact and passing them between a pair of backing surfaces, as in Figure 13.7(b).

b.   Complex encounters.   In complex reactions (e.g., exploiting pi-bond torsion, Section 8.5.6), a device must drive several coordinated, properly sequenced

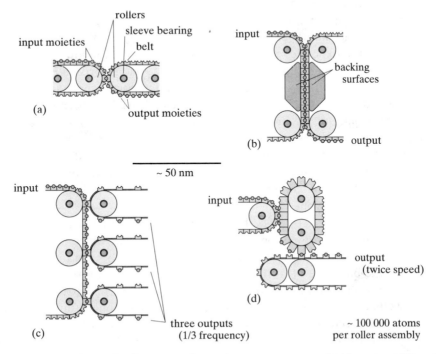

Figure 13.7.   Schematic diagrams of reactive encounter mechanisms, providing for simple contact and separation (a), prolonged contact with opportunity for substantial cam-driven manipulation (b), and transformations of moiety transit frequency (c) and speed (d).

motions. This can be accomplished by elaborating the structure of the reagent device to include moving parts actuated by auxiliary rollers or cam surfaces (Chapter 10), while extending the encounter time (if need be) as suggested in Figure 13.7(b).

This approach enables thorough control of molecular trajectories and reaction environments. As the two reagent devices come together, they can entirely surround the pair of moieties with a eutactic environment. Multinanometer-scale reagent devices containing multiple moving parts can manipulate this environment, forcing the moieties and adjacent structures to execute motions driven by stiff mechanisms capable of exerting large forces. Cam surfaces adjacent to the belts can interact with one or more protruding cam followers on the device to guide and power an arbitrarily long sequence of arbitrarily complex motions. Relatively small and simple mechanisms in this class are capable of exploiting pi-bond torsion effects, performing three-moiety encounters such as regeneration of alkynyl radicals assisted by radical displacement (Section 8.5.7), and so forth.

A complex encounter mechanism can also exploit nonbonded interactions to force the transfer of a group. For example, a bulky structure joined to another by a single bond can be mechanically clamped and then pulled to force cleavage of the bond and creation of a pair of radicals.

c. *Conditionally repeated encounters.*   Section 8.3.4f describes the use of conditional repetition to convert a reaction that on the basis of single-encounter thermodynamics and kinetics has a probability of failure P into a reaction process with a smaller probability of failure P^n, where n is the potential number of encounters. Conditional repetition can both increase reliability and (by permitting the use of less exoergic reactions) decrease energy dissipation.

Figure 13.8 schematically illustrates a reagent device incorporating a probe and a cam follower; the specific geometry and kinematics represent only one embodiment of a broad class of analogous mechanisms. For successful operation, the probability of the probe moving to its extended position [a state like (d)] must be negligible (e.g., $\leq 10^{-15}$) while the transferable group is still present. The driving force for this motion can be small (e.g., ~0.1 nN), and the formation of a bond between the probe and the site exposed by transfer of the group is among the acceptable outcomes, so long as that bond can be broken by the forcible retraction of the probe.

Figure 13.9 illustrates the use of a reagent device like that in Figure 13.8 to perform a conditional-repetition process, the net effect of which is to attempt the transfer reaction if and only if the group has not yet been transferred. As illustrated, the blockage of further attempts after a successful transfer can be direct, by blocking the reactive site; the motion of the probe structure could equally well achieve the same result though indirect mechanical linkages. Note that analogous mechanisms can be devised that make reaction attempts contingent on the *absence* of a group, hence the probe assembly need not be on the source side of a reactive encounter system.

When the cam follower passes through a location where two cam grooves merge, it undergoes a well merging process of the sort analyzed in Section 7.6. In any particular mechanism, the wells being merged have fixed probabilities of

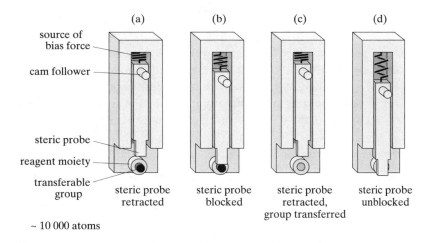

source of bias force

cam follower

steric probe

reagent moiety

transferable group

~ 10 000 atoms

(a) steric probe retracted

(b) steric probe blocked

(c) steric probe retracted, group transferred

(d) steric probe unblocked

Figure 13.8. Schematic diagram of a reagent device for implementing a conditional-repetition strategy to increase reaction reliability. In (a), the reagent moiety with its transferable group is exposed; the steric probe can be placed at any desired distance at this time. In (b), the probe is blocked by the transferable group and prevented from moving past it by a large (e.g., 150 maJ) energy barrier; the bias force need not be large. In (c), a successful reaction has transferred the group, removing the barrier. In (d), the bias force has driven the probe past its previous stopping point.

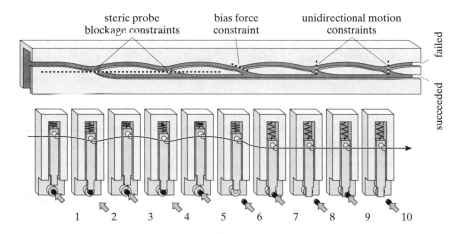

steric probe blockage constraints

bias force constraint

unidirectional motion constraints

failed

succeeded

1 2 3 4 5 6 7 8 9 10

Figure 13.9. Schematic diagram of a conditional-repetition system. Each small diagram below (numbered 1 to 10) represents the state of a mechanism like that illustrated in Figure 13.8 at a different time as it moves along one of the five possible paths through the system. The block above represents a structure with a network of cam grooves, shown from behind (with supporting material cut away). The cam-follower protrusions on the sliding mechanisms are (in the assembled configuration) constrained to follow the grooves. When a groove branches, the path taken is determined by the presence or absence of the transferable group at times 2, 4, 6, 8, and 10 (see Figure 13.8). So long as the group is present, the cam follower remains in the upper groove. Once it is absent (in the illustrated sequence, at time 6), the bias force moves the follower to the lower groove. The position of the probe then prevents further reactive encounters, as discussed in Section 13.3.1c. Several reactive devices can simultaneously be in transit through a cam mechanism of this sort, following different paths.

occupancy. If the sizes and depths of these wells are designed such that Eq. (7.73) holds, then the merging wells are in effect preequilibrated, and there is no fixed lower bound on the energy dissipation. Under these conditions, the work of compression done in merging wells directly increases the free energy of the product by reducing its entropy (i.e., increasing reaction reliability).

13.3.2. Interfacing mechanisms

a. Frequency and speed. System-level flexibility requires several kinds of transformations among streams of reagent moieties in molecular mill systems. Where several consumer subsystems require inputs of moieties at low frequencies, but a producer subsystem can efficiently produce them at high frequencies (or vice versa), it is desirable to transform frequencies by splitting and merging streams. Figure 13.7(c) diagrams how this can be done.

Differing subsystems can have widely differing optimal speeds of belt motion. Figure 13.7(d) diagrams how sites on a belt moving at one linear speed can be placed into smooth, nonsliding contact with sites on a belt moving at another linear speed while transferring materials.

b. Transfer to pallets, nondeterministic interfaces. In subsystems built using devices of the sorts described thus far, the rate of motion and frequency of operation at any given point fully determine the rates and frequencies at every other point (save for transient fluctuations). In building larger systems, it is desirable to surround subsystems of this sort with interfaces permitting greater flexibility of operation. In particular, this is required for fault-tolerant operation, which in turn is required in building relatively large and durable systems.

Transfer of a stream of moieties or product structures to a stream of pallets, moving along a track but not linked to form a belt, can provide the necessary flexibility. In regions where pallets are close packed, a stream of pallets can be treated like a moving belt, and similar transfer operations can be performed. In other regions, pallets can be more or less closely spaced (permitting the accumulation and use of buffer stocks), and the tracks bearing them can split or merge.

Two classes of track junction are useful in the design of fault-tolerant systems. These are fair, nonblocking merging junctions, and fair, nonblocking distribution junctions. A merging junction (1) accepts at least two input streams of pallets and produces a single stream of output pallets that includes contributions from both inputs (i.e., it is fair), and (2) continues to produce a stream of output pallets so long as one input stream has not failed (i.e., stopping inputs does not block the node). Similarly, a distribution junction (1) accepts a single input stream and produces one or more output streams, distributing the inputs across all the outputs, and (2) continues to distribute inputs to an output so long as one output stream remains unblocked. A *failed* junction of either kind is assumed to block one or more of its inputs; in the worst case, failure blocks all inputs (and thus outputs) by blocking all motion through the junction. These simple abstract behaviors can be implemented with active mechanisms, and perhaps without them.

A network of mechanosynthetic devices linked in this manner can behave as a demand-driven supply system. The combination of applied drive forces and mechanochemically derived forces from exoergic reactions can bias devices toward outputting product, but their output can be blocked when space for more product becomes unavailable. Production is then paced by the rate of removal of products at the end of the network, and no computational synchronization of production with demand is necessary. This somewhat resembles just-in-time delivery systems in conventional manufacturing, in which empty containers are the signal to produce more parts, but it more nearly resembles the delivery of water by a pipe, in which removal of material at one end creates a pressure gradient that moves material throughout the length.

13.3.3. Reagent preparation

Molecular mills can be used to transform low-entropy streams of feedstock molecules (Section 13.2.3) into low-entropy streams of reagent moieties covalently bound to reagent devices. Each of the group-transfer reactions in a process of this sort can occur in an enclosing environment designed to facilitate that reaction, to minimize °misreactions, and to minimize energy dissipation. Some of the feasible reactions and associated mechanochemical manipulations are surveyed in Chapter 8.

A typical input molecule might be ethyne; a typical reagent moiety might be a strained alkyne like **8.43**. The intermediate stages in this process might include two radical additions (Section 8.5.5) and hydrogen abstractions (Section 8.5.4), followed by the regeneration of the abstraction tools in a process that ends with the elimination of molecular hydrogen from a transition metal, Eq. (8.33). Each of these steps (save the last) requires an encounter between two reactive groups, and each results in the transfer of some number of atoms between one site and the other. This example suggests that cycles that prepare reagent devices by charging them with fresh reagent moieties may typically require on the order of 10 reactive encounters.

13.3.4. Reagent application

Molecular mills can be used to apply reagent devices to workpieces. Although programmable manipulators can do this with greater flexibility, producing many different products using one mechanism, mills can produce a single product with greater efficiency. Architectures for flexible manufacturing of macroscale products will likely include both mill-style and manipulator-style assembly mechanisms (Section 13.4).

Reagent application closely parallels reagent preparation, except that the workpiece-side structure in each reactive encounter is subject to the constraint that it be an intermediate in the construction of a desired device. This precludes thorough adaptation of the reaction environment for the purpose of facilitating reliable, low-dissipation operations. Nonetheless, the freedom to choose designs that can in fact be manufactured and synthetic sequences that generate relatively favorable intermediate structures provides considerable latitude for avoiding unfavorable reaction environments.

13.3.5. Size and mass estimates

The masses of components in a mill-style processing mechanism are significant both in estimating system productivity and in estimating radiation-damage lifetimes. As suggested by the discussion in Section 13.3.1, the structure (and hence the mass) of reactive encounter mechanisms can vary widely; to make a rough estimate of the typical system mass on a per-mechanism basis, a standard density can be assumed for all components (here, 2500 kg/m^3 or ~125 atoms/nm^3), and reference sizes can be allocated to provide for a moderately complex process and for transport of moieties over a moderately long distance between processes.

a.   Component sizes for the encounter mechanism.   Roller pairs and backing-surface pairs can have comparable masses; the roller option is considered here. A radius of 5 nm and a mean thickness of 2 nm provide ample room for a stiff bearing of the sort described in the sample calculations of Section 10.4.6 and shown in Figure 10.17. A pair of rollers, including axles, has a volume of ~310 nm^3.

A moderately complex encounter process can be driven by a mechanism occupying a volume 4 nm on a side, or 64 nm^3 (four times the volume of the exemplar interlock mechanism described in Section 12.3.3). This mechanism can be formed by the mating of reagent devices from each of two facing belts, each of which can then be $4 \times 4 \times 2$ nanometers in size. An auxiliary cam surface of substantial complexity can fit within a space $4 \times 2 \times 10$ nanometers in size, adding 80 nm^3.

These estimates are appropriate for reagent preparation processes, where only small moieties are handled. They can serve reasonably well for reagent application processes, so long as the workpiece is limited to a diameter of several nanometers.

b.   Component sizes for associated structure and transport mechanisms.   A mean belt length of 20 roller-radii (= 100 nm) per roller provides for substantial flexibility of system layout. If each belt is a strip with a cross sectional area of 0.5 nm^2, then the volume is 50 nm^3 plus that of one or more reagent-device pairs. If each belt is a close-packed chain of $4 \times 4 \times 2$ nanometer reagent devices, the volume of belt structure per roller-pair is 1600 nm^3. In the latter case, there are 50 reagent moieties per reactive encounter mechanism.

In a reasonably densely packed system, the distances between encounter mechanisms can be short; they can be stacked in parallel in near contact. A supporting structure with a mass equal to that of five struts of 4 nm^2 cross sectional area, each 15 nm (= 3 roller radii) in length seems ample; this amounts to 300 nm^3 of structural material per roller pair.

c.   Totals (per reactive encounter mechanism).   The total volume, mass, and atom-count per reagent device depend substantially on the choice of reagent-device density along the belts. At the upper extreme, where the belt consists of a chain of close-packed devices, the numbers are ~2300 nm^3, ~5.7×10^{-21} kg, and ~2.9×10^5 atoms. Systems of this sort have relatively high throughput and energy efficiency, and will serve as the basis for later calculations unless otherwise indicated. At the lower extreme, with only one pair of reagent devices per roller pair, the numbers are ~800 nm^3, ~2.0×10^{-21} kg, and ~1.0×10^5 atoms. Systems of this sort have lower throughput and efficiency, but less mass per kind of operation performed (i.e., greater versatility per unit mass). For comparison, 2.0×10^{-21} kg is ~40 times the mass of many enzyme molecules.

These estimates are substantially larger than would be appropriate for a simple encounter mechanism (Section 13.3.1a) and substantially smaller than would be appropriate for a conditionally repeated complex-encounter mechanism (Sections 13.3.1b and 13.3.1c), but should not grossly underestimate the size and mass required per encounter mechanism in a mill-style processing system handling small reagent moieties. Assuming an arbitrarily chosen filled-volume fraction for the more massive system of ~10% yields a crude estimate of the total volume per reactive encounter mechanism, $\sim 2 \times 10^{-23}$ m^3.

d.   Total mass of a reagent processing system.   As noted in Section 13.3.3, systems for transforming a stream of small feedstock molecules into a stream of small reagent moieties can be expected to require ~10 reactive encounters. Combining this estimate with those made in Section 13.3.5c yields $\sim 5.7 \times 10^{-20}$ and $\sim 2.0 \times 10^{-20}$ kg, respectively, or $\sim 2.9 \times 10^6$ and $\sim 1.0 \times 10^6$ atoms. These estimates indicate that high-throughput reagent processing systems can produce their own mass in deliverable moieties in ~3 s. Since reagent application adds a single step to a sequence of ~10, estimates of the mass and productivity of a system producing delivered moieties are little changed.

13.3.6.   Error rates and fail-stop systems

Errors in molecular mills can arise from damage to the mechanisms of the mill, from the instability of reagent moieties, and from failed reactions (i.e., °omitted reactions and °misreactions). Sections 8.3.3f and 8.3.4(d–g) discuss conditions that suffice to ensure that errors resulting from misreactions, omitted reactions, and reagent instabilities occur at rates $\leq 10^{-15}$ per operation.* The large stiffnesses feasible in mill mechanisms can help in meeting these conditions. Even with the exploitation of diamondoid, eutactic environments and stiff mechanochemical processes, however, these conditions exclude many potential reagents and transformations from use in reliable mechanosynthesis. It will here be assumed that choices compatible with these constraints are made, and that error rates in molecular mill systems are accordingly dominated by radiation damage. The following sections consider the consequences of damage, and describe how those consequences can, at the subsystem level, be made to result in a simple cessation of activity.

a.   Consequences of damage.   In a molecular mill, significant damage can be defined as damage that causes a failed reaction. In extreme instances, this may occur because damage has physically blocked the motion of the mill. In marginal instances, this may occur because damage has altered component geometries enough that misalignments increase the error rate.

In the former case, damage stops the mechanism directly; in all other cases, the mechanism continues to move for some time after the failed reaction. After

---

* An error rate of 10^{-15} corresponds to a mean time to failure of ~3000 years for a single device working at 10^4 operations per second. The failure rate of a mill module might be dominated by one device with an error rate of 10^{-15}, while a hundred others work with error rates of $\leq 10^{-18}$ (a factor of 10^3 in failure rate corresponds to a difference of <30 maJ in barrier height). The net mean time to failure then is still ~3000 years.

a failed reaction, however, one or more of the departing product moieties will have an incorrect structure. An incorrect structure will usually cause failures in further reactive encounters, causing a cascade of damage transmitted from reagent devices to other reagent devices. Accordingly, significant damage will rapidly result in widespread damage.

A damage cascade could be prevented by measurement and correction systems following each reactive encounter mechanism, but the complexity of doing so in general appears excessive. Conditional-repetition mechanisms (Sections 8.3.4f and 13.3.1c) are formally members of this class, but they correct only a single, well-defined kind of error: an omitted reaction.

b.   *Error detection and fail-stop subsystems.*   After a reaction, a processing system can gauge the shape of each moiety and product using a measurement mechanism based on a roller with a shape complementary to that of the correct structure. If the actual structure has a protrusion where the correct structure would not, then the roller (if it has a small contact area) is typically forced outward by a distance comparable to the size of the protrusion. Note that failed reactions will usually be signaled by a protrusion on one of the output structures: the atoms that enter the reactive encounter must leave again, and a misplaced atom will typically occupy space that would otherwise be empty (if the space is internal, detection of the event must be indirect). A single gauge mechanism can test a reagent device or a substantial fraction of the area of a workpiece. Gauge mechanisms can reliably discriminate between correct and incorrect structures (Sections 11.2.3 and 11.2.4), using iterated probing if necessary. Detection of an error can trigger a mechanism that blocks further activity in a molecular mill subsystem.

The number of gauge mechanisms necessary to detect an error cascade in a large molecular mill subsystem is a small fraction of the number of reactive encounter devices, provided that an error can safely be allowed to remain undetected for many operation cycle-times. If products from each subsystem must transit a relatively long belt before being delivered to a consumer subsystem, then a relatively tardy detection process can prevent the delivery of mismanufactured product structures. The resulting *fail-stop* subsystems make convenient building blocks in the design of redundant, fault-tolerant manufacturing systems (Section 14.3.3).

c.   *Acceptable subsystem masses.*   Redundant, fault-tolerant systems can be constructed from sets of subsystem units having identical functions, if these are linked in such a way that the essential input-output properties of each set are unchanged when one or more of its units fails. With cumulative damage (no repair is assumed here), system lifetime remains finite, limited by the lifetimes of the units.

It is convenient to consider units with a .01 probability of failure in a 10-year period. Assuming that failure is dominated by radiation damage, Eq. (6.54) implies that this degree of reliability can be achieved so long as the mass is $\leq 2 \times 10^{-18}$ kg, assuming a background radiation level of 0.5 rad/year. This mass corresponds to that of a solid block of material ~100 nm on a side, or to ~350 reactive-encounter mechanisms.

13.3.7. Estimates of energy dissipation

Energy dissipation in mechanochemical processes results from both mechanical and chemical steps. Mechanical energy dissipation results from the motions involved in transport and moietal manipulation; chemical energy dissipation results from the chemical transformations themselves (e.g., from merging shallow occupied wells with deep unoccupied wells, Section 7.6.4).

a. Energy dissipation caused by mechanical operations. Low energy dissipation is of greatest importance in systems designed for high throughput, and such systems will typically have belts with closely spaced reagent devices. Adopting a 4 nm spacing, a mechanism that delivers 10^6 moieties per second requires a 0.004 m/s belt speed; this speed is assumed throughout this section. Assuming a 5 nm roller radius (as in Section 13.3.5a), a 2 nm bearing radius and length, and 1000 N/m bearing stiffness (as in the examples of Section 10.4.6) yields an estimated power dissipation per roller of $\sim 7 \times 10^{-20}$ W, using the more adverse parameters for band-stiffness scattering. With ~ 20 rollers in the system, this corresponds to $\sim 1.4 \times 10^{-24}$ J per delivered moiety. This is negligible in comparison both to chemical energies and to kT_{300}.

Dissipation is larger where belts pass between backing surfaces. With assumptions analogous to those made for the bearing interface, a 16 nm^2 reagent device sliding over a surface at 0.004 m/s with an interfacial stiffness of 1000 N/m dissipates $\sim 10^{-19}$ W, or $\sim 10^{-24}$ J per pair sliding a distance of 20 nm. This remains quite small.

The motion of cam followers through cam grooves causes phonon scattering by the mechanism described in Section 7.3.4. Scaling from the sample calculation in that section yields an estimated energy dissipation of $\sim 10^{-27}$ J per nanometer traveled. Thermoelastic damping [Eq. (6.50)] is of the same order, assuming forces of ~ 1 nN applied to areas of ~ 1 nm^2. Since these values are $\sim 10^{-6} kT_{300}$ per nanometer of sliding motion, extensive use of cams is compatible with low energy dissipation per operation.

Reactive encounters can dissipate free energy through nonisothermal compression of potential wells (Section 7.5). With good design practice, typical values at the speeds assumed should commonly be ≤ 0.1 maJ per cycle. Dissipation from this source can easily exceed that from all other mechanical mechanisms combined. A reasonable estimate of the total energy dissipated by mechanical operations in a ten-step mechanochemical process is therefore ~ 1 maJ per moiety delivered; if nonisothermal compression losses are minimized, the energy dissipated can be reduced.

b. Energy dissipation caused by chemical transformations. A generous estimate of the energy dissipated in a series of chemical transformations can be made by assuming that each operation is as exoergic as the combustion of carbon, ~ 650 maJ per carbon atom. If each atom passes through ten such steps on the way to its final state, then the manufacture of a carbon-rich diamondoid product dissipates $\sim 3 \times 10^8$ J/kg in the form of waste heat. A less generous but still naively high estimate can be made by instead assuming that each chemical transformation must be sufficiently exoergic to make it highly reliable ($P_{err} \leq 10^{-15}$); this would require the dissipation of ~ 145 maJ per step, yielding an estimate of $\sim 7 \times 10^7$ J/kg. A more sophisticated estimate requires attention to ways in which good design practice can minimize energy dissipation. (Note that the naive estimate may in fact be reasonable for systems where low energy dissipation is not a design objective.)

As discussed in Section 8.5.2b, reliable mechanochemical operations can in some instances approach thermodynamical reversibility in the limit of slow motion. At 10^6 operations per second, many processes in this class dissipate energies in the range characteristic of mechanical motions, as discussed in Section 13.3.7a. In comparison to 145 maJ, these energies are negligible. Accordingly, *if* one assumes the development of a set of mechanochemical processes capable of transforming feedstock molecules into complex product structures using only reliable, nearly reversible steps, then the energy dissipated in mill-style manufacturing processes could approach zero, and in practice be ~ 1 maJ per atom, or $\sim 10^5$ J/kg. (The conversion of a disordered feedstock into an ordered product structure reduces entropy, producing waste heat without corresponding energy dissipation; see Section 14.4.8.) The conditions for combining reliability and near reversibility are, however, quite stringent: reagent moieties must on encounter have structures favoring the initial structure, then be transformed smoothly into structures that, during separation, favor the product state by ~ 145 maJ (to meet the reliability standards assumed in the present chapter).

The availability of conditional-repetition processes (Sections 8.3.4f, 13.3.1c) substantially loosens these constraints. A reliable, low-dissipation process can be implemented if the conditions of the preceding paragraph are met, but with a modulation in relative energies not of ~ 145 maJ, but of $\sim (5 + 145/N)$ maJ, where N is the maximum number of available repetitions, and 5 maJ is the energy bias favoring the initial structure. Under these conditions, the energy dissipated can be ~ 1 maJ per operation. For $N = 10$, the required energy difference (~ 20 maJ) can be induced by relatively modest manipulations of bond strain, overlap energy, electrostatics, or electronic structure. Further, a simple conditional-repetition process with $N = 10$ can achieve high reliability with an exoergicity (and energy dissipation) of ~ 15 maJ per operation, without need for well-depth manipulations.

In reagent preparation, the feasibility of tailoring the entire reaction environment should commonly permit the use of low-dissipation operations; but in reagent application, the prospects are less clear. Group-transfer reactions often proceed through a single transition state, without the formation and subsequent cleavage of stable bonds linking the reagent device to the workpiece. In reactions of this sort, it should commonly be feasible to choose a group-transfer reagent having bond strengths that make the reaction only moderately exoergic; in addition, it should commonly be feasible to modulate bond strengths so as to permit a close approach to thermodynamic reversibility. In these instances, reagent application causes an energy dissipation of ~ 15 and ~ 1 maJ, respectively.

Some reagents, however, form strong bonds to a workpiece without simultaneously sacrificing bonds to their carrier. These must pass though a second, bond-cleaving transition. Whether the energy liberated in the initial bond formation appears as heat or work depends on details of the reaction PES and the reagent device design. In particular, the net exoergicity of bond formation can often be reduced to low values by imposing mechanical constraints that force a temporarily unfavorable geometry on the product (e.g., large angle-bending strains in new single bonds, or large torsional strains in new double bonds). This tends to increase transition state energies, but large piezochemical forces can be used to reduce energy barriers (Sections 8.3.3c and 8.5) and drive the system along the reaction coordinate. In a suitably designed reagent device, the resulting strain energy can be relieved smoothly as cam-controlled motions modify the geometry of the surrounding structure, thus avoiding the production of thermal vibrational energy. The example of radical coupling (aside from spin-pairing considerations, the time reversal of homolytic bond cleavage, Section 8.5.3b) shows that a stiff mechanism can form highly exoergic bonds (~500 maJ) without substantial energy dissipation, even without exploiting the strained-intermediate strategy just described.

In light of these diverse strategies for reducing the required reaction exoergicity far below 145 maJ, and for reducing the energy dissipation far below the exoergicity, it seems that mechanosynthetic processes can be quite efficient. In estimating the energy dissipation of systems engineered for good efficiency, it seems reasonable to assume that nearly reversible pathways can be followed in reagent preparation, and that most reagent application steps are either nearly reversible or dissipate an amount of energy characteristic of a moderately exoergic reaction. In this picture, many reactions dissipate ~1 maJ, some dissipate ~15 maJ, and a few (owing to synthetic constraints requiring energetic reagents and precluding energy-recovery strategies) dissipate ~100 maJ or more. If the mean energy dissipated per operation is ~30 maJ, and the mean number of atoms transferred is ~1, then the energy dissipated per kilogram of product is ~1.5×10^6 J. If this estimate should prove to be wrong, the only conclusions affected are those regarding energy consumption and waste heat generation in manufacturing processes.

13.3.8.   Mechanochemical power generation

If a mechanochemical process is nearly thermodynamically reversible and is exoergic, then it converts chemical potential energy into mechanical work. Accordingly, mechanochemical systems can be designed to serve as sources of power; via electrostatic generators, they can provide electrical power. (Mechanochemical processes involving charge transfer can provide electrical power more directly, but are not considered here.)

The flexibility of transition-metal chemistry (Section 8.5.10) strongly suggests that a wide variety of reactions among small molecules can be carried out in a nearly reversible fashion via intermediate transition-metal complexes, including the reaction

$$2H_2 + O_2 \rightarrow 2H_2O \qquad (13.8)$$

which yields ~475 maJ per molecule of product (delivered to the liquid phase). If conducted at a rate of 10^6 product molecules per second with a mechanochemical system of the size and mass described in Section 13.3.5, the resulting power density relative to the mechanochemical apparatus itself is ~8×10^6 W/kg, or ~2×10^9 W/m^3. An overall efficiency >.99 is suggested by the energy dissipation estimates of Section 13.3.7. Similar results can presumably be obtained for the oxidation of small organic molecules.

In molecular manufacturing, the conversion of typical hydrogen-rich, small-molecule feedstocks into diamondoid structures liberates excess hydrogen. If this is combined with atmospheric oxygen, the yield of H_2O typically is on the order of one molecule per carbon atom in the product structure, and the associated net exoergicity is ~400 maJ per atom. This surplus energy can be made to appear as mechanical or electrical power.

13.4. Assembly operations using molecular manipulators

It is sometimes necessary to use a system to build other systems of equal or greater complexity (if this were not done, an infinite regress would make molecular manufacturing impossible). This can be achieved by applying reagent moieties using a multiple-degree-of-freedom positioning mechanism, controlled either by a local nanocomputer or by a more remote macroscale computer: a single mechanism of this kind can perform an arbitrarily complex sequence of operations, with no simple bound on the complexity of the resulting products. Whether directly or indirectly, mill-style systems are likely to be implemented using manipulator-style systems.

Industrial robots are examples of multiple-degree-of-freedom positioning mechanisms. With the availability of nanometer-scale digital logic systems, motors, gears, bearings, and so forth, analogous designs are feasible on a smaller scale (e.g., 100 nm). Aside from differences of scale and component properties, molecular manipulators differ from macroscale devices in that they must maintain positional accuracy despite thermal excitation. This problem can be minimized either (1) by operation at reduced temperatures, which receives no further attention here; or (2) by the use of a stiff mechanism, as described in Section 13.4.1; or (3) by use of local nonbonded contacts to align reagent devices to workpieces immediately before reaction, as discussed in Section 13.4.2.

Molecular manipulators typically execute many internal motions (e.g., rotations of drive shafts and gears, displacements of logic rods) per reagent moiety applied. Accordingly, their peak operating frequencies are reduced relative to mill-style systems, and their energy dissipation per operation (at a given frequency) is greater. Peak operating frequencies can exceed $10^6 \, s^{-1}$, although at a substantial (many kT_{300} per operation) penalty in energy consumption. Productivity on a per-unit-mass basis likewise is less than for a mill-style system, but still many orders of magnitude greater than that of conventional macroscale manufacturing systems.

13.4.1. A bounded-continuum design for a stiff manipulator

A sufficiently stiff manipulator can position a reagent moiety relative to a workpiece without needing to exploit interactions with the workpiece to ensure accurate alignment; this simplifies analysis. Completely general positioning requires control of six degrees of freedom (termed six axes): x, y, and z; and roll, pitch, and yaw. These functions could be divided among multiple loosely coupled mechanisms. For example, control of x and y could be provided by a mechanism moving the workpiece, with control of z provided by a mechanism moving the reagent device, and control of the roll, pitch, and yaw orientation of the reagent moiety could be provided by a mechanism that sets and locks the geometry of an adjustable reagent device before loading it into the z-axis mechanism. The present section, however, instead describes an integrated, armlike mechanism providing six-axis control. The range of choices in manipulator design is enormous; the specific set of choices explored here is intended to demonstrate lower bounds on capabilities without attempting to present an optimal design. In particular, little effort has been made to minimize energy dissipation, to maximize speed, or to simplify kinematics.

Section 8.5.5a concludes that positioning stiffnesses (for a reagent moiety with respect to a reaction site) of 5 to 10 N/m should be adequate to permit a wide range of reliable synthetic operations. The stiffness of an arm is best analyzed in terms of a series of (additive) compliances. A reasonable design goal is to keep the arm compliance small compared to 0.1 m/N (e.g., 0.04 m/N), thereby allowing most of the permissible compliance to be allocated to other parts of the system, such as the workpiece and the reagent moiety itself. This goal is, however, not trivial to meet: the scaling law of Eq. (2.3) suggests that a 100 nm long arm with a compliance of 0.04 m/N corresponds to a meter long arm with a compliance of 4×10^{-9} m/N. This is quite stiff compared to present practice in industrial robotics, motivating a clean-slate design rather than an attempt to scale and adapt existing designs.

a. General structure and kinematics.

A hollow tubular shape can combine low bending compliance with internal space for drive mechanisms. Motion requires joints, which add compliance. A telescoping joint with a threaded interface can combine good structural continuity with adjustable length, adding little compliance. Canted rotary joints (as in some experimental robot arms and hard-shell space suits) can permit angular motion of the arm with respect to the base and of the working end with respect to the arm, while again maintaining good structural continuity. The helical threads of the telescoping joint and the circular threads in the bearings of the rotary joints can form high-shear-stiffness interfaces like those in Section 10.4.7. Figure 13.10 illustrates the external form and range of motion of a device with these characteristics; it has a maximum extended length (with tool holder) of 100 nm, a diameter (save for the telescoping section) of 30 nm, and a typical wall thickness of 3.5 nm.

A drive system forces the tubular structural segments to rotate with respect to one another, and thus to rotate with respect to the local axis of the arm. In some instances (chiefly the two joints closest to the base, J1 and J2 in Figure 13.11) these rotational motions must be accomplished while maintaining high torsional stiffness. The approach chosen here exploits toroidal worm drives (Section 11.3.2) driven by mechanisms in a flexible core structure.

A stiff arm of this design has substantial dynamic friction, yielding relatively poor trade-offs between speed and energy dissipation. Speed and efficiency can be substantially improved if the arm need not move through a large distance between operations. Accordingly, the present design reserves a substantial volume of space for a transportation pathway from the base to the tip, permitting reagent moieties to be interchanged without arm motion by means of an internal conveyor mechanism.

b. Power supply and control. The power supply and control system is outside the arm itself, and hence is not subject to tight geometric constraints. It need only be assumed that shaft power is available, and that a control mechanism can engage and disengage smaller drive shafts (using clutches), causing them to turn by a set number of rotations between locked states. Reasonable implementations can be devised that use gear trains to produce an odometer-like encoding of the number of shaft rotations and use a corresponding device to encode a representation of the desired number of shaft rotations (this device can be written to like a register); a match between the two can be made to trigger a motion causing the disengagement of a clutch and the locking of the corresponding drive shaft. To initiate a motion, a digital logic device stores a suitable number in the register, then engages a clutch to cause shaft rotation in the appropriate direction.*

Drive shafts undergo longitudinal displacements as the arm moves (most obviously, when the telescoping joint J4 is actuated). A coupling that permits these motions while applying torque (e.g., a drive nut engaging longitudinal

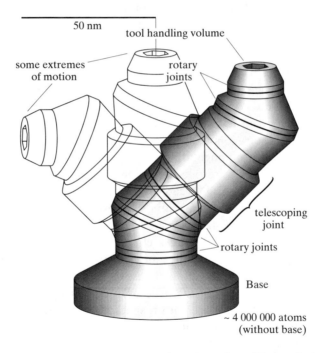

Figure 13.10. External shape and range of motion of a stiff arm design compatible with implementation on a 100 nm scale (internal mechanisms are described in Section 13.4.1 and Figures 13.11–13.). System compliance is analyzed in Section 13.4.1e.

---

* The ratios of the times and energies consumed by a typical nanomechanical computation operation to the times and energies consumed by a typical robotic arm motion are enormously greater than the corresponding ratios for microprocessor computation and macromechanical robotic arm motions. J. S. Hall has noted that this favors a shift away from moment-by-moment computer control of arm trajectories; the approach described here reflects this preference.

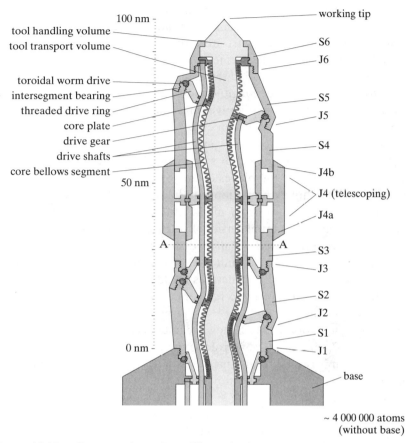

Figure 13.11. Cross section of a stiff manipulator arm and identification of parts (schematic); see also Figure 13.12.

grooves on a shaft segment) can interface these shafts to drive and control mechanisms with fixed positions.

c. Drive shafts, gears, and the core structure. Figure 13.11 shows a cross section of an arm, including all the moving parts required for arm motion save for three drive shafts (and gears) not in the plane of the section. As shown, the shafts have diameters of 1.5 nm; the compliance analysis of Section 13.4.1e assumes the use of a material with a modulus ~0.1 that of diamond, lowering the

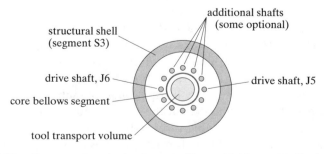

Figure 13.12. Cross section A–A of arm in Figure 13.11, perpendicular to the axis.

stresses caused by shaft bending. Surface strains in the shafts are well within the acceptable range even of high-modulus diamondoid structures. (Jointed and telescoping structures for drive shafts are also feasible.) One shaft, ending at joint J1, need not be flexible and is located outside the core structure.

The remaining shafts are threaded through bearing apertures in *core plates*, and are free to rotate and to slide longitudinally with respect to these plates. The rims of the core plates interlock with drive rings that in turn interlock with joints in the arm, forcing coplanarity between each core plate and its corresponding joint. Both segments linked by the joint can rotate with respect to the core plate.

Core plates are linked by core bellows segments. These flex to accommodate the angle and offset between the planes of the joints; the angle and offset have fixed magnitudes and relative orientation, but they can rotate with respect to the core structure. The core bellows segments link the core plates in a manner that gives the core structure substantial torsional stiffness. (Metal bellows often are used to transmit torque between misaligned shafts in macroscale engineering practice, despite the lower elastic limit and susceptibility to fatigue damage characteristic of macroscale metal structures.) Additional structures resembling the core plates could be introduced to further constrain and shape the paths of the bellows and drive shafts.

Within the envelope of the core bellows segments and core plates is a region with a diameter of 7 nm that plays no role in the arm structure or in driving its motions. This region is reserved for a mechanism to transport reagent moieties (or reactive clusters) to the tip of the arm, where a larger region is reserved for a mechanism to position and lock these moieties or clusters into position for use. An articulated conveyor system based on belts, rollers, and sliding-interface guideways can serve the transport function. Paths like those allocated for drive shafts can carry several shafts for actuating the reagent positioning and locking mechanism, or (in related applications) for transmitting signals from a sensor at the tip of the arm to a computational system outside the arm.

Drive shafts end in drive gears that engage the inner surface of a drive ring. As a drive shaft rotates, the drive gear forces rotation of the drive ring with respect to the core plate, which is itself restrained from substantial rotation about its axis by the torsional stiffness of the core bellows segments.

d. Joints and joint drive mechanisms.
The internal structure of the arm design is illustrated in Figure 13.11, including the components of the six joints. The joint nearest the tip of the arm, J6, plays no role in positioning and transverse compliance, affecting only the rotational angle of the tool. Its stiffness requirements are slight, hence it can be directly geared to a drive shaft.

The other rotary joints, J1, J2, J3, and J5 are implemented as toroidal worm drives (Section 11.3.2). Harmonic drives (Section 11.3.1) offer an alternative to the toroidal worm drive, but if repulsive interactions are used to implement their teeth, the large and time-varying radial forces applied by the wave generator are likely to result in larger energy dissipation at the interface between the wave generator and the flexspline than at the corresponding interface between the threaded drive tube and the triple-threaded torus. Further, the large asymmetrical loads would require additional structural analysis. The speed ratios and stiffnesses of these joints are examined in Section 11.3.2.

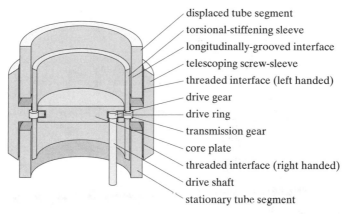

displaced tube segment
torsional-stiffening sleeve
longitudinally-grooved interface
telescoping screw-sleeve
threaded interface (left handed)
drive gear
drive ring
transmission gear
core plate
threaded interface (right handed)
drive shaft
stationary tube segment

Figure 13.13.   Structure and kinematics of a stiff telescoping joint. The drive shaft and attached drive gear rotate about their shared axis, forcing rotation of the drive ring about the shared axis of the core plate and the other illustrated components. The drive ring engages several transmission gears mounted in the torsional-stiffening sleeve. This sleeve shares longitudinally grooved interfaces with the upper and lower tube segments, preventing these segments and the torsional-stiffening sleeve from rotating with respect to one another, while permitting longitudinal displacements. The transmission gears engage an extension of the telescoping screw-sleeve, thereby coupling its rotational motion to that of the drive shaft. Rotation of the screw-sleeve drives longitudinal, telescoping motions of the tube segments via left- and right-handed helically threaded interfaces. The complexity of this scheme results from the placement of the screw sleeve outside the tube, which in turn is motivated by space limitations inside the tube and by advantages in bending stiffness resulting from a greater radius.

The telescoping joint assembly, J4 (including threaded interfaces J4a and J4b), can be directly geared to a drive shaft, since the thread angles (with a thread pitch of 0.5 nm) provide a speed reduction that provides ample stiffness [by Eq. (13.9)]. The structure and kinematics of this joint are described by Figure 13.13.

e. Estimated arm compliance.   The compliance of the arm (as in most beamlike structures) is greatest in bending modes. This section will estimate compliance in arms with the tip pointing upward or inward (relative to the orientation of Figure 13.10) in the worst-case geometry of full J4 extension and maximum bending (as in Figure 13.14). The compliance is then greatest for displacements perpendicular to the direction of bending (i.e., perpendicular to the plane of Figure 13.14). Table 13.1 summarizes the main contributions to total arm compliance in this geometry, and the physical assumptions behind the calculated contributions from structures outside the core and drive system.

The compliance contributed by the core and drive system can be reduced by using toroidal worm drives having large speed-reduction ratios. The compliance of the arm can be made almost independent of the core and drive compliances by exploiting the relationship

$$C_{s,2} = (v_2 / v_1)^2 C_{s,1} \qquad (13.9)$$

where $C_{s,1}$ and $C_{s,2}$ are the compliance of the system measured with respect to displacements at points 1 and 2, and v_1 and v_2 are the speeds of those points as

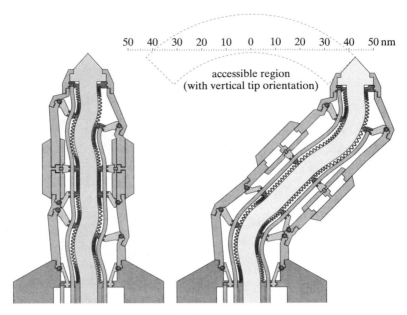

Figure 13.14. Cross section of a stiff manipulator arm, showing its range of motion (schematic).

the system executes some motion. (This expression assumes no compliance between points 1 and 2, i.e., that their velocities and displacements remain in a fixed ratio.) A mechanism providing a large speed-reduction ratio between the motion of the surface of the drive gear and the motion of the driven segment can tolerate large drive-shaft and power-supply compliances.

Drive shafts with a diameter of 1.5 nm and a shear modulus of 5×10^{10} N/m^2 will have a torsional compliance (for surface displacements) of ~6 m/N at a distance of 100 nm from a torsionally stiff locking mechanism. Torsional compliance in the core bellows structure adds directly to the drive rod compliance, but this increment can be made comparatively negligible.

For J6, direct gearing of the joint to the drive gear gives a standard deviation in joint angle of ~0.02 radians. This is orthogonal to the other compliances discussed here, and is negligibly small for most mechanochemical operations involving small reagent moieties.

For J4, if a drive gear is coupled to rotation of the threaded outer sleeve with a gear ratio of unity, it contributes a compliance in manipulator extension of ~0.018 m/N, assuming a threaded interface with a radius of 15 nm and threads of 0.5 nm pitch. This is orthogonal to the bending compliances considered elsewhere in this section.

In J2, J3, and J5, drive-rod compliance contributes to transverse-bending compliance. In each, the difference in radius between the geared inner surface and threaded outer surface of the drive ring contributes a speed increase of ~1.8, ~1.5, and ~1.8 respectively. The respective drive-rod lengths are ~25, ~35, and ~75 nm. The corresponding contribution of each of these joints to the transverse compliance at the tip can be limited to ≤0.001 m/N apiece if the speed-reduction ratios of the toroidal drive assemblies are 190, 70, and 65, respectively, measuring ratios from the speed across the threaded interface of the drive ring to the

Table 13.1.   Major contributions to transverse compliance of a manipulator.

Source	Coordinate	Magnitude (m/mN)	Multiplier[a]	Contribution (m/mN)
Tube bending[b]	End displacement	13.60	1	13.6
J1 rocking[c]	Max. local stretch	0.05	59	3.0
J2 rocking	Max. local stretch	0.05	45	2.3
J3 rocking	Max. local stretch	0.05	32	1.6
J4a rocking	Max. local stretch	0.05	19	1.0
J4b rocking	Max. local stretch	0.05	9	0.5
J5 rocking	Max. local stretch	0.05	4	0.2
J6 rocking[d]	Max. local stretch	0.07	3	0.2
J1 torsion[e]	Max. local shear	0.31	25	7.8
J2 torsion	Max. local shear	0.31	7	2.2
J3 torsion	Max. local shear	0.31	1	0.3
J5 torsion	Max. local shear	0.31	0.5	0.2
S1 torsion[f]	Max. local shear	0.11	25	2.8
Drive system	(see Section 13.4.1e)			4.0
Total				<40.0

[a] Values of $(v_2/v_1)^2$, as in Eq. (13.9).
[b] Tube bending: manipulator structure modeled as a tube of modulus $E = 10^{12}$ N/m^2, length = 100 nm, inner radius = 11.5 nm and outer radius = 15 nm (based on excluded volume), with a 0.1 nm surface correction for structural vs. excluded volume (Section 9.4.2).
[c] Rocking (joints J1, J2, J3, J4a, J4b, J5): compliance is measured with respect to the maximum stretching deformation across the joint (stretching varies from positive to negative around each ring). A bearing interface patterned on Figure 10.17, with a width of 2 nm and a radius of 13 nm, provides ~650 close interatomic contacts. Assuming a mean stiffness of 10 N/m per contact, the rocking compliance is ~0.00003 m/N. An additional compliance of ~0.00002 m/N results from the joint-associated thinning of the tube wall thickness. Each multiplier is the square of the ratio of the distance to tip and the radius of joint.
[d] Rocking (joint J6): As for other joints, but with a correction for smaller radius.
[e] Torsion (joints J1, J2, J3, J5): In these joints, torsional deformations resulting from shear across the threaded torus of the drive mechanism can result in tip motions (with differing lever arms). Threaded toruses with circumference = 80 nm and minor radius = 1 nm can have ~640 close interatomic contacts with threads on each of the two adjacent structural (or base) segments. A mean stiffness of 10 N/m per contact yields the stated compliance for the two interfaces taken together.
[f] Torsion (segment S1): Like J1 torsion, this contributes to tip motion. This segment is modeled as a tube (with dimensions as above) of length 15 nm and shear modulus $G = 5 \times 10^{11}$ N/m^2.

speed of the driven segment. (Toroidal worm drives of this size can provide ratios >3000.) In turning a tube segment through one full rotation, even the J3 and J5 drive shafts undergo >700 rotations.

For J1, no flexible drive rods are needed, hence the drive system compliance can be limited to a small value. Taking 0.1 m/N as a readily achievable target, the contribution to the tip compliance can be limited to ≤0.001 m/N with a speed-reduction ratio of 50. Drive system contributions to transverse compliance then total ≤0.004 m/N.

The total arm compliance estimated in Table 13.1 is 0.04 m/N, corresponding to a stiffness of 25 N/m. This permits the use of moderately compliant work-pieces and reagent moieties (~0.1 m/N) while keeping error rates $\leq 10^{-15}$ in directing reactions to chemically equivalent sites separated by a single bond length (Section 8.5.5a). This stiffness also permits the application of 1 nN forces with an elastic deflection in the arm of only 0.04 nm. (Elastic deflection under a planned load need not contribute to positional error.)

f. Speed, productivity, and magnitude of power dissipation.

To estimate the power dissipation of the manipulator system on a per-operation basis re-quires a model of a typical motion, which in turn requires consideration of a typical sequence of operations in the synthesis of diamondoid structures. To build such a model, let us assume a synthetic strategy based on the transfer of small moieties to a surface. The characteristic scale length of required motions can be taken as the 0.25 nm spacing of the bonds perpendicular to a diamond (111) surface. Efficiency favors small motions between operations. Consider an operation at a typical point (a), with subsequent operations required both at a set of ~6 neighboring points (b) and at a set of more distant points (c). After completing an operation at point (a), a large motion would be required if and only if synthetic constraints somehow demanded that each of the operations in set (b) be delayed until after executing an operation at a point (d) in set (c). Save in rare or contrived circumstances, this somewhat unusual constraint could arise only if the point (d) were adjacent to a point in set (b), thereby permitting a substantial electronic or steric interaction between them. Accordingly, the re-quired motion distance should often be one scale-length (0.25 nm), occasionally be two (0.5 nm), and only rarely be greater. Treating 0.5 nm as a typical dis-placement distance between operations should yield conservative estimates of energy dissipation.

The present arm design, however, causes substantial tip displacement in the course of executing a rotation in joints J3 and J5, which are provided chiefly to control tip orientation. Assuming that such motions are commonplace and re-sult in tip displacements of ~10 nm, it is conservative to assume that joints J1 and J2 will commonly have to execute motions able to cause comparable but compensating tip displacements. If such motions are performed in $\sim 10^{-6}$ s, the speed of the fastest sliding motion (that of the J2 drive ring) is ~1 m/s. Assuming that phonon drag at sliding interfaces (which scales with speed squared) is the dominant power-dissipation mechanism, motion in this joint can be expected to account for a substantial fraction of the total power dissipated.

To limit shear displacements of a threaded drive ring with respect to the tri-ple-threaded torus of the worm drive, an interfacial shear stiffness of ~100 N/m is ample; the interfacial compressive stiffness can be several times this value. The corresponding values of the phonon transmission coefficient, T_{trans} are $\sim 10^{-4}$ [from Eq. (7.41), assuming diamondlike materials]. Applying assump-tions like those used in the bearing analyses of Section 10.4.6 to a sliding-inter-face area of ~200 nm^2 yields an estimated energy dissipation per motion of ~10 maJ. An estimate of ~100 maJ per motion for the arm as a whole therefore

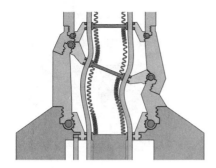

Figure 13.15.   A stiffer set of base joints for a manipulator arm.

appears generous. Note, however, that this estimate does not include energy dissipated in the power and control system, and is rather crude.

If an arm emplaces ~1 atom per motion, a dissipation of ~100 maJ is significant on the chemical energy scale. If an arm emplaces a cluster of ~2 nm size, containing ~1000 atoms, the energy dissipated per atom is a trivial ~0.1 maJ.

These calculations indicate that arms performing 10^6 operations per second need not incur excessive penalties from friction (although the energy cost is large compared to that of a mill-style mechanism). The total number of atoms in a mechanism of this size is ~5×10^6 (neglecting the base, power and control system, etc.); accordingly, the time required for a mechanism of this type to perform the motions necessary to build a product structure of similar mass and complexity is ~5 s. This suggests that the use of manipulator mechanisms need not preclude high overall system productivity.

g.   Toward better designs.   The design illustrated by Figure 13.11 is far from optimal. Although maximizing stiffness is a major design objective, and the major sources of compliance are in the region from joints J1 to J3, these joints and tube segments are no stiffer than the less critical structures further along the arm. A design based on a Stewart platform (Figures 16.4 and 16.5 illustrate the general structure and kinematics) might provide superior stiffness.

Minimizing energy dissipation is another major design objective. Increasing joint stiffness (as in Figure 13.15) and using several drive shafts for critical joints can lower the magnitude of required speed-reduction ratios, decreasing the sliding speeds in interfaces and lowering dissipation.

Other major improvements can doubtless be made. The present design exercise merely shows that suboptimal designs are adequate to enable reliable, reasonably efficient molecular manipulation.

13.4.2.   Self-aligning tips and compliant manipulators

A stiff manipulator can be used to apply small, simple reagent devices to workpieces without any special adaptation to the workpiece structure (aside from choice of moiety, position, orientation, force, etc.). At the cost of greater tip complexity, reactive encounters can exploit nonbonded interactions to achieve reliable alignment, permitting substantially greater arm compliance while maintaining high reaction reliability.

The intermediate states in the construction of a diamondoid structure can be limited to surface irregularities of atomic height on a nanometer length scale in the vicinity of each reaction site. To perform a reaction at such a site, a tip can be

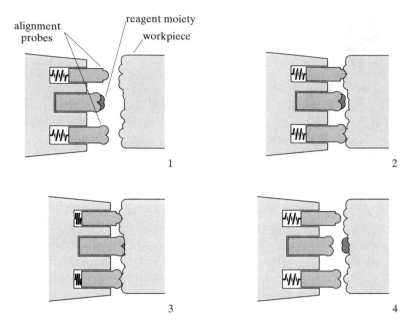

Figure 13.16. Schematic diagram showing alignment of a reagent moiety with respect to a workpiece aided by sterically complementary probes. The diagrams illustrate (1) a reagent device approaching the workpiece, (2) probes in contact with distinctive features (bumps and hollows) of the workpiece, bringing the reagent device into alignment before contact, (3) the reactive encounter, and (4) separation after reaction and transfer of a moiety. Note that the number of probes can be indefinitely multiplied, and that each complementary contact can increase alignment stiffness and specificity.

configured to have regions of specific steric complementarity with the workpiece surface, bringing the tip into stiff alignment (compliance ≤ 0.01 m/N) with the surface before the final reactive encounter (Figure 13.16). The required control of tip geometry can be achieved either by adding adjustable probes to the end of the arm (e.g., configured by geared mechanisms driven by rotating shafts), or by attaching probe structures of suitable geometry (selected from a diverse set) to the reagent device beforehand. Multiple-position detent mechanisms (Section 10.9.2) can be exploited to provide the necessary steric diversity.

13.4.3. Error rates and sensitivities

Manipulator-style mechanisms will tend to have lower stiffness and hence higher error rates than those of mill-style mechanisms. Nonetheless, the feasible stiffness permits errors stemming from thermal excitation to be kept to negligible levels ($\ll 10^{-15}$), given suitable choices of reagent, reaction site, and encounter geometry (Sections 8.3.3f and 13.4.1e). Self-aligning tips (Section 13.4.2) can increase stiffness, broadening the range of arm structures and reaction conditions that are compatible with low error rates.

Manipulators under programmable control can more easily be designed for fault tolerance than can mill systems. With the ability to vary the sequence of reagent devices and motions, and with the ability to make these sequences contingent on the results of measurements of the workpiece structure, it is feasible to continue operations after discarding damaged workpieces and tools. In some instances, flaws could be corrected, but this capability is not assumed in the following analyses.

13.4.4. Larger manipulator mechanisms

Within the size range where gravitational effects are negligible,* the design of manipulator arms becomes easier with increasing size. Assuming motions of constant speed (as in Section 2.3.2), structural stiffness scales with the characteristic dimension L, and joint stiffnesses and power dissipations scale as L^2. Since the mass being delivered commonly scales as L^3, and the time for its delivery scales as L, the energy per unit mass delivered is a constant. In practice, some of the increase in stiffness resulting from larger scale can be sacrificed in exchange for decreased friction, decreasing energy dissipated per unit mass. This is consistent with industrial experience, in which the energy required to move a macroscopic object over a short distance is trivial in comparison with its thermal or chemical energy content. Accordingly, the feasibility of stiff manipulator arms of sizes intermediate between 100 nm and macroscopic dimensions is assumed in the following sections without presenting additional design and analysis.

13.5. Conclusions

Nanomechanical systems can be used to acquire feedstock molecules from solution, sorting and ordering them to provide a flow of input materials to essentially deterministic molecule-processing systems. Exploiting mechanochemical processes, these systems can convert feedstock molecules into reactive moieties of the sorts discussed in Chapter 8. These moieties, in turn, can be applied to workpieces in complex patterns to build up complex structures.

Each of these operations can be made sufficiently tolerant of thermal excitation that damage to manufacturing mechanisms and product structures occurs chiefly as a result of ambient ionizing radiation. Molecular sorting operations can be quite efficient, with energy dissipation <1 maJ per molecule. Assembly operations using mills can be energetically efficient, with the energy dissipated in many operations being <1 maJ, and an estimated mean dissipation for a mix of efficient and inefficient operations being ~30 maJ per operation. The efficiency of manipulator motions is expected to depend heavily on the energy dissipated by their control mechanisms.

Each system has been characterized at a frequency of 10^6 operations per second. At this frequency, the sorting, processing, and manipulation subsystems each process a quantity of reagent moieties equaling its own mass within a few seconds. The next chapter describes how fast, reliable processing units like these can be combined with larger-scale assembly systems to build molecular manufacturing systems capable of delivering macroscopic products.

Some open problems.   As in the previous three chapters, the task of expanding the range of well-characterized, atomically detailed models for devices defines a large set of open problems. Modulated receptors can readily be designed within

---

* Note that the arm discussed in this Section 13.4 can easily apply 1 nN forces, enabling it to lift 10^{-10} kg against terrestrial gravity. This is ~10^9 times the mass of the arm itself.

the constraints of existing software and computers. A more detailed under-standing of purification processes could be developed by choosing a small feed-stock molecule, compiling an exhaustive list of similar molecules of smaller size, and characterizing the interactions of each with feedstock receptor structures of several designs.

Many of the problems of greatest interest in mill and manipulator design de-pend on the prior characterization of a compatible set of mechanochemical pro-cesses (as discussed at the close of Chapter 8); the discussion in this chapter highlights the interest in identifying sets of nearly thermodynamically reversible processes that can combine fuel and oxygen molecules. Various mechanical components of mill and manipulator systems (providing transport, power sup-ply, alignment, error detection) are independent of the specifics of the mecha-nochemical process being performed, hence their detailed design can proceed with existing knowledge (subject, as always, to limitations in the accuracy of mo-lecular mechanics methods).

14

Molecular
Manufacturing Systems

14.1. Overview

To bridge the gap between processing molecules and delivering macroscopic products, several issues must be considered. These include:

- Large-scale assembly operations
- Delivery to an external environment
- Cooling requirements
- Information sources
- Control mechanisms
- System architectures

Section 14.2 describes assembly operations at intermediate scales. Section 14.3 examines issues of system architecture, including overall organization, delivery of products, and reliability and redundancy. Section 14.4 puts these together to describe a system capable of manufacturing kilogram-scale product objects with a wide range of structures, including further manufacturing systems. Section 14.5 compares and contrasts present manufacturing technologies to molecular manufacturing technologies, considering their feedstocks, products, by-products, energy consumption, productivity, quality, and cost. Finally, Section 14.6 examines the complexity of macroscopic products of molecular manufacturing, including the manufacturing systems themselves, drawing on experience with complexity in computer design and software.

14.2. Assembly operations at intermediate scales

Several considerations (assembly speed, the mass of products in transit, the size of assembly units in fault-tolerant architectures) favor the use of convergent assembly over moiety-by-moiety assembly in making product structures of substantial size and complexity. In convergent assembly, small parts (initially individual transferable moieties) are combined to make larger parts, and these to make still larger parts, in a hierarchically organized process. Convergent assembly is familiar in macroscale manufacturing processes as well: in making an automobile, parts are combined to make engines before engines are joined to the chassis; electronic components are combined to make circuit boards before boards are installed into radios, and radios are assembled before being joined to the dashboard.

Section 14.4 will describe a flexible manufacturing architecture based on convergent assembly. Given moiety-level assembly, the chief physical questions raised by this approach involve assembly at intermediate scales.

14.2.1. Joining building blocks

a. Mechanisms for positioning and assembly.   Both mill-style and manipulator-style mechanisms can be used to transport and place building blocks of intermediate size. Section 13.4.4 discusses manipulators of sizes intermediate between the 100 nm and macroscopic scales; systems of belts, rollers, and cams can also be built at intermediate scales, again providing increases in mechanical stiffness.

The scale of these mechanisms depends on the scale of the parts being handled and of the product being assembled. In mill-style mechanisms, the transport system components must be scaled to the input parts. In manipulator-style mechanisms, the range of motion and hence the size of the device must be scaled to the output product.

Some useful assembly operations at intermediate scales may be quite specialized, such as stretching the sleeve of a grooved sleeve bearing in order to fit it over a shaft. Although specialized mechanisms are not explicitly considered here, they are likely to be commonplace in practice.

b. Interfaces between blocks.   The assembly of blocks to form larger structures spans a continuum of sizes between cluster-based synthesis strategies (Section 8.6.6) and joining of macroscopic blocks. At sufficiently large scales (e.g., ~100 nm) the use of snaps, screws, and the like is feasible. On all scales above the smallest, adhesive interfaces of the sort described in Section 9.7 are applicable. As noted, some adhesive interfaces can join to form a dense, continuous diamondoid structure (or diamond itself). As a consequence, designing systems to be built as compositions of many small parts need not entail any great sacrifice in structural strength and stiffness.

Systems of molecular machinery will commonly consist of many distinct, free-standing parts, joined by interlocking geometries rather than by covalent bonds. Systems of this sort can be built by methods resembling conventional robotic assembly, using grippers that do not bond to the part being moved. Surface forces (e.g., van der Waals attraction) must be considered in designing grippers and assembly sequences, but these present no great difficulty where diamondoid structures with nonreactive surfaces are concerned, given assembly adhesion. (The construction of free-standing parts can be accomplished by building up a workpiece on a substrate in a manner that allows clean cleavage at the workpiece-substrate interface; the cleaved surface can then be finished or extended by further mechanosynthetic operations. The forces required for cleavage and subsequent manipulation can be provided by nonbonding grippers.)

c. Energy efficiency.   The energy efficiency of intermediate-scale assembly can be high, on a per-atom basis. In moving large blocks, internal friction in positioning mechanisms is amortized over many atoms. A typical upper bound on the energy dissipated in joining surfaces is comparable to the surface energy, which for a large, thick block is small on a per-atom basis.

Energy dissipation is accordingly dominated by assembly of the smallest blocks. These can be approximated as one-nanometer cubes (consistent with the system description in Section 14.4). If the energy dissipated in joining surfaces were equal to the theoretical surface energy of diamond [~5 J/m^2 (Field, 1979)], this would amount to ~9×10^6 J/kg. It is reasonable to expect, however, that much or almost all of the potential energy of bonding can be recovered as mechanical work, reducing dissipation to a small fraction of this value. If adhesive interfaces were designed such that the exoergicity of C—C bond formation matches that in the polymerization of ethene, then the exoergicity of assembling a diamondoid solid from cubic nanometer blocks would be ~1×10^6 J/kg. This value probably overestimates the total energy dissipated in block assembly in a well-designed system producing a mix of diamondoid products; the estimate used here is 5×10^5 J/kg.

14.2.2. Reliability issues

In the assembly of intermediate-scale building blocks, ample exoergicity for stable, reliable assembly is no problem; even van der Waals attractions yield >150 maJ for typical interfaces of ≥ 1 nm^2. Further, the greater stiffness of larger structures makes it increasingly easy to limit positional uncertainty resulting from thermal excitation to low, acceptable values. The special problems arising on a larger scale are greater sensitivity to nonthermal vibration resulting from an increased ratio of mass to stiffness, and greater radiation damage rates resulting from increased mass directly.

a. Vibration and alignment. Under a fixed acceleration, inertial loads cause elastic deformations that scale in proportion to the square of the characteristic length of a system. Oscillating accelerations (vibrations) produce dynamic deformations that depend on resonant frequencies and damping. For perspective, a manipulator 10 cm long and scaled from the device described in Section 13.4 has a length and stiffness greater by a factor of 10^6 and a mass greater by a factor of 10^{18}. It deforms by ~10 nm under terrestrial gravity, and its lowest resonant frequency is >10^4 Hz.

Various combinations of (1) assembly tolerances, (2) measurement and control, and (3) precomputed corrections can compensate for static deformation. Various combinations of (1), (2), and mechanical isolation can correct for vibration, under ordinary circumstances. A surprise in the development of scanning tunneling microscope technology was the feasibility of achieving atomic-scale positional stability in centimeter-scale mechanisms mounted on ordinary laboratory tables, using 10-cm-scale vibration isolation systems consisting of ordinary elastomers and metal plates. Engineering molecular manufacturing systems for isolation from external vibrations and for control of internal vibrations should be feasible.

At larger scales, assembly tolerances can be increased without sacrificing precise control of the structure and alignment of the resulting interfaces. Figure 14.1 illustrates an adhesive interface structure in which nonbonded contacts occur as tapered pegs are inserted into matching holes. Tapering can provide substantial tolerance for initial misalignment, yet as the adhesive surfaces themselves approach contact, a stiff alignment can be achieved. As shown in the

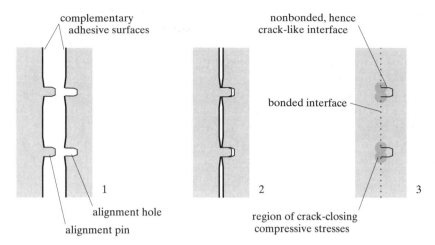

Figure 14.1. An adhesive interface with tapered alignment pegs and a geometry that prevents residual nonbonded interfaces from serving as crack initiation sites.

illustration, interfacial bonding can proceed outward from the alignment pegs, and the structure around the edge of the nonbonded contact surface can be placed in compression to inhibit crack initiation.

b. Radiation damage: beyond the single-point failure assumption. Typical nanometer-scale machine components have many locations where a point defect induced by ionizing radiation can cause failure; the default assumption throughout this work has been that a defect at any point will cause failure. Macroscale machines, however, function despite numerous defects. At some intermediate scale, machines can (and must) be designed such that a significant density of point defects is tolerable. This scale can be estimated by considering the effects of radiation damage on two basic characteristics of mechanical parts: their shapes and the properties of their interfaces. (Strength, stiffness, and the like are not at issue in the typical case.)

Applying an elastic continuum model, the effect of a point defect on the shape of a component can be described in terms of the resulting strain field. This is in general complex, but its longest-range component results from the local change in volume. Modeling the medium as homogeneous, isotropic, and elastic, the resulting radial strain Δr falls off as r^{-2}. An atomic-scale defect causes a strain of $\Delta r < 0.2$ nm at $r = 0.2$ nm, and hence a strain $\Delta r < 0.002$ nm at 2 nm, and so forth. Since geometrical errors of ~0.01 nm are tolerable in most mechanosynthetic systems, components having features of multinanometer or larger scale should usually prove able to tolerate radiation-induced defects in their interiors without failure. Large structural components with damage-sensitive interfaces (such as the segments of the stiff arm described in Section 13.4) can have substantial damage-tolerant interior and surface regions, in effect reducing their target sizes relative to the model described in Section 6.6.2.

Similar principles apply to misalignments caused by damage to bearing interfaces, although the nonlinearity of nonbonded interactions tends to increase the characteristic length scales. Moreover, damage occurring at a mobile interface can in some instances produce large static friction or adhesion, causing device failure. Bearings can be made damage tolerant by providing internal redundancy. For example, the surfaces of a sleeve bearing can be separated by multiple independent layers, each subdivided into parallel strips. A structure of this sort can tolerate several adhesion points among adjacent surfaces, strips, and layers without creating a connection between the sliding surfaces. Designs of this sort typically require interfacial thicknesses of several nanometers, which are geometrically compatible only with moving parts on a scale of tens of nanometers or more. Thus, systems large enough to employ internally redundant bearing interface structures commonly have structural components large enough to maintain good geometrical stability in the presence of point defects.

In a mill-style manufacturing process without fault-correction, a failed assembly operation can cause failure of a relatively large subsystem. The estimates made in this section regarding the scale at which machine components can tolerate point defects without failure suggest that assembly processes can tolerate damage in the parts themselves when these are of a similar scale. In the presence of point defects, assembly can proceed without disrupting the process; whether the resulting product is functional is a separate question. Accordingly, it will be assumed that when the scale of all components and building blocks in a manufacturing process is >10 nm, a moderate density of point defects can be tolerated without causing failure of the process.

14.3. Architectural issues

The previous sections of this chapter have chiefly described devices, subsystems, and operations, covering molecular sorting and orientation, transformation of molecules into reagent moieties, the application of reagent moieties to build small structures, and the hierarchical assembly of smaller structures into progressively larger structures. The present section outlines an architecture for molecular manufacturing systems capable of producing a wide range of macroscopic products and delivering them to an external environment. This requires attention to the spatial organization of material and energy flows, and to such system-level concerns as overall reliability and operational lifetime. These are addressed in substantial detail in the description of an exemplar architecture.

14.3.1. Combining parts to make large systems

Present manufacturing practice provides no precedent for processes that combine $\sim 10^{25}$ distinct parts to form a single object. A naive approach would use a mechanism to assemble parts one at a time; at 10^6 per second, however, assembling 10^{25} parts would take a time longer than standard estimates of the age of the universe. Viable approaches must coordinate many mechanisms in building a single object.

a. Construction-style assembly processes. A construction-style process uses small devices to work on or in a large structure. For example, many assembly mechanisms can work in parallel to build up a surface. If a product structure is

0.1 m thick, and each assembly mechanism occupies an area of $(100 \text{ nm})^2$, then an assembly rate of 10^6 small moieties per second implies a $\sim 10^8$ s construction time, more than a year. This strategy might be attractive for designs that lack thick solid regions: various porous, fibrous, foamlike, or honeycombed structures meet this criterion and have attractive mechanical properties.

Alternatively, assembly can proceed at higher rates through the use of larger building blocks. Keeping other parameters constant, increasing the block size to ~ 10 nm decreases the assembly time to 1000 s. Blocks of sufficient size and complexity (e.g., containing a computer, motor, and actuators) could be made self-assembling, although at a substantial penalty in properties such as strength-to-density ratio.

Some products can doubtless be assembled in a poorly controlled environment (e.g., in a solvent bath), simplifying problems of heat and mass transport, and of environmental control. Alternatively, eutactic environments of almost any desired size can be constructed by expanding a gas-tight barrier in the manner shown in Figure 14.2 or Figure 14.3. This style of assembly will not be considered further here.

b. *Manufacturing-style assembly processes.* In manufacturing-style processes, parts are manipulated and transported within larger mechanisms. If a construction-style process resembles the assembly of a building, a manufacturing-style process resembles the assembly of a computer. Processes of this sort will be described in greater depth.

The architecture described in Section 14.4 uses a convergent assembly sequence in which each structure is built from components within an order of magnitude or so of the structure's own linear dimensions. In one class of convergent assembly processes, the motion of components traces a tree in space: the trunk corresponds to the path traced by the final workpiece as the final components are assembled, the branches correspond to the paths traced by those components as they are assembled, and so forth. A simplified model (Figure 14.4) demonstrates that convergent assembly can be distributed in space in a manner that (1) provides an assembly volume proportional in size to the workpiece at each stage, and (2) requires only short-range transportation of parts between stages.

A system of the sort suggested by Figure 14.4 must operate all assembly stations at the same frequency, thereby slowing the smallest mechanisms to the frequencies of the largest. Varying the number of assembly operations per level in the hierarchy can alleviate this problem, as can allowing a small number of high-frequency processes to supply parts to a larger number of low-frequency processes. Both of these strategies are incorporated into the exemplar architecture summarized in Section 14.4.

The boundary between high-frequency and low-frequency processes in the exemplar architecture corresponds to a boundary between mill-style and manipulator-style assembly mechanisms. This, in turn, marks the boundary between purely repetitive operations producing standard building blocks and programmable operations that can stack these building blocks to make a wide variety of products. Although manipulators can handle blocks of ~ 10 nm scale with good energy efficiency, the exemplar architecture uses mill-style mechanisms up to a 1μ scale. This sacrifices flexibility, but the number of distinct kinds

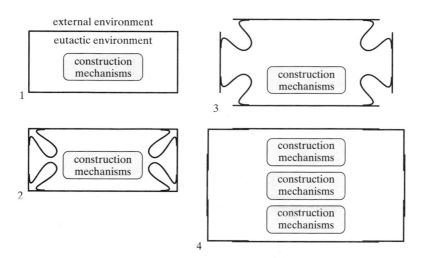

Figure 14.2.   Expansion of a eutactic environment by construction mechanisms operating inside it (ports for transfer of materials across the boundary not shown). In three
dimensions, the arrangement of folds at the corners where three edges meet is more
complex than suggested in steps 2 or 3.

of 1 μ blocks can be quite large. A 1 kg structure contains $\sim 10^{15}$ blocks. To manufacture them, the exemplar architecture uses $\sim 10^6$ separate systems, each
capable of producing a different structure. The demand for large numbers of a
few kinds tends to reduce diversity, but the option of running some lines at low
capacity (or only for producing certain products) substantially offsets this. The

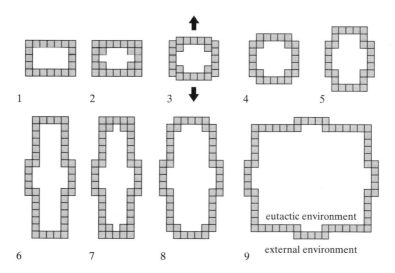

Figure 14.3.   Expansion of a eutactic environment as in Figure 14.2, using sliding
blocks rather than flexible pleats. This approach is based on the presumption that adequate seals can be maintained at all the junctions among the sliding blocks. Again,
geometries become more complex with expansion in three dimensions, but local block-
sliding relationships are unaltered. Note that the "blocks" could in principle be atoms
of a metallic material, with sliding occurring by dislocation movement.

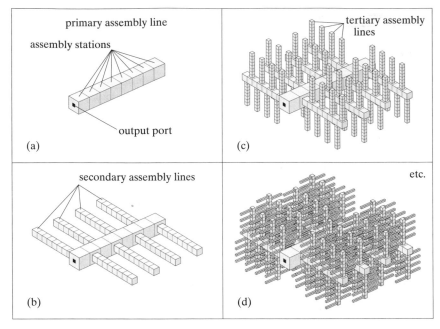

Figure 14.4. A simple model of a spatial arrangement for a hierarchical, convergent assembly process; panels (a), (b), (c), and (d) provide successively more detailed diagrams. In (a), a primary assembly line consisting of a series of 8 assembly stations (drawn as cubes) performs the final 8 assembly operations in a hypothetical manufacturing process. In (b), 8 secondary assembly lines provide parts to the final lines; (c) and (d) illustrate tertiary and quaternary assembly lines. Since each level of lines contains an equal volume, this pattern cannot be indefinitely extended without self-intersection (the maximum radius of expansion is bounded). With local rearrangements to postpone self-intersection until the available volume is nearly full, a branching pattern of this sort can be extended though >30 generations, enabling the assembly of objects from $>10^{27}$ pieces. This structure demonstrates that certain geometrical constraints can be met, but does not represent a proposed system.

range of systems that can be constructed from a catalogue of $\sim 10^6$ kinds of parts is large, particularly when the parts can be joined to make continuous volumes of high strength materials, and are each smaller than the usual tolerances in conventional manufacturing.

14.3.2. Delivering products to an external environment

In typical applications, the products of molecular manufacturing must be delivered to an external environment without permitting back contamination of the eutactic internal environment. Sections 11.4.2 and 11.4.3 discuss vacuum pumps and the use of sliding interfaces as seals; from these elements, various means can be devised for product delivery.

For example, a product object can be placed in a box bearing multiple external sealing rings and followed by a similarly ringed piston. The rings can provide a tight seal against the inside of an exit tube, and the piston can block the exit until another product-bearing box is ready for departure. Moderate failures in the seals could be rendered harmless by scavenging pumps operating on the spaces between them. Schemes that lack discardable boxes, pistons, and the like

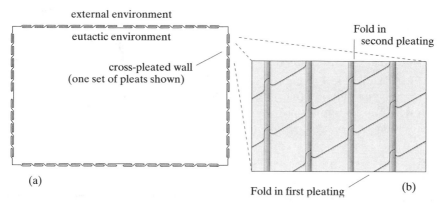

Figure 14.5. Schematic illustration of a gas-tight, expandable, doubly pleated enclosure for a eutactic environment (a). In three dimensions, a structure like that illustrated could undergo a volumetric expansion by a factor of ~27, unrolling one band of pleats at a time. Note that each face must have pleats running in two perpendicular directions, as shown in (b), an angled view of the surface of a doubly pleated enclosure. To satisfy constraints on elastic strain, the minimum radius of curvature in the first set of pleats must be several times the thickness of the gas-tight wall, and the radius of curvature in the second, superimposed set of pleats must be a further multiple of the radius of the first. Accordingly, the enclosure size must be orders of magnitude greater than the wall thickness, making this approach inapplicable to very small enclosures. Note that the wall system can include structural members and sturdy barriers both inside and outside of the comparatively delicate gas-tight wall.

can also be devised, with varying volume requirements, constraints on product object shape, and so forth. These will not be pursued further at present. Figures 14.5 and 14.6 illustrate how delivered products can be larger than the system that produced them.

14.3.3. Redundancy, reliability, and system lifetimes

A system of macroscopic scale and significant lifetime must tolerate localized, randomly distributed damage, which in nanomechanical systems will commonly cause failure of devices and subsystems. There is a large and diverse literature on fault-tolerant design for computers and aerospace systems, and similar approaches are applicable in many nanomechanical systems.

With a judicious choice of structures and reactions, damage to a molecular manufacturing system at ordinary temperatures is dominated by damage caused

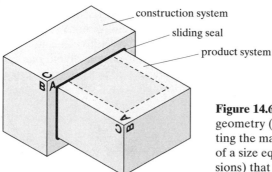

Figure 14.6. Schematic illustration of a geometry (suggested by R. Merkle) permitting the manufacture and delivery of objects of a size equaling or exceeding (in all dimensions) that of the manufacturing system.

by ionizing radiation. Accordingly, Eq. (6.54) can be taken as a damage model. Using this model, Section 13.3.6c concludes that mechanosynthetic units with a mass of $\leq 2 \times 10^{-18}$ kg have a probability of failure $\leq .01$ in a 10-year period in a typical terrestrial environment. The early sets of mill units described in Section 14.4 have overall operating frequencies of $\sim 10^6 \, s^{-1}$, but divided among 100 units in each set, yielding an operating frequency of $10^4 \, s^{-1}$ per unit. Accordingly, if the failure rate of a mill unit is dominated by a step with a failure rate of 10^{-15} (see Section 13.3.6), this will contribute $\sim .003$ to the probability of failure in a 10-year period.*

Section 13.3.6b shows that mechanosynthetic units can be built to exhibit fail-stop behavior, reliably ceasing operation rather than delivering damaged products. Section 13.3.2b outlines the characteristics of transportation components capable of transferring structures from belts to pallets, moving pallets along tracks through fair, nonblocking merging and distribution junctions, and transferring structures back to belts again. Figure 14.7 shows how fail-stop units

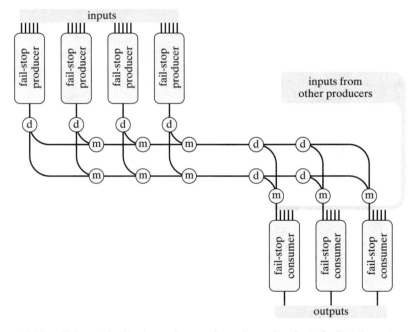

Figure 14.7. Schematic diagram of a portion of a redundant, fault-tolerant manufacturing architecture. The upper set of identical, fail-stop producers transforms inputs into intermediate products that become inputs to a set of identical, fail-stop consumers. The intermediate transportation network incorporates fair, nonblocking merging junctions (m) and distribution junctions (d). A system with this structure can continue to operate despite the failure of any component, including the failure of all but one producer and all but one consumer. Within the pattern illustrated, the degree of redundancy of producers, consumers, and transportation elements can be increased arbitrarily.

---

* Damage to a large fraction of units in a set increases the operation and damage rates of the remaining units, but this makes a negligible adverse contribution to overall system reliability.

can be joined by redundant tracks and junctions in such a manner that any single unit, track segment, or junction can fail without interrupting the transformation of inputs into outputs. In particular, the system can continue to work without error so long as one unit in any set remains operational. (The presence of redundant units decreases the operational frequency of each, decreasing energy dissipation per operation.)

In the typical case, failure rates are dominated by failures in the mechanosynthetic units rather than in the transportation network. In this case, the probability P_{syst} that a system remains operational is related to the probability P_{unit} that a unit remains operational by the expression

$$P_{\text{syst}} = \left[1-(1-P_{\text{unit}})^{N_{\text{unit}}}\right]^{N_{\text{set}}} \tag{14.1}$$

where N_{unit} and N_{set} are the number of units in a set and sets in a system, respectively. (Unit failures are assumed to be independent; substantial spatial separations can be provided.) A more readily evaluated expression is

$$P_{\text{syst}} \approx \exp\left[-N_{\text{set}}(1-P_{\text{unit}})^{N_{\text{unit}}}\right], \quad \text{where} \quad P_{\text{unit}} \approx 1. \tag{14.2}$$

These relationships correspond to Eqs. (6.58) and (6.59).

14.4. An exemplar manufacturing-system architecture

14.4.1. General approach

This section and Table 14.1 outline an architecture for a system capable of manufacturing macroscopic objects. The subsystem capacities are chosen to permit the conversion of a feedstock solution consisting of small organic molecules into ~ 1 kg product objects of ~ 0.2 m dimensions in a cycle time of ~ 1 hour. The flow of materials proceeds through molecule sorting and orientation (Section 13.2), preparation of reagent moieties (Section 13.3), several stages of convergent assembly using mill-style mechanisms (Sections 13.3 and 14.2), and several stages of convergent assembly using manipulator-style mechanisms (Sections 13.4 and 14.2).

This section focuses on mechanosynthesis and assembly, omitting the details of supporting systems. For example, although they would be essential in a detailed design, the specifics of a ~ 1 kW cooling system, a ~ 0.001 kg/s feedstock solution supply system, transportation and sealing mechanisms for product delivery, and so forth are peripheral to the central issues of molecular manufacturing. Figure 14.4 shows that convergent assembly processes can be performed without undue geometrical problems in material flow; accordingly, a reasonable estimate of overall system volume can be had by summing the volumes of the assembly workspaces without describing a particular three-dimensional layout. Maintenance of vacuum integrity has been discussed in Sections 11.4.2, 11.4.3, and 14.3.2. Compressive structures made of diamondlike materials can support terrestrial atmospheric pressures with masses of $\ll 1$ kg/m^3, and can be designed in many ways. Power conversion between mechanical and electrical forms can be performed with high efficiency and (for reasonable power levels) with negligible mass and volume (Section 11.7), imposing few constraints in this context.

Table 14.1. Manufacturing system parameters. See Section 14.4 for description.

Level	Number of sets	Number of units per set[a]	Number of inputs per product	Unit operating frequencies (Hz)	Product scale (nm)	Mechanism scale (nm)[b]	Unit radiation-sensitive mass (fg)[c]	Total mass of mechanisms (kg)
Input ordering[d]	1	10^{17}	~1	10^6	0.5	20	0.02	0.002
Reagent prep.[e]	1	10^{17}	~1	10^6	0.5	20	0.06	0.006
Stage 1 mill[f]	10^{15}	100	100	$\leq 10^6$	1	20	0.60	0.06
Stage 2 mill	10^{13}	100	100	$\leq 10^6$	5	20	0.60	<0.01
Stage 3 mill[g]	10^{11}	100	100	$\leq 10^6$	20	100	0.50	<0.01
Stage 4 mill[h]	10^9	100	100	10^4	100	500	–	<0.001
Stage 5 mill	10^6	100	1000	10^4	1000	5000	–	<0.001
Stage 1 manip.[i]	10^9	2	10^6	•500	10^5	2×10^5	–	<0.001
Stage 2 manip.	10^4	2	10^5	50	5×10^6	10^7	–	<0.001
Stage 3 manip.	1	2	10^4	5	10^8	2×10^8	–	0.020

[a] All numbers and frequencies are compatible with a ~1 gm/s throughput; within each stage, the combination of radiation-sensitive mass and number of units per redundant set is chosen to permit long operational lifetimes.

[b] Mechanism scale describes (e.g.) the size of a single rotor in a mechanosynthetic device, not of a multi-mechanism unit.

[c] The radiation-sensitive masses of units of each kind are stated in femtograms (1 fg = 10^{-18} kg).

[d] The mass of the input-ordering subsystem is estimated from Section 13.2.2.

[e] The mass of the reagent preparation subsystem is based on Section 13.3.5.

[f] Stage 1, 2 mill units: the mass per encounter mechanism is assumed to be the same as in reagent preparation.

[g] Stage 3 mill units: the moving parts are assumed to be large enough to permit designs tolerant of radiation damage (Section 14.2.2b), and the radiation-sensitive mass is that of product structures in transit through the unit.

[h] Stage 4, 5 mill units: the moving parts and product structures are assumed to be damage tolerant; the number of units per set is large chiefly to permit lower frequencies and speeds.

[i] Manipulator units: the arm structural mass is chosen to permit designs with stiffness >100 N/m, and with all bending-mode frequencies >100 times the unit operating frequency; the arm supporting structures are assumed to be 10 times more massive than the arms themselves.

The key issues in a manufacturing architecture of this sort center around the assembly mechanisms themselves: their numbers, kinds, reliabilities, operating frequencies, masses, volumes, and so forth. These issues are addressed in the next section; many parameters are summarized and explained in Table 14.1.

14.4.2. Products, building blocks, and assembly sequences

Using fixed, mill-style subsystems throughout a manufacturing system, from the moietal level to finished products, would limit it to making products of a single kind. Using flexible, manipulator-style subsystems throughout would permit independent control of every detail of a product, permitting an indefinitely large range of structures at all scales, but at the cost of relatively slow, inefficient manufacturing processes. Neither of these extremes is likely to prove desirable.

a. ***Building blocks and product diversity.*** Typical macroscopic products can conveniently be made from modular building blocks, many identical to one another. Micron-scale building blocks are small enough to make almost any macroscopic shape in ordinary use today within better tolerances than those provided by conventional machining. From sets of building blocks that can be assembled to make solids of differing strength, stiffness, conductivity, and so forth, objects can be made that exhibit material properties that equal or better those of present industrial products. Systems that take more direct advantage of nanoscale structures can still use extensively duplicated building blocks. Products containing a macroscopic quantity of computational hardware will typically contain many identical CPUs, memory arrays, and so forth. A product containing a macroscopic quantity of mechanical structure will typically contain many identical regions of structural material. Motors, gearboxes, vacuum pumps, data-bus segments, photovoltaic cells, storage batteries, conductors, struts, connectors, manipulator arms, reagent-processing subsystems—all can be constructed on a micron or submicron scale, and all are candidates for extensive duplication in macroscopic systems.

Accordingly, a wide range of macroscopic products can be made with good efficiency by using mills to make a diverse set of building blocks in the 10^{-7} to 10^{-6} m size range and then using manipulators to assemble them into macroscopic products. Inclusion of a secondary production capability that applies manipulators to smaller building blocks (at the moietal level where necessary), can provide unusual building blocks without requiring a dedicated mill mechanism. The exemplar system in Table 14.1 makes the simplifying assumption of a hand-off from mills to manipulators at a single block size, 10^{-6} m.

b. ***Alternative assembly sequences.*** The quantitative summary of the exemplar architecture assumes a strictly convergent assembly sequence in the mill mechanism: each moiety and each subassembly is destined for a unique product at the 10^{-6} m block level. The reagent preparation mill system, in contrast, produces standard moieties that can be used by any of a wide range of assembly mills. There is good reason to extend this pattern to the subassembly level, thereby permitting flexible allocation of the output of subassembly mill mechanisms to any of several higher-level assembly mills. Doing so enables a system to produce a greater range of 10^{-6} m scale blocks with a lesser increase in total mill system size.

A further assumption in the quantitative summary is that each assembly unit combines blocks of similar sizes to make a much larger structure. There is no reason to treat this as a constraint, and there are substantial advantages to assembling components of widely differing sizes. For example, after assembling two large blocks with an adhesive interface, moietal-scale modification of the edge of the interface region might be necessary to prepare a smooth surface spanning the two blocks. Macroscopic assembly processes frequently add small components to large structures.

c. *Adjustable blocks.* A large number of blocks will serve simple structural functions, holding active components such as motors, gearboxes, sensors, and the like in fixed geometrical relationships to one another. This requires struts of diverse lengths and joints of diverse angles. There is, however, no reason why structural elements of each needed length and angle must be manufactured by a distinct assembly mill, starting at the moietal level. It is straightforward to build struts and joints that contain internal sliding interfaces (Section 10.5) permitting them to be extended or twisted to assume a wide range of lengths or angles, and then twisted or compressed (respectively) to bring internal adhesive interfaces (Section 9.7) into contact, thus producing a stiff structural member with a particular geometry. In this manner, a mill system can make identical, adjustable members that each of several subsequent mills (or manipulators) can transform into one of many fixed, specialized structures.

14.4.3. Throughput, delays, and internal inventories

If all components operated at their stated frequencies, handling components with a density of 1000 kg/m^3, system throughput would be 0.001 kg/s. The five mill stages operate at high frequencies, performing relatively small numbers of operations per output object; the maximum delay for material passing through these five stages is ~ 0.1 s, implying a $\sim 10^{-4}$ kg inventory of materials in process. The three manipulator stages operate at lower frequencies, performing relatively large numbers of operations per output object; each requires 1000 s to complete its operations, and contains a time-average product mass of 0.5 kg. Allowing for feasible overlap in the operating times of the manipulator stages, the delay between process initiation and delivery of a 1 kg product object, starting from a feedstock solution, can be somewhat less than one hour.

14.4.4. Mass and volume

The volume of the first stages—input ordering, reagent preparation, and mill mechanisms—can be estimated by applying an estimated mean density to the tabulated masses. Using 100 kg/m^3 yields a volume of $\sim 1.8 \times 10^{-4} \text{ m}^3$. (System flexibility could be increased by devoting ~ 10 times this mass and volume to mill mechanisms, most idle in a typical production process; other system parameters would remain essentially unchanged.)

The volume of the manipulator stages depends on the size of the product structures, if these are to be totally enclosed in the workspace rather than constructed and extruded incrementally. A wide range of 1 kg products should fit (perhaps in a folded or partially disassembled configuration) within a 0.01 m^3 workspace. If the solid parts of a 1 kg product object are as dense as diamond, then these parts occupy ~0.03 of the workspace volume. Assuming that workspace volumes are nonoverlapping, but can be used for conveyance of parts from other workspace volumes, then the total volume required for the three manipulator stages is ~0.03 m^3. Multiplying by 1.66 to allow for packing inefficiencies, cooling channels, and so forth yields a total estimated volume of ~0.05 m^3.

The total mass of the manufacturing system components listed in Table 14.1 is <0.12 kg. Allowing 1 kg/m^3 for internal compressive structure to support external atmospheric pressure adds ~0.05 kg; allowing 0.1 kg/m^2 to provide for a sturdy case, rubber feet, and the like adds ~0.08 kg. These masses sum to <0.25 kg. Allowing as much as 0.75 kg for other subsystems yields an estimated total system mass <1 kg.

14.4.5. System lifetime

Radiation damage chiefly threatens the small yet relatively specialized devices in the first three stages of the mill system. Earlier stages are massively redundant; later stages use components large enough to permit radiation-insensitive designs (Section 14.2.2b). Applying the previous radiation-damage assumptions, units with a radiation-sensitive mass of $\leq 2 \times 10^{-18}$ kg have P_{unit} = .99 after ten years. The radiation-sensitive portions of mill stages 1–3 have a lower unit mass, with N_{unit} = 10, $N_{set} \approx 10^{15}$, and hence $P_{syst} \approx 1$ after ten years. Increasing N_{unit} to 20 raises the expected lifetime to over a century. All radiation-sensitive units are assumed to be fail-stop (Section 13.3.6) and linked by redundant transport networks (Section 13.3.2b). An enclosure of this size can easily be made UV opaque (Section 6.5.4).

14.4.6. Feedstock materials

A specific manufacturing mechanism will accept raw materials consisting of a specific set of small molecules in solution. Typical products require large quantities of C, moderate quantities of H, N, O, F, Si, P, S, and Cl, and lesser quantities of several other elements. These could be provided by a single solution, or by different solutions supplied through several ports. In designing a system, the choice of input compounds is largely a matter of cost and convenience.

14.4.7. By-products

Extended diamondlike structures have a smaller mass fraction of hydrogen than do typical small organic molecules: small molecules have a larger surface-to-volume ratio, and satisfaction of surface valences is most commonly achieved by inclusion of monovalent atoms, chiefly hydrogen. Fullerenes, cyanogen, perfluoroalkanes, and other unusual molecules are among the exceptions to this pattern. Aside from this, there is no strong reason why the elemental composition of the inputs should differ from that of the product, and hence no strong reason why there should be substantial by-products from a molecular manufacturing system. Excess hydrogen could be delivered as water, in a process that consumes atmospheric oxygen and (presumably) produces mechanical energy as a by-product (Section 13.3.8).

If a mechanism is supplied with a single solution of fixed composition, and must from this produce a variable mix of products, the (nonhydrogen) elemental composition of the products will in general fail to match that of the input solution, and the less-consumed compounds will accumulate. The resulting solutions can be recycled by adding the compounds in which they have become deficient.

14.4.8. Energy output and dissipation

The energy dissipated by mills is dominated by moietal operations, estimated by Section 13.3.7 as 1.5×10^6 J/kg; a further allowance of 5×10^5 J/kg is made for energy dissipated during block assembly (Section 14.2.1c). The mechanical aspects of manipulator operations result in comparatively negligible energy dissipation, as do the operations in molecular sorting and ordering (save for entropic effects included in the overall thermodynamic calculations that follow).

The energy dissipation resulting from computation used to direct manipulator operations is hard to estimate accurately, but can be made quite small in the exemplar architecture. The control of motions can be relatively direct: with compiled instructions, no run-time planning is necessary, hence the computations can be as simple as loading appropriate internal-coordinate settings into registers in controllers. A relatively computation-intensive process (perhaps analogous to the interpretation of a page-description language) might execute $\sim 10^6$ instructions per block placed. Section 12.7.4 derives an estimated energy dissipation of $\sim 10^{-16}$ J per instruction, and a 1 kg object contains $\sim 10^{15}$ cubic-micron blocks; hence executing even 10^6 instructions per block would increase the total energy dissipated by only 10^5 J/kg.

Transformation of hydrogen-rich organic feedstock molecules and oxygen into typical diamondoid products and water is an exoergic, entropy-reducing process. For example, a process with acetone as the primary carbon source and diamond as the chief product liberates a total energy equaling $\sim 1.7 \times 10^7$ J/kg of useful output, while decreasing local entropy by $\sim 5.7 \times 10^3$ J/kg·K (approximating the entropy of the product object as zero). The change in free energy accordingly is $\sim 1.5 \times 10^7$. Allowing for $\sim 3.1 \times 10^6$ J/kg of free energy dissipated (see preceding paragraphs), the system produces $\sim 1.2 \times 10^7$ J/kg of surplus energy in the form of mechanical work. This could be conveniently disposed of by delivery to an external power distribution system in the form of electrical energy.

Waste heat from the process results from both inefficiencies and computation ($\sim 3.1 \times 10^6$ J/kg) and from the reduced entropy of the products relative to the feedstocks ($\sim 1.7 \times 10^6$ J/kg). Thus, a system producing 1 kg/hr of diamondoid products dissipates an estimated 1.3 kW of waste heat. The required cooling capacity can be provided by fan-driven air flowing at a rate of 0.1 m^3/s with $\Delta T = $ <15 K between the intake and exhaust ports. Cooling channels can be placed in close proximity to the chief heat sources, permitting good thermal contact.

14.4.9. Information requirements

A typical product of the exemplar architecture consists of $\sim 10^{15}$ building blocks of $\sim 10^{-6}$ m linear dimensions. To specify a product requires specifying the type and position of each block. If there are $< 10^9$ different kinds of block, then specifying the type of a block requires <32 bits of information (substantially less, with efficient encoding). The total quantity of information required per block need seldom be much larger than this.

An analogy can be drawn to a pattern of type on a page: letter forms are complex, but standardized; with a little additional information (e.g., column widths, line spacings, superscript and subscript codes) a sequence of 8-bit codes specifying a sequence of letters determines the two-dimensional arrangement of those letters on a page. This may, however, depend on large auxiliary tables of information on letter widths, spacings for kerned pairs, and so forth. In a similar manner, a sequence of codes specifying blocks can be combined with auxiliary information describing the acceptable geometrical relationships between adjacent blocks to provide most of the information required to determine the three-dimensional arrangement of those blocks in space. This approach exploits stereotyped relationships among adjacent blocks to generate precise positioning information through simple computations.

Specification of an object should thus require substantially fewer than 100 bits per block, or $<10^{17}$ bits per product object. Since molecular tape storage systems can hold $\sim 10^{28}$ bits/m^3 (Section 12.6.4), a volume of $<10^{-5}$ m^3, containing <0.01 kg, can contain sufficient information to specify the structures of 10^6 product objects. (Note that the presence of repeated substructures above the block level, either within or between objects, can greatly reduce information requirements.)

14.4.10. Manufacture of manufacturing systems

Since most of the internal volume of the exemplar architecture is devoted to open workspaces for manipulators, it should be feasible to design a system that can be folded from linear dimensions of <0.4 m to linear dimensions of ~ 0.2 m, or (more to the point) that can be unfolded from the smaller dimensions to the larger; Figure 14.5 illustrates how a continuous, gas-tight wall can be expanded by a greater factor. This places the exemplar architecture within both the size and mass range of its own products.

With the use of programmable manipulators to build a diverse set of structures from a smaller set of building blocks, the output of a set of specialized mills can be used to build an identical set of mills, as well as many other structures. Accordingly, a suitably designed and programmed molecular manufacturing system, built along the lines of the exemplar architecture, can be used to build objects that, when unfolded, are substantially identical to itself.

The design of machines able to construct copies of themselves was described in von Neumann and Burks (1966), and many variations on this theme have been reviewed in Freitas and Gilbreath (1982) and the included references. It may seem somehow paradoxical that a machine can contain all the instructions needed to make a copy of itself, including those selfsame complex instructions, but this is easily resolved. In the simplest approach (as implicitly adopted here), the machine reads the instructions twice: first as commands to be obeyed, and then as data to be copied. Adding more data does not increase the complexity of the data-copying process, hence the set of instructions can be made as complex as is necessary to specify the rest of the system. By the same token, the instructions transmitted in a replication cycle can specify the construction of an indefinitely large number of other artifacts.

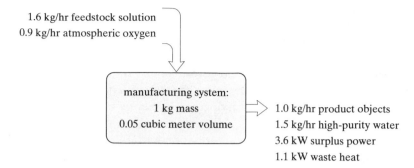

Figure 14.8.   The inputs and outputs of the exemplar manufacturing system discussed in Section 14.4.

14.5.   Comparison to conventional manufacturing

The characteristics of molecular manufacturing systems are best understood in comparison to the characteristics of conventional manufacturing systems. The following comparisons embrace required inputs of materials and energy, by-products, sizes of internal components, frequencies of internal operations, over-all productivity, characteristics of the resulting products, and costs of produc-tion. Differences in the technology base required for implementation are profound, and are implicitly addressed in Chapters 15 and 16, which describe strategies for developing molecular manufacturing capabilities.

   The discussion in this section is not a basis for further analysis of molecular machinery, manufacturing, or computation; rather, it attempts to evaluate some of the consequences of the previous analysis. In characterizing such broad topics as the entire range of product characteristics and the entire range of potential by-products, the descriptions are based on informed estimates. Most other con-clusions are direct consequences of the previous analysis, supplemented by commonplace knowledge of present technological capabilities. Figure 14.8 sum-marizes the inputs and outputs of the exemplar manufacturing system.

14.5.1.   Feedstocks and energy requirements

Defining what one means by a "feedstock" for conventional manufacturing can be difficult. A factory making products for household use may consume inputs as simple as sheet metal and wire, or as complex as integrated circuits and pre-fabricated assemblies. A system of factories considered as a whole typically includes smelters, steel mills, and the like, starting with crude feedstocks (iron ore, coal).

   In molecular manufacturing, a mixture of simple compounds can serve as a feedstock (Sections 13.2 and 13.3). Quantitatively, the largest requirement is for a source of carbon. Agricultural products (sugar, alcohol) or petrochemicals can serve this function at costs of ~0.1 \$/kg, comparable to the cost of raw steel. These inputs are converted into finished products without intermediate stages that require external handling or transportation.

   The discussion in Section 14.4.8 indicates that a molecular manufacturing process can be driven by the chemical energy content of the feedstock materials, producing electrical energy as a by-product (if only to reduce the heat dissipation burden). This contrasts with conventional manufacturing processes, which consume energy not contained in their inputs. Further, both the energy and material input requirements *per unit of product functional capacity* are substantially lower than those for conventional manufacturing processes, owing to greater functional capacity per unit product mass.

14.5.2. By-products and recycling

From the discussion in Sections 13.2 and 13.3, it should be clear that molecular sorting and processing mechanisms can perform transformations like those now performed in chemical plants, but with greater efficiency and control. The preparation of feedstock solutions need not generate by-products in uncontrolled forms; indeed, systems can be designed to produce no by-products at all, aside from some set of compounds containing any unwanted elements present in a crude input material. A general strategy for accomplishing this would be to extract and process small, soluble molecules from a mixture, then use conventional, nonspecific techniques (heat, oxidation, acids) to break down any residue into small, soluble molecules. Design studies to date suggest that relatively exotic and toxic materials (lead, mercury, cadmium) need play no role in molecular manufacturing processes or products.

Many products can be built as aggregates of modules joined by reversible means. These can presumably be recycled with little energy cost by disassembly and reassembly operating at the scale of micron-scale blocks. Failing this, almost any product can be recycled into small molecules subjecting it to chemical attack in a closed vessel (e.g., by oxidation in a sealed incinerator, followed by other processes to break down any residual solids).* These small molecules can then be sorted and reprocessed.

14.5.3. Internal component sizes and frequencies

In conventional manufacturing, human hands and other devices on a 0.01 to 1 m scale perform operations at frequencies that typically range from 10 to 0.1 Hz. In molecular manufacturing, the most numerous operations are performed by devices on a 10^{-9} to 10^{-7} m scale, operating at frequencies of $\sim 10^6$ Hz. In making macroscopic objects by convergent assembly, a modest number of operations are performed at frequencies of ~ 10 Hz.

14.5.4. Productivity

The physical productivity of a system can be measured in various ways. One is the time required for a system to produce outputs with a total mass equaling its own (by this standard, a short piece of pipe can be enormously productive).

---

* P. Barth has expressed concern that breaching the vacuum containment of a manufacturing system would cause its premature oxidation (i.e., that the finely-divided, carbon-rich nanomechanical components would burn). Note, however, that little of the mass of the system described serves a chemical or interfacial role; most of the mass serves as structure (including the interiors of moving parts). In systems to be used in environments where combustibility is a safety concern, much of this structural mass can be made from incombustible materials (e.g., aluminum oxide; see Table 9.1) with little change in overall design. Internal water reservoirs may also be useful.

Another basis, less precise but still significant, is the length of time required for a system to produce outputs with a total complexity equaling its own, starting with simple inputs. If this criterion is applied to the materials processing and manufacturing sector of an industrial economy, the result corresponds to the minimum doubling time for the total capital stock, and is presumably on the order of a year or more, $\sim 10^8$ s. For the exemplar molecular manufacturing system, the equivalent time is $\sim 10^3$ s. This disparity stems from the difference in the characteristic frequency of operations.

14.5.5. Some feasible product characteristics

a. Precision and reliability. Most products made by present manufacturing systems are poorly ordered: they have either amorphous microstructures or crystalline microstructures that contain a high density of atomic defects, impurities, dislocations, grain boundaries, and microcracks. Semiconductor substrates come closest to perfection, but even so, most devices are formed by imprecise processes such as diffusion of impurities, nonepitaxial deposition processes, etching, and the like. In addition to having poorly controlled microstructures, macroscale objects have shapes that are rarely precise to $<10^{-6}$ m. Without robust designs and careful process control, statistical variations in defect size can make products unreliable. Designs that tolerate material defects usually must sacrifice performance to do so.

Molecular manufacturing systems can make well-ordered products, save for scattered atomic-scale defects caused chiefly by background radiation. The materials-, component-, and system-level analyses in Chapters 9–12 have shown how this more precise control can yield systems of high performance. Designs that tolerate the residual flaws caused by radiation damage commonly sacrifice performance, but the lesser size and density of the flaws results in lesser sacrifices. For structures small enough to avoid damage, or large enough to be inherently tolerant of it, the reproducibility of properties and performance can approach perfection. This improves reliability.

b. Strength and stiffness. It is well known that intrinsically strong materials (e.g., ceramics such as aluminum oxide, silicon carbide, and diamond) are brittle, and that brittle materials placed in tension are sensitive to microcracks (Kelly, 1973). Accordingly, the ability of current engineering practice to exploit the high intrinsic strength of these materials is sharply limited by the inability of present manufacturing processes to make them without defects. The ability of molecular manufacturing to build structures with only atomic-scale defects permits these materials to be used at nearly their full theoretical strength. Diamond, in particular, has ~ 55 and 75 times the ratio of tensile strength to density of strong steel and aluminum alloys, respectively [assuming a conservative 50 GPa tensile strength for diamond, and tabulated values for AISI No. 9255 steel, hardest temper, and 7178-T6 aluminum (Tapley and Poston, 1990)].

The design of minimum-mass compressive structures often is constrained by stiffness, and the ratio of Young's modulus to density for diamond exceeds those for steel and aluminum by factors of 12 and 15, respectively. The design of compressive structures is often further constrained by buckling, and strategies for preventing buckling are often constrained by the scale of initial imperfections and by the difficulty of fabricating the optimum shapes (which can be intricate, and can pose minimum-gauge problems). Finally, it is worth noting that small moving parts, high power-density actuators, and fast sensing and control systems can be combined to make macroscopic objects that actively resist deformation, thereby simulating an extremely high modulus material within certain bounds of frequency response, length scale, and applied load.

c. *Other material properties.* Conventional manufacturing processes are constrained to work with materials that result from bulk (or biological) processes. Although it is difficult to characterize the limits of what such processes can make (particularly through organic synthesis), they control comparatively few variables: chiefly the time history of bulk composition, temperature, and pressure, augmented by shear rates, magnetic fields, and electric fields. Molecular manufacturing processes, in contrast, control a comparable number of variables *per moiety* during the fabrication process. This strongly suggests that the range of feasible materials (and therefore material properties) is broader. In some instances, however, bulk processes can presumably make materials that are at or near the limits of performance in particular respects.

The processes and exemplar architecture described in this chapter assume a high vacuum environment, permitting free use of highly active reagents with no special provision for local enclosures. This limits the output to product structures of high stability, and in particular, of negligible vapor pressure. Most high-strength materials and many polymers and elastomers meet these criteria, but many familiar products do not. Chapters 15 and 16, however, describe the use of mechanosynthetic processes in solution environments (with suitable constraints on reagents, mechanisms, and so forth). High vacuum is convenient and often desirable, but it should not be taken as a necessary condition for all molecular manufacturing processes; indeed, thorough control of the immediate environment of a reaction site can be provided in a liquid medium, as it is in enzymes. Convergent assembly via blocks of intermediate scale can impose constraints on structure, but the feasibility of joining a pair of blocks to form dense, continuous diamondoid structures (Section 9.7.3) suggests that these constraints are not severe, in practical terms.

d. *Size.* Manufacturing processes today rarely produce features less than 100 nm in size, and most products are macroscopic, having features on a scale of 0.001 m or more. Molecular manufacturing can construct objects with ~0.3 nm features, placing ~10^8 distinct atomic features in a volume of $(100 \text{ nm})^3$. At the opposite end of the size spectrum, both conventional and molecular manufacturing processes built along the lines of the exemplar system can construct objects of indefinitely large size (in both instances, by assembling macroscopic components to form larger structures).

e. *Energy conversion.* Section 11.7 describes a class of submicron electrostatic motors capable of converting between mechanical and electrical power in either direction at a power density of ~10^{15} W/m^3 and an estimated efficiency >.99. This power density is several orders of magnitude greater than that of electric motors produced by current manufacturing processes.

Section 13.3.8 uses previous calculations regarding mechanosynthetic processes to conclude that chemical and mechanical energy can be interconverted with an efficiency $>.99$ at a power density of $\sim 2 \times 10^9$ W/m^3, or $\sim 8 \times 10^6$ W/kg. If suitably low-dissipation designs can be found (this chiefly entails avoiding major nonisothermal compression losses), then higher speeds of motion (e.g., ~ 1 to 10 m/s, rather than ~ 0.004 m/s) will be readily achievable with good efficiency, and the power density values that are achievable will increase to $\sim 10^{12}$ W/m^3, or $\sim 10^{10}$ W/kg.

Conversion of optical to electrical energy has applications in power supply. Molecular manufacturing techniques can be used to make multilayer, multi-band-gap photovoltaic cells; structures of this sort can convert solar energy to electrical energy with efficiencies substantially greater than .3 (Hubbard, 1989).

Conversion of electrical to optical energy has applications in display and illumination. The use of quantum-well nanostructures is already of technological interest for improving the performance of light emitting diodes and solid-state lasers, hence the nanoscale structural control provided by molecular manufacturing techniques should lead to significant improvements. Isotropic light-emitting structures have no minimum scale, but owing to diffraction, good directionality requires a size that is large compared to an optical wavelength.

f. Energy storage.   Nearly reversible mechanochemical energy conversion can serve as a basis for energy storage. For example, a device that nearly reversibly transforms butane and oxygen into water and carbon dioxide can store $\sim 4 \times 10^7$ J/kg, based on the mass of the initial butane, or $\sim 9 \times 10^6$ J/kg, based on the total mass of the reactants. Rates of energy storage and release are as described in Section 13.3.8 (note that the power conversion rates possible in macroscopic systems will commonly be limited by cooling). Systems of this sort have substantially higher energy densities than those of storage cells that can be produced using present manufacturing technologies, and power densities that are orders of magnitude higher.

g. Computational capacity.   Molecular manufacturing processes can be used to construct computational systems like those described in Chapter 12 (or better, given the conservatism of the design rules employed there). In terms of component densities, some of the relevant system-level parameters include $\sim 10^{19}$ CPUs/m^3 (Section 12.7), $\sim 10^{25}$ bits/m^3 (for RAM storage, Section 12.6.3), and $\sim 10^{28}$ bits/m^3 (for tape storage, Section 12.6.4). Individual CPU speeds can equal or exceed 10^9 instructions per second (Section 12.5.2c) with an energy consumption of $\sim 10^{-16}$ J per instruction (Section 12.7.4). Aside from individual CPU speed, each of these parameters exceeds by several orders of magnitude the performance possible in systems made by conventional manufacturing processes.

14.5.6. Manufacturing costs

a. The feasibility of making cost estimates.   Estimating the cost of a conventional manufacturing process is difficult unless the product closely resembles one for which experience exists. Product costs depend on the costs of components which may or may not be in production. The costs of establishing, staffing, and managing a production process—including the controls necessary to ensure adequate product quality—can be hard to estimate. In the aerospace industry, in particular, the costs of testing and documentation for new technologies are huge. Where a technology differs greatly from existing practice, estimates of manufacturing costs have traditionally been little more than guesses.

The relationship between complexity and manufacturing costs highlights the unusual nature of cost estimation in molecular manufacturing processes. In conventional manufacturing processes, costs increase with the complexity of the product being made: more intricate systems require more parts and manufacturing operations. In molecular manufacturing processes, each moiety is treated as a distinct part, regardless of the apparent complexity or simplicity of the product. Accordingly, production cost will be essentially independent of complexity (Section 10.10.2d).

In molecular manufacturing, many of the traditional costs of building complex systems are avoided. Most of the costs that remain are sufficiently well defined that one can place upper bounds on them. These costs are estimated on a per-kilogram basis; for comparison, most manufactured products today fall in the cost range between ~ 10 \$/kg (e.g., automobiles, appliances) and $\sim 10^4$ \$/kg (e.g., aerospace vehicles).

The great exception to this generalization is development costs. These are hard to estimate, but are addressed in Section 14.6, and indirectly by the content of Chapters 15 and 16.

b.   Materials.   As has been discussed, the chief anticipated feedstock materials for molecular manufacturing (C, N, O, H) are available in bulk compounds for costs of ~ 0.1 \$/kg (at a substantial energy cost, they are available from atmospheric gases). Others that may play substantial roles (Si, P, S, F, Cl) are also available in reasonably inexpensive forms. Precious metals (e.g., of the platinum group) can be expected to be applied as catalysts in mechanochemical processes, but if used at only 10^4 cycles per second, a kilogram of platinum atoms (costing $\sim 10^5$ \$) can be used to process $>10^{10}$ kg of material per year, adding little to the product cost.

c.   Energy.   Using typical organic feedstocks, and assuming oxidation of surplus hydrogen, reasonably efficient molecular manufacturing processes are net energy producers (Section 14.4.8). At a typical price for electrical energy today, ~ 0.1 \$/kW·hr, the value of the by-product electrical energy would usually exceed the cost of the feedstock materials. The value of by-product energy will be ignored in the present cost estimate.

d.   Waste disposal.   The only wastes from the manufacturing process itself are high-purity water and a modest amount of heat rejected to the atmosphere (for a 1 kg/hr system, this is roughly equal to the heat produced by sunlight striking a square meter of dark surface). Neither of these would now be regarded as a significant form of waste. The cost of waste disposal from the preparation of feedstock materials is (within present environmental standards) reflected in the cost of those materials.

e.   Labor and distribution.   Between the input of raw materials and the removal of finished products, no labor is necessary. Both of these steps can be regarded as distribution costs falling outside the scope of manufacturing proper.

f.   Land.   The exemplar manufacturing system produces its own mass in product in less than an hour. Conventional manufacturing systems (which must in a fair comparison be taken to include factories that supply parts) take orders of magnitude longer to produce their own mass in product. Assuming comparable ratios of plant and product densities, this indicates that molecular manufacturing systems require orders of magnitude less space (and land) per unit of productive capacity. Costs of land will be reduced by the same factor.

g. Capital. In current manufacturing systems, the cost of capital goods is chiefly a manufacturing cost. In considering a new manufacturing system, this introduces a circular dependency in cost estimation: the cost of producing capital goods depends on the costs of materials, energy, waste disposal, labor, distribution, and capital goods. The equilibrium contribution of the cost of capital goods to the cost of later capital goods can be estimated as follows: Let r be the real interest rate (expressed as a fractional rate of return per year), c be the non-capital cost of producing a unit of capital goods, and t be the time (in years) for a unit of capital goods to produce another unit of capital goods (here treated as a batch process). The total cost C of a unit of capital goods is then

$$C = c + C[\exp(rt) - 1] \tag{14.3}$$

$$\approx \frac{c}{1 - rt}, \quad rt \ll 1 \tag{14.4}$$

In conventional manufacturing, $r \approx 0.1 \text{ yr}^{-1}$, and $t \approx 1$ yr (or more, particularly if system boundaries are drawn to include factories that supply parts), and the contribution of the cost of capital goods is substantial. In molecular manufacturing, $t \approx 10^{-4}$ yr, and these costs are negligible. In this connection, it might be noted that the cost of capital goods contributes to the cost of materials and energy, but the cost benefits of changes in the resulting circular dependencies will not be considered here.

h. Total costs of manufacturing. The preceding discussion indicates that the basic cost of production, here taken to exclude the costs of development and distribution (and of such imponderables as taxation, licensing agreements, and insurance), will be almost wholly determined by the cost of materials. The relevant materials presently cost ~0.1 to 0.5 \$/kg; this range can thus be taken as an upper bound on the basic cost of producing objects using a mature molecular manufacturing technology base. These costs are low enough for molecular manufacturing to be competitive in making a wide range of products.

14.6. Design and complexity

The design of systems that organize complex patterns of activity among reliable, submicron devices operating at millions of cycles per second is an everyday activity in computer science. Although molecular machines and manufacturing differ greatly from microelectronics and computation, many similar issues arise. It is reasonable to expect that computer scientists will play a leading role in the development of molecular manufacturing systems at levels above device design. Accordingly, this section repeatedly compares established concepts in computer science with the anticipated needs of molecular manufacturing.

Device design will likewise depend on computer science. Although experimentation is likely to play a large role, simulations and computer-aided design

(CAD) tools will be essential. Complex systems will become possible only when an adequate library of simple components and operations has been accumulated. The following discussion will assume the previous development of such a library, thereby sharpening the focus on issues arising from complexity.

14.6.1. Part counts and automation in design and computation

Many of the numbers that describe the manufacture of macroscopic objects by molecular means are enormous by most present standards. Building a cubic-micron block requires $\sim 10^{11}$ moietal assembly operations. Building a kilogram object from cubic-micron blocks requires $\sim 10^{15}$ block assembly operations. A team of a thousand people each making one design decision per second would require a century of 8-hour days to plan a single cubic micron, to say nothing of a kilogram: in a practical design process, most of these operations must be specified without direct human attention. Most operations must be as routine as painting a pixel in a raster output device or executing an instruction in a piece of code.

For comparison, typesetting this volume required that a complex pattern of $>10^{10}$ pixels be painted black or white; these operations were directed by a set of single-purpose PostScript programs that were automatically generated from $>10^8$ bits of data describing text and graphics. The software used to prepare this data executed $>10^{13}$ instructions (though mostly while waiting for inputs). The software itself consisted of $>10^7$ instructions, written by several teams to form separate programs that nonetheless cooperated to produce a single product. It is reasonable to assume that molecular manufacturing, which likewise generates complex patterns through repetitive operations, will involve broadly similar operations controlled by broadly similar software systems.

14.6.2. Design of components and small systems

Today, most machine code that directs computer operations is generated automatically from a more abstract description in a high-level language. The compiler that does this works by piecing together small, standardized fragments of machine code (consisting of one or more instructions) to implement the desired operations. These fragments and the rules for generating sequences of them are the result of human attention, but the results of this labor are used repeatedly without human attention. In a similar fashion, component-level nanomechanical designs and rules for their composition can be designed with human attention, then later combined by a *hardware compiler* to implement desired functions.

Examples of direct atom-by-atom design of molecular mechanical devices are scattered through this volume. Some designs, however, show the early results of automation, in which nanometer-scale components are described by languages that specify regular arrays of atoms and bonds. A good example is R. Merkle's prototype computer-aided design (CAD) software that starts with a single typed line of input parameters and generates the atomic coordinates of a regular, three-dimensional diamondoid structure that can contain many thousands of atoms and bonds (Merkle, 1991). Examples of its outputs include the shaft and sleeve in Figure 10.17 and the ring gear in Figure 10.32.

The design of less regular structures can also be automated. A precompiled library of Kaehler brackets (Section 9.5.5) can be used to choose bridging structures between surfaces. Simulated annealing methods (with change of valence and atom type as an operator) hold promise for generating regions of amorphous structure that meet particular boundary conditions.

Eventually, molecular CAD tools will combine principles like those applied in Merkle's and Kaehler's prototypes with features like those found in commercial CAD systems for macroscopic design, such as support for continuum descriptions of objects containing many atoms. With these CAD tools (and faster machines, better molecular potentials, and feedback from physical experiments) it will be possible to design components and systems of the complexity that can be managed today in the macroscopic world. These will include manipulator arms, computers, and molecular mill systems using multiple belts, rollers, and cam surfaces. Components and systems of this complexity can then serve as building blocks in the design of more complex systems.

14.6.3. Automated generation of synthesis and assembly procedures

a. Retrosynthetic analysis of chemical syntheses. Once the structure of a component is specified, a synthesis must be found or the component must be redesigned. Software for the automated generation of candidate chemical syntheses has been in use since the 1960s (see Corey and Cheng, 1989). These programs apply *retrosynthetic analysis,* the development and application of which won E. J. Corey the 1990 Nobel prize for chemistry. The central idea of retrosynthetic analysis to plan the assembly of a complex molecular structure by considering how it could be *disassembled* into small, available molecules using imaginary operations, each the *reverse* of a feasible synthetic step. The resulting disassembly process is therefore the reverse of a feasible synthesis. (Computer scientists will recognize this procedure as a form of backward chaining.)

Mechanosynthetic operations are equally subject to retrosynthetic analysis (but largely free of the complication of competing reactions). Each synthetic step can be regarded as the reverse of an etching reaction, and the synthesis designed (in reverse) as the etching away of the desired structure. Considering the mechanosynthesis of diamondoid structures, Soreff (1992) has suggested that a study of actual etching processes (e.g., of diamond by fluorine) can indicate geometries and transition states for synthetic steps; the forward-direction reagents would be analogous to the hypothetical etching products, but with bond strengths (etc.) favoring deposition over etching.

b. Hierarchical decomposition of larger structures. If an object is to be made from an array of parts, it (and its assembly procedure) can be designed by dividing it into regions, each made of subregions, and so on down to the level of the individual parts. The divisions chosen during the design phase must create interfaces that permit assembly during the manufacturing phase. If the final product were fully specified at the outset, this hierarchical decomposition approach would be a form of retrosynthetic analysis.

In conventional engineering, however, the specification of the final product is influenced by the nature of the available parts and assembly operations. Abstractly, this means that the hierarchical decomposition is developed together with the design, rather than afterward. In dividing a system into regions, the interfaces can be altered as the design becomes more detailed. A design process that includes an assembly-oriented hierarchical decomposition can directly generate the outlines of an assembly plan, and can provide a framework for dividing the computational load of the design process over a large number of processors. (For other purposes, designing in terms of functional rather than geometrical decompositions will be important, requiring multiple representations of the design.) For an early essay on the general applicability of hierarchical decompositions, see Simon (1981).

14.6.4. Shape description languages and part arrays

A *shape description language* can describe a shape so as to enable an interpreter driving an output device to make a corresponding physical structure. In a program in the PostScript language, which describes two-dimensional shapes, a set of instructions can describe a circle by its radius, color, and border. A language for describing three-dimensional shapes might similarly describe a sphere by its radius, material, and surface structure. With PostScript driving a printer, the circle is approximated by printing an array of dots, fixed with respect to a page; with a three-dimensional language driving a molecular manufacturing system, a sphere would be approximated by assembling an array of blocks, and, unless specified as joined to another object, would remain mobile. In either case, the complexity of the description is independent of the size of the object and hence of the number of pieces from which it is made.

Three-dimensional shape description languages are standards in computer-aided design systems for mechanical engineering. An interpreter operating on a CAD description could generate instructions for a molecular manufacturing system, directing it to make a corresponding set of solid objects. Using cubic-micron blocks that can be composed to make solid regions of widely varying properties, the output structures could form a wide variety of high-performance systems with submicron control of component shapes. Not all CAD descriptions will correspond to feasible assembly sequences, but many structures can be made to fall within the required design rules (for example, that each object be supported while it is being built, and that each object be subject to a hierarchical decomposition forming suitable interfaces).

14.6.5. Compilers

a. Assembly-process compilers.   Given a fully specified design that meets the constraints necessary to permit assembly, there remains the task of specifying the operations of the assembly process. If the product is to be made from micron-scale blocks, this amounts to specifying the transport of blocks of the correct types to the correct assembly workspaces in the correct order, and specifying the sequence of motions to be executed by manipulators in putting the blocks together. As mentioned in Section 14.6.3b, the design of structures and assembly procedures by hierarchical decomposition directly generates a tree of assembly steps. If this tree has been chosen to generate parts of the appropriate sizes and numbers, then it can be mapped onto a manufacturing layout (perhaps one not *entirely* unlike that in Figure 14.4). The vast number of manipulator motions to be specified at the finest, earliest, and most-dispersed levels of the assembly process can be planned in parallel by identical software systems running almost independently on separate processors. The result of this parallel assembly-process compilation is a set of instructions that, when executed by the transportation system and manipulator controllers of a manufacturing mechanism, will result in the assembly of an object corresponding to the initial design.

b. Design compilers. A conventional software compiler generates a set of machine instructions from a more abstract specification: a program written in a high-level language. An assembly-process compiler likewise generates a set of machine instructions from an abstract specification, the hierarchically decomposed structure of a particular object (which could itself be specified by a shape-description language). Like a software compiler, it acts as a compiler *with respect to the machine instructions,* although starting with a different kind of specification. A modern *silicon compiler*, in contrast, generates a complex pattern of transistors and conductors from an abstract specification of the properties of a digital circuit. It thus acts as a compiler *with respect to the design of a machine.* The implementation of artifacts as complex as the exemplar architecture of Section 14.4 seems likely to require the use of analogous design compilers for nanomechanical systems.

There can be no fully general design compilers, capable of converting an arbitrary design specification into a design: not all specifications can be met, and of those that can, not all can be met by the application of compiler-style rules guiding the composition of stereotyped solutions. Like a high-level language compiler or a silicon compiler, a design compiler will accept only specifications for a restricted class of mechanisms in a restricted format.

A compiler for the design of manufacturing systems can only be developed after solutions have been developed to each of the stereotyped, smaller-scale problems that will be encountered. These include:

- Software modules for performing retrosynthetic analyses of components.
- Specifications of reaction geometries and environments for effecting standard synthesis steps in mill systems.
- Designs for subsystems that assemble special components (Section 14.2.1).
- Designs for structural elements, reagent devices, mill mechanisms, gauges, manipulator arms, computers, communication channels, and so forth.
- Parameterized descriptions of subsystems for transportation, power distribution, communication networks, and so forth.
- Design rules for composing subsystems and selecting spatial layouts.

A further requirement, of course, is experimental feedback to permit debugging of system and subsystem designs that fail to work. Note that compiler-aided design need not all be performed by a single tool with seamless end-to-end automation.

The compilation of manufacturing systems will have similarities to the compilation both of digital circuits and of machine code: The product of the compilation process must cause large numbers of operations to be performed in the correct sequence. In designing this product, trade-offs involving time, space, and power must be made, taking into account the relative frequencies and costs of different operations. Preexisting solutions developed for small-scale problems must be described and organized so that they can be applied by algorithms. Throughput, latency, buffers, storage, and transport paths must all be considered. Compact arrangement of parts in space is important (albeit in three dimensions, rather than the two of silicon compilers).

c. The economics of compiler development. Compilers have not been developed for macroscopic mechanical systems,* presumably because either (1) the cost-benefit ratio in developing and using them has been adverse, or (2) they are impossible to develop within the constraints of present computer hardware and software technology. If (2) were true because of software technology constraints, this would cast doubt on the feasibility of developing compilers for nanomechanical systems. The history of software and chip-design compilers sheds light on this situation.

When processors were orders of magnitude slower and memory orders of magnitude more expensive, programs were short, and it was crucial to minimize both size and execution time. Small programs could be described in assembly language, providing instruction-by-instruction control of the machine, and there were great incentives to do so. The first computers became operational in the mid-1940s, but the first high-level language to achieve widespread use (Fortran II) was introduced over a decade later, in 1958. Through the 1970s, machine cycles and memory space were scarce enough, and the quality of compiled code poor enough, that assembly language continued to find widespread use in critical programs. Today, faster machines and cheaper memory have enabled the development of programs too large and complex for practical assembly language coding, and improvements in speed, cost, and compilers have combined to make assembly-language coding a shrinking part of software development.

To summarize the abstract lesson: When resources are scarce, systems must be relatively small and simple, and detailed human planning is both feasible and desirable. The incentive to develop compilers is low, and the evolution of compilers good enough to compete with human designers does not even begin. When resources are abundant and inexpensive, detailed human planning cannot organize them on a sufficient scale to exploit their potential. Even inefficient compilers become attractive. They are implemented, and evolve toward greater efficiency.

In macroscopic hardware design today, parts are few and production costs are large. Relatively crude compilers could not compete with human designers in the quality and cost effectiveness of their designs. People can do the job, and do it better. With molecular manufacturing, however, part counts can grow into the trillions and beyond, and per-part costs will fall to minuscule fractions of a dollar. Limiting designs to the number of parts that an unaided team of human designers can manage will be unattractive, if a compiler-aided design team can specify a system a million times more complex and capable. This preference would often remain true even if each compiler-specified system wastes twice as much space, mass, and energy as would a comparable system specified in detail by a competent human designer.

---

* CAD systems are automating more functions, and research in entirely automated design is underway. Accordingly, this statement is becoming debatable.

In a design process following the present compiler model, human design will remain dominant at the level of parts and subsystems (in the form of knowledge built into the compiler) and at the level of overall system organization and purpose (in the form of specifications given to the compiler when it is used). The intermediate levels will be designed, with considerable inefficiency, using algorithms and heuristics that represent a workable subset of human knowledge of design principles.

This pattern is, in the early 1990s, already familiar in one hardware domain. Since the early 1960s, the number of components in an integrated circuit has climbed by a factor of $\sim 10^6$, and the manufacturing cost per transistor has fallen to $\sim 10^{-5}$ dollars or less. If a 10^6 transistor design has an expected market of 10^5 units, then every dollar of design cost per transistor adds ten dollars to the price of each chip, yet a dollar cannot buy much time from a human design team. The design cost is often a large fraction of the product cost. In response to this pressure, silicon compiler technology emerged in the 1980s. At first considerably less efficient than a human designer in maximizing speed and minimizing circuit size and cost, silicon compilers nonetheless gained a foothold, then steadily improved, becoming an integral part of the design process.

This section has given only a rough sketch of past developments. A more detailed study of the evolution from lower-level CAD systems to silicon compilers could yield further insights into the likely path from today's molecular modeling and mechanical CAD packages toward future compiler-style support for the engineering of nanomechanical systems.

14.6.6.   Relative complexities

As suggested by the discussion in Section 14.6.4, many nanomechanical systems are relatively simple. Highly regular structures provide an obvious set of examples. Among more complex structures, mechanical nanocomputers are of essentially the same complexity as microelectronic computers, once a modest library of logic-element designs has been accumulated (indeed, some CAD software for digital logic design would be applicable without modification). Regular arrays of nanocomputers with local interconnection will not be substantially more complex than single computers, hence the design of extremely complex systems can eventually rely on such arrays to solve computationally expensive problems.

Small molecular manufacturing systems are comparable in complexity to modern automated factories, both in parts count and in organization. Since macroscopic automated factories fall within the range of present human design capabilities, their nanomechanical equivalents should as well. The focus in this chapter on large systems of great complexity should not be taken to imply that *all* useful manufacturing systems need be so complex.

Even the most complex systems described here are arguably less complex than some modern software systems. Although their part counts greatly exceed the number of lines of code in any program, parts and lines of code are not analogous. Mechanisms in a manufacturing system will appear in many copies; a particular mechanism (and its parts) is less like a unique subroutine or object class (and its code) than like one of many invocations of that subroutine, or instances of that object. Indeed, it is often like an invocation or instance *with variables identical to those already used,* existing in multiple copies in order to produce identical "constants," or building blocks, for repeated use in a physical product.* Writing a program with 10^{15} lines of code would be out of the question today on grounds of complexity. Running a program that executes 10^{15} instructions is less extraordinary, amounting to a month of supercomputer time. The difference between these cases lies in the massive repetition of standard operations. Nanomechanical systems can exhibit a similarly massive repetition of physical structures, without necessarily entailing unprecedented complexity.

14.7. Conclusions

Molecular mill and manipulator mechanisms can be combined in architectures capable of transforming solutions of small organic molecules into complex macroscopic artifacts built of diamondoid materials and containing components of the sorts described in Chapters 9–13. A 1 kg system can build a 1 kg product object in ~1 hr, dissipating ~1.1 kW of heat. The range of products that can be produced is large, encompassing high-performance structures, massively parallel supercomputers, and additional molecular manufacturing systems. A review of costs (other than those of design, licensing, taxation, and the like) indicates that the cost of producing objects using such systems can approach the cost of the required feedstocks, for exoergic reactions. Assuming present materials prices, the anticipated product costs would be ~0.10 to 0.50 \$/kg. The complexity of macroscopic molecular manufacturing systems will require extensive automation of the design process.

Some open problems. Far more detailed system-level descriptions of molecular manufacturing could be provided by studies that first attempt to clarify the relationship between (1) the number of distinct building-block types and (2) the range of capabilities that can be delivered by products made from those building blocks. Alternative architectures are worth considering, including architectures that incorporate greater flexibility and more extensive use of general-purpose mechanisms. Section 14.6 can be regarded as a partial survey of open problems in automating the design of products and the specification of production processes. The literature on analogous macroscale problems is a natural point of departure for work in this area.

---

* The computational analogue would be a comparatively simple massively parallel SISD machine, pointlessly producing many instances of the same answer. In manufacturing, applying the same operation ("single instruction") to identical objects ("single datum") is more useful, producing multiple instances of a physical object.

Part III

Implementation Strategies

15

Macromolecular Engineering

15.1. Overview

Earlier chapters describe a family of technologies *assuming the availability of advanced manufacturing systems to build other advanced systems:* the analysis is based on present science, but not on present technology. This chapter and the next attempt to show how the gap between present technology and advanced molecular manufacturing can be bridged. Unlike the chapters of Part II, the present chapter adopts the constraint that the objects and capabilities described be within the reach of research programs that exploit only presently available laboratory capabilities.

Present technology cannot build complex structures containing billions of atoms, each occupying a predictable location. In seeking paths toward this ability, it is natural to start with technologies that can provide precise, molecular control: these include biotechnology, synthetic chemistry, and (more recently) proximal probe systems such as the scanning tunneling and atomic force microscopes (STMs and AFMs). This chapter describes how each of these fields can evolve into a more advanced technology for macromolecular engineering. The next chapter shows how these macromolecular engineering capabilities can be a starting point for a series of steps leading to advanced molecular manufacturing.

This chapter suggests objectives and approaches in several existing fields, aiming to identify lines of development that avoid insurmountable obstacles and offer large long-term rewards. This analysis, however, does not constitute a detailed proposal and cannot be compared to hard-won experimental results. It is offered as an aid in choosing experimental objectives, not as a substitute for achieving them.

15.2. Macromolecular objects via biotechnology

15.2.1. Motivation

In introducing the concept of molecular manufacturing, examples of molecular devices and processes from biology were used to make the case for physical feasibility, and an analysis of the feasibility of protein engineering was used to make the case for technological feasibility (Drexler, 1981). Although this use of examples from molecular biology (Table 15.1) has since been superseded by

Table 15.1. A comparison of macroscale and biomolecular components and functions (from Drexler, 1981).

Device	Function	Molecular example(s)
Struts, beams, casings	Transmit force, hold positions	Microtubules, cellulose
Cables	Transmit tension	Collagen
Fasteners, glue	Connect parts	Intermolecular forces
Solenoids, actuators	Move things	Conformation-changing proteins, actin/myosin
Motors	Turn shafts	Flagellar motor
Drive shafts	Transmit torque	Bacterial flagella
Bearings	Support moving parts	Sigma bonds
Containers	Hold fluids	Vesicles
Pumps	Move fluids	Flagella, membrane proteins
Conveyor belts	Move components	RNA moved by fixed ribosome (partial analogue)
Clamps	Hold workpieces	Enzymatic binding sites
Tools	Modify workpieces	Metallic complexes, functional groups
Production lines	Construct devices	Enzyme systems, ribosomes
Numerical control systems	Store and read programs	Genetic system

more direct design and analysis, these examples still illustrate how the engineering of biomolecules could enable the construction of a first generation of molecular machines.

15.2.2. DNA, RNA, and protein

The macromolecules of central concern in biotechnology and molecular biology are DNA, RNA, and protein. Genetic engineering now includes powerful and general techniques for constructing DNA molecules; via RNA, these can direct the molecular machinery of bacterial cells to make protein molecules.

DNA ordinarily functions as an information molecule, but N. Seeman has demonstrated that it can also be used to build structures with both branches and loops (Seeman, 1991), including a molecule with the connectivity of a cube. The ability of complementary segments of DNA to bind to one another provides a powerful, readily controllable mechanism for guiding the °Brownian assembly of molecules. Hybrid structures of DNA and other macromolecules may prove particularly attractive.

RNA in living cells has diverse functions; some are structural and chemical (e.g., in the ribosome) and some informational (e.g., in the messenger RNA that directs ribosomal synthesis of proteins). Like DNA, it can fold and bind to match up complementary segments, forming a double helix.

Molecular machines in living cells are chiefly made of protein. Unlike pure DNA and RNA structures, protein structures can be dense, compact, and relatively stiff; further, their 20 amino acids give them greater chemical and structural versatility than is found in structures consisting only of the four nucleotides of DNA or of RNA. These characteristics suggest that proteins will play a central role if biotechnology is used to build nanomechanical systems.

Unlike DNA and RNA, however, protein chains do not bind to one another by pairwise matching of their monomers: no simple notion of complementarity can be used to predict or design the folding of a protein. Although nature demonstrates that proteins are adequate building blocks for systems of molecular machinery, and although biotechnology demonstrates that novel proteins can be constructed, the problem of protein design—of relating amino acid sequence to three-dimensional structure and function—has slowed progress toward protein-based macromolecular systems engineering.

15.2.3. Protein folding: prediction vs. design

The "protein folding problem" has classically been defined as a *fold-prediction* problem, that is, predicting the three-dimensional folded conformation of a naturally occurring protein molecule given only its amino acid sequence (and no knowledge of the fold of a similar sequence). Despite continuing progress (Bowie et al., 1991; Jaenicke, 1987; Ponder and Richards, 1987b), no general solution to this problem has been found.

It was long assumed that solving the fold-prediction problem was a prerequisite for solving the fold-design problem; this impression lingers, discouraging efforts in *de novo* protein engineering. Actually, these problems are quite distinct. In solving a design problem, one is free to bias all manipulable characteristics to favor a particular result, but in predicting a natural phenomenon, one faces whatever ambiguities nature happens to present. Based on this elementary observation, together with evolutionary considerations that predict marginal stability for natural proteins, (Drexler, 1981) split the "protein folding problem" into the distinct problems of fold design and fold prediction, arguing that fold design is easier. *De novo* design of globular proteins has since been pursued (Eisenberg et al., 1986; Erickson et al., 1988; Hecht et al., 1989; Ho and DeGrado, 1987), yielding a recent landmark success (DeGrado et al., 1987; Regan and DeGrado, 1988) which confirmed predictions that design could yield structures of unusual stability (Drexler, 1981). Meanwhile, the ongoing failure to predict the folds of novel natural proteins confirmed the conclusion that fold design is both distinct from and easier than fold prediction. [These suggestions regarding the design of protein folds have been cited and used as a basis for attacks on the fold-prediction problem itself by Ponder and Richards (1987b); see also Lim and Sauer (1989).]

15.2.4. Rational design and evolutionary approaches

At present, the field of protein engineering divides into several main areas. Most protein engineering experiments modify one or more amino acid residues in a natural structure to alter protein stability or function (e.g., Bowie et al., 1990; Imanaka et al., 1986; Pantoliano et al., 1987; Pantoliano et al., 1988; Wilks et al., 1988). More ambitious experiments splice portions of known structures together to make new proteins (Blundell et al., 1988). The most ambitious experiments, termed *de novo* protein engineering, make protein molecules that have no close resemblance to any found in nature (e.g., DeGrado et al., 1987; Hecht et al.,

1989). Experiments of the first sort have become routine; *de novo* protein engineering remains exceptional.

Recent years have seen the emergence of several techniques that develop new proteins not by design, but by an evolutionary process of variation, selection, and replication. The first group, monoclonal antibody technologies, exploits animal immune systems to develop protein molecules that bind a specific ligand molecule; by developing antibodies that bind a molecule resembling the transition state of a particular chemical reaction, it has proved possible to develop antibodies (*abzymes*) that catalyze the reaction (e.g., Janda et al., 1989; Janda et al., 1988). More recently, techniques for *in vitro* evolution of monoclonal antibody molecules manufactured by bacteria have been demonstrated (Huse et al., 1989). Systems that select among molecular objects by their affinity to a chromatographic column have been used to evolve specific binding affinities both in protein fragments exposed on viral surfaces (Scott and Smith, 1990) and in isolated RNA fragments (Ellington and Szostak, 1990; Tuerk and Gold, 1990).

Evolutionary techniques can be used to develop protein molecules capable of binding (and altering) other molecules, so long as the desired property permits differential selection and does not interfere with the replication mechanism. If protein molecules are to be used as self-assembling building blocks in larger structures, developing proteins that bind to the correct surface sites is a basic requirement. Evolutionary techniques may provide means for developing such molecules, if methods can be found to select for strong binding at desired sites on a target structure. Methods for masking alternative sites, or for selecting *against* binding to alternative sites may prove useful in evolving the desired specificity.

15.2.5. Material and device properties

In designing machinery for molecular manufacturing, stiffness is a central concern. Different protein structures have widely differing moduli of elasticity (and are inhomogeneous and anisotropic as well); a Young's modulus of ~10 GPa nonetheless "seems to be common to various kinds of polymers of globular protein molecules" (Oosawa and Asakura, 1975). This falls within the range of moduli of polymeric materials such as polyethylene, 1 GPa; polystyrene, 3 GPa; polyphenylene sulfide, 8 GPa; poly(chlorotrifluoroethylene), 14 GPa; and most kinds of wood, 8 to 12 GPa (Tapley and Poston, 1990). Although a modulus of 10 GPa is adequate for many purposes, it is only 1% of the modulus of diamond; there is thus ample motivation to move to better materials.

Existing protein devices demonstrate forces, speeds, and frequencies that can be achieved by machinery constructed from biomolecules working in solution. The enzyme having the highest known operating frequency, catalase, decomposes hydrogen peroxide molecules at 4×10^7 s^{-1} (Creighton, 1984), bacterial DNA replication machinery can add nucleotides at 750 s^{-1}, and bacterial ribosomes can add amino acids at about 40 s^{-1} (Watson et al., 1987). The bacterial flagellar motor can rotate at almost 300 Hz, and can develop a torque of ~2.5×10^{-18} N·m [= 2.5 nN·nm (Blair, 1990)]. Several molecular linear motors (based on myosin, kinesin, and dynein) exert forces that have been estimated at 0.0005 to 0.0026 nN (Ashkin et al., 1990).

Development paths starting with biomolecular systems can exploit existing molecular machines and components. This advantage is reduced by the complexity, fragility, and specialization of existing biomolecular systems, and by the limited knowledge of many structures (for example, no motor mechanism has yet been characterized in atomic detail). Structural knowledge will accumulate (aided by new instruments), and protein engineering techniques might be used to adapt these mechanisms to other applications and make them more robust. Developments along several other pathways, however, will be compatible with hybrid strategies in which biomolecular and wholly synthetic components are combined. Accordingly, it seems unlikely that a *purely* biomolecular approach will be carried through.

15.3.   Macromolecular objects via solution synthesis*

15.3.1.   Motivation

Design becomes easier in systems that exhibit good functional modularity, and becomes more reliable when ample safety margins are possible; greater freedom of design can further both of these objectives. Alternatives to standard, genetically encoded proteins can provide a more tractable medium for the design of folded molecules with proteinlike functions, chiefly by providing better modularity and safety margins. Greater design freedom will permit novel strategies for stabilizing folds; estimates based on observed correlations between structure and free energy suggest that order-of-magnitude increases in the stability of the folded state can be achieved. Increased stability and rigidity can be expected to mitigate errors in molecular modeling and design, increasing the odds of success in a particular design. The family of approaches outlined in this section adheres closely to the protein model; attractive approaches can likely be found that are quite dissimilar from any biological model.

Constructing molecular objects poses challenges both in design and in synthesis, but one challenge can be traded off against the other. Alternatives to standard proteins typically sacrifice the convenience of biosynthesis (but see Noren et al., 1989) in exchange for greater stability and reduced design difficulty. Protein design remains challenging chiefly because the structure of a globular protein emerges from a cooperative folding process dependent on a balance of weak, noncovalent forces in a large, flexible structure. At the other extreme of molecular size and flexibility, proteinlike receptors have been constructed from relatively small, rigid, °polycyclic molecules (e.g., Cram, 1988; Lehn, 1988). These molecules are comparatively easy to design: many dozens have been made, vs. the few proteins designed *de novo*. Their polycyclic structures, however, usually present synthetic challenges greater than those of °peptides of similar molecular weight. Their molecular masses typically fall in the 200 to 1200 dalton range (Cram, 1988; Lehn, 1988), vs. ~2000 daltons for commercially available custom peptides and 5000 to 15000 daltons for experimental products (Kent, 1988).

---

* Save for editing and partial updating of citations, the material in this section is drawn from Drexler (1989b).

These two approaches, growing out of protein chemistry and more general chemical synthesis, represent extremes on a spectrum. In the first, synthesis is relatively straightforward, but design is difficult; in the second, design is relatively straightforward, but synthesis is difficult. Intermediate approaches may avoid the worst problems of each, by exploiting peptide-style synthetic strategies to construct molecular objects more rigid, stable, and designable than natural proteins. In examining the potential of this approach, fundamental principles of design are relevant.

15.3.2. Basic design principles

Designers exploit *modularity:* In his classic essay, "The Architecture of Complexity," Herbert Simon notes the utility of modular structure in understanding, designing, and building complex systems (Simon, 1981). Organizing a system as an assembly of relatively distinct, independent building blocks enables the use of divide-and-conquer strategies in design and construction. Early successes in protein design have used α helices as modular units in design and development (DeGrado et al., 1987; Mutter et al., 1988).

Designers exploit *safety margins:* Evolutionary considerations suggest that evolved proteins can be expected to have minimal stability while designed proteins can have a greater margin of stability (Drexler, 1981); recent experience shows that protein designers can indeed build molecules of increased stability (Matthews et al., 1987; Pantoliano et al., 1988; Perry and Wetzel, 1984; Regan and DeGrado, 1988). Safety margins in stability will improve the odds of success in designing molecular functions.

Finally, designers exploit *design freedom:* In design, unlike the analytical sciences, one is free to seek components having the best available combinations of stability, predictability, manufacturability, and performance. Design freedom in protein engineering enables the successful pursuit of modularity and safety margins, as has been demonstrated (Mutter et al., 1988; Regan and DeGrado, 1988).

15.3.3. Alternatives to standard proteins

Structures based on sequence-specific polymers have advantages worth keeping. To assemble complex macromolecular objects from small molecules, linkages must be introduced in stages. Ease of synthesis favors starting by building linear, polymeric structures. Established synthetic techniques, the entropic advantages of rotationally rigid bonds, and the extensive knowledge resulting from protein science all favor further exploitation of peptide chemistry to build those polymers. (Note, however, that a similar case could be made for exploiting nucleic acid chemistry.)

Noncovalent linkages are well suited for the subsequent steps: they can form under conditions that preserve existing covalent structures; they can link complementary, polyatomic surfaces, enabling specific binding; and they can allow equilibration, breaking and reforming linkages, thereby enabling the system to "search" for a cooperative binding configuration planned by the designer. (Disulfide bonds also allow equilibration under mild conditions and are of similar utility.) Accordingly, proteinlike structures have engineering advantages quite apart from their biological interest.

Within this framework, design freedom can be expanded by choosing from a broader range of parts and chain topologies. It may be that linear sequences of the 20 genetically encoded amino acids of *standard proteins* will provide adequate design freedom, eventually making *de novo* protein design practical and routine. Nonetheless, it may be useful to examine the advantages and disadvantages of *augmented proteins* made using (1) a set of monomers augmented by inclusion of nonstandard amino acids and (2) a set of chain topologies augmented by branched (e.g., Mutter et al., 1988) and cyclic structures. Augmented structures have repeatedly proven their utility in peptide design (DeGrado, 1988), and nonstandard amino acids have appeared in proposals and experiments in protein design (Erickson et al., 1988; Mutter et al., 1988). Nature's use of nonstandard amino acids, phosphorylation, glycosylation, and so forth (via enzymatic modification of proteins) also indicates the utility of a broader set of building blocks and structures than the 20 genetically encoded amino acids.

The following sections build on current understanding of protein folding and stability to explore the utility of augmented proteins for designing stable folded structures. Many of these remarks apply equally well to nonpeptide structures, which would further expand design freedom but are not explicitly considered here. [The following discussion in this section makes free use of terms and concepts from protein chemistry; an excellent introduction is (Creighton, 1984).]

15.3.4. Strategies for stabilizing specific folds

Achieving a stable fold entails destabilizing (or excluding) both unfolded and misfolded states. Achieving a stable *and predictable* fold requires that the designed fold be clearly stable and that alternative folds be clearly improbable; this requires a margin of safety that exceeds modeling uncertainties. Both of these objectives can be pursued by (1) minimizing conformational entropy in the unfolded chain, and by (2) maximizing the strength and specificity of interactions favoring a chosen folded state.

Multiple stabilizing strategies can be applied, but quantifying their value in a design context raises questions regarding the choice of reference state. The *stability* of a protein can be defined as $-\Delta\mathcal{G}_f$, the negative of the change in Gibbs free energy of folding. The change in stability resulting from a given residue substitution is then $-\Delta(\Delta\mathcal{G}_f)_{sub}$. Employing a stabilization tactic in a design, however, will typically involve not a single substitution, but a coordinated set of changes. Consider two protein molecules, A and B, having grossly similar dimensions and properties. Assume that each has been designed for maximum stability, subject to similar functional constraints, but that use of a tactic X (e.g., use of a dialkylglycine residue, a cross-link, etc.) is allowed only in the design of B, which may differ from A in any manner needed to exploit X effectively. Tactic X should be considered stabilizing if B can *usually* be made more stable than A. More generally, one can define the *expected stabilization* resulting from the application of a tactic X as the average change in $-\Delta\mathcal{G}_f$ resulting from the comprehensive redesign of typical molecules to take advantage of X; this can be represented as $-\Delta(\Delta\mathcal{G}_f)_{design}$. The magnitude of $-\Delta(\Delta\mathcal{G}_f)_{design}$ cannot be measured directly without an enormous amount of design and experimentation. Since good designs will avoid unfavorable interactions in both A and B, however, an estimate of the value of $-\Delta(\Delta\mathcal{G}_f)_{design}$ is the value of $-\Delta(\Delta\mathcal{G}_f)_{sub}$ that is expected when a single-site substitution (cyclization, etc.) introduces no adverse interactions, such as strain or steric crowding. This will be referred to as a *compatible modification.*

a. *Backbone rigidity.* Decreasing the conformational freedom of the backbone decreases the entropy of unfolding, thus stabilizing a compatible folded state. (In the configuration space picture of Section 4.3.3, folding is equivalent to compression; less-flexible molecules expand into a smaller region of configuration space when unfolded, hence the work of compression required to fold them is reduced). In accord with theoretical predictions, the compatible replacement of Gly with Ala (or, it is expected, with any amino acid having a β carbon) or of Ala with Pro each increase stability by ~7 maJ* (Matthews et al., 1987); placing Pro in an α helix, however, would illustrate an *incompatible* replacement. In augmented proteins, various cyclic amino and imino acids (Figure 15.1) could be used to constrain conformations in a similar manner, but with varied geometries.

Structures **15.1–15.4** suggest how cyclic moieties can alter and constrain backbone conformation. Cubane derivatives (suggested by J. Bottaro) are examples of rigid, saturated polycyclic structures. Structure **15.5** and derivatives can strongly constrain backbone ϕ and ψ angles (Toniolo, 1989). Structure **15.6** is

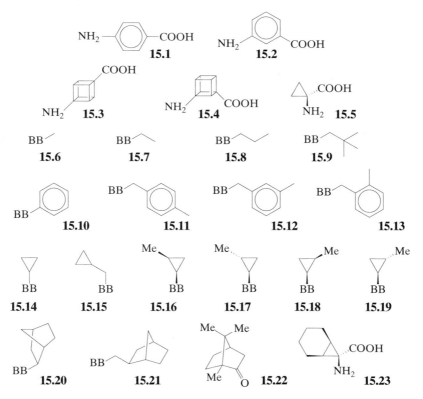

Figure 15.1. Structures illustrating rigidity and steric diversity; BB represents the backbone of a polypeptide

---

* In the following sections, values cited as ~7 maJ are to be regarded as round numbers, representing ~1 kcal/mole.

that of alanine; **15.7–15.9** are those of small, nonstandard R-groups of evident use in designing core packings. Structures **15.10–15.13** illustrate known phenylalanine-like structures, formally omitting a methylene [in dialkylglycines (Cotton et al., 1986)] or adding a methyl group (Porter et al., 1987). Removing the methylene increases R-group rigidity, but freely rotating methyl groups can, if constrained in the folded state, adversely affect stability. Structures **15.14–15.19** can be formally derived from standard amino acids by converting a methyl group into the methylene group of a cyclopropane ring, yielding a small change in steric bulk. Structures **15.14** and **15.15** correspond to valine and leucine; **15.16–15.19** correspond to isoleucine (and a side-chain diastereomer) and exhibit reduced conformational freedom. Structures **15.20** and **15.21** illustrate the use of cyclic moieties (which can bear rotationally hindered methyl groups) to provide rigid, diverse structures; the related structure of camphor, **15.22**, points to the usefulness of natural products as a source of diverse, inexpensive, chirally pure precursors. Structure **15.23** exemplifies the use of cyclic moieties for rigidly fixing bulky R-groups to the backbone. In synthesis, difficult monomers can be sandwiched between others with more tractable properties before incorporation into chains (Cotton et al., 1986). Selection of synthetically accessible amino acids and chain sequences is a legitimate application of design freedom.

Peptide researchers have exploited nonstandard amino acids and characterized their conformational effects (DeGrado, 1988; Toniolo, 1989). In particular, dialkylglycine residues impose severe restrictions on ϕ and ψ angles; relative to Ala, they should yield a $-\Delta(\Delta\mathscr{G}_f)_{design}$ on the order of ~7 maJ per use.

b. R-group rigidity. Folding flexible R-groups into a tightly packed, hydrophobic protein core reduces their conformational entropy. This adverse effect can be reduced by making greater use of relatively rigid R-group structures, such as the rings of phenylalanine, tryptophan, or nonstandard amino acids (Figure 15.1). On entropic grounds, each elimination of a rotational degree of freedom from an aliphatic chain should yield a $-\Delta(\Delta\mathscr{G}_f)_{design}$ equaling ~4 maJ.

Like other constraints compatible with a planned fold, increased R-group rigidity inhibits unplanned folds. The potential for unplanned folding is suggested by the work of Ponder and Richards on defining sets of residue sequences and R-group conformations compatible with a given backbone conformation (Ponder and Richards, 1987b). A geometric test for sequences of core residues compatible with the fold of rubredoxin yielded 44 sequences, but 236 sets of compatible conformations—that is, an average of over 5 plausible core packings per sequence. Allowing small variations in backbone conformation would presumably have enlarged this number. Both lamprey hemoglobin and crambin have internal residues with alternative conformations (Smith et al., 1986).

Greater use of rigid R-groups makes alternative packings less likely, increasing the specificity of packing and folding. However, it will also increase the difficulty of finding a sequence that has even one acceptable packing.

c. Steric diversity. More rigid R-groups provide fewer shapes per structure, but this can be countered by providing more structures. How many shapes are needed to make tight packings possible? The nonpolar amino acids Ala, Val, Leu, Ile, Pro, Phe, and Trp assume 18 common rotameric conformations in proteins (Ponder and Richards, 1987b); in a sample protein population they make up 37% of the residues (Creighton, 1984), and a larger fraction of the interior residues. This suggests that use of a few tens of relatively rigid R-groups can enable the design of packings having a large fraction of rigid structure.

Natural protein interiors have a mean volumetric packing density of about .74 (Ponder and Richards, 1987b) with local variations from <.6 to >.85 (Richards, 1977). A cavity the size of a methylene group decreases the stability of the folded state by about 8 maJ (Kellis et al., 1988), and various unfavorable interactions in the folded state (e.g., adverse van der Waals contacts) reduce stability by an estimated 14 maJ per residue (Creighton, 1984). Expanding design freedom by including D-amino acids and nonstandard R-groups (Figure 15.1) should increase achievable packing densities while reducing adverse van der Waals contacts—a wider choice of pieces will help designers find better-fitting patterns.

d. Pairwise matching. An unworkable—yet instructive—way to think of protein folding is to regard each interior R-group as binding to a receptor site formed by the rest of the fold. In this entirely nonmodular picture of cooperative folding, every part fits because of the way the rest fits, offering no point of departure for design and analysis. In standard proteins, regularities at the level of secondary structure provide a partial escape from this quandary.

In augmented proteins, a large R-group could provide a preformed, fold-independent binding site for another R-group, yielding modularity at the level of monomer pairs. Supramolecular complexes (Cram, 1988; Lehn, 1988) and nucleic-acid base pairing in folded RNA and in DNA junctions (Robinson and Seeman, 1987) provide models for such pairwise specificity. Even weak pairwise binding could aid folding specificity, while stronger interactions could strongly direct folding by forming complexes stable enough to mimic covalent cross-links.

e. Cyclic and branched backbones. Fold-compatible cross-links decrease the entropy of unfolding, stabilizing the folded state [by about 2.8 maJ per residue in a 10-residue loop at room temperature (Poland and Scheraga, 1965)]. Engineered disulfide cross-links have been used to stabilize natural proteins (Perry and Wetzel, 1984). Since looped structures (closed, for example, by °amide bonds between R-groups) can be made using solution chemistry and then treated as monomers during solid-phase synthesis, augmented proteins can incorporate cyclic backbone segments.

Studies of omega loops (Leszczynski and Rose, 1986) show that small looped structures (6 to 16 residues) can be globular and densely packed. Analogous cyclized structures can be regarded as small domains or large monomers, serving as modular elements in design and folding. These elements can be large enough to exhibit pairwise selectivity and binding, strongly guiding the folding process.

Branched structures decrease the entropy of unfolding through excluded-volume effects. This augmentation strategy (Mutter et al., 1988) has brought striking success in constructing stable tertiary structures, including a 73-residue molecule with enzymatic activity (Hahn et al., 1990).

f. Staged thiol pairing. Many standard proteins form cycles postsynthetically, through oxidation of pairs of cysteine °thiol groups to form disulfide bonds. Cross-links exposed to a suitable mix of thiols and disulfides in solution undergo rapid thiol-disulfide exchange through S_N2 displacement (Saxena and Wetlaufer, 1970); this can anneal the disulfide bond system toward a pairing pattern

Table 15.2. Model thiol compounds and their pK_a values in aqueous solution.

Thiol	pK_a	ref.
Thiobenzoic acid	2.48	[a]
Thiolacetic acid	3.5	[b]
2-Nitro-5-mercaptobenzoic acid	4.53	[c]
4-Mercaptobenzenesulfonic acid	5.73	[b]
Thiophenol	6.5	[b]
Methyl mercaptoacetate	7.68	[b]
Cysteine	8.30	[b]
1-Thioglycerol	9.51	[b]
Ethanethiol	10.61	[b]
2-Methyl-2-butanethiol	11.35	[b]

[a] Whitesides et al., 1977.
[b] Danehy and Parameswaran, 1968.
[c] Riddles et al., 1979.

of minimal free energy (Creighton, 1978; Creighton, 1984). In augmented proteins, a wider choice of thiol R-groups can broaden the range of disulfide cross-link geometries. Further, bulky monomers (or cyclic backbone segments) can provide local steric constraints to guide cross-link formation.

Formation of disulfides under conditions of increasing oxidation and (initially) high pH can favor pairing of thiols in stages corresponding to their pK_a values, from highest to lowest (Table 15.2), making pairing more specific. Data from studies of the kinetics and equilibria of thiol-disulfide exchange (Szajewski and Whitesides, 1980; Whitesides et al., 1977) indicate that, although equilibration among disulfides would destroy pK_a-specific pairing, lower pK_a thiols can be paired and equilibrated among themselves rapidly enough to limit destructive equilibration with established, less-labile disulfides formed from thiols of substantially higher pK_a. Preliminary Monte Carlo simulations of staged thiol pairing in this approximation indicate that molecules having solely within-class pairing (for 12 to 18 thiols in three pK_a classes) can be obtained with yields of tens of percent.

In this sequential annealing process, early-forming bonds can sterically constrain later bond formation [as in standard proteins (Creighton, 1978)]. This mechanism can hinder incorrect pairings within a set of thiols of a particular pK_a, enabling the formation of multiple specific cross-links per stage. The chief disadvantage of disulfides formed from low-pK_a thiols is extreme sensitivity to alkaline degradation (Danehy, 1971).

g. *Metal complexation.* Suitable R-groups can form constraining cross-links when metal ions are introduced into solution. In standard proteins, these constraints can force folding of structures as small as 30-residues [e.g., zinc fingers (Párraga et al., 1988)]. Engineering of metal binding sites has increased the stability of subtilisin (Pantoliano et al., 1988). Work in supramolecular chemistry suggests that single R-groups (e.g., based on α,α' bipyridine) can serve as preformed halves of metal binding sites, undergoing pairwise matching when metal

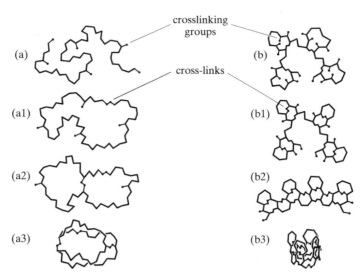

Figure 15.2. Staged crosslinking and cyclic backbone segments.

ions are added to a solution (Lehn, 1988). As in staged thiol pairing, the time at which crosslinking occurs can be controlled by manipulation of solution composition. Thus, metal complexation can add a further stage to the strategy of forming cross-links in locations determined by constraints established by previous cross-links. Metal binding can also drive the quaternary assembly of separately folded structures (Frankel et al., 1988); disulfide formation can do likewise. Figure 15.2 suggests how staged crosslinking could strongly constrain a folded state.

Series (a) shows in schematic form how six crosslinking groups in three reactivity classes (based on thiol pK_a values, Table 15.2, or on distinct binding specificities) might be used to constrain conformation through staged crosslinking. In stage (a1), the molecule is cyclized; in (a2) and (a3) the initial loop is further crosslinked. A theory of entropy for multiply-looped chains (Poland and Scheraga, 1965) indicates that a fold based on (a3) would be stabilized by roughly 105 maJ relative to the uncrosslinked structure (at 30 °C). More speculatively, series (b) suggests an extensive application of staged crosslinking in which multiple crosslinking groups pair at each stage and selectivity of pairing within a class (if achievable) is dependent on constraints resulting from prior structure. In this sketch, a chain is synthesized with cyclic backbone segments designed to sterically and entropically favor the pattern of pairing shown in stage (b1). Selectivity in stage (b2) depends on adequate steric constraints established in (b1), and stage (b3) similarly depends on the results of (b2). This suggests the possibility of structures in which cross-links virtually preclude unfolding.

15.3.5. Consequences for design

Table 15.3 compares a noncrosslinked standard protein to an augmented protein of similar size, but designed to exploit the tactics described in Section 15.3.4; relative stabilizations exceeding 450 maJ seem achievable in an augmented protein of only 50 residues. This increment is large compared to the

Table 15.3. Stability increment expected for a 50-residue augmented protein.[a]

Mechanism	Stability improvement (maJ)
a. Increased backbone rigidity	115
b. Increased R-group rigidity	50
c. Improved packing density	70
d. Reduction in bad van der Waals contacts	?
e. Cyclic backbone segments	110
f. Disulfide cross-links, metal complexation	110
Sum:	>455

[a] Differences between augmented and standard proteins can contribute to increased stability of the folded state. The basis of the estimates listed above is as follows: (a) 7 maJ per three residues, resulting from use of dialkylglycines and cyclic structures in one third of all residues. (b) 3 maJ per three residues, resulting from elimination of one rotational degree of freedom in one third of all R-groups. (c) Exploitation of steric diversity is assumed to increase packing density by 4%, from the mean for proteins, .74, to a high value for organic crystals, .78, which still falls short of densely packed regions of proteins (>.85); see Richards (1977). The associated stability increment is estimated from the 8 maJ stabilization resulting from filling a cavity the size of a methylene group (Kellis et al., 1988). (d) Expected to be substantial; not estimated here. (e) Based on change in conformational entropy from forming four 10-residue loops (Poland and Scheraga, 1965). (f) Taken as equal to (e); as with (a) and (b), the effect of this tactic will vary with the extent of its application.

stability of small evolved proteins (typically ~35-70 maJ), or even of α_4 [estimated at ~150 maJ (Regan and DeGrado, 1988)]. It may well be impractical (and unnecessary) to apply all these tactics to a single molecule. Margins of safety of this magnitude can expand design freedom in functionally critical parts of a molecule: if crafting a binding site or a flexible joint demands the sacrifice of several hydrogen bonds or creation of 0.1 nm^3 of internal space, this need not jeopardize the stability of the folded state.

Mechanisms for imposing conformational constraints—including backbone rigidity, R-group rigidity, cyclic segments, and staged thiol- and metal-based crosslinking—will not only stabilize the desired fold, but will destabilize (or simply block) an indefinitely large set of alternative folds. The destabilization of unwanted alternatives can be quantified: since ~275 maJ of the stability increment results from decreases in conformational entropy imposed before folding begins, the number of energetically accessible alternative configurations (including both unfolded, misfolded, and misaggregated states) is reduced to ~10^{-29} of the unconstrained value. By making results more predictable, this should increase the success rate in the trial-and-error process of designing macromolecular objects. Several groups have developed software to aid protein design (Blundell et al., 1988; Pabo and Suchanek, 1986; Ponder and Richards, 1987b); these systems appear adaptable to augmented proteins. Further, as indicated by the discussion in Section 4.4.3, stability achieved through rigidity reduces the sensitivity of designs to the errors in molecular mechanics models; these errors pose a serious problem where natural proteins are concerned.

This discussion has focused on backbone and hydrophobic core structures, but surface properties largely determine molecular behavior. These, too, are targets for design. For example, solubility—though aided by structures that favor the folded state and thus disfavor alternative, aggregated states (Mutter et al., 1988)—can be further increased by the generous use of polar and charged R-groups on the molecular surface; these can include groups bearing polar (Gehrhardt and Mutter, 1987) or multiply-charged polymeric chains (large, solubilizing side chains that can be cleaved and discarded after a molecular object has bound to a larger assembly may be of particular value). Broadened residue options will facilitate the design of surface geometry, electrostatics, reactivity, and so forth.

15.3.6. Trade-offs and applications

By expanding design freedom, the use of augmented proteins can aid the design of stable, predictable, functional molecules. Several of the tactics proposed here for augmented proteins may prove difficult to implement, but avoidance of those that prove too difficult is a proper use of design freedom. Commercial solid-phase synthesis of standard peptides now offers products with costs of ~$10 per milligram for $1000 orders of crude oligopeptides in the 10-mer to 20-mer range. Prices climb steeply with chain length and purity; they fall steeply with increased quantity. Lengths near the 50-mer range are approaching routine practice (Kent, 1988).

Feasible tactics should yield stable folds in chains of readily accessible length. Small size aids characterization (without crystallization) via solution nuclear magnetic resonance techniques (Clore and Gronenborn, 1987); these have been extended to structures several times as large as those contemplated here (Kay et al., 1990). Larger molecular systems can then be designed as quaternary structures self-assembled from these building blocks; a review of relevant principles appears in Lindsey (1991), and a more general survey appears in Whitesides et al. (1991). Various high-value molecular products might be worth making, despite the cost of chemical synthesis. Among these are systems of molecular machinery like those discussed in Section 16.4.

15.4. Macromolecular objects via mechanosynthesis*

15.4.1. Motivation

The atomic force microscope [AFM (Binnig and Quate, 1986) also termed a scanning force microscope] enables surfaces to be positioned relative to one another in a solution environment with ~0.01 nm positional precision and tip compressive loads as low as 0.01 nN (Hansma, 1990; Weisenhorn et al., 1989). AFMs are most often used to form images of surfaces by probing mechanical interactions at many points. This section suggests how AFM mechanisms could be exploited to enable solution-phase mechanosynthesis; instruments of the sort described also have more immediate applications in improving AFM imaging.

---

* Portions of the discussion in this section are based on Drexler (1991b, 1992).

The lack of reproducible, well-characterized, atomically sharp tips in standard AFM systems causes difficulties in image resolution, interpretation, and reproducibility. It has been proposed (Drexler and Foster, 1990) that AFMs with molecular tips could both alleviate these problems and provide an approach to achieving positional control of chemical synthesis. Single molecular tips, however, present problems of fabrication, yield, and damage during use.

Molecular tip arrays as proposed here can permit screening and interchange of tips during operation, reducing sensitivity to yield and damage. They can enable sequential use of differing tips to image a sample and to perform solution-phase mechanosynthesis. A penalty of this flexibility is limitation to small substrates. R. Merkle has proposed an analogous approach based on scanning tunneling microscope mechanisms working with adsorbed reagent molecules in vacuum.

15.4.2.   Tip-array geometry and forces

a.   Tip-array geometry.   The geometry of a tip-array system is illustrated in Figures 15.3 and 15.4. On a large scale, Figure 15.3(d), a *primary bead* can be viewed as an AFM tip, imaging an array of *flat-side beads*. On a small scale, Figure 15.3(a), a single molecular tip on the flat-side *secondary bead* images the primary bead. The terms *substrate* and *tip* are thus ambiguous, but under the conditions of greatest interest, the primary bead acts as the substrate and a molecule on the secondary bead acts as the tip. Key geometrical parameters are defined by Figure 15.4. Reasonable dimensions for a device with this geometry

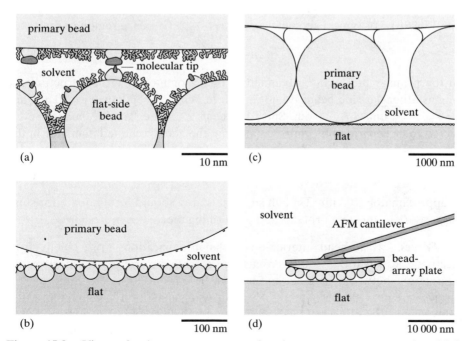

Figure 15.3.   Views of a tip-array geometry using tip support structures and multiple substrate beads on an AFM cantilever. Each view differs from the next by a factor of ten in scale.

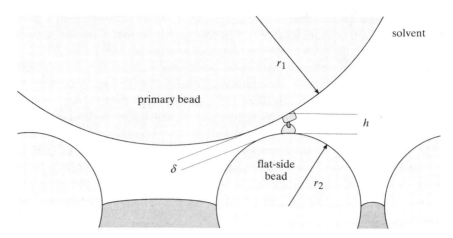

Figure 15.4. Sketch illustrating the relationships among the bead, flat, and molecular tips, including basic geometric parameters.

are $r_1 \approx 500$ nm, $r_2 \approx 50$ nm, $h \approx 4$ nm, and $\delta \approx 6$ nm. Commercial AFMs (Prater et al., 1990) enable operation in a fluid medium with atomic resolution and scan ranges of 25 μ; they seem suitable as basic mechanisms. The following assumes that the fluid medium is an aqueous solution, but anhydrous solvents are feasible and sometimes desirable.

If bead-bead forces and stiffnesses are small relative to the required imaging forces and stiffnesses, they will permit imaging based on interactions between the tip and the target structure. To limit bead-bead interactions, a tip must have a height h greater than some minimum imaging separation δ (discussed in Section 15.4.2b). For typical geometries with $r_1 > r_2 \gg (h - \delta)$, a region of diameter

$$d \approx \sqrt{8(2h-\delta)(r_1+r_2)r_1/r_2} \qquad (15.1)$$

on the primary bead can be scanned to image or modify target structures, so long as other flat-side beads do not interfere. With $r_1 = 500$ nm, $r_2 = 50$ nm, and $(2h - \delta) = 2$ nm, $d \approx 300$ nm. Since the centers of the flat-side beads are expected to be separated by approximately r_2, the actual scanned diameter in this instance is

$$D \approx 2r_2 \qquad (15.2)$$

or approximately 100 nm. Though small, this area should be adequate for some purposes (e.g., imaging proteins and performing mechanosynthesis).

b. Forces. Bead-bead interactions at small separations arise chiefly from electrostatic, solvation, van der Waals, and so-called steric forces. Electrostatic interactions between the bead and flat can be minimized by ensuring that the surfaces are near electrical neutrality, or by using a solution in which ions effectively screen the effects of surface charge. Moderate concentrations of salt in water (≥ 0.1 M NaCl) result in a Debye screening length <1 nm, small compared to $\delta \approx 6$ nm. Solvation interactions can operate over multiple molecular diameters: in water, the repulsion between hydrophilic surfaces (Israelachvili, 1992) falls off exponentially, becoming acceptably small at separations of 1 to 3 nm (hydrophobic forces would cause strong adhesion between surfaces and must be avoided). Oscillating solvation forces associated with molecular size effects become small at separations greater than ~2 nm in water (Israelachvili, 1992), and are reduced by surface roughness and mixed-solvent systems. With rough bead surfaces and $\delta \approx 6$ nm, oscillating forces should not be observed.

Bead-flat van der Waals interactions can be described using continuum models based on the Hamaker constant (Section 3.5.1a). The results are thought to give a reasonably accurate picture of the forces and their dependence on geometrical parameters. One family of approaches to bead fabrication would use polystyrene-like materials; for polystyrene interacting through water, the Hamaker constant $\mathcal{A} \approx 1.3 \times 10^{-20}$ (Israelachvili, 1992). (This neglects Debye screening of zero-frequency contributions to the Hamaker constant.) With the given geometrical parameters and an intersurface separation $s = \delta = 6$ nm the force F is attractive and ≈ -0.003 nN (from the relationship stated in Figure 3.10(c), which neglects retardation effects). This is smaller than the minimum tip contact force reported in AFM work, 0.01 nN, and is associated with a stiffness of only ~ -0.001 N/m, which is negligible in comparison to the stiffness of a typical AFM cantilever or of an interaction between two molecules in contact. Accordingly, this force should cause little difficulty.

Previous papers (Drexler, 1991b; Drexler, 1992; Drexler and Foster, 1990) have described less-symmetrical structures, with support molecules attached on a single side to avoid a requirement for two distinct attachment technologies. In those geometries, intersurface separations are smaller, requiring material and solvent choices that minimize the net Hamaker constant. The present approach has greater symmetry, enabling a single attachment technology to mount both target-supporting and tip-supporting molecules. The resulting increase in intersurface separation relaxes constraints on the Hamaker constant, and hence on the choice of materials and solvents.

Figure 15.3(a) illustrates the use of short solvophilic chains (e.g., polyoxyethlyene) on the bead surfaces to provide a short-range repulsive force caused chiefly by entropic effects. This mechanism is termed *steric stabilization* and is widely used to prevent the coagulation of colloids. Attachment of these chains renders the bead surfaces hydrophilic and inhibits the bead-bead adhesion that would otherwise occur though short-range hydrophobic and van der Waals attractions.

c.　Multiple beads.　Figures 15.3(c) and 15.3(d) illustrate larger-scale geometrical features of a system that permits primary beads to be interchanged during operation by tilting the flat relative to the cantilever. Except for molecular structures at the finest scales, Figure 15.3(a), and platelike structures at the largest scales, Figure 15.3(d), each of the shapes illustrated can be generated by the solidification either of a droplet or of a fluid forming a meniscus between two objects. With the same exceptions, all close contacts occur between spherical surfaces and are insensitive to angular misalignment. Accordingly, the most severe fabrication challenge appears to be that of attaching suitable molecules to beads, and doing so with adequate stiffness.

15.4.3. Molecular tips and supports in AFM

a. Protein-ligand tips.   Potential tip structures include proteins or particles bearing small adsorbed molecules. A particularly versatile approach exploits the capabilities of both organic synthesis and biotechnology by using synthetic ligands as tips and protein molecules as supporting structures. Many natural proteins bind partially exposed ligands. Ligand analogs could be synthesized with protruding moieties having steric properties suiting them for use as AFM tips. Extensions of monoclonal antibody technology (e.g., Huse et al., 1989) can rapidly generate proteins able to bind almost any selected small molecule. Single-chain Fv proteins [combining antibody V_L and V_H sequences (Bird et al., 1988)] are compact, ~3 nm in height, and lack hinge regions. Use of these antibody-derived proteins can allow broad freedom in ligand design, if their stability is (or can be made) adequate for use in an AFM mechanism (see Section 15.4.3c). Techniques for attaching proteins to surfaces are noted in Section 15.4.4.

b. Stiffness.   To be used as AFM tip-supports and as imaging targets, proteins and protein-ligand complexes must have adequate mechanical stiffness. In protein crystals [which have been used as models for bound protein-protein complexes (Finkelstein and Janin, 1990)], individual atoms in the protein interior typically undergo a 0.03 to 0.05 nm root-mean-square displacement caused by thermal vibration (Creighton, 1984). The atomic-displacement stiffness, k_s, can be derived from the root mean square displacement via the relationship for a thermally excited harmonic oscillator [Eq. (5.4)], which implies $k_s \approx 1.6$ to 4.6 N/m for typical atomic displacements in the interior (k_s for a side chain on a surface often is far lower). Proteins in crystals typically are anchored to neighbors by a few side chain contacts; comparably stiff attachment of proteins to surfaces seems achievable.

Rigid, polycyclic structures of substantial size can be made by organic synthesis (e.g., Webb and Wilcox, 1990), and their internal stiffnesses can exceed those of proteins. Such ligands can be anchored relative to the protein by numerous van der Waals interactions of substantial stiffness (Section 3.3.2e), yielding ligand atomic-displacement stiffnesses toward or beyond the upper end of the range characteristic of proteins: the following sections will assume a value of 3 N/m. The stiffness of the interaction in imaging a protein by a protein/ligand complex can thus be ≥ 1 N/m; AFM cantilevers with $k_s < 1$ N/m should provide good responsiveness.

c. Stability of proteins under compressive loads.   Applied forces can destabilize protein folding and ligand binding. The free energy required for unfolding or unbinding (Creighton, 1984) typically is ~50 to 100 maJ; at the larger energy, the unfolding half-life is on the order of 1000 years (assuming a 1 s folding time; typical values range from 0.1 to 1000 s). From a kinetic perspective, the destabilizing energy increment resulting from a force equals the work it performs as the molecule moves from equilibrium to a transition state for unfolding (where the location of the transition state is affected by the force); estimating this energy requires an estimate of the atomic displacements associated with such a transition state.

In a linear elastic system, the strain energy $\mathcal{V}_s$ is related to a displacement Δx by

$$\Delta x = \left(2\mathcal{V}_s / k_s\right)^{1/2} \tag{15.3}$$

Large $\mathcal{V}_s$ and low k_s yield a large (adverse) value of Δx. For $\mathcal{V}_s = 100$ maJ and $k_s = 1.6$ N/m, $\Delta x = 0.35$ nm. Localized displacements of this magnitude (an atomic diameter) might independently be expected to disrupt the tight core packing necessary for stability (Richards, 1977). Over this displacement, 0.1 nN performs 35 maJ of work, which is small compared to the energetic differences between more and less stable proteins. Thus, on kinetic stability grounds, this estimate suggests that 0.1 nN forces should be compatible with imaging of proteins of moderate stability (occasionally interrupted by denaturation), and compatible with the use of tips based on proteins of good stability incorporating well-bound ligands.

Considering only stiffness and acceptable elastic deformations (taken to be 0.01 nm), and calculating from a continuum model, a maximal tip force of 0.01 nN had been suggested for protein imaging (Persson, 1987). Note, however, that with $k_s \approx 1.6$ N/m (a low value from experimental data), a 0.1 nN force would yield $\Delta x \approx 0.06$ nm and $\mathcal{V}_s < kT_{300}$. These values are compatible with imaging yielding useful structural data.

In AFM systems of standard geometry, repulsive interatomic tip forces (in the presence of net long-range attraction) have been reduced to low values. It had been suggested that forces as low as 0.01 nN should be within reach for systems operating in solution (Weisenhorn et al., 1989), and this was subsequently achieved (Hansma, 1990). Comparably low forces should likewise be possible in multiple-tip systems, providing a substantial margin of safety when using proteins as tip supports or as imaging targets.

15.4.4. Attachment of supporting molecules

The immobilization of molecules (and proteins in particular) is a well-established technique in biotechnology (Scouten, 1987; Uy and Wold, 1977), often used to retain molecules in a vessel as fluid flows past. Immobilization commonly relies on the formation of cross-links between reactive moieties on a protein and on a surface. Molecular tip arrays, however, will require more than immobilization against macroscopic displacement: they will require relatively stiff attachment of molecules to a surface.

Immobilization requires only a molecular tether; stiff attachment requires three (or more) points of contact providing tripodlike stability. The stiffness of a tip can be no greater than the stiffness of the interface attaching it to a surface. A stiffness of ~1.6 N/m combined with a tip contact load of ~0.01 nN can produce an effective concentration of ~145 nm^{-3} for a tip moiety relative to a point on a substrate, in the presence of AFM jitter and thermal noise (Section 15.4.6a). This stiffness is compatible with the mechanical properties of proteins, assuming good attachment. It also suffices for strong positional control of chemical reactions.

A previous paper (Drexler, 1992) suggests strategies for achieving stiff attachment of proteins to surfaces, several of which exploit the engineering of protein-surface side chains to provide specific binding interactions. One family of approaches would attach proteins to surfaces directly, via the bonding of side chains (e.g., the thiols of cysteine) or chemically-modified side chains. Another family of approaches would use techniques akin to the freeze-cleave-etch methods used to prepare specimens for electron microscopy. A final family of approaches discussed there would link proteins to molecules embedded in fluid Langmuir–Blodgett films, which could then be polymerized to increase their stability; this has the advantage of forming (in effect) a binding site for the protein, thereby increasing its stability. A more recent approach, suggested by the Langmuir–Blodgett film approach combined with ideas suggested by work on imprinting small molecules in polymers (e.g., Fischer et al., 1991; Robinson and Mosbach, 1989) would link proteins to surfactant molecules embedded in a droplet that is later polymerized to form a bead. In light of these possibilities, suitable attachment techniques appear feasible.

15.4.5. Imaging with molecular tips

On a large scale, scanning the primary bead provides a low-resolution image of the array of flat-side beads. In regions where the primary bead and a secondary bead interact through steric repulsion forces between the polymers, the image is expected to be essentially featureless. In regions where a molecular tip on the secondary bead interacts with a target structure on the primary bead, the image is expected to show structural features at high resolution. Systematic probing in these regions (moving in and out by several nanometers at a series of points on a grid) can generate a map of force vs. distance for the interaction between tip and target. A surface in that map corresponding to a large, constant stiffness should in many instances resemble an atomic-resolution topographic map.

For a system with a $\sim$25 μ scan range and $d \approx 100$ nm, the number of immediately available tips can be $\sim 10^5$. A tip array can advantageously include tips that have diverse structures and orientations.

On an isotropic substrate, supporting molecules (and their bound tips) will be attached at diverse azimuthal angles. Bead curvature and structural irregularities will increase the diversity of orientations in the other two rotational degrees of freedom.

Tips can be polar, nonpolar, hydrogen bonding, positively charged, or negatively charged. Information gained by imaging a single molecule with tips of multiple types may help fulfill the goal (Drake et al., 1989) of determining folded protein structures through AFM and computational modeling. It may be that features can be reproducibly mapped on the surface of a protein, yet fail to indicate which residues occupy which surface locations. If so, then imaging and comparing a series of proteins (prepared via site-directed mutagenesis) that differ only by the substitution of a distinctive residue at a known position in the sequence could reveal much about the structure of the folded state.

Tips with specific biochemical affinities (e.g., enzyme substrate analogues, candidate drug molecule analogs, etc.) can yield data of special biological interest. A tip can be moved along a trajectory aligned with the flexure direction of the AFM cantilever, carrying it from solution to (for example) a protein surface and back again. Integrating the force curve (with suitable corrections for bead-bead interactions) then yields an intermolecular mean-force potential, that is, a

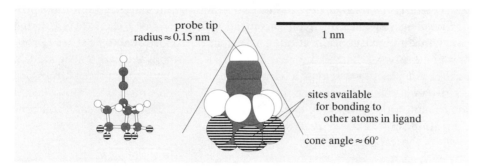

Figure 15.5. A candidate ligand framework with a sharp tip.

high-resolution spatial map of the free energy of the intermolecular interaction. For probes with tip structures of biological interest, this information can be of direct biological relevance regardless of the apparent image resolution.

To facilitate topographical imaging, organic synthesis can be used to prepare ligands with stiff tips of considerable geometric acuity. For example, molecular mechanics calculations (MM2/CSC) indicate that the H atom of $R-C\equiv C-H$ has a transverse bending stiffness of ~2 N/m relative to an sp^3 carbon in the R-group (e.g., in a polycyclic ligand structure like that in Figure 15.5); the longitudinal k_s is ≥ 100 N/m. The effective tip radius for such a probe is ~0.13 nm.

Tips made by current microfabrication techniques have irregular structures with characteristic radii estimated to be ~10 nm (Park Scientific Instruments, 1992). The molecular tips proposed here can have well-characterized structures of diverse types, attached in diverse orientations. Imaging specimens of known structure using multiple tips should permit discrimination and mapping of the tip types and orientations in a particular tip array. To take a simple example, positive, negative, and neutral tips will have distinct, contrasting responses to a bound charge.

Chemists frequently prepare derivatives of molecules to identify the original structure. Ligands used as molecular tips can include chemically reactive moieties that can be used to map sites of differing reactivity. This mode of operation will usually require prolonged dwell times at a candidate site, as opposed to steady scanning or rapid probing.

15.4.6. Solution-phase mechanosynthesis

Organic synthesis today offers powerful techniques for building molecular structures. Use of maneuverable molecular tips can add a fundamental novelty: flexible, positional control of sequences of synthetic steps based on control of local effective concentration. Receptors that bind reagent ligands can serve as tool holders for mechanosynthesis.

Molecular tip array systems enable approaches substantially different from those previously proposed (Drexler and Foster, 1990): Tip arrays and antibody technologies will enable the use of distinct receptors for reagents of different types, permitting a series of reactions without cycling the composition of the solution. With rapid, spontaneous dissociation no longer necessary for interchange, reagent ligands can be bound tightly; this enables larger effective concentration ratios and hence better site specificity.

a. Effective concentrations.   Reagent reaction rates are by definition proportional to reagent effective concentrations (Section 8.3.3a). All else being equal, the effective concentration of a reagent is proportional to its spatial probability density. Modeling a reagent ligand as a thermally excited object, subject to a force F pressing it against a barrier, to constraint by a transverse stiffness k_s, and to an additional Gaussian jitter in the AFM mechanism with a root mean square amplitude = σ_{AFM}, yields a probability density having a peak local concentration (m^{-3})

$$c_{local} = \frac{F}{2\pi kT}\left(\frac{kT}{k_s} + \sigma_{AFM}^2\right)^{-1} \qquad (15.4)$$

Neglecting energetic and orientational effects (which can, when favorable, greatly increase the effective concentration), and provided that k_s is substantially less than the transverse stiffness in the reaction transition state (as is the case here), local concentration corresponds to effective concentration. Displacements of well under 0.01 nm are now routinely measurable in stationary tips (Albrecht, 1990), but jitter from ambient vibration is of this order (Drake et al., 1989). Assuming σ_{AFM} = 0.01 nm, T = 300 K, F = 0.01 nN, and k_s = 1.5 N/m (between two structures with stiffnesses of 3 N/m) yields $c_{local} \approx 135$ nm^{-3} (~225 M) at 300 K. For comparison, mobile surface-residue thiol groups in proteins can exhibit effective concentrations exceeding 60 nm^{-3}; interior residues often exhibit much larger values (Creighton, 1984).

Rate constants for ligand binding commonly exceed 10^8 $nm^3 s^{-1}$. Antibody technologies commonly yield proteins that bind a chosen ligand with equilibrium °dissociation constants on the order of 10^{-9} nm^{-3} (Bird et al., 1988; Huse et al., 1989). On exposure to a solution with a 10^{-6} nm^{-3} ligand concentration, reagent receptors with these properties will become occupied in ~10^{-2} s, and at equilibrium will remain occupied with probability ~.999 (with these parameters, the mean waiting time for dissociation to occur is ~10 s). Application of mechanical load to a tip tends to increase its equilibrium probability of dissociation while decreasing its rate of dissociation; this behavior is compatible with the tip applications described here.

Under the conditions just specified, the peak enhancement in effective concentration of a reactive moiety on a molecular tip relative to the background solution concentration is ~$(100$ $nm^{-3})/(10^{-6}$ $nm^{-3}) = 10^8$. This localized, positionable enhancement in effective concentration can direct site-specific reactions on molecular structures attached to the bead. The transverse probability density distribution is Gaussian; with these parameters, the enhancement falls to background levels at a radius of ~0.3 nm, or about one atomic diameter. Thus, this mechanism for positional control can produce reaction rate differentials on the order of 10^8 between otherwise chemically equivalent sites separated by two to three bond lengths.

b. Anticipated synthetic capabilities.   The choice of reagent ligands is constrained by requirements for chemical compatibility. Protein molecules have been suggested as reagent receptors; they have reactive functional groups and ordinarily exist in an aqueous medium. Modifications (including immobilization itself) may help stabilize proteins, permitting their use in polar, nonaqueous media.

Reagent ligands also must not react with one another at substantial rates in solution, unless the resulting by-products are harmless to the process. Low concentrations reduce this problem, and a system that steadily flushes the AFM cell with freshly mixed reagent solution (ensuring a low residence time) can further reduce unwanted reactions. If the residence time is ~100 s, the effective-concentration ratio is ~10^8, and a typical mechanosynthetic reaction time is ~1 s (at an effective concentration of 100 nm^{-3}), then the concentration of side-reaction products will be ~10^{-12} nm^{-3}, or ~10^{-6} that of the reagents.

A mature AFM-based mechanosynthetic technology could potentially perform syntheses involving >10^5 sequential steps in the presence of ~100 chemically equivalent sites with only a small probability of failure resulting from reactions with uncontrolled reagent molecules in solution. In an operating system, each reaction step removes a reagent ligand from a receptor that is promptly recharged by a molecule from the solution. Each reaction step can be tested for success by AFM imaging, to ensure completion. The extension of tip-array technology to encompass solution-phase mechanosynthesis will require a multidisciplinary collaboration with synthetic organic chemistry playing a prominent role. A major objective is the identification of building blocks that are (1) accessible via conventional solution-phase synthesis, and (2) adequate for use in building molecular machine systems, given positional control of the expected reliability; initial studies have been undertaken by M. Krummenacker.

15.4.7. Summary

Analysis suggests that AFMs with arrays of molecular tips may have substantial advantages over conventional AFM systems for the intensive study of molecular-scale specimens, including protein molecules. The chief sacrifice is a drastic reduction of the effective substrate area, which is limited to a fraction of the bead diameter. The opportunity to select among ~10^6 tips, guided by imaging results and without interrupting system operation, sidesteps the requirement for good tip yield and durability imposed by single-tip systems. Selection of tips with desirable mechanical and orientational properties from a population comprising well-defined molecular structures will enable more thorough and unambiguous characterization of samples. Use of reactive molecules as tips will enable nanofabrication by positional control of chemical synthesis. The estimated reliability is sufficient to enable >10^5 sequential steps, enabling the construction of molecular objects ~10^3 times larger than typical protein molecules and existing products of chemical synthesis.

15.5. Conclusions

Present technology can be extended to build macromolecular objects and systems in any of several ways:

- Biotechnology (and protein engineering in particular) can be extended to make ~300 monomer molecular objects capable of forming larger structures of indefinite size and complexity by Brownian assembly. Problems with design are the chief obstacles to this path.

- Solution-phase synthesis can be used to make ~50 monomer molecular objects with greater cost per unit mass, but with lessened design difficulty. Again, with suitable design, these objects can be made to self-assemble to form larger and more complex objects. Problems on this path seem smaller in total, but are more interdisciplinary, requiring innovations both in synthesis and design.
- Solution-phase mechanosynthesis guided by AFM mechanisms appears feasible, and could be used to build molecular objects containing ~10^5 highly crosslinked monomers without need for Brownian assembly. This path promises the greatest speed and flexibility of molecular construction, but its implementation requires still broader interdisciplinary collaboration.

Opportunities for hybrid approaches are evident. Biotechnology (and the stockpile of molecular components found in nature) can provide devices that can be incorporated into other structures. Solution-phase synthesis of medium-sized building blocks could provide components useful in mechanosynthesis. Molecular-tip AFM mechanisms could be used to characterize the products of biotechnological or chemical experiments, reducing the time required to detect and correct errors.

Each of the three approaches surveyed in this chapter grows out of experimental research already in progress; each can yield short-term scientific rewards while developing capabilities useful in building solution-phase molecular machine systems. Making new systems work in the laboratory is far more difficult than merely surveying possible approaches. By the same token, however, it is important to identify areas where successful experimental work can yield capabilities that open new fields of endeavor. If work is difficult, then it is best applied where the rewards seem large.

Some open problems. Much of this chapter can be taken as a description of open experimental problems. On the theoretical side, it would be useful to identify sets of compatible building-block molecules for use in solution-phase synthesis or in mechanosynthesis. This is a system design problem with numerous chemical constraints, not a problem of organic synthesis per se; it will not be solved by the traditional research strategy of synthesizing individual molecules having interesting structures or useful interactions with living organisms. Molecular CAD systems for the design of structures and syntheses based on such building blocks would be of great value.

16

Paths to Molecular Manufacturing

16.1. Overview

Chapter 15 describes extensions of present capabilities that would facilitate the construction of complex macromolecular structures, working in a solution environment. The chapters of Part II describe systems able to perform high-speed, high-efficiency, high-reliability mechanosynthesis of macroscopic objects, working in a vacuum environment. The present chapter attempts to describe how the gap between these technologies could be bridged by the development of a series of small mechanosynthetic devices. Section 16.2 describes the overall strategy. Sections 16.3 and 16.4 describe intermediate technologies, planning backward from advanced objectives toward present capabilities. Section 16.5 discusses factors affecting the pace of development.

The author apologizes in advance to experimentalists who may be offended by so quick a summary of so large a set of objectives. The work outlined here could readily absorb researcher-centuries of effort: many of the steps described will, if attempted, spawn a host of subproblems, each demanding long, hard, and creative work. It would, however, become burdensome to point this out at every turn. Developments that will one day make molecular manufacturing fast and easy will result from efforts that are slow and difficult.

16.2. Backward chaining to identify strategies

16.2.1. Forward vs. backward chaining

In a *forward chaining* search (as the term is used in computer science), one pursues a goal by taking steps that may lead toward it, sometimes exploring blind alleys. If, however, all possible destinations are considered equally good, then there is no real goal, hence (by this loose standard) forward chaining never fails. In science, it is common to pursue experimental programs opportunistically, choosing next steps based on immediate prospects and a sense of what is interesting, important, and fundable. This process resembles forward chaining with abundant, unranked goals, and it routinely produces incremental advances in knowledge and capabilities.

In *backward chaining,* one first describes a goal, then searches for intermediate situations one step removed from the goal, then for situations one step removed from *those,* and so forth, planning backward toward situations that are immediately accessible—that is, toward potential first steps on an implementation pathway. If there are many potential first steps, then backward chaining can be particularly attractive. In technology, it is common to select a goal based on its near-term feasibility and economic attractiveness and then plan backward to select the necessary parts and procedures. The enormous range of modern technological capabilities often provides many possibilities, hence engineers are more often concerned with cost and performance than with feasibility alone.

16.2.2.   Evaluating paths to molecular manufacturing

Molecular manufacturing is a technological goal (presenting many scientific challenges), but it cannot be achieved in a few steps or a few years. Accordingly, one cannot expect to succeed by combining just existing parts and procedures. Backward chaining is still appropriate, but the links in the chain are intermediate technologies, not mere parts and assembly procedures.

Opportunistic, forward-chaining research strategies have moved toward increasing molecular control, and continued progress will eventually encompass the methods of molecular manufacturing (unless other methods can deliver equivalent capabilities). These opportunistic steps have been guided by perceived short-term rewards, but molecular manufacturing can be achieved more directly if long-term strategies and rewards guide short-term decisions. In a society that heavily discounts its future, however, long-term research directions must deliver short-term rewards. Early steps are described in Chapter 15, and some mention has been made of their immediate applications.

Short-term rewards aside, the chief criteria for development paths are directness and reliability. A path with fewer sequential steps can likely be traversed more rapidly. A path (or family of paths) that offers many choices at each step is more reliable, since it is unlikely that any single problem can block all choices. Assuming adequate short-term rewards, the total cost of following a development path is of little significance. Since 1950, the total cost of developing computers has been enormous, but the industry has earned steady returns. The present case seems similar.

16.2.3.   Overview of the backward chain

Table 16.1 summarizes the characteristics of several technologies (Stages 1–3) that might be developed on the way to advanced molecular manufacturing systems (Stage 4). The descriptions of Stages 2 and 3 assume that the first molecular machine systems work in a liquid medium. The description of Stage 1 assumes that the process begins with AFM-based mechanosynthesis.

This table omits many alternatives. Sections 15.2 and 15.3 have described (Brownian) self-assembly strategies that could replace Stage 1 as a basis for developing Stage 2. R. Merkle has suggested vacuum-based mechanosynthesis systems; the most likely subsequent steps would differ greatly from Stages 2 and 3 as described here.

The following sections trace a path backward through the stages of Table 16.1. Section 16.3 attempts to describe the simplest molecular manufacturing

Table 16.1.   Differences between proposed generations of mechanosynthetic assembly systems, taking Stage 1 as AFM-based molecular manipulation (Section 15.4, but see also Section 15.2 and 15.3); Stage 4 embraces a range of systems, including those described in Chapter 14.

Characteristic	Stage 1	Stage 2	Stage 3	Stage 4
System size	macroscopic	submicron	submicron	up to macro
Initial production	present tech.	Stage 1 syst.	Stage 2 syst.	Stage 3 syst.
Molecular parts	$\sim 10^1$ to 10^2	$\sim 10^5$ to 10^6	$\sim 10^7$ to 10^8	$\sim 10^8$ to 10^9+
System assembly	surface chemistry	folding block-chain	positional assembly	positional assembly
Structural materials	folded polymers	crosslinked polymers	crosslinked polymers	diamondoid solids
Instructions	external	external	external	internal
Internal comp.	no	no	no	yes
Cycle times	~ 1 s	~ 100 ms	~ 10 ms	~ 1 μs
Error rates	$\sim 10^{-6}$	$\sim 10^{-9}$	$\sim 10^{-12}$	$\sim 10^{-15}$
Working medium	solution	pure liquid	pure liquid or vacuum	vacuum
Reagent transport	diffusion	minimal conveyance	simple conveyance	conveyance
Stray molecules	solvent, reagents	solvent, contaminants	solvent, contaminants	rare
Feedstocks	complex	complex	complex	simple
Reagent processing	none	reactive activation?	reactive activation?	extensive
Reagents	moderate reactivity	high reactivity	high reactivity	extreme reactivity
Applied forces	small	moderate	moderate	large
Energy efficiency	negligible	low	low	good

systems able to use highly active reagents (Chapter 8) to produce diamondoid materials (Part II). Section 16.4 then describes how systems of this sort could themselves be built by systems working with more conventional reagents in a liquid medium, and then how *those* systems could be built by smaller and simpler systems working in the same medium. These, in turn, could be built using one of the macromolecular engineering approaches described in Chapter 15.

16.3.   Smaller, simpler systems (stages 3–4)

16.3.1.   Macroscopic via microscopic manufacturing systems

With suitable instructions, the exemplar manufacturing system described in Chapter 14 could build a similar system of its own size, or larger. Conversely, it could be built by a similar system smaller than itself. What is the smallest, simplest system of this kind, able to build similar systems that are larger and more complex? Since both CPUs and manipulators can be of submicron size, an initial system can evidently be of microscopic dimensions, but more radical simplifications are possible.

One strategy for simplifying early nanoscale systems is to transfer complexity to the macroscopic world. Section 16.3.2 describes how onboard computing, information storage, and power supply systems can be replaced by transducers that respond to externally imposed acoustic signals. Section 16.3.5 describes how feedstocks consisting of pure solutions of large molecules can enable one-step mechanisms to replace complex molecule acquisition and purification systems. Sections 16.3.3 and 16.3.4 describe tactics for building simpler manipulators and for achieving the most important benefits of vacuum operation in a fluid medium. Combining these features, Section 16.3.6. describes a small mechanosynthetic device.

16.3.2. Acoustic power and control

a. Broadcast instructions. Replacing computers and stored instructions with broadcast instructions can simplify molecular manufacturing systems. If all operations are sufficiently reliable, then sensors and logic for directing conditional operations can be omitted. (Reliability will result from some combination of computational modeling, model-error-tolerant design, experiment, and redesign.) Broadcast instructions can enable system replication without internally stored instructions.

Signals can be broadcast to molecular mechanical systems in several ways (e.g., by light via photochemical isomerization), but signaling by modulated pressure via mechanical transducers is both simple and adequate. It is thus of interest to consider pressures and pressure histories that can be produced by the application of present technologies. The discussion that follows is generic, but some of the specific parameters and conclusions developed are used in later sections.

b. External control of the pressure history. The pressure on a device can vary over (at least) the range in which the surrounding solution remains liquid. At ~300 K, the solidification pressures of several pure solvents are >2 GPa (for example, >3 GPa for methanol); using mixed solvents can expand the available range. The following calculations assume that pressure can vary over the range $0 \leq p \leq 2$ GPa.

Available laboratory equipment can subject 50 cm^3 of solution to static pressures of up to 2 GPa (Isaacs and George, 1987). Broadcasting of instructions, however, requires varying the pressure. At low frequencies, suitable designs include small cells modulated by pistons driven by relatively large actuators. If the diameter of the cell and piston are ~10^{-3} m (permitting volumes >$10^9 \mu^3$), then the maximum required force is ~2×10^3 N, and electromechanical actuator dimensions <5 cm appear reasonable. It appears feasible to sweep the ambient pressure in a cell through its full range in ~ 1 ms. The approach outlined below uses the ambient pressure in the cell to select transducers of a particular class, and acoustic pressure fluctuations of lesser magnitude (± 1 to 100 MPa) and greater frequency (~10^6 to 10^7 Hz) to drive them. (Acoustic wave amplitudes can be increased by focusing.)

At pressures up to 2 GPa, the speed of sound in organic liquids is ~1000 to 2000 m/s, with negligible dispersion of phase velocities at megahertz frequencies

(Gray, 1972). The amplitude of acoustic plane waves in a liquid decays exponentially with distance, and is usually well described by the expression

$$A = A_0 \exp\left(-C_{\text{atten}} f^2 x\right) \tag{16.1}$$

where A is the amplitude, A_0 the initial amplitude, f the frequency (Hz), x the propagation distance (m), and C_{atten} a measure of the attenuation (s²/m). For organic liquids at ~300 K and low pressures, the value of C_{atten} ranges from ~3×10^{-14} to ~6×10^{-12} s²/m, declining with increasing pressure (Gray, 1972). For a typical value, $C_{\text{atten}} = 10^{-13}$ s²/m, a 10 MHz signal can propagate ~1 cm with ~10% attenuation; sizes and frequencies in this range are compatible.

A final concern is conversion of high-amplitude sound waves to shock waves by nonlinear properties of the fluid. Shock-wave formation, however, requires a substantial propagation distance and does not occur in systems having sizes, frequencies, and amplitudes in the range considered here.

At 10 MHz, a 100-pulse train lasts 10^{-5} s, during which the ambient pressure (if modulated over its full range in 1 ms) changes by only ~1% of its peak value. It will here be assumed that, at a particular point in the cell, the pressure can be varied over a 2 GPa range in 1 ms, and that acoustic pulse trains with frequencies of up to 10 MHz can be superimposed. For comparison, pulse-echo diagnostic ultrasound systems used in medicine can produce peak-to-trough pressure differentials of ~10 MPa, with frequencies of 1 to 20 MHz and pulse lengths measured in microseconds (ter Haar, 1988).

c. Pressure-threshold actuators. To convert a pattern of pressure fluctuations directly into a series of mechanical operations requires transducers that function as pressure-driven actuators. Devices like that abstractly represented in Figure 16.1 prove convenient. Each actuator of this kind is characterized by a pressure threshold p_{thresh}, a volumetric change ΔV, and a corresponding length change $\Delta \ell$. The Gibbs free energy difference between the extended and retracted configurations is

$$\Delta \mathcal{G}_{\text{extend}} = \left(p - p_{\text{thresh}}\right)\Delta V \tag{16.2}$$

and the outward force applied by the actuator at lengths between its limiting values is

$$F_{\text{extend}} = -\left(p - p_{\text{thresh}}\right)\Delta V / \Delta \ell \tag{16.3}$$

liquid medium

constant-force spring
(schematic)

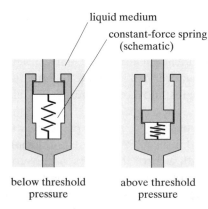

below threshold
pressure

above threshold
pressure

Figure 16.1. Schematic diagram of a pressure-threshold actuator, here illustrated with a constant-force spring (Section 10.5.3) and a piston moving between two limit stops. Actuators of this abstract class can be constructed with widely varying structures.

neglecting the effects of compliance. By adding a mechanism such as a lever, the effective value of $\Delta\ell$ can be changed without varying other parameters. The discussions that follow will use one of two sample sets of parameters:

1. High-pressure actuator parameters. With $\Delta p = p - p_{thresh} = -0.1$ GPa and $\Delta V = 5$ nm^3, $\Delta\mathcal{G}_{extend} = -500$ maJ, providing reliable avoidance of the disfavored position despite thermal excitation. With $\Delta\ell = 1$ nm, the maximum force delivered at $\Delta p = -0.1$ GPa is 0.5 nN.

2. Low-pressure actuator parameters. With $\Delta p = -0.001$ GPa, $\Delta V = 500$ nm^3, $\Delta\mathcal{G}_{extend}$ is again -500 maJ/nm^3, and with a suitable geometry and mechanical advantage, the delivered force can again be 0.5 nN.

In a system described by the high-pressure parameters, a 2 GPa pressure range permits use of actuators divided into 20 distinct classes, with $p_{thresh} = 0.1$, 0.2, 0.3,..., 1.9, 2.0 GPa. With the low-pressure parameters, 20 classes can be accommodated in a pressure range of 20 MPa, or 2000 in a pressure range of 2 GPa. (This is conservative; closer analysis of the mechanism described in Section 16.3.2d indicates that thresholds can be spaced at intervals less than Δp.)

Actuators in any one pressure class can be cycled through a reliable length transition repeatedly without activating those of other pressure classes. Shifting the ambient pressure monotonically from the value of p_{thresh} for one class to that of another class, then back again, will cycle all actuators of intermediate p_{thresh} exactly once. If the pressure is swept over the range only once, each set of actuators becomes active only once, each in a fixed sequence but with an independently controllable number of cycles including at least one more extension than retraction.

d. Control via acoustic signals. Two actuators with different thresholds could provide clock and data signals, permitting the input of strings of bits to a nanomechanical controller with data rates on the order of 10^7 bits per second. The present application, however, requires control of the motion of several different parts, which would require mechanisms to decode the instructions and then apply the results at different locations.

A more direct approach can reduce complexity. Figure 16.2 illustrates how pressure-threshold actuators can drive a ratchet mechanism that moves a rod. When the *latching pawl* is engaged, the rod will not move left or right spontaneously, but can readily be driven to the right by an applied force. The *driving pawl* provides this force: every downward pressure transition from p_{thresh} to $p_{thresh} -\Delta p$ pushes the rod one increment to the right. (A reasonable spacing for the ratchet teeth is ~1 nm.) When both ratchets are disengaged, a small bias force retracts the rod leftward to a fixed position. For a mechanism of this kind to work reliably, the ratchet barrier heights must be adequate (e.g., 250 maJ); actuators with $\Delta\mathcal{G}_{extend}$ on the order of 500 maJ are more than adequate to surmount these barriers. Incrementing a ratchet dissipates substantial energy, but efficiency here is not a concern.

A typical operational cycle (Figure 16.2) would begin at high pressure, with both ratchets disengaged and the rod to the left. Lowering the pressure through the first range engages the latching ratchet with a force that is essentially constant at all lower pressures, owing to the limit stop. After lowering the pressure further (possibly much further), the driving rachet extends and moves the rod to

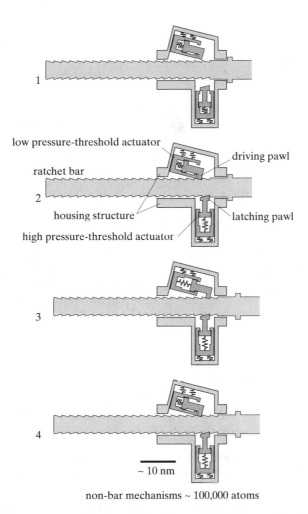

low pressure-threshold actuator

driving pawl

ratchet bar

housing structure

latching pawl

high pressure-threshold actuator

~ 10 nm

non-bar mechanisms ~ 100,000 atoms

Figure 16.2. A ratchet mechanism driven by a pair of pressure-threshold actuators. At high pressures (1), both pawls are retracted. As the pressure falls, the latching pawl and the driving pawl extend and engage (2). As the pressure drops through the active range of the driving pawl, it extends, forcing the rod to the right and momentarily lifting the latching pawl against the force of a spring, leading to state (3). An increase in pressure crossing the active range of the driving pawl resets it, preparing to move the rod a further step during the next downward pressure transition. Raising the pressure to the level in state (1) retracts both pawls again, allowing a bias force to reset the rod position to the left. Springs outside the two actuator cylinders represent mechanical compliances not directly involved in actuation.

the right. The pressure then can be lowered further with no effect on this mechanism, or it can be cycled up and down across the active range of the driving ratchet, moving the rod by further increments. After an operation has been performed, a return to high pressure resets the rod to its initial position. All latching ratchets can have actuators of a single pressure class, hence a system with N distinct classes can directly control $N - 1$ distinct motions.

Simple ratchet-driven rods of this sort would, however, provide inadequate positional resolution (~1 nm) for mechanosynthesis of diamondoid structures, if

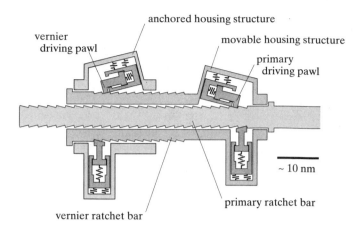

Figure 16.3. A vernier ratchet mechanism driven by four pressure-threshold actuators. Ratchets with differing tooth spacings can provide positional resolution equal to the difference in spacing (Section 16.3.2d). This schematic illustration places both sets of actuators in the same plane, suggesting a mechanical interference that need not occur in a noncoplanar geometry. Note that the two latching pawls can be driven by actuators of the same pressure threshold.

their displacements were applied in a 1:1 ratio. Use of levers, screws, or gears to provide mechanical advantage is one option; use of additional vernier displacements of a slightly different increment size (Figure 16.3) is another (for example, a total displacement can be the sum of 1 nm displacements at one interface and 1.1 nm displacements at another, yielding 0.1 nm resolution).

e. Power via acoustic signals.   A single actuator described by either of the sets of parameters described in Section 16.3.2c can yield $\sim 5 \times 10^{-12}$ W of mechanical power when cycled at 10 MHz. This power level could support 10^6 operations per second at 5 aJ/operation (far larger than the comparable value estimated for the exemplar system in Chapter 14).

· In systems of the sort considered here, control and power supply are inseparable. The power supplied by an actuator is sufficient so long as the forces it applies are sufficient to drive component motions. A force of ~ 0.5 nN (Section 16.3.2c) is ample to overcome both the inertial and viscous drag forces caused by displacing a 10 nm diameter, 100 nm long rod by 1 nm in 10^{-7} s; these values are chosen to roughly model the struts of a molecular manipulator like that discussed in Section 16.3.6d.

16.3.3. Simpler manipulators

The manipulator described in Section 13.4 requires stiff materials and intricate parts, making it difficult to implement. An alternative design (Figure 16.4) uses a *Stewart platform,* widely used in aircraft flight simulators and occasionally in robotics (its use in this context was suggested by Professor D. Taylor of Cornell). The Stewart platform has a roughly octahedral geometry, with a fixed triangular base linked to a mobile triangular platform by six struts of adjustable length. Figure 16.5 shows a two-dimensional analogue with three struts and indicates a portion of the available range of motion. Strut lengths can be altered with

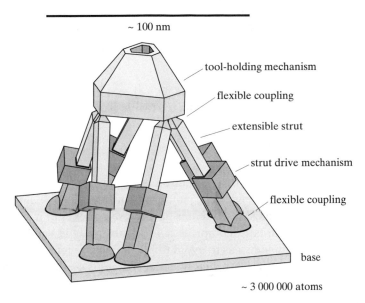

~ 100 nm

tool-holding mechanism

flexible coupling

extensible strut

strut drive mechanism

flexible coupling

base

~ 3 000 000 atoms

Figure 16.4. A positioning mechanism based on the Stewart platform (schematic). Each strut has a joint at each end, shown for simplicity as ball-and-socket joints at the base and as flexible point attachments at the top. Each strut has a set of actuators able to adjust its length. See Figure 16.5.

~0.1 nm accuracy by means of pressure-actuated vernier ratchet mechanisms like those described in Section 16.3.2d.

A manipulator can replace many other control mechanisms. With a "hand" in the workspace, various devices can be "hand operated"—for example, a manipulator could push any one of a dozen levers to select and move a particular transport mechanism, or it could move a partially finished part from a clamp to a

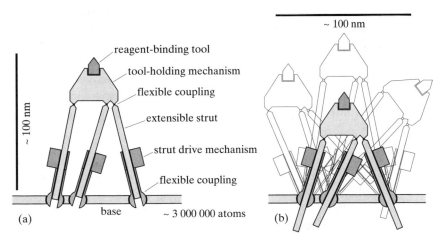

~ 100 nm

reagent-binding tool

tool-holding mechanism

flexible coupling

extensible strut

strut drive mechanism

flexible coupling

~ 100 nm

base ~ 3 000 000 atoms

(a) (b)

Figure 16.5. A two-dimensional schematic of a positioning mechanism based on the Stewart platform (a); the kinematics of this structure correspond to three of the six degrees of freedom of a three-dimensional platform (Figure 16.4). Panel (b) illustrates a portion of the available range of motion. Note that the lateral range exceeds the length of the arm in a retracted position.

special jig for a series of operations, and then to its destination. To minimize parts and physical complexity, it may be desirable to use a manipulator for many purposes beyond the narrowly mechanochemical.

16.3.4. Inert internal environment

The previous chapters describe nanomechanical systems working in vacuum. This environment has three basic advantages (aside from analytical simplicity):

- No viscous drag
- No mechanical interference
- No chemical reactivity

Viscous drag is acceptable here because energy dissipation is of negligible concern. Mechanical interference (i.e., jamming) could pose problems, but only if a motion (e.g., of gears meshing) creates an enclosed pocket of molecular size or larger and then attempts to compress it to a substantially smaller volume; mechanisms can be designed so as to avoid this. Chemical reactivity is a concern in many media (particularly with highly active reagents), but not in a nearly inert medium such as helium.

A helium-filled structure can equalize interior and exterior pressures to minimize wall flexure. More precisely, an enclosure can include an expansion bellows, thereby concentrating flexure in a region designed to accept it. Evacuated regions would be needed (if at all) only inside actuators, and these could be pumped: the presence of a few molecules would be inconsequential. Adopting the low-pressure actuator parameters (Section 16.3.2c), a pressure range of 0.02 GPa is adequate to control many distinct devices. If the total mean pressure is 0.2 GPa, the pressure (and gas-filled volume) fluctuations are then ~10%. Helium can be admitted (and large molecules excluded) by helium-permeable pores. A suitable internal helium pressure will result if the surrounding liquid is in contact with gaseous helium, and if the mean fluid pressure does not vary too widely or too fast.

If the interior volume is $\sim(100 \text{ nm})^3$, then so long as the contained helium in a set of devices has a mean mole-fraction impurity concentration $<10^{-9}$, the probability of the presence of a stray nonhelium molecule in any one device is $<.1$, hence most lack any contamination. Many reactive contaminants could be neutralized by *gettering*, that is, by providing sacrificial reactive surfaces with which they can combine without harmful effect. One such contaminant is hydrogen, the molecule most likely to enter through helium-permeable pores.

16.3.5. Sorting and ordering molecules

To simplify cost estimation (and to reduce estimated production costs), Chapter 14 describes molecular manufacturing systems able to use low-purity solutions of small-molecule feedstocks. The present objectives differ; to simplify the systems themselves (and to reduce estimated development costs), it is better to assume larger-molecule feedstocks of high purity.

Large feedstock molecules of several kinds can differ greatly both from solvent molecules and from one another. Reliable acquisition of a large molecule of a particular kind can then be straightforward, in part because larger molecules can have enormous binding affinities. For example, a 31-atom molecule (biotin) binds to a protein (avidin) in water with a dissociation constant K_d of

$\sim 10^{-15}$ nm^{-3}. Assuming solute concentrations of ~ 1 nm^{-3}, a dissociation constant this low provides more than adequate sorting and ordering in a single step, assuming that no closely similar molecules are present to compete for the receptor. This can be ensured by careful purification of the large-molecule components of the feedstock solution (e.g., by repeated affinity chromatography). Accordingly, it seems that a judicious choice of high-purity feedstock components can enable a manufacturing device to acquire feedstock molecules of one kind with good reliability in a single step. A mechanism with a single receptor could then replace complex systems like those shown in Figures 13.4 and 13.5.

16.3.6. Minimal diamondoid-material systems

Any attempt to describe a "minimal" system must be considered speculative. On one hand, a general description may overlook some minor but complex and necessary feature; on the other hand, clever designs may result in smaller, simpler systems overall. The mechanosynthetic system described here has passed through several cycles of simplification, and there is no reason to regard it as ideal or final. This sketch, like those that follow for other stages, is offered only to provide a stimulus and point of reference for further work.

a. Preparing and applying reagents. Chapter 14 describes systems using specialized mill mechanisms to prepare reagents and to perform several stages of subsequent assembly; this specialization improves throughput and energy efficiency. Parts count can be reduced, however, by applying manipulator mechanisms more directly to feedstock molecules.

Conventional chemical synthesis can prepare feedstock molecules of several kinds, each having a reagent-precursor moiety at one end and a distinctive "handle" of several dozen atoms at the other. Differences between handles can enable the selective binding assumed in Section 16.3.5. Forces applied by a manipulator can drive the rotation of a receptor from the exterior to the interior of the enclosure, and can force the release of a bound feedstock molecule; the handle can then bind to a larger block with a standardized, grippable surface.

A pair of manipulators (one can be an immobile gripper) can be situated so that their tips can interact. Bound feedstock molecules can then be moved, placed, and made to interact with flexible control of encounter sequence and geometry. The resulting mechanochemical transformations can generate a wide range of reagent moieties and subassemblies from a smaller range of feedstock molecules. In particular, relatively unreactive precursors (stable in solution) can be transformed into highly active reagents in the inert environment within the enclosure.

Once prepared, reagents can be applied, transferring groups to (or abstracting atoms from) a workpiece. The degree of control available is essentially that described in Chapter 8.

b. Waste disposal. The preparation and application of reagent moieties will generate wastes in the form of discarded handle structures. These will typically have a volume greater than that of the product structure, and must be cleared from the workspace. The manipulator can deliver them to a relatively unselective positive-displacement pumping mechanism for expulsion into the surrounding solution.

Spent handles might resemble feedstock molecules enough to compete for receptors. This problem can be avoided by either (1) ensuring that their use modifies structures important to binding, (2) ensuring that their concentration in solution remains very small, or (3) trapping them in an expandable external compartment (e.g., a bag).

*c. **Wall design and product delivery.***   Pressure equalization minimizes loads on walls, enabling the use of a flat-walled enclosure scheme like that illustrated in Figure 14.7. This geometry relies on a sliding interface to provide a seal between the enclosing wall of the manufacturing system and the emerging wall of the (possibly larger) product structure. Section 11.4.2 describes suitable seals.

*d. **Estimated size, complexity, and replication time.***   A system of the sort described in this section could be packaged in a volume $\sim(100 \text{ nm})^3$ (although not equidimensional). The structural components of a Stewart-platform manipulator (Section 16.3.3) having 6 struts ~100 nm long and ~5 nm wide contribute $\sim2 \times 10^4$ nm^3 of solid structure. Each strut has 4 actuators, including verniers. If an additional 4 actuators are provided to control operations at the manipulator tip, 8 to control a second, more rudimentary arm (Section 16.3.6a), 4 to control overall movement of the product structure, and 2 to drive a waste-expulsion pump, then the total is 42. If each actuator is approximated as a box 10 nm on a side with walls 1 nm thick, actuators contribute $\sim2.5 \times 10^4$ nm^3 of solid structure (actuators this large permit operation in the low-pressure regime described in Section 16.3.2c). Allowing for 10 reagent-acquisition mechanisms, each ~5 nm wide and ~10 nm tall, adds a negligible 0.25×10^4 nm^3. Walls 1 nm thick contribute $\sim6 \times 10^4$ nm^3. The total volume, including a 50% allowance for additional components and structures, is then $\sim1.6 \times 10^5$ nm^3, or $\sim2 \times 10^7$ atoms. The estimated number of moving parts is, according to this estimate, ~100.

With the parameters assumed here, motion frequencies are limited by the frequency at which the pressure in the cell is swept over its full range. At ~1 ms per sweep and 5 motion cycles per reagent application cycle, reagent application operations can be performed at $\sim200 \text{ s}^{-1}$. If the mean number of atoms transferred per operation is ~1, then the replication time for the system is $\sim10^5$ s, or ~1 day. Since there is reason to regard the atom count as generous and the motion frequencies as conservative, this time estimate seems likely to be generous. (Note that the acceptable mean error rate per operation can be larger than the goal discussed in Chapter 8, $\sim10^{-8}$ rather than $\sim10^{-15}$.)

*e. **Instrumentation.***   Techniques for probing product structures will be central to experimentation and process development. One approach would exploit modulation of the fluorescence yield, lifetime, and spectrum of a chromophore by changes in its mechanical and electronic environment. In the solution phase, changes in solvent polarity can produce dramatic shifts in fluorescence spectra (Wayne, 1988). Changes in the distribution of charge near a bound chromophore or direct mechanical deformation of a conjugated structure can produce changes of similar magnitude.

Existing instrumentation using subnanosecond excitation pulses and spectrally filtered, nanosecond time-gated detection to suppress background photons permits the detection of single molecules (each with a single chromophore) by their fluorescence in a ~10 ms interval (Shera et al., 1990). With accessible improvements (e.g., reduction of photobleaching by tailoring the molecular environment) more photons can be collected, and changes in the properties of the chromophore should be readily detectable. This can enable readout of data from an individual mechanosynthetic device. A chromophore can be manipulated like a feedstock molecule, and its fluorescent properties (frequency, quantum yield) can be modulated by direct contact with the structure being probed, causing electrostatic or mechanical perturbation of the chromophore. Since the chromophore is a proximal probe, repeated positioning at different locations can be used to build up an image of a workpiece structure, as in AFM and STM. If measurements can be performed in ~10 ms or less, atomic-resolution images can be acquired at ~1 nm^2/s.

Unless their photostability is excellent, probe chromophores will require regular replacement. This can be accomplished by mechanisms like those used to provide reagent moieties. To avoid interference from unbound chromophores, however, the feedstock molecules must initially have negligible fluorescence at the signal wavelengths, and must be activated after binding to the mechanism. Alternatively, the mechanosynthetic device could be tethered to a surface and fluid flow could be used to purge the surrounding volume before gathering data from a proximal fluorescent probe.

f. Summary. An inert workspace environment in which highly active reagent moieties can be positioned relative to workpieces with high precision and reliability (the essential requirements of qualitatively advanced molecular manufacturing) can be provided by a mechanism with ~10^2 moving parts and ~2×10^7 atoms. Pressure-threshold actuators enable substantially independent control of several tens of mechanisms within a single manufacturing system; an indefinitely large number of identical systems can be driven through the same sequence of operations at the same time. Using a manipulator mechanism able to reach a large portion of the interior space, complex behaviors can be implemented by complex control sequences, rather than by complex hardware. With 10^7 Hz actuator stepping rates, manufacturing can proceed at ~200 operations per second, enabling a system to build an object of its own complexity in about a day. Molecular probes can be used to determine the outcome of operations within a single device, facilitating experimentation and the development of new operations.

These system characteristics assume the use of diamondoid materials, and are chosen to permit the fabrication of diamondoid materials. The following steps in the backward-chaining analysis will move away from these materials and conditions toward more familiar polymeric structures and solution-based chemistry.

16.4. Softer, smaller, solution-phase systems (stages 2–3)

16.4.1. Diamondoid via nondiamondoid systems

Diamondoid materials with diamondlike stiffness and atom number density are challenging targets for synthesis (Section 8.6.1), and are here assumed to require the use of highly active reagents. These, in turn, are assumed to be incompatible with solution-phase synthesis conditions, owing to the potential for reaction with solvents. (If either of these assumptions proves incorrect, development will be easier.*) Along development paths that start with solution-phase chemistry, diamondoid structures must be built by nondiamondoid systems.

A nondiamondoid system can build diamondoid structures if it can execute suitable motions with adequate stiffness in an environment compatible with highly active reagents. A mechanism resembling that described in Section 16.3.3 will suffice, using thicker struts to compensate for lower modulus. Highly oriented polymers can exhibit a Young's modulus >100 GPa; examples include polyethylene and Kevlar 49. Similar materials can be made by solution-phase mechanosynthesis. A strut ~10 nm in diameter and ~100 nm long would then have a stretching stiffness $k_s \approx 100$ N/m.

Highly crosslinked organic structures of good stiffness can presumably be made by positioning multifunctional monomers of conventional reactivity; extensive use of cycloaddition reactions, for example, can enable the formation of relatively dense bond networks. The number of atoms required for a given functional capability will usually be larger for such structures than it would be for more diamondlike structures, but the number of (relatively large) monomers placed, and hence the number of mechanosynthetic operations, should typically be lower.

These crosslinked organic structures can be built by mechanosynthetic systems of still lower stiffness. Required manipulator stiffness varies as the inverse square of the site separation. The large size of multifunctional monomers in comparison to simple radicals and alkynes tends to increase the separation between reactive sites; the alignment requirements of cycloaddition reactions can further widen the separation in configuration space (Section 8.3.3f).

The picture that emerges from these considerations is of a progression from relatively compliant mechanosynthetic systems that position large monomers in solution to relatively stiff systems that position small, highly active moieties in an inert environment. Monomer-based structures built by compliant mechanisms using solvent-compatible chemistry can themselves be stiff enough to build diamondoid structures.

16.4.2. Inert environments from solvent-based systems

The transition from solvent to an inert environment can be achieved by a manufacturing system that (1) tolerates but does not require an internal solvent, and (2) can replicate while maintaining a barrier against solvent entry. Consider a system meeting this specification that initially contains 10^9 molecules of solvent. After one replication cycle, the mean number of solvent molecules in each of the resulting pair of systems is $10^9/2$. After 40 replication cycles, the mean number is $10^9/2^{40} \approx 10^{-3}$; hence most systems are solvent-free by virtue of extreme dilution.

If all solvent molecules were sufficiently mobile, then replication-based dilution could be replaced by simple pumping. Problems that might be caused by

---

* J. Bonaventura has suggested the use of liquid xenon (e.g., at 16 C, 58 atm) as a solvent (Rentzepis and Douglass, 1981); this may avoid some of these constraints. Xenon has little interaction with (e.g.) reactive organic radicals (Cook and Roberts, 1983).

operating in the presence of a helium-solvent phase boundary can be avoided by pumping out all mobile solvent molecules, then using replication cycles to produce solvent-free mechanisms by dilution.

If enclosures made of polymeric materials are permeable to small reactive contaminant molecules (e.g., hydrogen), then either the external concentration of such contaminants must be kept low (by avoiding sources or providing sinks) or the internal mechanisms must tolerate their presence. Note that dangerously reactive contaminants should be susceptible to chemical scavenging.

16.4.3. Solution-synthesized pressure-threshold actuators

Relatively advanced solution-phase mechanosynthesis systems can presumably make (and use) pressure-threshold actuators based on piston mechanisms like the one illustrated in Figure 16.1. An alternative approach, presenting less synthetic difficulty, exploits pressure-driven changes in volume analogous to solid-state phase transitions in bulk materials.* Various salts and organic crystals exhibit phase changes at pressures between 0.05 and 2 GPa with fractional volumetric changes of ~0.05 to 0.1 (Gray, 1972). A block of material with this behavior can serve as a pressure-threshold actuator: the nature of the relationship among force, displacement, and pressure is identical. The chief disadvantage is the smaller value of $\Delta V/V$; this can be accommodated by roughly doubling the linear dimensions of the device.

16.4.4. Smaller liquid-based mechanisms

Sections 16.3.6 and 16.4.2 assume that synthesis occurs inside a ~100 nm scale enclosure, with each moiety (or monomer) emplaced by a manipulator of comparable size. The enclosure must surround the system, and the manipulator (to enable replication) must span its diameter. A rough estimate (Section 16.3.6d) places most of the system mass in these components. To add a further link in the backward chain from advanced manufacturing systems, it is desirable to keep the chief benefits of enclosures (exclusion of stray reagent molecules) and of manipulators (use of positional control to build complex structures) while eliminating most of their mass. The result is an easier target for construction.

a. Stray reagent exclusion without a wall. If reagent monomers contain dozens of atoms (or have discardable handles), then it is not a severe constraint to require that they carry enough polar or charged groups to make them substantially soluble only in a polar medium. If the workspace is located within a droplet of an unreactive, nonpolar solvent (e.g., an aliphatic hydrocarbon), then reagent monomers carried by an external polar solvent will be largely excluded from the interior. Droplets can be stabilized against coalescence by polymeric surfactant structures of low mass and complexity; their equilibrium size can be fixed (for example) by the colligative properties of their contents, or by the

---

* Martensitic transitions have desirable properties: they transform one solid structure into another, are cooperative, occur through local structural displacements, require no diffusion, cause little change in entropy, and can propagate at speeds near that of sound.

surfactant structures. Reagent monomers can be transported into the workspace by mechanically rotatable receptors like those discussed in Section 16.3.5.

b. Folding-assisted assembly.   Biological molecular machines are manufactured by devices capable of controlling only the sequence of a chain of small monomers. A manipulator-based assembly system with even a small range of motion (e.g., 5 nm) could build relatively rigid, highly crosslinked molecular objects of substantial size—more like a protein than like an amino acid residue. A chain of these objects could be designed to undergo a specific, predictable folding sequence as it exited the assembly system, guided by the binding of complementary surfaces with areas of several square nanometers. Note that identical pairs of complementary surfaces can be used repeatedly, so long as only one of any particular kind is ever exposed within reach of a tethered complementary structure.

Accordingly, a manipulator with a range of motion ~1/20 that of the device described in Section 16.3.6d can build structures far larger than itself, including larger manipulators. A single ratchet-driven bar can be used to move any of several reagent receptor mechanisms within reach. As in the proposed AFM-based mechanism (Section 15.4), the workpiece can be mounted on the manipulator, enabling monomers to be bound to it directly from a receptor. The comments in Section 16.3.6e regarding chromophores as probes are applicable here as well.

The advantages over an AFM-based mechanism include (1) isolation of the workpiece from stray reagent monomers, (2) motion control in six degrees of freedom, (3) a "substrate" that can include mechanisms for product clamping and release, (4) reagent receptors in known locations that need not be mapped, and (5) the feasibility of directing a replication process that generates large numbers of low-cost mechanosynthetic devices. The disadvantages include (1) inability to correct for reaction failures—at least when many devices are sharing one stream of broadcast signals—and (2) the requirement for much more complex nanomechanisms.

c. Estimated size.   To minimize the scale of the pressure-threshold actuators, pressures and sizes can be chosen in the range suggested by the high-pressure actuator parameters in Section 16.3.2c. With $\Delta V = 5$ nm^3, the required volume per actuator is ~50 nm^3, and a set of ~40 actuators (of ~20 different pressure classes) requires ~2000 nm^3 of material. The size of the struts is negligible in comparison, and the wall is nonexistent. In systems with large pressure swings, structural deformations resulting from ordinary compressibility (rather than discrete phase changes) will be an important consideration in design.

Based on these estimates, a total structural volume of ~4000 nm^3 appears ample, suggesting an atom count of ~4×10^5. Assuming ~20 atoms per monomer, the number of synthetic operations required for construction is ~2×10^4. This is within the >10^5 operations estimated to be feasible using AFM-directed mechanosynthesis (Section 15.4); alternatively, a self-assembled structure might be made from a set of ~400 molecular objects each consisting of ~50 monomers. Once fully operational, a device of this sort could build another device of comparable size and complexity in ~1 hour, working at only 5 operations per second.

16.5. Development time: some considerations

The preceding sections outline an ambitious effort. Reaching the first stage described in this chapter (Stage 2, Section 16.4.4) will require the construction of precise molecular structures $\sim 10^2$ to 10^3 times larger than those previously achieved. Subsequent steps, building on this base, will move still farther from present laboratory experience.

Naive assessments of feasibility often implicitly assume that *future* developments will be limited (in scope and pace) by *today's* tools, instruments, and techniques. This mistake is natural, because only today's capabilities are widely and concretely known. Where the developments under consideration involve several stages of tool development, however, this static picture can be quite misleading. To formulate a more accurate understanding, one must instead consider each step in light of the capabilities provided by previous steps. The balance of this section makes a preliminary effort in this direction, beginning with a conceptual framework of more general applicability.

16.5.1. Determinants of the development time

a. The cycle of experimentation. Development of new technologies involves experimentation leading to the accumulation of reliable systems and tested capabilities. The cycle of experimentation has several stages: (1) choosing an objective, (2) choosing an approach, (3) trying the approach, and (4) evaluating the result. So long as the result is unsatisfactory, stages (2) through (4) are repeated until success is achieved or the objective is altered or abandoned.

b. The number of sequential experimental steps. The time required to develop a system equals the sum of the durations of the required sequential steps. Good strategy can reduce the number of steps (and their difficulty); there is no reason to believe that the best strategy for developing molecular manufacturing has yet been identified. Further, many steps can be taken in parallel. For example, multiple teams can work with similar mechanosynthetic systems, each doing experiments that add to the repertoire of reproducible synthetic operations. Low instrument and materials costs can enable more teams to work, increasing parallelism. Concurrent development of a system and of plans for using it (via computational modeling) can further increase parallelism.

c. The number of teams working on a step. Even without cooperation, work by multiple teams can speed completion of a single step, because this happens as soon as the fastest (or luckiest) group succeeds. The number of teams funded to work on a problem will depend in part on the perceived importance of solving it.

d. Computational modeling. If computational modeling were perfect (fast, inexpensive, and accurate), then it would enable all correctly executed experiments to succeed on the first try. In reality, it does far less, yet modeling can help researchers avoid unworkable approaches without rejecting too many workable approaches. This tends to increase the fraction of experiments that can, after repeated tries, be made to work, thereby decreasing the mean delay in achieving an objective.

Computational modeling is more useful in planning mechanosynthesis than it has been in solution synthesis. Mechanosynthesis can impose a particular interaction geometry on a pair of reagents, thereby defining the system to be modeled and rendering small errors in transition-state energy relatively unimportant (Section 4.4.3). Further, relatively rigid, crosslinked structures are less sensitive to modeling errors than are conformationally flexible peptides and typical products of organic synthesis. As inexpensive computers become faster and software becomes more sophisticated, the utility of computational modeling will increase.

e. Time for synthesis. If the cycle of experimentation is to be short, the time required to synthesize a structure must be short. This consideration favors active reagents, high effective concentrations, and the use of software control to direct long sequences of operations.

f. Time for evaluation. Short experimental cycles likewise demand short evaluation times. This tends to favor intermediate technologies that combine sensors and manipulators in a single package.

g. Willingness to limit evaluation and abandon subgoals. In routine science, an unexpected result often provides a problem to study. In building novel systems, however, an unexpected result often gives reason to seek a more predictable system. To evaluate a result, success and failure must be distinguished, but the nature of a failure need not always be understood—an attempt to study it in detail seldom yields fundamental knowledge, and may merely yield detailed knowledge of something useless. Even a consistent pattern of failures may motivate, not a prolonged halt for study, but instead a redesign that avoids the difficulty entirely. Analyzing unexpected failures can be of great value, producing rules to guide future design; but quickly trying a different approach can sometimes speed development. (Note that evolution developed sophisticated molecular machine systems without analyzing failures.)

16.5.2. Stage 1a: Brownian assembly of medium-scale blocks

Building a solution-phase mechanosynthesis system of the sort described in 16.4.4c by Brownian assembly (rather than by AFM-directed mechanosynthesis) might involve developing ~400 self-assembling building blocks of ~50 monomers apiece (Section 15.3). Monomers can be developed in parallel with one another. With good CAD software to define their functions and interfaces, building blocks can also be developed in parallel with one another. Further, design of building blocks can be overlapped with the design and synthesis of monomers.

Designing ~400 building blocks need not require the design of a similar number of unique complementary interfaces. If building blocks are added to solution sequentially (and removed afterward), each will have a unique destination so long as the assembly sequence exposes only one interface of each kind at any one time; supramolecular variants of protective-group strategies may be useful here. The quality of the tools available for computer-aided design of folded structures that meet geometrical and functional specifications appears critical to the pace of development along this pathway.

Blocks of ≥150 monomers can be characterized by NMR spectroscopy (Kay et al., 1990). Interpretation of NMR data is simpler if the desired fold is known, and if structures that differ widely from the objective are discarded without detailed study. Molecular-tip AFM techniques could aid in evaluating both building blocks and assembled structures.

In past protein design efforts, the time required for synthesis and characterization has often been measured in months or more, and success rates per trial have been moderate. Augmented proteins, designed to make design and synthesis easier, will presumably permit somewhat faster cycles.

16.5.3. Stage 1b: Mechanosynthetic assembly of small building blocks

AFM-directed mechanosynthesis can sidestep the problem of designing folding polymer structures and enable immediate characterization of product structures using the AFM mechanism itself. The required monomers are likely to be complex (bearing multiple functional groups) but can be designed to permit synthesis by conventional means. Monomers can be developed in parallel, and to avoid investing synthetic effort in chemical blind alleys, they are best developed in parallel with designs intended to use them.

Effective exploitation of AFM-directed mechanosynthesis systems will require substantial software development: image interpretation software able to determine the types, positions, and orientations of specific sets of reagent binding molecules; control software to automate reagent positioning and sensing, thereby automating the execution of long sequences of reactions; and computer-aided design software to plan both what to build and how to build it.

The cost of an AFM-based mechanosynthesis system can be estimated today: it will likely be less than the price of an AFM system in the early 1990s, $\sim \$10^5$. Reasonably low prices will enable many teams to enter the field, some working in parallel to develop and expand a set of reliable synthetic operations. With suitable reagents, reaction times can be ~ 1 s, enabling the construction of moderately complex objects in minutes and of objects containing $\sim 10^5$ monomers in about a day. Once good tools are in hand, the time required to develop mechanical systems with $\sim 10^2$ moving parts (as many as in the system in Section 16.3.6d) consisting of $\sim 2 \times 10^4$ monomers (estimated to be sufficient to build a minimal solution-based molecular manufacturing system, Section 16.4.4c) could be a fraction of a decade.

16.5.4. Stage 2: First-generation solution-based systems

Stage 2 manufacturing systems as described in Section 16.4.4 have capabilities resembling those of AFM-based mechanosynthetic systems, hence they can use identical or similar monomers, and their applications can be planned using similar CAD software. Since these systems are precisely constructed, they create no requirement for image-interpretation software to locate and discriminate among reagent receptors. Evaluation of the results of an operation (including imaging) can be achieved as described in Section 16.3.6e.

The estimated time required for a system of this kind to build a copy of itself (a key threshold in development) is several hours. If the replication time is ~ 1 hour (Section 16.4.4c), then the time required to produce a macroscopic quantity of systems is several days. Subsequent production of other devices can supply relatively low-cost, disposable laboratories for mechanosynthesis. The cost of required reagents is negligible on a per-device basis. For example, a cost of 10^4 \$/gm amounts to ~\$$10^{-12}$ per complex device. New reagents, operations, and devices can be developed in parallel in many laboratories. From this technology base, the direction of advance is toward enclosed systems with larger manipulators, then toward inert interiors, then toward more active reagents. Experimental cycle times can remain short throughout this process.

16.5.5.   Stage 3: Inert environments, diamondoid materials

After the transitions enabling the use of highly active reagent moieties and the construction of devices made from diamondoid materials, development can proceed more or less directly to advanced molecular manufacturing systems. CAD systems for diamondoid materials will differ from those used with larger monomers, but will be in some respects simpler. The frequency of synthetic operations can be increased, in part by the use of more active reagents and in part by the construction and use of decoding subsystems that more fully exploit the capacity of information input channels. Further, once systems are built with internal control and data storage devices, brief instructions can activate complex subroutines, greatly relaxing constraints associated with data transmission rates. A single manufacturing system can then contain multiple production lines, each working at high frequency. Construction of molecule sorting and preparation subsystems can enable the use of simpler, less pure, less expensive reagents, reducing product costs. With scale-up, integration, and suitable software for design and control, these features provide the basic capabilities described in Chapter 15.

16.6.   Conclusions

A backward-chaining analysis indicates that feasible developmental pathways link our present technology base to the technology base described in Part II. Alternative pathways are known, and others can be devised.

To reach the first stage discussed in this chapter requires the assembly of molecular machine systems from ~10^4 to 10^5 monomers. Chapter 14 describes several ways to achieve this ability, via Brownian assembly or the development of AFM-directed mechanosynthesis systems. Once assembled, small mechanosynthetic devices can be made to execute complex patterns of motion powered and controlled by imposed pressure fluctuations. Relatively small and simple devices can be used to construct larger and more complex devices. A series of steps can enable a relatively smooth transition from solution-phase assembly of monomers with conventional reactivity to the assembly of diamondoid mechanisms in an inert (eventually, vacuum) environment using highly active reagents.

Each step along this pathway will present great practical challenges, but each step will also bring valuable new capabilities. The long-term rewards, measured in terms of scientific and technological capabilities, appear large.

Methodological Issues in Theoretical Applied Science

A.1. The role of theoretical applied science

This appendix discusses issues at a higher (but less mathematical) level of abstraction than the rest of this volume. It considers methodological issues associated with efforts to understand technological possibilities, examining how trade-offs in goals and intellectual effort can make possible surprisingly robust and fault-tolerant reasoning. It offers a perspective from which to judge studies of molecular nanotechnology; this perspective may be useful to researchers exploring technological possibilities in other fields.

Part II of this volume is an exercise in *theoretical applied science,** a mode of research which aims to describe technological possibilities as constrained not by present-day laboratory and factory techniques, but by physical law. (To provide a definite basis for analysis, this is taken to be physical law *as presently understood.*) Within conventional (experimental) applied science, theoretical studies are used to design realizable instruments and experiments, and then to interpret their results. This use of theory thus centers around the experimental demonstrations of useful new phenomena. Theoretical applied science, however, like theoretical physics, produces no experimental results: a typical product is instead a theoretical analysis demonstrating the possibility of a class of as-yet *unrealizable* devices, including estimated lower bounds on their performance. Theoretical applied science is likewise distinct from *engineering,* which pursues the economical, near-term production of physical devices. Theoretical applied science fills a gap in the matrix resulting from the familiar distinctions of theoretical vs. experimental and pure vs. applied (Figure A.1).

There are several reasons for undertaking research of this kind: First, theoretical applied science can be viewed as a branch of theoretical physics, studying certain time-independent consequences of physical law; it is thus of basic

---

* This has previously been called *exploratory engineering* (Drexler, 1991a), a term that does not adequately convey the theoretical nature of the studies. The term *theoretical applied science* could be applied to theoretical studies performed in direct support of experimental applied science, but *applied theoretical science* seems a better term. No clearer alternative name for the present topic has yet been suggested.

	Pure	Applied
Theoretical	pure theoretical science	theoretical applied science
Experimental	pure experimental science	experimental applied science

Figure A.1.   Theoretical applied science in a matrix of familiar distinctions.

scientific interest. Second, theoretical applied science can expose otherwise-unexpected rewards from pursuing particular research directions in laboratories today; it can thus improve the allocation of scientific resources. Finally, a better understanding of physical possibilities, when linked to feasible development programs (as in Part III), can yield a better-informed picture of future capabilities; this is of broad importance if one wishes to make realistic plans.

The study of technological possibilities is closely allied with—yet distinct from—more familiar forms of research in science and engineering. Its technical content (drawing extensively from physical theory and experimental results) and the nature of its product (knowledge, rather than hardware) link it closely to scientific research. Yet it is also closely akin to engineering: studying technological possibilities poses problems of design and analysis. The products of theoretical applied science can be termed *exploratory designs,* although some take the form of a rather abstract analysis.

In some instances, the systems under study may be so many steps removed from present fabrication technologies (or their capabilities may be of so little practical value) that the research is as theoretical and noncommercial as the purest science. In other instances, the systems under study may be so useful and accessible that exploratory studies can promptly transition into development. Part III concentrates on more accessible systems; Part II concentrates on useful but less accessible systems. This appendix examines methodological issues common to both. It proceeds largely by examining similarities and differences among engineering, pure science, and theoretical applied science (some of the remarks regarding engineering could, with some adaptation, be applied to experimental applied science).

As illustrated in this volume, familiar principles of science and engineering can be used to formulate and evaluate technological concepts that are beyond the reach of implementation by present fabrication techniques. In addition to the concrete technical issues raised by molecular nanotechnology, studies of this sort raise basic issues of objective and methodology. To understand the requirements for successful reasoning in theoretical applied science, one must understand what it does and does not attempt to achieve, and how the sacrifice of one set of goals can facilitate the achievement of another. Theoretical applied science operates under stringent constraints stemming from the inability to make and test the objects it studies: It is hardly economical to develop a complex, detailed design if it is known beforehand that a system cannot be built (who would fund the work, and why?). More basically, if a system cannot be built, it cannot be tested. In the context of standard engineering, either of these constraints would be fatal to manufacturing and to marketplace success. The following sections show how these and related constraints can be accepted—and exploited—where the objective is knowledge rather than competitive manufactured products.

A.2.   Basic issues

A.2.1.   Establishing upper vs. lower bounds

Theoretical applied science, as pursued here, sets *lower* bounds on the *maximum possible* performance of devices by establishing the possibility of certain capabilities. Every existing device sets a lower bound on what is possible: it constitutes a proof by example. To go beyond this direct form of proof, one must apply physical theory. A brief examination of some issues associated with the study of upper and lower bounds on possible device performance may clarify the nature of theoretical applied science.

The performance of any device for some well-specified purpose is given by some mathematical function of its physical properties (the existence of such a function can be taken to define a "well-specified purpose"). The maximum possible value of this function is determined by physical law (Section A.2.2), and one can attempt to estimate this value by bracketing it between upper and lower bounds. Upper bounds—which are beyond the scope of the present work—are typically sought by attempting to determine whether performance at or beyond some specified level would or would not contradict a given set of physical laws. Studies of the theoretical physical limits of computation (Toffoli, 1981; Bennett, 1982; Fredkin and Toffoli, 1982; Landauer, 1982; Likharev, 1982; Feynman, 1985; Landauer, 1987) provide notable examples of this kind of work (which yields surprisingly many null results). When such studies analyze a hypothetical device, they resemble theoretical applied science; they frequently differ, however, in their neglect of certain properties of real materials and systems. Studies of computation have assumed (for example) perfect initial conditions (Fredkin and Toffoli, 1982), idealized quantum Hamiltonians (Feynman, 1985), or mechanisms made from physics-textbook solids and components [in which, for example, rigid volumes exclude other rigid volumes (Bennett, 1982; Toffoli, 1981)]. These idealized models have shown, for example, that neither the basic principles of thermodynamics nor those of classical or quantum mechanics prohibit the construction of devices that perform arbitrarily complex combinational logic operations without dissipation of free energy, so long as the input can be determined from the output: logical reversibility permits (but does not guarantee) thermodynamic reversibility. (This is illustrated in Section 12.3.8b.)

Successful research in theoretical applied science sets lower bounds to possible capabilities. This demands that capabilities be consistent with *all* relevant physical constraints, not just those needed to illustrate basic consequences of physical law. To show this, one typically engages in a process of design and analysis paralleling that of engineering. Indeed, the tested products of engineering make good points of departure for theoretical applied science: by holding dimensionless numbers constant in a design based on an existing product, a complex design process can sometimes be reduced (in part) to a simple scaling analysis (Chapter 2 discusses scaling laws and their limitations).

A.2.2. Are there objective, physical limits to device performance?

The concept of upper and lower bounds to maximal achievable device performance presumes the existence of a limiting, maximal value. Does this make sense?

According to the present understanding of physical law, a finite spherical volume with a bounded total energy, containing a finite number of particles chosen from a finite number of kinds, can exhibit only a finite number of quantum states; its possible contents are therefore selected from a finite set. This implies that the set of devices of finite volume (and so forth) is itself finite. (One usually regards multiple quantum states as corresponding to a single device, but each such many-to-one mapping reduces the size of the set of devices).

Given a function on a domain consisting of a finite set of elements, there exists a single maximum value (treating infinity as a value, for present purposes) that corresponds to one or more elements. Thus, for every clearly defined objective function (and bounding volume), there must exist one or more devices that maximize the function (unlike the set of integers, in which for every integer N there is a larger integer $N + 1$). The capabilities of the function-maximizing device mark the limit of the possible, relative to the given measure of performance. To assert that a limit exists implies knowledge neither of its numerical value nor of nature of the device that exhibits it.

The only constraints on possibility assumed in the preceding paragraph are those of physical law, excluding limitations on design and fabrication. These have a different character. Design limitations, though real, are never absolute: given a formal device-specification language, the output of a random number generator may correspond to the solution of even the hardest design problem, though typically with a negligible probability. Likewise with fabrication limitations: in the approximation that physical law is time reversible, any structure that could be destroyed can be created, though perhaps with vanishingly small probability (Drexler, 1986a). (More rigorously, given the charge-parity-time invariance of present physical theory, any device can be created with some finite probability so long as a geometrical mirror-image made of antiparticles could be destroyed.) The observation that thermal fluctuations have a finite probability of creating silicon wafers from sand, however, is of no use to semiconductor technology. To describe practical possibilities, theoretical applied science must take account of limitations on design and fabrication, even though these are not absolute in a purely physical sense.

A.2.3. Certainties, probabilities, and possibilities

Research in theoretical applied science can produce results of differing confidence levels. Statements regarding simple systems can be virtually certain: "A flawless ring of diamond can be spun with a rim speed in excess of two kilometers per second without rupturing" (this follows from elementary mechanics and from the density and tensile strength of diamond). Statements regarding ill-defined complex systems can likewise be virtually certain: "A physically-possible system of molecular machines can build a copy of itself from simple chemical compounds" (this follows from the existence of bacteria). Less certain propositions can also be of interest; indeed, the theory of rational decisions (Raiffa, 1968) indicates that information of the form "One cannot have confidence $>.95$ that capability X is impossible" may in some circumstances be valuable information indeed.*

In practice, the nature of theoretical applied science places a premium on establishing results with near certainty, because these results frequently serve as foundations for further design and analysis. In building a complex analysis, one must take care to ensure that each required assumption is reliable, lest the overall analysis prove weak. Relative to engineering, theoretical applied science favors designs with *increased* safety margins and *decreased* commitment to a single detailed specification (i.e., retaining redundant options); both differences can make conclusions more reliable (Sections A.4.4 and A.4.6). Likewise, the standards of criticism should be stringent: if weak evidence suggests a failure mechanism, and no strong evidence counters it, then failure should be assumed to occur. Engineering complex, reliable analyses has much in common with engineering complex, reliable systems.

A.3. Science, engineering, and theoretical applied science

A.3.1. Science and engineering

Science and engineering differ in their goals. Science strives to understand how things work; engineering strives to make things work. Science takes an object as given and studies its behavior; engineering takes a behavior as given and studies how to make an object that will exhibit that behavior. Science modifies objects to probe them; engineering modifies objects to improve them. ("Science" in sentences like these can be read as shorthand for a phrase like "scientists, when acting in *direct* pursuit of *purely* scientific goals"; "engineering" and other impersonal abstractions treated as grammatical subjects can be read similarly.) This sharp conceptual distinction between science and engineering does not imply a sharp distinction among researchers or among research programs: when designing experimental apparatus, scientists do engineering; when studying novel physical systems, engineers do science. Knowledge guides action, and action yields knowledge.

Science and engineering differ radically in their ability to describe their own future accomplishments. Knowledge of the content of a future scientific discovery is logically impossible: if one were to *know* today what one will "discover" tomorrow, it would not be a discovery. Knowledge of feasible engineering achievements is a different matter: since engineering aims to *do* rather than to *discover*, no logical problem arises. No contradiction arose in the early 1960s

---

* For example, if a war were in progress, and X could be used to destroy the country, and a defense against X would cost one billion dollars, then the expected value of this information could be on the order of 5% of the value of the country, minus one billion dollars. If determining the possibility or impossibility of X with near certainty can be achieved promptly at a cost of only one million dollars, so much the better.

when engineers reiterated the feasibility of landing a person on the Moon. When scientists predict future knowledge, they describe not what they will *learn,* but what they will learn *about* (e.g., that science, aided by engineering, would learn about the composition of the lunar surface).

In the following comparisons, the chief model of science is theoretical physics. This has two motivations: (1) physical theory is the foundation for much engineering work and (2) theoretical physics has been a popular model for philosophers of science. Indeed, it has been objected that much "philosophy of science" might better be termed "philosophy of theoretical physics" and that other fields of science often—quite properly—have distinct aims and methodologies (Bartley, 1987a).

Ideally, theoretical physics produces a set of foundational principles supporting precise models of well-specified physical systems. These models are subject to rigorous experimental testing and (again, ideally) pass all tests because they provide a uniquely correct description of observable physical reality.

Engineering produces devices, not theories. Again, experimental testing and thorough specification of physical systems play key roles. Engineering, however, faces requirements of manufacturability and competitive performance that have no parallel in physics. Further, no rigid standard of correctness demands unique answers in engineering, although competitive pressures sometimes force convergence toward a unique optimum.

A.3.2. Engineering vs. theoretical applied science

Like physics, theoretical applied science produces theoretical results, but these describe the capabilities of specific classes of devices rather than universal laws of nature. The following section explores the requirements for sound reasoning in this domain, where experiments are, for the moment, impossible. A major theme will be advantages that can be had during the design process by sacrificing one or more of the usual objectives of engineering, such as precise specification, present-day manufacturability, and competitive performance. Theoretical applied science offers no free lunch, no way to achieve physical engineering objectives without applying standard engineering methods. Instead, it exploits tradeoffs among objectives to achieve a general understanding of what can be made to work. Table A.1 provides a compact overview of how various issues interact in different domains.

A.4.   Issues in theoretical applied science

A.4.1.   Product manufacturability

Engineering attempts to produce functional products, usually within a one-to-ten-year planning horizon. Engineers must accordingly adapt their designs to the limitations of current or near-term manufacturing technologies. If a design cannot be translated into hardware (perhaps after some adjustment), it is a failure.

Successful theoretical applied science produces reliable analyses that establish the possibility of a class of devices. In theoretical applied science, planning horizons and near-term manufacturing capabilities are of no fundamental significance. For a class of devices to be truly possible, however, it must be possible to

Table A.1. Engineering, theoretical applied science, and theoretical physics.

Characteristic	Engineering	Theoretical applied science	Theoretical physics
Sound analysis	○ Not necessary: can demonstrate	• Necessary: is itself the product	• Necessary: is itself the product
Direct experiment	• Part of standard procedure	- Not possible (a central problem)	• Part of standard procedure
Accurate modeling	○ Experiment often can substitute	○ Margin for error often can compensate	• Ideally, complete and precise
Physical specification	• Thorough, to enable manufacture	○ Partial specification often can suffice	• Ideally, complete and precise
Product manufacture	• Must be possible in the present or near future	○ Must be possible, but no time limit	- Not relevant
Product performance	• Must compete with current systems	• Must (easily) surpass current systems	- Not relevant
Unique answers	○ Competition promotes convergence	- Unimportant	• Goal is to formulate a uniquely true theory

• Denotes an issue or activity that is crucial in this area.
° Denotes an issue or activity that is *relatively* unimportant.
- Denotes an issue or activity that is irrelevant or impossible.

construct instances of the class with possible tools. Molecular nanotechnology falls within the scope of theoretical applied science because it will require (and provide) manufacturing capabilities that do not yet exist.

A.4.2. Product performance

The products of engineering often face competitive tests in markets or battle-fields. Competition forces engineers to pursue small advantages, though this may incur considerable expense and development risk. Competitive pressures affect engineering in far-reaching ways.

The requirements of theoretical applied science, in contrast, give little reason to seek small advantages. The calculated performance of an exploratory design may far exceed that of analogous existing technologies, but this is typically a direct consequence of a powerful new technology, not of cleverness in stretching an existing technology past its present limits. Sacrificing a factor of two in this high performance—sometimes a factor of a thousand—may well be worthwhile if it makes an analysis simpler or more robust. (Accordingly, Chapter 12 examines relatively simple, mechanical systems for logic and computation, despite good reasons for expecting electronic nanocomputers to be far faster.)

In engineering, one can seldom afford to make an enormous sacrifice in performance merely to simplify analysis, therefore the potential gains from doing so are largely unfamiliar. An architectural example shows the principle. It may be difficult to design an economically viable five-kilometer-tall office complex, but it is easy to show that, with lavish use of steel, one could build a habitable, five-kilometer-tall structure. Indeed, by relaxing concerns regarding cost, aspect ratio, and the volume fraction of habitable space by a factor of a thousand, the basic question of feasibility can be answered with little more than a reference to the strength-to-density ratio of steel. The cost of analysis thus drops from that of a major design project to that of a few keystrokes on a calculator.

In theoretical applied science, the threat is not that a marginally superior competitor will dominate the market, but that a flawed analysis will bring down a large (and costly) intellectual structure. Sacrificing performance can help by permitting simpler and hence more reliable analyses. By attempting to leave room for advance, it can reduce the likelihood of being forced to retreat.

These considerations also decrease the value of finding and solving hard problems. By solving hard problems, an engineer is apt to gain an edge in performance, leaving competitors behind *precisely because of the problem's difficulty*. In theoretical applied science, this advantage disappears: the simplest analysis may still entail solving hard problems, but there is no such systematic reward for seeking them.

A.4.3. Direct experimentation

In theoretical applied science, one cannot build systems and hence cannot experiment with them. This presents a key challenge.

The absence of direct experimentation does not isolate theoretical applied science and exploratory designs from contact with empirical data. Lacking the ability to build a system, one may still be able to build (or find) examples of its components, and test those. In studying molecular nanotechnology, experimental data on model compounds, materials, and surfaces answers many questions. Exploratory designs can also be built on (and judged in terms of) tested scientific theories. Testing proposals against theories does not constitute direct experimentation, yet it ties proposals to the world of experimental reality (albeit via possibly inaccurate models). Computational experiments have become widespread in science and are equally applicable here.

The inability to experiment with systems nonetheless has a profound effect on the practical scope of theoretical applied science. Again, engineering provides a natural comparison.

Competitive pressures drive engineers to seek and exploit any genuine, reliable technical advantage, regardless of whether its physical basis is fully understood. For example, engineers must minimize manufacturing costs, confronting them with the complexities of manufacturing systems. They must seek the best materials, confronting them with the complexities of metallurgy, ceramics, and polymer chemistry. They sometimes must push the limits of precision, cleanliness, purity, and complexity, as in state-of-the-art microprocessor production. Almost any production process is likely to exploit many tested, reproducible, but inexplicable tricks: cleaning a surface with detergent of one brand results in a good adhesive bond in the next step, cleaning with another brand doesn't, no one knows why, and no one has any good reason to find out. Success in manufacturing demands experience more than theory, and experience means experimentation.

Almost any attempt to plan and execute a major innovation in engineering will require experimentation. Usually, one must either (1) exploit phenomena that are poorly understood and characterized, or (2) sacrifice performance and lose to the competition (or both). But relying on such phenomena without testing one's assumptions through experiment risks outright failure, and taking several such gambles in a single system would virtually assure failure. Accordingly, experimentation is universally recognized as necessary.

This case for the necessity of experimentation does not apply to theoretical applied science: if competitive pressures do not force reliance on phenomena that are poorly understood and characterized, the preceding considerations do not make experimentation a condition of success. In theoretical applied science, it is legitimate and often practical to choose well-understood components that can be combined into well-understood systems. If a device or process is too poorly understood, then it cannot be used in exploratory designs. (A related argument for experimentation—the likelihood of bugs in a design based on well-understood components—is addressed in Section A.4.5.)

A.4.4. Accurate modeling

Theoretical applied science often relies on mathematical models of physical systems. As suggested previously, the evolution of competitive systems usually requires more than just modeling, because modeling alone is less powerful than modeling combined with experimentation. Competitive pressures forbid engineers the intellectual luxury of staying on well-understood ground.

What constitutes "well-understood ground" depends in large measure on the required accuracy of the model, and accuracy varies widely. Shortcomings are of several kinds: Fundamental physical theory has known shortcomings (e.g., at high energies). Where physical theory is adequate, practical computational models may still have shortcomings (e.g., in solid-state physics). Where computational models are in principle adequate, they may in practice be too slow and expensive for a problem of interest (e.g., in *ab initio* quantum chemistry). Finally, a model may represent a substantial (or dramatic) simplification of physical reality, made for the sake of clarity and convenience.

Since mathematical models of complex physical systems are always inaccurate to some degree, it is important to consider tolerance for error. This, too, varies widely, depending on the objective. (Variable tolerance for error helps explain the use of different models to describe the same phenomena.)

In theoretical physics, approximations are valued in practice, yet the ideal of a complete, accurate description of nature leaves—at least in principle—no tolerance for error. Physics deals in equations, that is, in precise equalities.

Engineering requirements, in contrast, are expressed by inequalities. For example, to know that an aircraft is safe against wing failure, one need know neither the precise strength of its wings, nor the precise stresses they will encounter: the requirement is that strength exceed stress—or, more generally, that capacities exceed requirements. Consequently, engineers can tolerate inaccurate models if the extent of their inaccuracy is (approximately) known. If stresses and strengths are both known within 10%, then it suffices to make components having 1.3 times the strength demanded by the model. If stresses are uncertain by a factor of 5, and strengths by a factor of 2, then it suffices to make components having 15 times the strength demanded by the model. In practice, however, competitive pressures seldom allow engineers the luxury of 15-fold safety margins. It usually makes more sense to pay for a better analysis (or for a series of experiments) than to accept the cost and performance penalties of enormous safety margins.

In theoretical applied science the incentives differ: analysis is the main cost, experiments are unavailable, and safety margins are, in a sense, free of cost. Accordingly, simple models and large safety margins often make sense, even where engineering would demand a more efficient design based on more accurate data. In a specific instance, of course, physical constraints may preclude designs having ample safety margins to cover errors in the model; this is one of many ways in which a design may become unreliable. (If constraints impose a negative safety margin, then the analysis suggests an impossibility.)

In an unknown environment, uncertainties themselves may be unknown. Accordingly, theoretical applied science is easier when one can assume a simple, well-defined environment. The assumption of a sealed environment containing no loose molecules, for example, simplifies the design and analysis of nanomechanisms. This assumption is made throughout most of Part II and is supported by the analysis of walls, seals, and pumps in Chapter 11.

As H. Simon discusses in "The Architecture of Complexity" (Simon, 1981), structuring a system as a hierarchy of relatively independent subsystems simplifies design and analysis. One reason is that this architecture places each subsystem in a simpler, better-understood environment, reducing the difficulty of modeling its interactions with adequate accuracy.

A.4.5. Physical specification

Theoretical applied science deals with the feasibility of *classes* of devices rather than with the implementation of specific designs, hence it need not provide a full and detailed specification for every proposal. In a superficial paradox, sacrificing commitment to a detailed specification can make for a more reliable conclusion.

Returning momentarily to the issue of tolerance for modeling errors, it is important to remember that not all constraints are of kinds to which large safety margins can be applied. Engineers sometimes require not a simple inequality (e.g., $B > A$), which in principle permits an arbitrarily large safety margin, but satisfaction of a tolerance (e.g., $C > B > A$). If $C \approx A$, then $B \approx A$, a constraint which approximates that of equality. It may be that the best available model relating designs to values of B leaves an uncertainty in this parameter greater than $C - A$. If so, then showing that some designs yield large values of B no longer suffices to show that the constraint can be satisfied.

One can instead show that choosing from a *range of designs* allows B to be adjusted within a *range of values* from $D > C$ to $E < A$, with a distribution of choices that assures the existence of a design with $C > B > A$. Thus, modeling inaccuracies may both permit the conclusion that one or more choices are guaranteed to satisfy the constraint, yet preclude the conclusion that any particular design will do so. For example, a design constraint might require that a rotor be almost perfectly balanced, and yet be made of two parts of differing materials and uncertain density, located on opposite sides of the shaft. Any specified size and shape for the two parts would, for most values of their densities, yield an unbalanced rotor. Yet so long as the designer can adjust the dimensions of at least one part to bring the assembly into balance, the unknown densities permit confidence in satisfying the constraint. The required *range of choice* in dimensions depends directly on the *range of uncertainty* in the ratio of the densities, and the adequacy of a given range of choice can be verified or rejected. The result of such an analysis would be an exploratory design in which neither the densities nor the dimensions of the parts are specified, but in which satisfaction of the balance criterion can be guaranteed. A design exercise incorporating such a rotor (in a housing that, inefficiently, provides room for the full range of possible dimensions) could then proceed with confidence.

A particular design for a complex system may have a bug, that is, an unanticipated interaction that causes failure. Engineers commonly detect bugs by testing (that is, by experimentation), then remove them by redesign. If the bugs in systems of some class can consistently be removed by redesign, then a description of such systems *at a level of abstraction that omits features of the sort changed during routine bug removal* can be regarded as bug free (it requires no change). Although it may be impossible to design a working system in full detail without testing, one may nonetheless have confidence that the available range of choice includes designs that work.

In considering nanotechnology, there is a temptation to demand atom-by-atom specifications of structures. This is necessary for sufficiently small structures, but unnecessary for many others (Section 9.5). In engineering macroscale objects today, one never specifies the positions of all the atoms, even though these objects are subject to the full constraints of engineering practice, including manufacturability and success in a competitive marketplace. In molecular nanotechnology (at least during the theoretical applied science stage), the requirements for detailed specification need not *always* be more strict.

A.4.6. Confidence despite reduced detail

Several forms of reasoning illustrate how a partial (even nonexistent) physical specification can establish the physical feasibility of certain classes of devices. For example:

a. Biological analogy. Biological systems can demonstrate capabilities without revealing their underlying physical structures and mechanisms. The original case for programmable molecular°assemblers (Drexler, 1981) exploited analogies with enzymes (which guide chemical reactions) and with ribosomes (which execute a programmed series of operations). Likewise, the simplest case for the possibility of systems of molecular machinery able to build similar systems of molecular machinery is the existence and replication of bacteria.

b. Continuum models. In studying nanotechnology, useful design calculations can sometimes be performed based on a model of components made of homogeneous materials of a certain density, stiffness, surface stiffness (Sections 9.3 and 9.4), and so forth. The chief hazard is the use of such an approximation on the wrong scale (Chapter 2). A continuum model constitutes a partial specification that often can give an adequate description of the engineering possibilities.

500 Methodological Issues in Theoretical Applied Science

c. ***Encompassing options.*** In this approach, one attempts to establish that a range of options is broad enough and rich enough to include a solution, without specifying which option will actually satisfy the requirements (Section 10.3 provides an example). For a nanomechanical example, any of over 10^{75} different structures can be built within a single cubic nanometer (Section 9.5.2c); this range of options should suffice to join two slim rods of diamond with a bend of any desired angle and twist, satisfying tight tolerances on both. The chief hazard in this sort of argument is that a combination of constraints (angle *and* twist *and* offset *and* modulus *and* surface structure), each easily satisfied in isolation, may each eliminate a large fraction of the apparent options, perhaps leaving none that satisfy all the conditions simultaneously.

d. ***Engineering analogy.*** The possibility of parts that are analogous to those composing a system X will often make possible a system analogous to X itself. For example, given mechanical parts analogous to conduction paths, transistors, and so forth, mechanical computers will be possible (Chapter 12). Given parts equivalent to conventional gears, bearings, motors, and so forth (Chapters 10 and 11), robotic arms will be possible (Chapter 13). To reach these conclusions, one need not specify every detail of a particular computer or robotic arm; accordingly, one can postpone enormous efforts in design, analysis, and debugging to the engineering phase. In this mode of reasoning, the chief hazard is an inadequately close analogy between the sets of parts. For example, if one proposes devices intended to be analogous to transistors, then one must consider not only the basic logic operations that each gate performs, but other characteristics important in building systems: noise tolerance, signal fan-out, and logic-level restoration. Shortcomings in these areas have caused the demise of many proposed computer devices (Keyes, 1985; Keyes, 1989). To be reliable, analogies must either be sufficiently exact, or be buttressed by an analysis of the functionally significant differences.

A.4.7. Unique answers (and confidence from "uncertainty")

As outlined in Section A.3, theoretical applied science is less concerned with uniquely correct answers than is theoretical physics or even engineering. In physics, only one set of theoretical predictions can be completely true; discrepancies between two theories imply that at least one must be false. In engineering, many designs may work, hence there is no uniquely correct answer in any strict sense. If each design differs in performance, however, then in a hypothetical world of idealized competition and unbounded experimentation, engineering development would converge on uniquely optimal designs.

In engineering, different designs that serve the same purpose compete for the same market; in theoretical applied science, they cooperate in supporting the same conclusion. If a capability can seemingly be realized in multiple ways, then no single error is likely to invalidate the conclusion that this capability is indeed possible.

If a design is incompletely specified, this leaves uncertainty regarding how it is to be completed. Since "uncertainty" is an antonym of "confidence," it might naively seem that uncertainty regarding how a design will be completed must erode confidence in whether the design (when completed) will work. Yet we have seen (Section A.4.5) that uncertainty in one area (e.g., the accuracy of a parameter in a mathematical model) can be neutralized by freedom of choice elsewhere (e.g., choice of a parameter in a design). Uncertainty in the model leads to uncertainty in the design, but the two uncertainties more nearly cancel than add. This kind of uncertainty reflects not a risky gamble but an opportunity for a deferred problem-solving choice. Uncertainty (of the right kind) can thus increase confidence (in a distinct but related conclusion).

a. Uncertainties in different areas.   Uncertainties play different roles in science, engineering, and theoretical applied science. The intuitive rule regarding uncertainty in large sets of ideas or proposals is simple: if a conclusion or design rests on layer upon layer of shaky premises, it will surely fall. But this intuition sometimes misleads. To see where it works and where it fails, consider an imaginary proposition P in science ("Theory Y is true"), and a proposition P' in theoretical applied science ("A machine can do Z"). Each proposition is assumed to have N essential constituents P_a or P'_a ($a = a_1, a_2,..., a_N$), and each constituent is assumed to be selected from a set of $M = 10$ equally plausible possibilities P_{ab} or P'_{ab} ($b = b_1, b_2,..., b_M$). For a wide range of parameters in which both N and M are large, the result would be an absurdly speculative theory in science, but a robust proposal in theoretical applied science.

In the imaginary theory, each of the five constituents is a hypothesis regarding a distinct question. For example, a theory regarding a geological structure might comprise $N = 5$ hypotheses regarding (1) the interpretation of a seismogram, (2) the cause of bands in certain mineral grains, (3) the species identification of associated microfossils, (4) the origin of carbonaceous material, and (5) the source of a particular trace element. For each question, only one hypothesis can be true, and the stated assumption of M equally plausible possibilities implies than any given hypothesis will have (at best) a $1/M$ probability. If $M = 10$, the probability of making 5 correct choices is $.1^5 = .00001$ (assuming independent probabilities). For a theory to be true, each part must be true, and this becomes unlikely as low-confidence assumptions multiply. (Figure A.2 provides a graphical representation of this situation.)

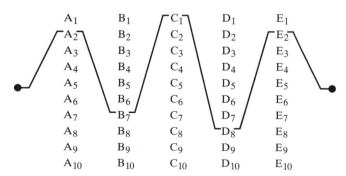

Figure A.2.   One path through an array of many choices: if only one choice is correct at each step, then only one of the 10^5 possible paths can be correct. If none of the known choices for some step is correct, then no path can be correct.

Consider a superficially similar problem in theoretical applied science: analyzing the possibility of a nanomechanical system that consists of 5 essential subsystems, each serving a particular function: (1) a motor, (2) a power supply, (3) a vacuum pump, (4) a pressure sensor, and (5) a gas-tight wall. Again we assume that for each there are 10 equally plausible possibilities—but unlike alternative scientific hypotheses, design options in theoretical applied science are not mutually exclusive. Thus, we can assume (for concreteness) that each of the $M = 10$ options for a subsystem has an arbitrarily chosen .5 probability of being a workable design (rather than a $1/M = .1$ probability of being the "one true design"). Further, the problem is to determine the possibility of a mechanism with a particular function, not to specify a single detailed design. (Figure A.3 shows a graphical representation.)

Assuming (for the moment) independent probabilities, the probability that all ten options will fail, leaving no workable choice for a particular subsystem, is $.5^{10} \approx .001$. Taking account of the risk of unworkability associated with each of the 5 necessary subsystems, the overall probability that a successful combination exists is $(1 - .5^{10})^5 \approx .995$.

In this example, a near certainty emerges from a combination of possibilities, each of which is as likely to fail as to succeed. Real examples can yield still more confidence for at least two reasons: First, one or more options may be essentially sure bets. Second, the probabilities of the various options may not be independent. For example, a set of options taken together may encompass a range that is guaranteed to contain a workable solution, even though any individual option, taken alone, is improbable. (A lack of independence caused by all options sharing a dubious assumption would have an adverse effect.)

Thus, uncertainties in theoretical applied science need not combine adversely, as do the superficially similar uncertainties of science. Engineering, however, has a somewhat closer resemblance to science: one must propose a single, specific design, build it, and live with the consequences. Time and budgets are limited, and the failure of a large system may leave no resources for another try. In an adaptation of the preceding model, this would mean making five choices with a .5 probability of success in each, yielding an overall probability of success $\sim.03$. Concerns of this sort motivate engineers to analyze and test components with care before building complex, expensive systems.

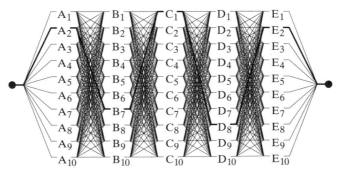

Figure A.3.   Many paths through an array of many choices: if many choices are likely to be correct at each step, then many (though perhaps a small fraction) of the 10^5 possible paths are likely to be correct.

A.4.8. Reliable reasoning

Different kinds of errors have different effects on an analysis. Errors of optimism regarding performance have no obvious upper bound, but errors of conservatism plainly have a lower bound: present capabilities. Further, errors of optimism place costly intellectual structures at risk, since work built on such assumptions is apt to be undermined by their failure. Errors of conservatism, in contrast, strengthen intellectual structures because they can often provide a margin of safety able to compensate for inadvertently optimistic assumptions made elsewhere. A systematic bias toward errors of conservatism can make analyses more robust.

Several considerations combine to make theoretical applied science more feasible than it might seem, chiefly by relaxing certain constraints of standard engineering practice. Exploratory designs need not be manufacturable using available technologies: they need only be physically possible. They need not compete with other, similar designs: they need only be workable (P. Morrison notes that this has parallels in the design of scientific instruments and elsewhere in experimental applied science). They can be grossly overdesigned to compensate for uncertainties. Since their purpose is to establish a possibility, not to guide the setup of a manufacturing process, they can omit details and include room for corrections. Finally, forms of uncertainty whose closest analogues in science and engineering would be intolerable prove to be perfectly acceptable in theoretical applied science. All these considerations aid in constructing reliable chains (and networks) of reasoning.

It may be objected that relaxing these constraints sacrifices most of the value of standard engineering, but human activities occur in a web of trade-offs that routinely forces the sacrifice of one goal to further another. This is true within an engineering design space and is equally true in intellectual work, such as design itself. The sacrifice of standard engineering goals is precisely what makes far-ranging exploratory design in theoretical applied science a feasible enterprise.

Theoretical applied science cannot substitute for experimental applied science or for engineering. Even the most reliable reasoning about a system is inferior to a physical example: real products both prove their own feasibility and enable physical accomplishments. The virtue of disciplined research in theoretical applied science is its ability to provide a partial survey of a field before experimentation and engineering become possible, offering some measure of knowledge when the alternative is ignorance.

A.5. A sketch of some epistemological issues

A.5.1. Philosophy of science (i.e., of physics)

Considerable attention has been given to the problem of knowledge in science; a few notes on the problem of knowledge in engineering and theoretical applied science may be useful here. These amount to no more than a sketch, taking the views of certain philosophers of science as a point of departure.

The philosophical view that an exact, general physical theory could be proved by experiment was dealt a mortal wound by the displacement (after long success) of Newtonian mechanics. As K. Popper points out, experiments cannot verify such theories; they can at best (provisionally) falsify them (Popper, 1963). More generally, concord between theory and experiment cannot show either to be correct, but discord between them shows a defect somewhere in the system of ideas (Lakatos, 1978).

W. W. Bartley III, Popper's student and biographer, has described a generalization of the Popperian position, termed *pancritical rationalism* (Bartley, 1987b). This holds that views cannot be proved in any ultimate sense, but that they can be criticized in terms of background assumptions that are for the moment considered nonproblematic. This seems to reflect actual practice in the scientific community, in which theories are themselves criticized in terms of their consistency with experiments and other theories, while experimental results are criticized in the same fashion. Ideally, nothing is taken as dogma, everything is open to criticism, and ideas are winnowed in the resulting Darwinian competition. Science thus is viewed as an evolutionary process.

A.5.2. Philosophy of engineering

Epistemological issues in engineering appear to have received little attention (though the practicalities of gathering and using knowledge in engineering have received massive attention). These epistemological issues appear to differ from those raised by physical theories.

A general physical theory characteristically makes precise statements about all forms of matter everywhere, asserting the truth of an equation. Engineering, in contrast, characteristically makes a statement about the behavior of a specific device, expressible as the satisfaction of a set of inequalities and tolerances. Although no finite set of measurements can show that a general theory holds in all instances, and no real physical measurement can show that an exact theory holds true even in a single instance, a single observation *can* show that a particular device sometimes works. Further, a series of observations can provide good evidence that devices of a particular type will work with high reliability, given certain conditions.

This comparison (1) neglects the problem, common to physics and engineering, of defining devices and conditions, including the *ceteris paribus* problem, (2) treats the notion of experimental observations as nonproblematic, (3) neglects the logical possibility that the laws of the universe might change in any manner at any time, and so forth. Nonetheless, it shows that typical propositions in engineering (e.g., "Device D can accomplish goal G") can, in an important sense, be better supported by experiment than can typical propositions in physics. This is a direct consequence of their lower precision and lesser generality.

A.5.3. Philosophy of theoretical applied science

Theoretical applied science, like engineering, makes assertions of lesser generality than does physics. A typical assertion might be of the form "*Some* device D can accomplish goal G." In engineering, such assertions can be verified by demonstration (in the approximation that the experimental results are nonproblematic), but the conditions of theoretical applied science do not permit this (when demonstrations become possible, theoretical applied science is over). Further, statements of this form cannot be experimentally falsified under any circumstances, since no finite set of experiments can test all possible devices.

How can propositions in theoretical applied science be tested against reality, if they cannot be tested experimentally? A close parallel occurs in the design phase of an engineering project. Here, too, experimentation is deferred, but this does not leave engineers adrift in fantasy. They test designs against generally accepted facts and theories about physical systems, which have themselves been tested (though not proved) by experimentation. Further, because engineering works with inequalities and tolerances, its conclusions can be more reliable than are its theoretical premises. For example, most engineering calculations are based on Newtonian mechanics, and can yield reliable engineering conclusions even though Newtonian mechanics is false. The accuracy of these conclusions is well insulated from most plausible revolutions in theoretical physics. Even in the quantum domain, current theories regarding the behavior of electrons on a molecular scale seem well insulated from uncertainties regarding systems involving (for example) neutrinos, subnuclear dimensions, or extreme energies.

To summarize, theoretical applied science takes the body of generally accepted fact and theory amassed by science and engineering as nonproblematic background knowledge. By testing propositions in theoretical applied science for consistency with this body of knowledge, they can be criticized and their likelihood judged, even when they are formally unfalsifiable by direct experimentation.* *Propositions in theoretical applied science are falsifiable if one adopts the rule that any conflict with established scientific knowledge constitutes falsification.* (Note that ambiguities like those in "established scientific knowledge" also appear in "experimental results": either can be clear or disputed, and both depend on theory and interpretation.) For example, all classes of device that would violate the second law of thermodynamics can immediately be rejected. A more stringent rule, adopted in the present work, rejects propositions if they are inadequately substantiated, for example, rejecting all devices that would require materials stronger than those known or described by accepted physical models. By adopting these rules for falsification and rejection, work in theoretical applied science can be grounded in our best scientific understanding of the physical world.

A.6. Theoretical applied science as intellectual scaffolding

Theoretical applied science can provide intellectual scaffolding for further study of a field. A scaffold serves the goals of architecture and must be structurally sound, yet it is judged by criteria different from those used to judge the ultimate architectural product. Like scaffolding, theoretical applied science analyses are adapted for rapid construction, and their parts may all be removed and replaced as work progresses.

A.6.1. Scaffolding for molecular manufacturing

The case for molecular manufacturing is today an example of theoretical applied science. Like a scaffold, it consists of parts that join to form a structure. These

---

* The idea that experimentally unfalsifiable statements can be tested against established theory has been discussed in the philosophy of science literature by Wisdom (1963).

parts support the physical feasibility of various capabilities. They include reasons for expecting that we can, given suitable tools and effort:

- Engineer complex molecular objects
- Assemble more complex systems from these molecular objects
- Build and control molecular machine systems
- Use molecular machine systems to perform molecular manufacturing
- Use molecular manufacturing to build nanocomputers
- Use nanocomputers to control molecular manufacturing
- Use manufacturing systems to build more manufacturing systems
- With the preceding, achieve thorough control of the structure of matter

These arguments interlock: establishing the feasibility of the whole requires support for the pieces, and the apparent feasibility of the whole then motivates greater scrutiny of each piece, either to criticize it or to improve it. Given molecular manufacturing, improved nanocomputer designs become more interesting. Given nanocomputers, improved molecular manipulator designs become more interesting. Given the feasibility of the whole, implementation strategies become more interesting.

As exploratory designs grow more detailed, they come to resemble descriptions of experiments in applied science, or of engineering prototypes. As the tools required for fabrication become available (perhaps speeded by a better understanding of what they can build), engineering practice will encroach on theoretical applied science, and theoretical studies will give way to experiment, production, and use. The scaffolding will then have been replaced with brick.

A.7. Conclusions

Theoretical applied science draws on the enormous body of knowledge amassed by science and engineering, but exploits that knowledge for different purposes using different methodologies. Its aim is neither to describe nature nor to build devices, but to describe lower bounds to the performance achievable with physically possible classes of devices.

Theoretical applied science can achieve its goals only by sacrificing many of the goals of experimental science and of engineering. It produces analyses, not devices, and thus avoids the stringent requirement that its designs be fully specified, manufacturable, and competitive. This latitude can be exploited to mitigate the problems posed by inaccurate models and the infeasibility of direct experimentation. Research in theoretical applied science typically makes no pretense of designing systems that can be built today, or that will be built tomorrow. Today we lack the tools; tomorrow we will have better designs.

In an ideal world, theoretical applied science would consume only a tiny fraction of the effort devoted to pure theoretical science, to experimentation, or to engineering. The resulting picture of technological prospects can nonetheless be of considerable value: it can indicate areas of science and technology that are likely to prove especially fruitful, and it can help us understand the opportunities and challenges that technological development is likely to bring.

Appendix B

Related Research

B.1. Overview

The present volume analyzes the capabilities of nanomechanical systems, including computers and manufacturing systems able to construct additional nanomechanical systems. There has been an enormous amount of relevant work in chemistry, statistical mechanics, solid-state physics, mechanical engineering, and so forth; portions of that work are cited throughout the previous chapters and form their foundation. Nonetheless, despite prior discussions of biological-style mechanisms (i.e., molecules and polymer-based molecular machines interacting in solution), and numerous proposals, efforts, and successes in building microscale, nanoscale, and molecular devices, there has until recently been little published work by other authors that parallels the direction taken here.* Accordingly, much of this appendix discusses adjacent fields.

Section B.2 describes how these fields have been divided in their methods and objectives. Sections B.3 to B.6 survey relevant work in the fields of mechanical engineering, microtechnology, chemistry, molecular biology, and protein engineering. Section B.7 discusses proximal probe experiments that have demonstrated limited forms of molecular manipulation. Finally, Section B.8 discusses the remarkably foresighted suggestions made by R. Feynman in 1959.

B.2. How related fields have been divided

B.2.1. Scientific goals vs. technological goals

The construction of molecular manufacturing systems, like the construction of conventional manufacturing systems, is a technological goal. In many countries (e.g., the U.S.), the study of molecules is usually taught as chemistry and defined

---

* Diverse work in "nanotechnology" is reviewed by Franks (1987), but the then-new term was taken by that author as possibly including glass polishing and fine-powder technologies. Taniguchi (1974) applied 'Nano-technology' (in this form, with quotes) to processes such as ion sputtering. A broad and uneven survey of pre-1980 speculations regarding small devices of various kinds appears in Schneiker (1989). Recent collections of papers organized around molecular nanotechnology in the present sense appear in the proceedings of the First and Second Foresight Conferences on Molecular Nanotechnology, held in 1989 and 1991 (Crandall and Lewis, 1992; Teague, 1992).

as a natural science. As materials science is to integrated circuit design, and as physics is to mechanical engineering, so chemistry is to molecular engineering: a distinct but intimately related discipline. It would be remarkable if materials scientists developed computers, or if physicists developed automobiles; likewise, it is not surprising that chemists have not developed molecular manufacturing systems.

B.2.2. Top-down vs. bottom-up approaches

Engineering systems today range from macroscopic to microscopic, with active research on building electronic and mechanical systems on ever-smaller scales using microtechnologies. The micron scale, however, is volumetrically 10^9 times larger than the nanometer scale, and existing microtechnologies provide no mechanism for gaining precise, molecular control of the surface and interior of a complex, three-dimensional structure. A wide gap has separated the top-down path of microtechnology (starting with large, complex, and irregular structures) from the bottom-up path of chemistry (starting with small, simple, and exact structures). In microtechnology, the challenge is to make imprecise structures smaller; in chemical synthesis, the challenge is to make precise structures larger. An engineering discipline is only now forming around the latter goal.

B.2.3. Immediate goals vs. long-term prospects

Science and technology are united by a focus on what can be made or tested within a few years. The molecular sciences are centered around the laboratory, and hence around current capabilities. Engineering is centered around the workshop, again binding creative thought to current capabilities. In most sciences, theoretical work is scarce relative to experimental work (the abundant output of theoretical physics is an outstanding exception). The study of prospects in technology (Appendix A), though based on present scientific knowledge, is a theoretical discipline that falls outside the usual scope of physics, of laboratory science, and of practical engineering.

B.3.   Mechanical engineering and microtechnology

When miniaturization is viewed as an incremental, top-down process, as in mechanical engineering and microtechnology, working at the molecular size scale appears to be a distant goal. Current research in top-down miniaturization (usually termed "microtechnology" but sometimes termed "nanotechnology") offers no obvious way to achieve the goals of molecular manufacturing. Precise control of mechanical structures at the molecular scale has only recently begun to be considered as a goal for mechanical engineering.

B.4.   Chemistry

Chemistry is usually regarded as a laboratory-centered natural science.* Lacking the means to design and synthesize complex molecular machines, and also lacking an engineering tradition urging them in that direction, chemists have yet

---

* The Department of Synthetic Chemistry at the University of Tokyo, however, is in the Faculty of Engineering.

to place the design and analysis of *systems* of molecular machinery high on their research agenda. Chemical research has nonetheless made progress toward complex molecular systems; the following is a highly fragmentary review.

The early 1980s saw a wave of interest in molecular electronics. In his Nobel lecture (Lehn, 1988), J.-M. Lehn stated:

> Components and molecular devices such as molecular wires, channels, resistors, rectifiers, diodes, and photosensitive elements might be assembled into nanocircuits and combined with organized polymolecular assemblies to yield systems capable ultimately of performing functions of detection, storage, processing, amplification, and transfer of signals and information by means of various mediators (photons, electrons, protons, metal cations, anions, molecules) with coupling and regulation.

This paragraph closes with a list of references, of which the earliest are Nagle et al. (1981), which describes experiments on charge-transfer excited states in synthetic molecules; Drexler (1981), which suggests molecular computers as possible products of molecular manufacturing; and an address by J.-M. Lehn (1980). Molecular rectifiers had been proposed by Aviram and Ratner (1974), and other molecular electronic systems had been proposed by F. Carter (e.g., in Carter, 1980).

Most experimental research in molecular electronics has focused on the development of molecules that exhibit useful electronic properties in thin films or in microscale aggregates; some proposals, however, have focused on the construction of computational devices in which individual molecules or moieties would serve as signal carrying and switching elements [e.g., papers in Carter (1982, 1987); also (Robinson and Seeman, 1987) and many others]. These have suggested various combinations of chemical synthesis, protein engineering, and DNA engineering to make self-assembling systems on a broadly biological model. This objective is a form of molecular systems engineering (though not of machines or manufacturing systems) and the capabilities required would resemble those discussed in Chapter 15.

Chemists have constructed molecular devices including sophisticated reagents and catalysts, molecules in which two rotating sections are coupled in a gear-like manner (Mislow, 1989), molecules that self-assemble to form small structures (for examples, see Cram, 1986; Cram, 1988; Diederich, 1988; Lehn, 1988; Rebek, 1987), molecules that self-assemble in a planned manner to form crystals (for example, Fagan et al., 1989), and a molecule that catalyzes the synthesis of copies of itself (Tjivikua et al., 1990). Recent years have seen the development of molecular systems including components that join covalently in a process described as "structure-directed synthesis" (Ashton et al., 1989). Efforts of this kind have been described as steps toward "molecular LEGO" (Kohnke et al., 1989) or "molecular Meccano" (Anelli et al., 1992); synthetic rods have been described as steps toward "molecular Tinkertoys" (Kaszynski et al., 1992). These descriptions, however, have yet to be supported by the necessary system-level analysis: in the cited toy systems, well-known means (e.g., hands) join the building blocks to form complex, aperiodic, functional patterns, but comparable means remain to be described in connection with these molecular structures. Lindsey (1991) provides a useful discussion of the self-assembly principles that might be used to fill this gap (in the absence of direct molecular manipulation).

Chemistry has traditionally followed a forward-chaining strategy (Section 16.2.1), taking solution-phase phenomena and synthetic capabilities as a point of departure. The concepts in this volume result from a backward-chaining analysis that first explores deterministic molecular machine systems as an objective, and then examines present laboratory capabilities as a means of achieving that objective. Since machine-phase systems are quite unlike molecules undergoing Brownian motion in solution, and since they are not immediately realizable in the laboratory, it is natural that molecular manufacturing has been slow to appear on the research agenda of chemistry.

B.5. Molecular biology

Molecular biologists study and modify systems of molecular machines. Genetic engineers reprogram them, sometimes to build novel multinanometer-scale molecular objects with complex functions. Molecular biologists, however, work within the traditions of natural science, and they are seldom system builders. Although observations at the level of "biomolecules might be used as components in some device" appeared sporadically in earlier years, there are apparently no publications that predate Drexler (1981) and argue that devices resembling biomolecular motors, actuators, and structural components could be combined to build molecular machine systems analogous to machine systems in the macroscopic world. (Proposals patterned on living systems, coupling catalytic and regulatory molecules by diffusive transport, while of considerable interest, are in a distinct category of no direct relevance to the goal of molecular manufacturing.)

The chief forms of molecular engineering to emerge from molecular biology have been protein engineering (usually to produce isolated catalysts in solution or immobilized on surfaces), and the engineering of three-dimensional structures from branched DNA (Chen and Seeman, 1991; Seeman, 1982; Seeman, 1991). The use of complementary DNA sequences provides one answer to the question of how to form complex patterns from self-assembling molecular objects, and proposals have been advanced for the application of this work to molecular mechanical and electronic devices (Robinson and Seeman, 1987).

B.6. Protein engineering

Protein molecules constitute much of the molecular machinery found in living systems, and protein engineering has in the last decade become a substantial area of research. It has been suggested that protein engineering can provide a path for developing molecular manufacturing (Drexler, 1981), if pursued with the objective of constructing self-assembling systems of molecular machines.

The journal *Protein Engineering,* in its "Instructions to Authors," gives a sense of how the field has developed. The MIT School of Engineering (MIT Bulletin, 1988–89) defines engineering as "a creative profession concerned with developing and applying scientific knowledge and technology to meet societal needs," but *Protein Engineering* states that "the objectives of those engaged in this area of research are to investigate the principles by which particular structural features in proteins relate to the mechanisms through which biological function is expressed, and to test these principles in an empirical fashion by introduction of specific changes followed by evaluation of any altered structural and/or functional properties." In short, the stated objective is scientific knowledge, not the construction of useful new proteins.

Considerable progress has been made in engineering novel objects from proteins. Outstanding examples include what is generally regarded as the first *de novo* structure (DeGrado et al., 1987), the development of synthetic, branched proteinlike structures that depart substantially from biological models (Mutter et al., 1988), and the engineering of a branched, nonbiological protein with enzymatic activity (Hahn et al., 1990).

B.7. Proximal probe technologies

Proximal probe instruments—scanning tunneling microscopes (STMs), atomic force microscopes (AFMs) and their relatives—provide a means for positioning and maneuvering tips near surfaces with atomic precision. The possibility of modifying surfaces with these tips was evident from the earliest years of STM research, since inadvertent contact between tips and surfaces routinely caused such modifications. Suggestions for controlled surface modification have appeared (Farrell and Levinson, 1985), including control based on the precise application of molecular tools (Drexler, 1986a). Experiments have since demonstrated (for example) atomic-scale surface modifications on germanium (Becker et al., 1987), pinning of organic molecules to graphite (Foster et al., 1988), arrangement of 35 xenon atoms to spell "IBM" on a nickel surface (Eigler and Schweizer, 1990) and subsequent positioning of carbon monoxide and platinum atoms on platinum surfaces in the same laboratory. Much of this work has been conducted at IBM, where the objective of developing atomic-scale mechanisms for data storage and processing has been explicitly articulated. Japan's Science and Technology Agency in 1989 initiated the Aono Atomcraft Project to pursue research in this area.

B.8. Feynman's 1959 talk

The mechanical construction of molecules was suggested by R. Feynman in an after-dinner speech, "There's Plenty of Room at the Bottom," given at the 1959 annual meeting of the West Coast Section of the American Physical Society and later published (Feynman, 1960). Most of the talk focused on miniaturization and microtechnology: this section anticipated capabilities like those that emerged in the microelectronics industry, and then proposed an alternative approach to miniaturization that would use machines to build smaller machines, which would build still smaller machines, and so forth. Toward the close of this discussion are four paragraphs that comprise the clearest prior discussion of molecular manufacturing:

> At the atomic level, we have new kinds of forces and new kinds of possibilities, new kinds of effects. The problems of manufacture and reproduction of materials will be quite different. I am, as I said, inspired by the biological phenomena in which chemical forces are used in a repetitive fashion to produce all kinds of weird effects (one of which is the author).

The principles of physics, as far as I can see, do not speak against the possibility of maneuvering things atom by atom. It is not an attempt to violate any laws; it is something, in principle, that can be done; but, in practice, it has not been done because we are too big.

Ultimately, we can do chemical synthesis. A chemist comes to us and says, "Look, I want a molecule that has the atoms arranged thus and so; make me that molecule." The chemist does a mysterious thing when he wants to make a molecule. He sees that he has got that ring, so he mixes this and that, and he shakes it, and he fiddles around. And, at the end of a difficult process, he usually does succeed in synthesizing what he wants. By the time I get my devices working, so that we can do it by physics, he will have figured out how to synthesize absolutely anything, so that this will really be useless.

But it is interesting that it would be, in principle, possible (I think) for a physicist to synthesize any chemical substance that the chemist writes down. Give the orders and the physicist synthesizes it. How? Put the atoms down where the chemist says, and so you make the substance. The problems of chemistry and biology can be greatly helped if our ability to see what we are doing, and to do things on an atomic level, is ultimately developed—a development which I think cannot be avoided.

These remarks pointed in the direction explored in this volume. Why was there so little response? Presumably because these long-term goals appeared to lack near-term consequences: they neither defined an accessible scientific problem nor suggested an immediate engineering project. They might have sparked work in theoretical applied science, but the study of long-term technological possibilities has had few serious practitioners.

B.9. Conclusions

Prior work in physics, chemistry, molecular biology, and engineering forms an adequate and essential foundation for the work presented in the previous chapters. Nonetheless, the pre-1991 literature appears to contain few research reports describing studies of molecular mechanical engineering and molecular manufacturing. Although these ideas are, in a sense, an obvious extrapolation of present knowledge and abilities, several circumstances have discouraged sustained analytical efforts in this area. These include the usual time horizons for research funding, the strong laboratory orientation of the molecular sciences, and the gap between the goals of conventional chemistry and those of theoretical applied science.

In light of emerging capabilities in chemistry, protein engineering, and proximal probe technologies, the time appears ripe for design and experimentation aimed at the goal of molecular manufacturing; indeed, a research community has begun to form. The coming effort will draw on the knowledge and skills of disciplines as diverse as chemistry, physics, mechanical engineering, and computer science, and its products will contribute to fields spanning the whole of science and technology. This volume has drawn on fundamental principles from several disciplines to assemble a set of conceptual and mathematical tools adequate for designing and analyzing a limited class of molecular mechanical systems. If it has shown some of the potential of molecular nanotechnology, and given some help to those wishing to enter the field, then it has achieved its major objective.

Afterword

Readers seeking further information on molecular nanotechnology may wish to contact the Foresight Institute regarding publications, conferences, and relevant research such as that sponsored by the Institute for Molecular Manufacturing. The present volume focuses on physical principles and basic technologies; the Foresight Institute also addresses the broader issues of applications and policy raised by the prospect of molecular nanotechnology.

Foresight Institute
Nanosystems information
P.O. Box 61058
Palo Alto, CA 94306 USA

telephone: 415-324-2490
fax: 415-324-2497
email: foresight@cup.portal.com

Reports of errata and inquiries regarding problems for use in courses can be sent to the author at the above address.

Symbols, Units, and Constants

Equations are stated in terms of base units such as meters, kilograms, and seconds. Physical quantities in the text and tables are commonly described using scaled units such as nanometers, femtograms, and nanoseconds.

A	ampere
A	amplitude; rate of change of amplitude; preexponential factor
$\mathscr{A}$	Hamaker constant (J)
a	atto- (10^{-18})
a	acceleration (m/s^2)
amu	atomic mass unit $(= 1/12$ mass of $^{12}C \approx 1.661 \times 10^{-27}$ kg)
aJ	attojoule $(= 10^{-18}$ J)
B	boron
Be	beryllium
Br	bromine
b	exponential factor
c	speed of light $(\sim 3 \times 10^8$ m/s); concentration (m^{-3})
C	coulomb; carbon
Cl	chlorine
C	London dispersion force coefficient $(J \cdot m^6)$; constant of integration; compliance (m/N); capacitance (F)
C_{atten}	acoustic attenuation coefficient (s^2/m)
C_{vol}	heat capacity per unit volume at constant volume $(J/K \cdot m^3)$

$\mathscr{C}_n$	ratio of longitudinal contraction to transverse vibrational energy (m/J)
D	radiation dose (rad)
D_0	binding energy of bond, with zero-point correction
D_e	binding energy of bond, without zero-point correction
D_{sr}	coefficient of shear-reflection drag
d	distance (m); thickness (m)
d'	dimensionless measure of interfacial stiffness
d_a	interatomic distance along a line (m)
E	Young's modulus (N/m^2); electric field strength (V/m)
E_ℓ	rod elastic modulus (N)
$\mathscr{E}$	total energy (J)
eV	electron-volt $(\approx 1.602 \times 10^{-19}$ J)
e	elementary charge $(\approx 1.602 \times 10^{-19}$ C)
F	fluorine; farad
F	force (N)
$\mathscr{F}$	Helmholtz free energy (J)
f	femto- (10^{-15})
f	fractional quantity; frequency (cycles/s)
f_{TST}	frequency factor in transition state theory
$f_x(x)$	probability density function with respect to a variable x
f_x	spatial frequency along axis x (cycles/m)

G   giga- (10^9)

G shear modulus (N/m^2)

G_ℓ rod shear modulus (N)

$\mathcal{G}$ Gibbs free energy (J)

gcd   greatest common denominator

H   hydrogen

He   helium

Hz   Hertz (= cycle/s)

$|\mathrm{H}(f_x)|$ amplitude spectral density of a potential along axis x

h   henry

h height (m)

$\hbar$ Planck's constant ($\approx 1.055 \times 10^{-34}$ J·s)

$\hbar\omega$ vibrational energy quantum (J)

I   iodine

I moment of inertia (kg·m^2)

i an integer (often an index); $(-1)^{1/2}$

J   joule

J exchange integral (J)

j an integer (often an index)

K   degree kelvin

K_d dissociation constant (m^{-3})

kg   kilogram

k Boltzmann's constant ($\approx 1.381 \times 10^{-23}$ J/K); optical extinction coefficient

k_{12} rate of transitions from state 1 to state 2, given occupancy of state 1 (s^{-1})

k_a stiffness per unit area of interface (N/m^3)

k_b bending stiffness of a beam (J·m/rad^2)

k_{cubic} MM2 cubic bond stretching constant (m^{-1})

k_{err} erroneous reaction rate (s^{-1})

k_{isc} intersystem crossing rate (s^{-1})

k_ℓ transverse stiffness per unit length (N/m^2)

k_{react} correct reaction rate (s^{-1})

k_s mechanical stiffness (N/m^2)

k_{sextic} MM2 sextic bond angle-bending constant (rad^{-4})

$k_{s\theta}$ stretch-bend parameter (N/rad)

$k_{s\perp}$ mechanical stiffness perpendicular to a bond or axis (N/m^2)

kT characteristic thermal energy (J)

$k_{t,b}$ transverse-displacement stiffness of a bending beam (N/m)

k_n stiffness of vibrational mode n (N/m = J/m^2)

k_θ angular spring constant (J/rad^2)

$\boldsymbol{k}$ wave vector (m^{-1}); magnitude of wave vector (m^{-1})

$\boldsymbol{k}_D$ Debye radius (in reciprocal space) (m^{-1})

L   ligand

Li   lithium

LP   lone pair

L length in scaling relationships (m); inductance (h)

lcm   least common multiple

ℓ length (m)

M   molar (10^3 mole/m^3); metal atom

MIPS   million instructions per second

M modulus of elasticity (N/m^2)

M_s Mach number (= v/v_s)

m   meter; milli- (10^{-3})

maJ   milli-attojoule (10^{-21} J)

mole   gram mole (= 6.022×10^{23})

m mass (kg); integer

N   newton (kg·m/s^2); nitrogen

N integer

n   nano- (10^{-9})

nm   nanometer (10^{-9} m)

n integer; refractive index

n_a areal number density (m^{-2})

n_v volumetric number density (m^{-2})

O   oxygen

P   phosphorus

Pa   pascal (N/m^2)

P probability; power (W)

P_{err} probability of error

$P(x)$ probability of condition x

p   pico- (10^{-12})

pK_a pH at which 50% of the molecules of an acidic species are dissociated

p pressure (N/m^2); magnitude of momentum (kg·m/s)

p_i momentum coordinate i (kg·m/s)

$\mathbf{p}$ momentum vector (kg·m/s)

Q Coulomb integral (J)

q electrical charge (C); generalized coordinate

q partition function

R alignment band velocity ratio; Reynolds number; concentration ratio; resistance (Ω)

R_{temp} compression temperature ratio

$\mathscr{R}_n$ nth root of Eqs. (5.45) and (5.46) describing beam vibrations

rad   radians; radiation dose (10^{-2} J/kg)

r distance or radius (m); interest rate (yr^{-1})

r_0 reference distance or radius

r_i position coordinate i (m)

r_{vdw0} MM2 van der Waals radius

$\mathbf{r}$ coordinate vector in configuration space (m)

S   sulfur

Si   silicon

S_N2 nucleophilic substitution by direct displacement

S area (in real space, m^2; in n-dimensional configuration space, m^{n-1})

$\mathscr{S}$ entropy (J/K)

s   second

s separation (m)

T   tesla

T temperature (K); transmittance; torque (N·m)

T' T/T_D

T_{300} 300 K

T_D Debye temperature (K)

T_{trans} phonon transmission coefficient

$\mathscr{T}$ kinetic energy (J)

t time (s)

t_{act} actuation time

t_{trans} transformation time

V   volt

V volume (in real space, m^3; in n-dimensional configuration space, m^n)

$\mathscr{V}$ potential energy (often a function of position coordinates) (J)

$\mathscr{V}'$ potential energy of a well, corrected for zero point v speed (m/s)

$\mathscr{V}^\dagger$ potential energy including z-axis elastic constraint (J)

$\Delta\mathscr{V}^\ddagger$ barrier height (transition-state energy minus well energy) (J)

$\Delta\mathscr{V}'^\ddagger$ barrier height based on corrected well energy (J)

$\mathscr{V}_\omega$ potential energy of bond torsion

$\mathscr{V}_\theta$ potential energy of bond angle-bending

v speed (m/s)

v_s speed of sound (m/s)

W work (J)

w width (m)

x spatial coordinate (m)

yr   year ($\approx 3.154 \times 10^7$ s)

$y(x)$ displacement as a function of x

y spatial coordinate (m); transverse displacement (m); general-purpose variable

z spatial coordinate (m); general purpose variable

α accommodation coefficient; selective transport coefficient

β Morse function scaling parameter (m^{-1}); volume coefficient of thermal expansion (K^{-1})

Γ^* Wigner tunneling correction factor

γ Shear stress (N/m^2)

γ_G Grüneisen number

γ_ℓ tension in rod (N)

δ positional tolerance (m); minimum imaging separation (m)

δ_{surf} distance between structural- and excluded-volume boundaries (m)

$\partial x/\partial y$ derivative of x with respect to y, holding other variables constant

ε zero-frequency dielectric constant; phonon energy density (J/m^3); surface contour tolerance (m)

ε_0 electrical permittivity of free space (8.854×10^{-12} F/m)

ε_{vdw} MM2 van der Waals energy

η viscosity (N·s/m^2)

η_{phonon} phonon viscosity (N·s/m^2)

θ angle (rad); bond angle (rad)

θ_0 reference angle (rad)

θ_{sym} minimum rotational symmetry angle (rad)

κ_n spatial frequency factor in beam bending equations

λ wavelength (m)

μ reduced mass (kg); micro- (10^{-6}); micron (10^{-6} m)

ν Poisson's ratio

ρ mass density (kg/m^3)

ρ_ℓ linear mass density (kg/m)

σ standard deviation in position (m); scattering cross section (m^2)

ϕ phase angle (rad); spherical coordinate angle (rad)

ϕ_w work function (eV)

ψ spherical coordinate angle (rad)

$\psi(x)$ wave function (commonly complex-valued)

$\psi^*(x)$ complex conjugate of $\psi(x)$

Ω ohm

ω angular frequency (rad/s); torsion angle (rad)

ω_0 reference torsion angle (rad)

ω_{rc} imaginary frequency of motion along reaction coordinate (s^{-1})

$\$$ U.S. dollar, 1990

Energy conversion factors:

kJ/mole = kilojoule per mole = 1.661×10^{-21} J = 1.661 maJ
  1 maJ = 0.6022 kJ/mole

kcal/mole = kilocalorie per mole = 6.952×10^{-21} J = 6.952 maJ
  1 maJ = 0.1438 kcal/mole

eV = electron volt = 1.602×10^{-19} J = 160.2 maJ
  1 maJ = 6.242×10^{-3} eV

a.u. = Hartree = 4.360×10^{-18} J = 4360 maJ
  1 maJ = 2.295×10^{-4} a.u.

kT_{300} = thermal vibrational energy at 300 K = 4.142×10^{-21} J = 4.142 maJ
  1 maJ = $0.2414 kT_{300}$

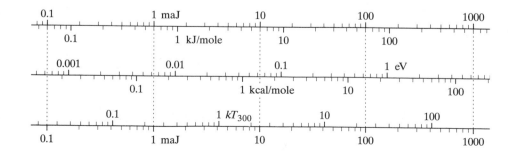

Comparison of chemical symbols:

CH_2ClF

$C_6H_5CH_2COOH$

Glossary

This glossary is intended to offer help in crossing disciplinary boundaries, especially for those beginning to explore the terminology-rich field of chemistry. The entries are neither exhaustive in scope nor definitional in quality. A small fraction of the entries define terms introduced in the present work. Cross-references are marked with a small circle (for example, °abstraction reaction); an alphabetical search will occasionally yield a different form of the indicated word.* Some words in the main text are marked in the same way.

* This system is modeled on Atkins (1974).

Abstraction reaction   A °reaction that removes an atom from a structure.

Acid   In the Brønsted definition, an acid is a chemical species that can donate a proton to another species (a °base). In the Lewis definition, an acid is a chemical species that can accept (and share) a pair of electrons from another species. Hydrochloric acid is a Brønsted acid; the proton it donates is a Lewis acid. A neutral Lewis acid and a neutral Lewis base can commonly form a °dipolar bond.

Activation energy   Distinct states correspond to minima of a °potential energy surface in a °configuration space. In this classical picture, the activation energy for transforming state A into state B is the maximum increase in energy (relative to the °ground state of A) encountered on a minimum-energy path from A

to B. °Energy here refers to potential energy; an analogous definition based on °free energy can be constructed. When tunneling is considered, lower energy paths become possible, but an activation energy can be associated with the °reaction (at a given °temperature) via the relationship between temperature and reaction rate.

Acyclic   Not °cyclic.

Affinity constant   The reciprocal of the °dissociation constant; a measure of the °binding energy of a °ligand in a °receptor.

AFM   An °atomic force microscope.

Alkane   A °saturated, °acyclic hydrocarbon structure; usually quite inert.

Alkene   A °hydrocarbon containing a °double bond; often rather reactive.

Alkyne   A °hydrocarbon containing a °triple bond; often rather reactive.

Amide   A molecule containing an °amine bonded to a carboxyl group; the resulting bond has substantial double-bond character. Also termed a °peptide; °amide bonds link °amino acids in °proteins.

Amine   A molecule containing N with a single bond to C and two other single bonds to H or C (but not an °amide); the amine group or moiety.

Amino acid   A molecule containing both an °amine and a °carboxylic acid group; in the 20 genetically encoded amino acids in biology, both groups are bound to the same C. Amino acids joined by °amide bonds form °peptides

519

and°proteins; these do not contain amino acids as such, and are instead said to contain *amino acid residues*.

Anion   A negatively charged °ion.

Aromatic   A term used to describe °cyclic °pi-bonded structures of special stability.

Assembler   In recent popular usage, any °nanomachine, usually assumed to offer magical, universal capabilities in an atom-sized package. In the author's usage, any programmable nanomechanical system able to perform a wide range of mechanosynthetic operations. See °molecular manipulator, °molecular mill.

Atomic force microscope   A device in which the deflection of a sharp stylus mounted on a soft spring is monitored as the stylus is moved across a surface. If the deflection is kept constant by moving the surface up and down by measured increments, the result (under favorable conditions) is an atomic-resolution topographic map of the surface. Also termed a scanning force microscope.

Barrier height   Roughly synonymous with °activation energy.

Base   In the Brønsted definition, a base is a chemical species that can accept a proton from another species. In the Lewis definition, a base is a chemical species that can donate (and share) a pair of electrons with another species. See °acid.

Bearing   A °mechanical device that permits the motion of a component (ideally, with minimal resistance) in one or more degrees of freedom while resisting motion (ideally, with a stiff restoring force) in all other degrees of freedom.

Binding energy   The reduction in the °free energy of a system that occurs when a °ligand binds to a °receptor. Generally used to describe the total energy required to remove something, or to take a system apart into its constituent particles—for example, to separate two atoms from one another, or to separate an atom into electrons and nuclei.

Binding site   The active region of a °receptor; any site at which a chemical °species of interest tends to bind.

Binding   The process by which a molecule (or °ligand) becomes bound, that is, confined in position (and often orientation) with respect to a °receptor. Confinement occurs because structural features of the receptor create a °potential well for the ligand; °van der Waals and electrostatic interactions commonly contribute.

Bond   Two atoms are said to be bonded when the energy required to separate them is substantially larger than the van der Waals attraction energy. Ionic bonds result from the electrostatic attraction between ions; °covalent and metallic bonds result from the sharing of electrons among atoms; °hydrogen bonds are weaker and result from dipole interactions and limited electron sharing. When used without modification, "bond" usually refers to a covalent bond.

Brownian assembly   °Brownian motion in a fluid brings molecules together in various positions and orientations. If molecules have suitable complementary surfaces, they can bind, assembling to form a specific structure. Brownian assembly is a less paradoxical name for self-assembly (how can a structure assemble itself, or do anything, when it does not yet exist?).

Brownian motion   Motion of a particle in a fluid owing to thermal agitation, observed in 1827 by Robert Brown. (Originally thought to be caused by a vital force, Brownian motion in fact plays a vital role in the assembly and activity of the molecular structures of life.)

CAD   Computer-aided design.

Cam   A component that translates or rotates to move a contoured surface past a °follower; the contours impose a sequence of motions (potentially complex) on the follower.

Carbanion   A highly reactive °anionic chemical species with an even number of electrons and an unshared pair of electrons on a tetravalent carbon atom.

Carbene   A highly reactive chemical species containing an electrically neutral, divalent carbon atom with two nonbonding °valence electrons; a prototype is CH_2.

Carbonium ion A highly reactive °cationic chemical species with an even number of electrons and an unoccupied orbital on a carbon atom.

Carbonyl A chemical moiety consisting of O with a double bond to C. If the C is bonded to N, the resulting structure is termed an °amide; if it is bonded to O, it is termed a °carboxylic acid or an °ester linkage.

Carboxylic acid A molecule that includes a C having a °double bond to O and a °single bond to OH.

Catalyst A chemical °species or other structure that facilitates a chemical °reaction without itself undergoing a permanent change.

Cation A positively charged °ion.

Classical mechanics Classical mechanics describes a °mechanical system as a set of particles (which in a limiting case can form continuous media) having a well-defined geometry at any given time, and undergoing motions determined by applied forces and by the initial positions and velocities of the particles. The forces themselves may have electromagnetic or °quantum mechanical origins. Classical °statistical mechanics uses the same physical model, but treats the geometry and velocities as uncertain, statistical quantities subject to random thermally-induced fluctuations. Classical mechanics and classical statistical mechanics give a good account of many mechanical properties and behaviors of molecules; but for describing the °electronic properties and behaviors of molecules, they are often useless.

CMOS An acronym for *complementary metal-oxide-semiconductor,* as in CMOS transistor and CMOS logic.

Col In describing landforms, a pass between two valleys is sometimes termed a col. In describing molecular °potential energy functions, this term is commonly used to describe analogous features of the PES; a col is the region around a saddle point having negative curvature along one axis and positive curvature along all orthogonal axes.

Compliance The reciprocal of °stiffness; in a linear elastic system, displacement equals force times compliance.

Configuration space A mathematical space describing the three-dimensional configuration of a system of particles (e.g., atoms in a °nanomechanical structure) as a single point; the configuration space for an N particle system has $3N$ dimensions.

Conformation A molecular geometry that differs from other geometries chiefly by rotation about single or triple bonds; distinct conformations (termed conformers) are associated with distinct °potential wells. Typical biomolecules and products of organic synthesis can interconvert among many conformations; typical °diamondoid structures are locked into a single potential well, and thus lack conformational flexibility.

Conjugated A conjugated °pi system is one in which pi bonds alternate with single bonds. The resulting electron distribution gives the intervening single bonds partial double-bond character, the pi electrons become delocalized, and the energy of the system is reduced.

Conservative In design and analysis, a conservative model or a conservative assumption is one that departs from accuracy in such a way that it reduces the chances of a false-positive assessment of the feasibility of the system in question. Conservative assumptions overestimate problems and underestimate capabilities.

Covalent bond A °bond formed by sharing a pair of electrons between two atoms.

Covalent radius Given a set of N elements that can form °covalent °single bonds in molecules, with $N(N-1)$ possible elemental pairings, it has proved possible to define a covalent radius for each element such that the actual bond length between any two elements that form a covalent single bond is roughly equal to the sums of their covalent radii.

CPU The central processing unit of a computer, responsible for executing instructions to process information.

Cyclic A structure is termed cyclic if its °covalent bonds form one or more rings.

Cycloaddition A °reaction in which two unsaturated molecules (or moieties within a molecule) join, forming a ring.

Dative bond A °dipolar bond.

Diamondoid As used in this volume, this term describes structures that resemble diamond in a broad sense: strong, stiff structures containing dense, three-dimensional networks of °covalent bonds, formed chiefly from first and second row atoms with a °valence of three or more. Many of the most useful diamondoid structures will in fact be rich in tetrahedrally coordinated carbon. *Diamondoid* is used more narrowly elsewhere in the literature.

Dipolar bond A °covalent bond in which one atom supplies both bonding electrons, and the other atom supplies an empty orbital in which to share them. Also termed a dative bond.

Dissociation constant For systems in which °ligands of a particular kind bind to a °receptor in a solvent there will be a characteristic frequency with which existing ligand-receptor complexes dissociate as a result of thermal excitation, and a characteristic frequency with which empty receptors bind ligands as a result of Brownian encounters, forming new complexes. The frequency of binding is proportional to the concentration of the ligand in solution. The dissociation constant is the magnitude of the ligand concentration at which the probability that the receptor will be found occupied is 1/2.

Double bond Two atoms sharing electrons as in a °single bond (that is, a °sigma bond) may also share electrons in an orbital with a node passing through the two atoms. This adds a second, weaker bonding interaction (a °pi bond); the combination is termed a double bond. A twisting motion that forces the nodal plane at one atom to become perpendicular to the nodal plane on the other atom eliminates the (signed) °overlap between the atomic orbitals, destroying the pi bond. The energy required to do this cre-

ates a large barrier to rotation about the bond (see °triple bond).

Doublet The electronic state of a molecule having one unpaired spin is termed a doublet (see °radical). This term is derived from spectroscopy: an unpaired spin can be either up or down with respect to a magnetic field, and these states have different energy, resulting in field-dependent pairs, or doublets, of spectral lines. (*See* °triplet, °singlet.)

Effective mass In a vibrating system, a particular vibrational mode can be described as a harmonic oscillator with some mass and stiffness. Given some measure of vibrational amplitude, there exists a unique choice of mass and stiffness that yields the correct values for both frequency and energy; these are the effective mass and effective stiffness.

Effective stiffness See °effective mass.

Elastic An object behaves elastically if it returns to its original shape after a force is applied and then removed. (If an applied force causes a permanent deformation, the behavior is termed *plastic*.) In an elastic system, the internal potential energy is a function of shape alone, independent of past forces and deformations.

Electron density The location of an electron is not fixed, but is instead described by a probability density function. The sum of the probability densities of all the electrons in a region is the electron density in that region.

Electronegativity A measure of the tendency of an atom (or moiety) to withdraw electrons from structures to which it is bonded. In most circumstances, for example, sodium tends to donate electron density (it has a low electronegativity) and fluorine tends to withdraw electron density (it has a high electronegativity).

Electronic Pertaining to the energies, distributions, and behaviors of electrons; see °mechanical.

Endoergic A transformation is termed endoergic if it absorbs energy; such a °reaction increases molecular °potential energy. (Sometimes wrongly equated to the narrower term °endothermic.)

Endothermic   A transformation is termed endothermic if it absorbs energy in the form of heat. A typical endothermic °reaction increases both °entropy and molecular °potential energy (and is thus analogous to a gas expanding while absorbing heat and compressing a spring).

Energy   A conserved quantity that can be interconverted among many forms, including °kinetic energy, °potential energy, and electromagnetic energy. Sometimes defined as "the capacity to do °work," but in an environment at a uniform nonzero °temperature, °thermal energy does not provide this capacity. (Note, however, that all energy has mass, and thus can be used to do work by virtue of its gravitational potential energy; this caveat, however, is of no practical significance unless a *really* deep gravity well is available.) See °free energy.

Enthalpy   The enthalpy of a system is its actual °energy (termed the *internal energy*) plus the product of its volume and the external pressure. Though sometimes termed "heat content," the enthalpy in fact includes energy *not* contained in the system. Enthalpy proves convenient for describing processes in gases and liquids in laboratory environments, if one does not wish to account explicitly for energy stored in the atmosphere by work done when a system expands. It is of little use, however, in describing processes in °nanomechanical systems, where work can take many forms: internal energy is then more convenient. Enthalpy is to energy what the °Gibbs free energy is to the °Helmholtz free energy.

Entropy   A measure of uncertainty regarding the state of a system: for example, a gas molecule at an unknown location in a large volume has a higher entropy than one known to be confined to a smaller volume. °Free energy can be extracted in converting a low-entropy state to a high-entropy state: the (time-average) pressure exerted by a gas molecule can do useful work as a small volume

is expanded to a larger volume. In the classical °configuration space picture, any molecular system can be viewed as a single-particle gas in a high-dimensional space. In the quantum mechanical picture, entropy is described as a function of the probabilities of occupancy of different members of a set of alternative quantum states. Increased information regarding the state of a system reduces its entropy and thereby increases its free energy, as shown by the resulting ability to extract more work from it.

An illustrative contradiction in the simple textbook view of entropy as a local property of a material (defining an entropy per mole, and so forth) can be shown as follows: The third law of thermodynamics states that a perfect crystal at absolute zero has zero entropy;* this is true regardless of its size. A piece of disordered material, such as a glass, has some finite entropy $\mathscr{S}_0 > 0$ at absolute zero. In the local-property view, N pieces of glass, even (or especially) if all are atomically identical, must have an entropy of $N\mathscr{S}_0$. If these N pieces of glass are arranged in a regular three-dimensional lattice, however, the resulting structure constitutes a perfect crystal (with a large unit cell); at absolute zero, the third law states that this crystal has zero entropy, not $N\mathscr{S}_0$. To understand the informational perspective on entropy, it is a useful exercise to consider (1) what the actual entropy of such crystal is as a function of N, with and without information describing the structure of the unit cell, (2) how the third law can be phrased more precisely, and (3) what this more precise statement implies for the entropy of well-defined aperiodic structures. Note that any one unit cell in the crystal can be regarded as a description of all the rest.

Enzyme   A °protein molecule that acts as a specific °catalyst, binding to other

---

* Some textbooks state a slightly weaker form: a reaction that converts perfect crystals at absolute zero into other perfect crystals at absolute zero results in no change in entropy. This is essentially equivalent.

molecules in a manner that facilitates a particular chemical °reaction.

Equilibrium   A system is said to be at equilibrium (with respect to some set of feasible transformations) if it has minimal °free energy. A system containing objects at different °temperatures is in disequilibrium, because heat flow can reduce the free energy. Springs have equilibrium lengths, reactants and products in solution have equilibrium concentrations, thermally excited systems have equilibrium probabilities of occupying various states, and so forth.

Ester   A molecule containing an ester linkage, a °carbonyl group bonded to an O that is in turn bonded to a C.

Ether   A molecule containing a C—O—C structure, termed an ether linkage (unless one of the C atoms has a °double bond to another O, making this part of an ester linkage, or some other exception holds).

Eutactic   Characterized by precise molecular order, like that of a perfect crystal, the interior of a °protein molecule, or a machine-phase system; contrasted to the disorder of bulk materials, solution environments, or biological structures on a cellular scale. Borderline cases can be identified, but perfection is not necessary. As a crystal with sparse defects is best described as a crystal (rather than as amorphous), so a eutactic structure with sparse defects is best described as (imperfectly) eutactic, rather than as disordered.

Excluded volume   The presence of one molecule (or moiety) reduces the volume available for other molecules (or moieties); resulting reductions in their °entropy are termed *excluded volume effects*.

Exoergic   The opposite of °endoergic; describes a transformation that releases energy.

Exothermic   The opposite of °endothermic; describes an °exoergic transformation in which energy is released as °heat. Exoergic °reactions in solution are commonly exothermic.

Fail-stop   Describes a component or subsystem that, in the event of a failure, produces no output (e.g., of material or data) rather than producing a damaged or incorrect output.

Fault-tolerant   Describes a system that can suffer failure in a component or subsystem, yet continue to function correctly.

Follower   A component in a °cam system that is driven through a pattern of displacements as it rests against a moving contoured surface.

Free energy   Free energy is a measure of the ability of a system to do work, such that a reduction in free energy could in principle yield an equivalent quantity of work. The °Helmholtz free energy describes the free energy *within* a system; the °Gibbs free energy does not.

Free radical   A °radical.

Gate   In digital logic, a component that can switch the state of an output dependent on the states of one or more inputs.

Gibbs free energy   The Gibbs °free energy is the °Helmholtz free energy plus the product of the system volume and the external pressure. Changes in the Gibbs free energy at a constant pressure thus include work done against external pressure as a system undergoes volumetric changes. This proves convenient for describing equilibria in gases and liquids at a constant pressure (e.g., at one atmosphere), but is of little use in describing machine-phase chemical processes. Changes in the Gibbs free energy caused by a change in the applied pressure (at constant volume) have no direct physical significance. (See also °enthalpy.)

Ground state   The lowest-energy state of a system. The electronic ground state of a system cannot reduce its energy by an electronic transition, but may contain vibrational energy (°kinetic and °potential energy associated with the motions and positions of its atoms); extended systems at ordinary temperatures are always vibrationally excited, and so "ground state" is often taken to mean "electronic ground state."

Group   A set of linked atoms in a molecule; a defined substructure. Typically, a set that is usefully regarded as a unit in chemical °reactions of interest.

Group velocity   In wave propagation, the speed of the waveform (e.g., of a peak) can be different from the speed of a group of waves (e.g., of a set of ripples in water). The latter is the *group velocity,* and is the speed of propagation of information and wave energy. The waveform speed is the *phase velocity.*

Harmonic oscillator   A system in which a mass is subject to a linear restoring force, like an ideal spring. A harmonic oscillator vibrates at a fixed frequency, independent of amplitude.

Heat   As defined in thermodynamics, heat is the energy that flows between two systems as a result of °temperature differences (a system *contains* neither heat nor °work, but can *produce* heat or *do* work). Heat thus differs from °thermal energy.

Heat capacity   The ratio of the °heat input to the °temperature increase in a system. Note that this definition does not imply that a system *contains* heat, despite the name *heat capacity.*

Helmholtz free energy   The internal energy of a system minus the product of its °entropy and °temperature; see °free energy.

Hydrocarbon   A molecule consisting only of H and C.

Hydrogen bond   A hydrogen atom °covalently bound to an electronegative atom (e.g., nitrogen, oxygen) has a significant positive charge and can form a weak bond to another electronegative atom; this is termed a hydrogen bond.

Hydrophobic force   Water molecules are linked by a network of °hydrogen bonds. A nonpolar, nonwetting, surface (e.g., wax) cannot form hydrogen bonds. To form their full complement of hydrogen bonds, the nearby water molecules must form a more orderly (hence lower °entropy) network. This both increases °free energy and causes forces that tend to draw hydrophobic surfaces together across distances of several nanometers.

Intermolecular   Describes an interaction (e.g., a chemical °reaction) between different molecules.

Internal energy   The sum of the °kinetic and °potential energies (including electromagnetic field energies) of the particles that make up a system.

Intramolecular   Describes an interaction (e.g., a chemical °reaction) within a single molecule. Intramolecular interactions between widely separated parts of a molecule resemble °intermolecular interactions in most respects.

Ion   An atom or °molecule with a net charge.

Ionic bond   A chemical °bond resulting chiefly from the electrostatic attraction between positive and negative °ions.

Isoelectronic   Two molecules are described as isoelectronic if they have the same number of valence electrons in similar orbitals, although they may differ in their distribution of nuclear charges (e.g., $H-C\equiv N$ and $H-N^+\equiv C^-$ are isoelectronic).

Kinetic   Pertaining to the rates of chemical °reactions. A fast reaction is said to have fast kinetics; if the balance of products in a reaction is controlled by reaction rates rather than by thermodynamic equilibria, the reaction is said to be kinetically controlled.

Kinetic energy   Energy resulting from the motion of masses.

Ligand   In °protein chemistry, a small molecule that is (or can be) bound by a larger molecule is termed a ligand. In organometallic chemistry, a moiety bonded to a central metal atom is also termed a ligand; the latter definition is more common in general chemistry.

Linear   Aside from its geometric meaning, *linear* describes systems in which an output is directly proportional to an input. In particular, a linear °elastic system is one in which the internal displacements are (at equilibrium) directly proportional to applied forces.

London dispersion force   An attractive force caused by quantum-mechanical electron correlation. For example, a neutral spherical molecule (such as a single argon atom) has no charge and produces no external electric field, yet a pair of molecules has a distribution of electron configurations weighted toward those with lesser electron-electron repulsions; this creates a small net attraction.

Lone pair   Two valence electrons of an atom that share an orbital but do not participate in a bond.

Machine-phase chemistry   The chemistry of systems in which all potentially reactive moieties follow controlled trajectories (e.g., guided by molecular machines working in vacuum).

Mechanical   Pertaining to the positions and motions of atoms, as defined by the positions of their nuclei; see °electronic. A purely mechanical device can be described in terms of atomic positions and motions without reference to electronic properties, save through their effect on the °potential energy function.

Mechanochemistry   In this volume, the chemistry of processes in which °mechanical systems operating with atomic-scale precision either guide, drive, or are driven by chemical transformations. In general usage, the chemistry of processes in which energy is converted from mechanical to chemical form, or vice versa.

Mechanosynthesis   Chemical °synthesis controlled by °mechanical systems operating with atomic-scale precision, enabling direct positional selection of °reaction sites; synthetic applications of °mechanochemistry. Suitable mechanical systems include °AFM mechanisms, °molecular manipulators, and °molecular mill systems. Processes that fall outside the intended scope of this definition include reactions guided by the incorporation of reactive °moieties into a shared °covalent framework (i.e., conventional °intramolecular reactions), or by the binding of reagents to °enzymes or enzymelike °catalysts.

Metastable   A classical system is metastable if it is above its minimum-energy state, but requires an energy input before it can reach a lower-energy state; accordingly, a metastable system can act like a °stable system, provided that energy inputs (e.g., thermal fluctuations) remain below some threshold. Systems with strong metastability are commonly described as stable. Quantum mechanical effects can permit metastable states to reach lower energies by tunneling, without an energy input; an associated, broader definition of *metastable* embraces all systems that have a long lifetime (by some standard) in a state above the minimum-energy state.

Misreaction   A chemical °reaction that fails by yielding an unwanted product.

MM2   A molecular mechanics program developed by Norman Allinger and coworkers; the "MM2 model" is the molecular °potential energy function described by the equations, rules, and parameters embodied in that program.

MM2/CSC   A molecular mechanics program developed by Cambridge Scientific Computing that closely follows the MM2 model, adding a graphical user interface and other features.

Modulus   Any of several measures of °strain versus applied °stress. See °shear modulus, °Young's modulus.

Moiety   A portion of a molecular structure having some property of interest.

Mole   A number of instances of something (typically a molecular species) equaling $\sim 6.022 \times 10^{23}$. *Mole* ordinarily means gram-mole; a kilogram-mole is $\sim 6.022 \times 10^{26}$.

Molecular machine   A °mechanical device that performs a useful function using components of nanometer scale and defined molecular structure; includes both artificial °nanomachines and naturally occurring devices found in biological systems.

Molecular manipulator   A programmable device able to position molecular tools with high precision, for example, to direct a sequence of °mechanosynthetic steps; a molecular °assembler.

Molecular manufacturing   The production of complex structures via nonbiological °mechanosynthesis (and subsequent assembly operations).

Molecular mechanics models   Many of the properties of molecular systems are determined by the molecular °potential energy function. Molecular mechanics models approximate this function as a sum of 2-atom, 3-atom, and 4-atom terms, each determined by the geometries and bonds of the component atoms. The 2-atom and 3-atom terms describing bonded interactions roughly correspond to linear springs.

Molecular mill   A °mechanochemical processing system characterized by limited motions and repetitive operations without programmable flexibility (see °molecular manipulator).

Molecular nanotechnology   See °nanotechnology.

Molecule   A set of atoms linked by °covalent bonds. A macroscopic piece of diamond is technically a single molecule. (Sets of atoms linked by bonds of other kinds are sometimes also termed molecules.)

Nanomachine   An artificial °eutactic °mechanical device that relies on nanometer-scale components; see °molecular machine.

Nanomechanical   Pertaining to °nanomachines.

Nanoscale   On a scale of nanometers, from atomic dimensions to ~100 nm.

Nanosystem   A °eutactic set of nanoscale components working together to serve a set of purposes; complex nanosystems can be of macroscopic size.

Nanotechnology   In recent general usage, any technology related to features of nanometer scale: thin films, fine particles, chemical synthesis, advanced microlithography, and so forth. As introduced by the author, a technology based on the ability to build structures to complex, atomic specifications by means of mechanosynthesis; this can be termed molecular nanotechnology.

NMOS   An acronym for n-*channel metal-oxide-semiconductor,* as in NMOS transistor and NMOS logic.

Nucleus   The positively charged core of an atom, an object of ~0.00001 atomic diameters containing >99.9% of the atomic mass. Nuclear positions define atomic positions.

Olefin   An °alkene.

Omitted reaction   A chemical °reaction that fails by not occurring (see °misreaction).

Orbital   In the approximation that each electron in a molecule has a distinct, independent °wave function, the spatial distribution of an electron wave function corresponds to a molecular orbital. These, in turn, can be approximated as sums of contributions from the orbitals characteristic of the isolated atoms. An electron added to a molecule—or, similarly, one excited to a higher-energy state within a molecule—would occupy a state with a different wave function from the rest; an unoccupied state of this kind corresponds to an unoccupied molecular orbital. Orbital-symmetry effects on reaction rates arise when a reaction requires °overlap between two lobes of the orbitals on each of two °reagents: if the algebraic signs of the wave functions in the facing lobes do not match, bond formation between those orbitals is prohibited.

Overlap   °Orbitals lack sharply defined surfaces, declining in amplitude exponentially in their surface regions. When two orbitals are brought together, regions of substantial amplitude overlap. The resulting system can be described as two new orbitals, one formed by joining the two original orbitals without introducing a node in the °wave function, and the other formed with a node between them. The nodeless joining reduces the energy of the electrons relative to the separate orbitals, resulting in a bonding interaction; joining with a node raises the energy, producing an antibonding interaction. If both new orbitals are occupied, antibonding forces dominate, resulting in °overlap repulsion. °Molecular mechanics models give an approximate description of overlap (and other) forces for a certain range of atoms and geometries.

Overlap repulsion   A repulsive force resulting from the nonbonding °overlap of two atoms.

Partition function   A function determined by the probability distribution (over °phase space in the °classical treatment; over °quantum states in the quantum treatment) describing a thermally equilibrated system; many thermodynamic quantities can be expressed in terms of the partition function and its derivatives.

PDF   See °probability density function.

Peptide   A short chain of amino acids; see °protein.

PES   See °potential energy surface.

Phase space   A classical system of N particles can be described by its $3N$ position and $3N$ momentum coordinates. The phase space associated with the system is the $6N$ dimensional space defined by these coordinates.

Phonon   A quantum of acoustic energy, analogous to the quantum of electromagnetic radiation, the photon. Thermal excitations in a crystal or in an elastic continuum can be described as a population of phonons (analogous to blackbody electromagnetic radiation). In highly inhomogeneous solids, a description in terms of phonons breaks down and localized vibrational modes become important.

Pi bond   A °covalent bond formed by °overlap between two p orbitals on different atoms (see °sp). Pi bonds are superimposed on °sigma bonds, forming °double or °triple bonds.

Poisson's ratio   A bar of an isotropic, elastic material ordinarily shrinks laterally when it is stretched longitudinally. The lateral contracting strain divided by the applied tensile strain is Poisson's ratio, which varies from material to material.

Polycyclic   A °cyclic structure contains rings of bonds; a structure having many such rings is termed polycyclic. In the polycyclic structures of interest in this volume, a large fraction of the atoms are members of multiple small rings, resulting in considerable rigidity.

Potential energy   The energy associated with a configuration of particles, as distinct from their motions. In macroscopic experience, potential energy can be increased (for example) by stretching a spring or by lifting a mass against a gravitational force; in molecular systems, potential energy can be increased (for example) by stretching a bond or by separating molecules against a °van der Waals attraction.

Potential energy surface   The °potential energy of a °ground-state molecular system containing N atoms is a function of its geometry, defined by $3N$ spatial coordinates (a °configuration space). If the energy is imagined as corresponding to a height in a $3N + 1$ dimensional space, the resulting landscape of hills, hollows, and valleys is the potential energy surface.

Potential well   In a °potential energy surface, the region surrounding a local energy minimum. Typically taken to include at least those points in °configuration space such that a path of steadily declining energy can be found that leads to the minimum in question, and such that no similar path can be found to any other minimum. If the PES were a landscape, this would be the region around the minimum that could be filled with water without any flowing down and away toward another minimum.

Probability density function   Consider an uncertain physical property and a corresponding space describing the range of values that the property can have (e.g., the configuration of a thermally excited N particle system and the corresponding $3N$ dimensional configuration space). The probability density function associated with a property is defined over the corresponding space; its value at a particular point is the probability per unit volume that the property has a value in an infinitesimal region around that point.

Protein   Living cells contain many molecules that consist of amino acid polymers folded to form more-or-less definite three-dimensional structures; these are termed proteins. Short polymers lacking definite three-dimensional structures are termed °peptides. Many proteins incorporate structures other than amino acids, either as °covalently attached side chains or as bound °ligands. Molecular objects made of protein form much of the molecular machinery of living cells.

Quantum mechanics   Quantum mechanics describes a system of particles in terms of a °wave function defined over the °configuration space of the system. Although the concept of particles having distinct locations is implicit in the °potential energy function that determines the wave function (e.g., of a °ground-state system), the observable dynamics of the system cannot be described in terms of the motion of such particles from point to point. In describing the energies, distributions, and behaviors of electrons in nanometer-scale structures, quantum mechanical methods are necessary. Electron °wave functions help determine the °potential energy surface of a molecular system, which in turn is the basis for °classical descriptions of molecular motion. °Nanomechanical systems can almost always be described in terms of classical mechanics, with occasional quantum mechanical corrections applied within the framework of a classical model.

Radiation damage   Chemical changes (bond breakage, ionization) caused by high-energy radiation (e.g., x-rays, gamma rays, high-speed electrons, protons, etc.).

Radical   A structure with an unpaired electron (but excluding certain metal ions). In organic molecules, a radical is often associated with a highly reactive site of reduced valence (see °doublet). The term radical is sometimes used to describe a substructure within a molecule; the term *free radical* then describes a radical in this sense, viewed as the result of cleaving the bond linking the substructure to the rest of the molecule.

Reaction   A process that transforms one or more chemical °species into others. Typical reactions make or break °bonds; others change the state of ionization or other properties taken to distinguish chemical species.

Reagent   A chemical °species that undergoes change as a result of a chemical °reaction.

Reagent device   A large °reagent structure (or a large structure that binds a smaller reagent) serving as a component of a °mechanochemical system. A reagent device exists chiefly to hold, position, and manipulate the environment of a °reagent moiety.

Reagent moiety   The portion of a °reagent device that is intimately involved in a chemical °reaction.

Receptor   A structure that can capture a molecule (often of a specific type in a specific orientation) owing to complementary surface shapes, charge distributions, and so forth, without forming a °covalent bond. See °dissociation constant.

Reconstruction   A crystal consists of a regular array of atoms, and the simplest model of a crystal surface would be generated by simply discarding all atoms to one side of a surface without changing the positions of the rest. In reality, however, the positions of the remaining atoms do change. A pattern of displacements that lowers the symmetry of the surface (relative to the ideally terminated crystal) is termed a surface reconstruction; some reconstructions alter the pattern of bonds.

Reduced mass   Many dynamical properties of a system consisting of two interacting masses, m_1 and m_2, are equivalent to those of a system in which one mass is fixed in space and the other has a mass (the reduced mass) with the value $m_1 m_2/(m_1 + m_2)$. The reduced mass description has fewer dynamical variables.

Register   A temporary storage location for an array of bits within a digital logic system.

Relaxation time   A measure of the rate at which a disequilibrium distribution decays toward an °equilibrium distribution. The electron relaxation time in a metal, for example, describes the time required for a disequilibrium distribution of electron momenta (e.g., in a flowing current) to decay toward equilibrium in the absence of an ongoing driving force and can be interpreted as the mean time between scattering events for a given electron.

Representative point   The point in a °configuration space that represents the geometry of a system.

Rigid structure   As used in this volume, a °covalent structure that is reasonably stiff. In a typical rigid structure, all modes of deformation encounter first-order restoring forces resulting from some combination of bond stretching and angle bending; such a structure cannot undergo deformation by bond torsion alone. Meeting this condition usually requires a °polycyclic °diamondoid structure.

Salt bridge   An ionic bond between charged °groups that are part of larger °covalent structures; salt bridges occur in many °proteins.

Saturated   An organic molecule is described as saturated if it is a closed shell species lacking °double or °triple bonds; forming a new bond to a saturated molecule requires the cleavage of an existing bond.

Scanning tunneling microscope   A device in which a sharp conductive tip is moved across a conductive surface close enough to permit a substantial tunneling current (typically a nanometer or less). In a common mode of operation, the voltage is kept constant and the current is monitored and kept constant by controlling the height of the tip above the surface; the result, under favorable conditions, is an atomic-resolution map of the surface reflecting a combination of topography and electronic properties. The STM has been used to manipulate atoms and molecules on surfaces.

Self assembly   A term commonly used for °Brownian assembly.

Shear modulus   Shear °stress divided by shear °strain; has the units of force per unit area.

Shear   A shear deformation is one that displaces successive layers of a material transversely with respect to one another,

like a crooked stack of cards. Shear is a dimensionless quantity measured by the ratio of the transverse displacement to the thickness over which it occurs.

Sigma bond   A °covalent bond in which °overlap between two atomic °orbitals (e.g., of °sp, sp^2, or sp^3 hybridization) produces a single bonding orbital in which the distribution of shared electrons has a roughly cylindrical symmetry about the axis linking the two atoms; see °pi bond, °single bond, °double bond, °triple bond. By themselves, sigma bonds present little barrier to rotation of one substructure with respect to another, although °steric effects and °cyclic structures may hinder or block rotation.

Single bond   A °sigma bond having no associated °pi bonds.

Singlet   An electronic state of a molecule in which all spins are paired; see °doublet, °triplet.

sp, sp^2, sp^3   An isolated carbon atom has four °valence °orbitals: three mutually perpendicular p orbitals, each with a single nodal plane, and one spherically symmetric s orbital. A carbon atom in a typical molecule can be regarded as bonding with four orbitals consisting of weighted sums (termed hybrids) of these s and p orbitals. One common pattern has four equivalent orbitals, each formed by combining the three p orbitals with the s orbital; this is sp^3 hybridization. An sp^3 carbon atom forms four °sigma bonds, usually in a roughly tetrahedral arrangement. Another common pattern has three equivalent orbitals formed by combining two p orbitals with the s orbital; this is termed sp^2 hybridization. An sp^2 carbon atom forms three roughly coplanar sigma bonds, usually separated by ~120°, and one °pi bond (or several fractional pi bonds). If a single p orbital is combined with the s orbital, the result is sp hybridization, forming two sigma bonds and two pi bonds (usually in a straight line). Atoms of other kinds (e.g., N and O) can hybridize in an analogous manner.

Species   In chemistry, a distinct kind of °molecule, °ion, or other structure.

Stable Strictly speaking, a system is termed stable if no rearrangement of its parts can form a system of lower °free energy. In practice, the term is used with an implicit proviso regarding the transformations to be considered. Hydrogen is not considered unstable merely because it is subject to nuclear fusion at extreme temperatures. A system is usually regarded as stable (more precisely, as kinetically stable) if its rate of transformation to a state of lower free energy is negligible (by some standard) under the ambient conditions. In °nanomechanical systems, a structure can commonly be regarded as stable if it has an extremely low rate of transformations when subjected to its intended operating conditions.

State A physical system is said to be in a particular state when its physical properties fall within some particular range; the boundaries of the range defining a state depend on the problem under consideration. In a °classical world, each point in °phase space could be said to correspond to a distinct state. In the real world, time-invariant systems in °quantum mechanics have a set of discrete states, particular superpositions of which constitute complete descriptions of the system. In practice, broader boundaries are usually drawn. A molecule is often said to be in a particular excited °electronic state, regardless of its state of °mechanical vibration. In °nanomechanical systems, the °PES often corresponds to a set of distinct °potential wells, and all points in configuration space within a particular well can be regarded as one state. Definitions of state in the thermodynamics of bulk matter are analogous, but extremely coarse by these standards.

Statistical mechanics Statistical mechanics treats the detailed °state of a system (its °quantum state or, in °classical models, its position in °phase space) as unknown and subject to statistical uncertainties; °entropy is a measure of this uncertainty. Statistical mechanics describes the distribution of states in an °equilibrium system at a given °tempera-

ture (describing either the distribution of probabilities of quantum states or the °probability density function in °phase space), and can be used to derive °thermodynamic properties from properties at the molecular level. These equilibrium results are useful in °nanomechanical design.

Steric Pertaining to the spatial relationships of atoms in a molecular structure, and in particular, to the space-filling properties of a molecule. If molecules were rigid and had hard surfaces, *steric properties* would merely be an opaque way of saying "shape"; a flexible sidechain, however, has definite steric properties but no fixed shape. °Nanomechanical systems make extensive use of the steric properties of relatively rigid molecules, for which the term "shape" has essentially its conventional meaning so long as one remembers that the surface interactions are soft on small lengthscales.

Steric hindrance Slowing of the rate of a chemical °reaction owing to the presence of structures on the °reagents that mechanically interfere with the motions associated with the reaction, typically by obstructing the reaction site.

Stiffness The stiffness of a system with respect to a deformation (e.g., the stiffness of a spring with respect to stretching) is the second derivative of the energy with respect to the corresponding displacement; this measures the curvature of the °potential energy surface along a particular direction. Positive stiffness is associated with stability, and a large stiffness can result in a small positional uncertainty in the presence of thermal excitation. Negative stiffnesses correspond to unstable locations on the potential energy surface. Alternative terms for stiffness include *force gradient* and *rigidity*.

STM A °scanning tunneling microscope.

Strain In mechanical engineering, strain is a measure of the deformation resulting from °stress (that is, force per unit area); the displacement of one point with respect to another, divided by their °equilibrium separation in the absence of stress. In chemistry, a molecular fragment generally has some equilibrium geometry (bond lengths, interbond angles, etc.) when the rest of the molecular structure does not impose special constraints (e.g., bending bonds to form a small ring). Deviations from this equilibrium geometry are described as strain, and increase the energy of the molecule. Strain in the mechanical engineering sense causes strain in the chemical sense.

Stress   Force per unit area applied by one part of an object to another. Pressure is an isotropic compressive stress. Suspending a mass from a fiber places it in tensile stress. Gluing a layer of rubber between two plates and then sliding one over the other (while holding their separation constant) places the rubber in °shear stress.

Structural volume   The interior of a °diamondoid structure typically consists of a dense network of °covalent bonds; a larger °excluded volume, however, is determined by nonbonded repulsions at the surface. The structural volume corresponds to a region smaller than the excluded volume, chosen to make properties such as the strength and °modulus nearly size independent by correcting for surface effects.

Synthesis   The production of a specific °molecular structure by a series of chemical °reactions.

System   In scientific usage, usually equivalent to "a collection of matter and energy being analyzed as a unit." In engineering usage, usually equivalent to "a set of components working together to serve a set of purposes."

Temperature   A system in which internal vibrational modes have equilibrated with one another can be said to have a particular temperature. Two systems A and B are said to be at different temperatures if, when brought into contact, °heat flows from (say) A to B, increasing the °thermal energy of B at the expense of the thermal energy of A.

Thermal energy   The °internal energy present in a system as a result of the energy of thermally equilibrated vibrational modes and other motions (including both °kinetic energy and molecular °potential energy). The mean thermal energy of a classical °harmonic oscillator is kT.

Thermal expansion coefficient   The rate of change of length with respect to °temperature for a particular material.

Thermal fluctuations   The °thermal energy of a system (or of a particular part or mode of motion in a system) has a mean value determined by the °temperature and by the structure of the system. Statistical deviations about that mean are termed thermal fluctuations; these are of great importance in determining both rates of chemical °reactions and error rates in °nanomechanical systems.

Thermodynamics   A field of study embracing energy conversion among various forms, including °heat, °work, and °potential and °kinetic energy.

Thermoelastic   Both °stress and °temperature changes alter the dimensions of an object having a finite °stiffness and a nonzero thermal expansion coefficient. Applying a stress then produces a temperature change; this can result in a °heat flow which then changes the stress: these are thermoelastic effects, and result in losses of °free energy.

Thiol   An SH group, or a molecule containing one. Also known as a sulfhydryl or mercapto group.

Tight-receptor structures   A receptor structure in which a bound °ligand of a particular kind is confined on all sides by repulsive interactions (note that favorable binding energies are compatible with repulsive forces). A tight-receptor structure discriminates strongly against all molecules larger than the target.

Transition state   At the saddle point of a °col linking two °potential wells, the direction of maximum negative curvature defines the reaction coordinate; the transition state is a hypothetical system of reduced dimensionality, free to move only on a hypersurface perpendicular to the reaction coordinate at its point of maximum energy.

Transition state theory   Any of several theories that give approximate descriptions of chemical °reaction rates based on the °PES of the system, and in particular, on the properties of two °potential wells and a °transition state between them.

Triple bond   A °double bond is formed when a pi bond is superimposed on a °single bond; adding a second pi bond results in a triple bond. The two pi bonds have perpendicular nodal planes, and their sum has roughly cylindrical symmetry, permitting rotation in much the same manner as a single bond.

Triplet   An electronic state of a molecule in which two spins are aligned. This term is derived from spectroscopy: a system of two aligned spins has three possible orientations with respect to a magnetic field; each has a different energy, resulting in sets of three field-dependent spectral lines (see °doublet, °singlet.)

TST   °Transition state theory.

Tunneling   A classical particle or system could not penetrate regions in which its energy would be negative, that is, barrier regions in which the °potential energy is greater than the system energy. In the real world, however, a °wave function of significant amplitude may extend into and beyond such a region. If the wave function extends into another region of positive energy, the barrier is crossed with some probability; this process is termed tunneling (since the barrier is penetrated rather than climbed).

Unimolecular   Occurring to or within a single molecule; like °intramolecular, but can refer to fragmentation reactions.

Unsaturated   Possessing °double or °triple bonds.

Valence electrons   Electrons that can participate in bonds and in chemical reactions; °lone-pair electrons are valence electrons, although not participating in a bond.

Valence   In °covalent compounds, the valence of an atom is the number of bonds it forms to other atoms.

Van der Waals force   Any of several intermolecular attractive forces not resulting from ionic charges; in this volume, only °London dispersion forces are described by this name. Descriptions of the potential energy of nonbonded interactions follow the convention of expressing van der Waals attractive forces and overlap repulsion forces as a single *van der Waals potential.*

Wave function   In °quantum mechanics, a complex function extending over the °configuration space of a system; its complex conjugate yields the °probability density function, and other mathematical operations yield other physical quantities.

Work   Energy transferred by applying a force over a distance; lifting a mass does work against gravity, and stores gravitational °potential energy.

Young's modulus   A °modulus relating tensile (or compressive) °stress to °strain in a rod that is free to contract or expand transversely in accord with its °Poisson's ratio. The relevant measure of strain is the elongation divided by the initial length.

References

Adam, W., P. Hössel, W. Hümmer, and H. Platsch. (1987) UV-Laser Photochemistry: Diffusion-Controlled Trapping of Cyclic 1,3-Diradicals by Molecular Oxygen. Conformation Effects on Triplet Lifetimes. *J. Am. Chem. Soc.* **109:**7570–7572.

Alberts, B., D. Bray, J. Lewis, M. Raff, K. Roberts, and J. D. Watson. (1989) *Molecular Biology of the Cell.* Second Edition. New York: Garland.

Albrecht, T. R. (1990) Personal communication.

Allinger, N. L. (1977) MM2. A Hydrocarbon Force Field Utilizing V1 and V2 Torsional Terms. *J. Am. Chem. Soc.* **99:** 8127–8134.

Allinger, N. L. (1986) Personal communication.

Allinger, N. L., F. Li, and L. Yan. (1990) Molecular Mechanics. The MM3 Force Field for Alkenes. *J. Comp. Chem.* **11:** 848–867.

Allinger, N. L., and A. Pathiaseril. (1987) A Force Field for Allenes and for Nonlinear Acetylenes within the MM2 Approximation. *J. Comp. Chem.* **8:**1225–1231.

Allinger, N. L., Y. H. Yuh, and J. Lii. (1989) Molecular Mechanics. The MM3 Force Field for Hydrocarbons. 1. *J. Am. Chem. Soc.* **111:**8551–8566.

Alshits, V. I., and L. V. Indenbom. (1986) Mechanisms of Dislocation Drag. *Dislocations in Solids.* F. R. N. Nabarro, ed. Amsterdam: North-Holland.

Amdur, I., and J. E. Jordan. (1966) Elastic Scattering of High-Energy Beams: Repulsive Forces. *Molecular Beams.* J. Ross, ed. New York: John Wiley & Sons.

Anelli, P. L., P. R. Ashton, R. Ballardini, V. Balzani, M. Delgado, M. T. Gandolfi, T. T. Goodnow, A. E. Kaifer, D. Philp, M. Pietraszkiewcz, L. Prodi, M. V. Reddington, A. M. Z. Slawin, N. Spencer, J. F. Stoddart, C. Vincent, and D. J. Williams. (1992) Molecular Meccano. 1. [2]Rotaxanes and a [2]Catenane Made to Order. *J. Am. Chem. Soc.* **114:**193–218.

Aoyagi, Y., K. Katano, H. Suguna, J. Primeau, L. Chang, and S. M. Hecht. (1982) Total Synthesis of Bleomycin. *J. Am. Chem. Soc.* **104:**5537–5538.

Arnaud, R., and R. Subra. (1982) Substituent Effects and Radical Reactivity: Addition of *p*-Substituted Phenyl Radicals to Ethylene. *Nou. J. Chim.* **6:**91–95.

Ashcroft, N. W., and N. D. Mermin. (1976) *Solid State Physics.* New York: Holt, Rinehart and Winston.

Ashkin, A., K. Schütze, J. M. Dziedzic, U. Euteneuer, and M. Schliwa. (1990) Force generation of organelle transport measured *in vivo* by an infrared laser trap. *Nature.* **348:**346–348.

Ashton, P. R., N. S. Isaacs, F. H. Kohnke, J. P. Mathias, and J. F. Stoddart. (1989) Stereoregular Oligomerization by Repetitive Diels–Alder Reactions. *Angew. Chem. Int. Ed. Engl.* **28:**1258–1261.

Ashurst, W. T., and W. G. Hoover. (1975) Dense-fluid shear viscosity via nonequilibrium molecular dynamics. *Phys. Rev. A.* **11:**658–678.

Atkins, P. W. (1970) *Molecular Quantum Mechanics.* Oxford: Clarendon Press.

Atkins, P. W. (1974) *Quanta: a handbook of concepts.* Oxford: Oxford University Press.

Aviram, A. (1988) Molecules for Memory, Logic, and Amplification. *J. Am. Chem. Soc.* **110:**5687–5692.

Aviram, A., and M. A. Ratner. (1974) Molecular Rectifiers. *Chem. Phys. Lett.* **29:**277–283.

Badziag, P., and W. S. Verwoerd. (1987) MNDO Analysis of the Oxidised Diamond (100) Surface. *Surf. Sci.* **183:** 469–483.

Barner, J. B., and S. T. Ruggiero. (1987) Observation of the Incremental Charging of Ag Particles by Single Electrons. *Phys. Rev. Lett.* **59:**807–810.

Baron, W. J., M. R. DeCamp, M. E. Hendrick, M. Jones, Jr., R. H. Levin, and M. B. Sohn. (1973) Carbenes from Diazo Compounds. *Carbenes.* M. Jones, Jr., and R. A. Moss, eds. New York: John Wiley & Sons.

Bartley, W. W., III. (1987a) Philosophy of Biology versus Philosophy of Physics. *Evolutionary Epistemology.* G. Radnitzky and W. W. Bartley III, eds. .La Salle, Illinois: Open Court.

Bartley, W. W., III. (1987b) Theories of Rationality. *Evolutionary Epistemology.* G. Radnitzky and W. W. Bartley III, eds. La Salle, Illinois: Open Court.

Bates, R. B., and C. A. Ogle. (1983) *Carbanion Chemistry.* Berlin: Springer-Verlag.

Beauregard, G., and M. Potier. (1985) Temperature Dependence of the Radiation Inactivation of Proteins. *Anal. Biochem.* **150:**117–120.

Becker, R. S., J. A. Golovchenko, and B. S. Swartzentruber. (1987) Atomic-scale surface modifications using a tunneling microscope. *Nature.* **325:**419–421.

Bell, R. P. (1959) The Tunnel Effect Correction for Parabolic Potential Barriers. *Trans. Farad. Soc.* **55:**1–4.

Bennett, C. H. (1982) The Thermodynamics of Computation—a Review. *Int. J. Theor. Phys.* **21:**905–940.

Bérces, T., and F. Márta. (1988) Activation Energies for Metathesis Reactions of Radicals. *Chemical Kinetics of Small Organic Radicals.* Z. B. Alfassi, ed. Boca Raton: CRC Press.

Bigeleisen, J. (1949) The Relative Reaction Velocities of Isotopic Molecules. *J. Chem. Phys.* **17:**675–678.

Binnig, G., and C. F. Quate. (1986) Atomic Force Microscope. *Phys. Rev. Letters.* **56:** 930–933.

Bird, R. E., K. D. Hardman, J. W. Jacobson, S. Johnson, B. M. Kaufman, S. Lee, T. Lee, S. H. Pope, G. S. Riordan, and M. Whitlow. (1988) Single-Chain Antigen-Binding Proteins. *Science.* **242:**423–426.

Blair, D. F. (1990) The bacterial flagellar motor. *Sem. in Cell Bio.* **1:**75–85.

Blundell, T., D. Carney, S. Gardner, F. Hayes, B. Howlin, T. Hubbard, J. Overington, D. A. Singh, B. L. Sibanda, and M. Sutcliffe. (1988) Knowledge-based protein modelling and design. *Eur. J. Biochem.* **172:**513–520.

Bockris, J. O., and A. K. N. Reddy. (1970a) *Modern Electrochemistry.* Vol. 1. New York: Plenum.

Bockris, J. O., and A. K. N. Reddy. (1970b) *Modern Electrochemistry.* Vol. 2. New York: Plenum.

Bowie, J. U., R. Lüthy, and D. Eisenberg. (1991) A Method to Identify Protein Sequences That Fold into a Known Three-Dimensional Structure. *Science.* **253:**164–170.

Bowie, J. U., J. F. Reidhaar-Olson, W. A. Lim, and R. T. Sauer. (1990) Deciphering the Message in Protein Sequences: Tolerance to Amino Acid Substitutions. *Science.* **247:**1306–1310.

Brooke, J. H. (1985) Organic Chemistry. *Recent Developments in the History of Chemistry.* C. A. Russell, ed. London: Royal Society of Chemistry.

Brookes, C. A., V. R. Howes, and A. R. Parry. (1988) Multiple slip in diamond due to a nominal contact pressure of 10 GPa at 1,000 °C. *Nature.* **332:**139–141.

Brooks, B. R., R. E. Bruccoleri, B. D. Olafson, D. J. States, S. Swaminathan, and M. Karplus. (1983) CHARMM: A Program for Macromolecular Energy, Minimization, and Dynamics Calculations. *J. Comp. Chem.* **4:**187–217.

Brown, F. B., and D. G. Truhlar. (1985) Dissociation Potential for Breaking a C—H Bond in Methane. *Chem. Phys. Lett.* **113:**441–446.

Brown, R. F. C. (1980) *Pyrolytic Methods in Organic Chemistry.* Organic Chemistry, Vol. 21, H. H. Wasserman, ed. New York: Academic Press.

Burkert, U., and N. L. Allinger. (1982) *Molecular Mechanics.* ACS Monograph 177. Washington, DC: American Chemical Society.

Candlin, J. P., K. A. Taylor, and D. T. Thompson. (1968) *Reactions of Transition-Metal Complexes.* Amsterdam: Elsevier.

Carey, F. A., and R. J. Sundberg. (1983a) *Advanced Organic Chemistry, Part A: Structure and Mechanisms.* Second Edition. New York: Plenum Press.

Carey, F. A., and R. J. Sundberg. (1983b) *Advanced Organic Chemistry, Part B: Reactions and Synthesis.* Second Edition. New York: Plenum Press.

Carter, F. L. (1980) NRC Memorandum Report 4335. National Research Council.

Carter, F. L. (1982) *Molecular Electronic Devices.* New York: Marcel Dekker.

Carter, F. L. (1987) *Molecular Electronic Devices II.* New York: Marcel Dekker.

Caruthers, M. H. (1985) Gene Synthesis Machines: DNA Chemistry and Its Uses. *Science.* 230:281–285.

Casale, A., and R. S. Porter. (1978) *Polymer Stress Reactions.* New York: Academic Press.

Ceperley, D., and B. Alder. (1986) Quantum Monte Carlo. *Science.* 231:555–560.

Chen, C. J. (1991) Attractive interatomic force as a tunnelling phenomenon. *J. Phys.: Condens. Matter.* 3:1227–1245.

Chen, J., and N. Seeman. (1991) Synthesis from DNA of a molecule with the connectivity of a cube. *Nature.* 350:631–633.

Chu, C. H., and S. K. Estreicher. (1990) Similarities, differences, and trends in the properties of interstitial H in cubic C, Si, BN, BP, AlP, and SiC. *Phys. Rev. B.* 42:9486–9495.

Chu, J. G., and Z. Y. Hua. (1982) The statistical theory of turbomolecular pumps. *J. Vac. Sci. Tech.* 20:1101–1104.

Clark, T. (1985) *A Handbook of Computational Chemistry.* New York: John Wiley & Sons.

Clore, G. M., and A. M. Gronenborn. (1987) Determination of three-dimensional structures of proteins in solution by nuclear magnetic resonance spectroscopy. *Protein Eng.* 1:275–288.

Cook, M. D., and B. P. Roberts. (1983) Liquid Xenon as a Solvent for E.S.R. Studies. *J. Chem. Soc., Chem. Commun.* Issue 6, 264–266.

Corbridge, D. E. C. (1990) *Phosphorus: An Outline of its Chemistry, Biochemistry, and Technology.* Fourth Edition. Amsterdam: Elsevier.

Corey, E. J., and X. Cheng. (1989) *The Logic of Chemical Synthesis.* New York: John Wiley & Sons.

Cotton, R., P. M. Hardy, and A. E. Langran-Goldsmith. (1986) Peptides containing dialkylglycines. *Int. J. Peptide Protein Res.* 28:245–253.

Cowan, D. O., and R. L. Drisko. (1976) *Elements of Organic Photochemistry.* New York: Plenum Press.

Crabtree, R. H. (1987) *The Organometallic Chemistry of the Transition Metals.* New York: John Wiley & Sons.

Cram, D. J. (1986) Preorganization—From Solvents to Spherands. *Angew. Chem. Int. Ed. Engl.* 25:1039–1134.

Cram, D. J. (1988) The Design of Molecular Hosts, Guests, and Their Complexes. *Science.* 240:760–767.

Cram, D. J., M. E. Tanner, and R. Thomas. (1991) The Taming of Cyclobutadiene. *Angew. Chem. Int. Ed. Engl.* 30:1024–1027.

Crandall, B., and J. B. Lewis. (1992) *Nanotechnology: Research and Perspectives.* Cambridge, Massachusetts: MIT Press.

Creighton, T. E. (1978) Experimental studies of protein folding and unfolding. *Prog. Biophys. Molec. Biol.* 33:231–297.

Creighton, T. E. (1984) *Proteins.* New York: W. H. Freeman and Company.

Critchley, J. P., G. J. Knight, and W. W. Wright. (1983) *Heat Resistant Polymers.* New York: Plenum Press.

Danehy, J. P. (1971) The alkaline decomposition of organic disulfides and related nucleophilic displacements of sulfur from sulfur. *Int. J. Sulfur Chem., B.* 6:103–114.

Danehy, J. P., and K. N. Parameswaran. (1968) Acidic Dissociation Constants of Thiols. *J. Chem. Eng. Data.* 13:386–389.

Davis, J. H., W. A. Goddard III, and L. B. Harding. (1977) Theoretical Studies of the Low-Lying States of Vinylidene. *J. Am. Chem. Soc.* 99:2919–2925.

DeGrado, W. F. (1988) Design of Peptides and Proteins. *Adv. Protein Chem.* 39:51–124.

DeGrado, W. F., L. Regan, and S. P. Ho. (1987) The Design of a Four-helix Bundle Protein. *Cold Spring Harbor Symposia on Quantitative Biology, Volume 52.* Cold Spring Harbor Laboratory.

Diederich, F. (1988) Complexation of Neutral Molecules by Cyclophane Hosts. *Angew. Chem. Int. Ed. Engl.* 27:362–386.

Diederich, F., Y. Rubin, C. B. Knobler, R. L. Whetten, K. E. Schriver, K. N. Houk, and Y. Li. (1989) All-Carbon Molecules: Evidence for the Generation of Cyclo[18]carbon from a Stable Organic Precursor. *Science.* 245:1088–1090.

Dötz, K. H., H. Fischer, P. Hofmann, F. R. Kreissl, U. Schubert, and K. Weiss. (1983)

Transition Metal Carbene Complexes. Weinheim: Verlag Chemie.

Drake, B., C. B. Prater, A. L. Weisenhorn, S. A. C. Gould, T. R. Albrecht, C. F. Quate, D. S. Cannell, H. G. Hansma, and P. K. Hansma. (1989) Imaging Crystals, Polymers, and Processes in Water with the Atomic Force Microscope. *Science.* **243:** 1586–1589.

Drexler, K. E. (1981) Molecular engineering: An approach to the development of general capabilities for molecular manipulation. *Proc. Natnl. Acad. Sci. U.S.A.* **78:**5275–5278.

Drexler, K. E. (1986a) *Engines of Creation: The Coming Era of Nanotechnology.* New York: Anchor Press/Doubleday. (Drafts circulated in 1982, 83, and 84.)

Drexler, K. E. (1986b) Molecular engineering: Assemblers and Future Space Hardware. American Astronautical Society: AAS-86-415.

Drexler, K. E. (1987a) Molecular Machinery and Molecular Electronic Devices. *Molecular Electronic Devices II.* F. L. Carter, ed. New York: Marcel Dekker.

Drexler, K. E. (1987b) *Nanomachinery: Atomically precise gears and bearings.* IEEE Micro Robots and Teleoperators Workshop. Hyannis, Massachusetts: IEEE, Cat. # 87TH0204-8.

Drexler, K. E. (1988a) Nanotechnology and future supercomputing. *Proceedings of the Third International Conference on Supercomputing.* S. P. Kartashev and S. I. Kartashev, eds. International Supercomputing Institute.

Drexler, K. E. (1988b) Rod logic and thermal noise in the mechanical nanocomputer. *Molecular Electronic Devices.* Amsterdam: North-Holland.

Drexler, K. E. (1989a) Biological and Nanomechanical Systems: Contrasts in Evolutionary Capacity. *Artificial Life.* Reading, Massachusetts: Addison-Wesley.

Drexler, K. E. (1989b) *Engineering Macromolecular Objects.* (Draft circulated for comment, submitted to the journal *Protein Engineering* in 1989, and withdrawn in early 1992 owing to editorial inaction.)

Drexler, K. E. (1989c) Machines of Inner Space. *1990 Yearbook of Science and the Future.* D. Calhoun, ed. Chicago: Encyclopædia Britannica.

Drexler, K. E. (1991a) Exploring Future Technologies. *Doing Science.* J. Brockman, ed. New York: Prentice Hall.

Drexler, K. E. (1991b) Molecular tip arrays for molecular imaging and nanofabrication. *JVST-B.* **9:**1394–1397.

Drexler, K. E. (1992) Molecular directions in nanotechnology. *Nanotechnology.* **2:** 113–118.

Drexler, K. E., and J. S. Foster. (1990) Synthetic tips. *Nature.* **343:**600.

Drexler, K. E., and M. S. Miller. (1988) Incentive Engineering for Computational Resource Management. *The Ecology of Computation.* B. A. Huberman, ed. Amsterdam: Elsevier Science Publishers, B. V.

Drexler, K. E., C. Peterson, and G. Pergamit. (1991) *Unbounding the Future: The Nanotechnology Revolution.* New York: William Morrow.

Dürig, U., O. Züger, and D. W. Pohl. (1988) Force sensing in scanning tunnelling microscopy: observation of adhesion forces on clean metal surfaces. *J. Microscopy.* **152:**259–267.

Eaton, P. E., and G. Castaldi. (1985) Systematic Substitution on the Cubane Nucleus. Amide Activation for Metalation of "Saturated" Systems. *J. Am. Chem. Soc.* **107:**724–726.

Eaton, P. E., and M. Maggini. (1988) Cubene (1,2-Dehydrocubane). *J. Am. Chem. Soc.* **110:**7230–7332.

Eggers, D. F., Jr., N. W. Gregory, G. D. Halsey, Jr., and B. S. Rabinovitch. (1964) *Physical Chemistry.* New York: John Wiley & Sons.

Eigler, D. M., and E. K. Schweizer. (1990) Positioning single atoms with a scanning tunnelling microscope. *Nature.* **344:**524–526.

Eisenberg, D., W. Wilcox, S. M. Eshita, P. M. Pryciak, S. P. Hu, and W. F. DeGrado. (1986) The Design, Synthesis, and Crystallization of an Alpha-Helical Peptide. *Proteins Struct. Funct. Genet.* **1:**16–22.

Eisenthal, K. B., R. A. Moss, and N. J. Turro. (1984) Divalent Carbon Intermediates: Laser Photolysis and Spectroscopy. *Science.* **225:**1439–1445.

Ellington, A. D., and J. W. Szostak. (1990) *In vitro* selection of RNA molecules that bind specific ligands. *Nature.* **346:** 818–822.

Emch, R., J. Nogami, M. M. Dovek, C. A. Lang, and C. F. Quate. (1988) Characterization and local modification of atomically flat gold surfaces by STM. *J. Microscopy.* **152:**129–135.

Erickson, B. W., S. B. Daniels, P. A. Reddy, M. L. Higgins, J. S. Richardson, and D. C. Richardson. (1988) *Chemical Synthesis of Designed Beta Structures.* Advances in Gene Technology: Protein Engineering and Production. Miami, Florida: ICSU.

Fagan, P. J., M. D. Ward, and J. C. Calabrese. (1989) Molecular Engineering of Solid-State Materials: Organometallic Building Blocks. *J. Am. Chem. Soc.* **111:**1698–1719.

Farrall, G. A. (1980) Electrical Breakdown in Vacuum. *Vacuum Arcs: Theory and Application.* J. M. Lafferty, ed. New York: John Wiley & Sons.

Farrell, H. H., and M. Levinson. (1985) Scanning tunneling microscope as a structure-modifying tool. *Phys. Rev. B.* **31:**3593–3598.

Feynman, R. P. (1960) There's Plenty of Room at the Bottom. *Eng. and Sci.* **23:**22–36.

Feynman, R.P. (1961) There's Plenty of Room at the Bottom. *Miniaturization.* H. D. Gilbert, ed. New York: Reinhold.

Feynman, R. P. (1985) Quantum Mechanical Computers. *Optics News.* **11:**11–20.

Feynman, R. P., R. B. Leighton, and M. Sands. (1963) *The Feynman Lectures on Physics.* Reading: Addison-Wesley.

Field, J. E. (1979) Strength and Fracture Properties of Diamond. *The Properties of Diamond.* J. E. Field, ed. London: Academic Press.

Finkelstein, A. V., and J. Janin. (1990) The price of lost freedom: entropy of biomolecular complex formation. *Prot. Eng.* **3:**1–3.

Fischer, L., R. Müller, B. Ekberg, and K. Mosbach. (1991) Direct Enantioseparation of β-Adrenergic Blockers Using a Chiral Stationary Phase Prepared by Molecular Imprinting. *J. Am. Chem. Soc.* **113:**9358–9360.

Fogarasi, G., and J. E. Boggs. (1983) Theoretical equilibrium geometry, vibrational frequencies and the first electronic transition energy of HCC. *Mol. Phys.* **50:**139–151.

Foster, J. S., J. E. Frommer, and P. C. Arnett. (1988) Molecular manipulation using a tunnelling microscope. *Nature.* **331:**324–326.

Frankel, A. D., D. S. Bredt, and C. O. Pabo. (1988) Tat Protein from Human Immunodeficiency Virus Forms a Metal-Linked Dimer. *Science.* **240:**70–73.

Franks, A. (1987) Nanotechnology. *J. Phys. E: Sci. Instr.* **20:**1442–1451.

Fredkin, E., and T. Toffoli. (1982) Conservative Logic. *Int. J. Theor. Phys.* **21:**219–253.

Freitas, R. A., Jr., and W. P. Gilbreath, editors. (1982) Advanced Automation for Space Missions. National Aeronautics and Space Administration: NASA CP-2255

Frenklach, M., and K. E. Spear. (1988) Growth Mechanism of Vapor-Deposited Diamond. *J. Mater. Res.* **3:**133–140.

Gehrhardt, H., and M. Mutter. (1987) Soluble polymers in organic chemistry. 5. Preparation of carboxyl- and amino-terminal polyethylene glycol of low molecular weight. *Polymer Bull.* **18:** 487–493.

Gimzewski, J. K. (1988) Scanning tunneling microscopic techniques applied to roughness of silver surfaces. *Surface Measurement and Characterization.* Bellingham, Washington: Society of Photo-Optical Instrumentation Engineers.

Goodman, F. O. (1980) Thermal Accommodation Coefficients. *J. Phys. Chem.* **84:**1431–1445.

Goodman, F. O., and H. Y. Wachman. (1976) *Dynamics of Gas-Surface Scattering.* New York: Academic Press.

Gray, D. E. (1972) *American Institute of Physics Handbook.* Third Edition. New York: McGraw-Hill.

Hagler, A. T., J. R. Maple, T. S. Thacher, G. B. Fitzgerald, and U. Dinur. (1989) Potential energy functions for organic and biomolecular systems. *Computer Simulations of Biomolecular Systems.* W. F. van Gunsteren and P. K. Weiner, eds. Leiden: Escom Science Publishers.

Hahn, K. W., W. A. Klis, and J. M. Steward. (1990) Design and Synthesis of a Peptide Having Chymotrypsin-Like Esterase Activity. *Science.* **248:**1544–1547.

Hamza, A. V., G. D. Kubiak, and R. H. Stulen. (1988) The role of hydrogen on the diamond C(111)-(2 × 1) reconstruction. *Surf. Sci.* **206:**L833–L844.

Hansma, P. K. (1990) Personal communication.

Hecht, M. H., D. C. Richardson, and J. S. Richardson. (1989) Design, Expression, and Preliminary Characterization of Felix: A Model Protein. *J. Cell. Biochem.* **13A:**86.

Hehre, W. J., L. Radom, P. v. R. Schleyer, and J. A. Pople. (1986) *Ab Initio Molecular Orbital Theory.* New York: John Wiley & Sons.

Heinicke, G. (1984) *Tribochemistry.* Berlin: Akademie-Verlag.

Hemley, R. J., and H. K. Mao. (1990) Critical Behavior in the Hydrogen Insulator-Metal Transition. *Science.* **249:**391–393.

Hendricks, M. B., H. E. Bergeson, R. B. Coats, J. W. Keuffel, M. O. Larson, J. L. Osborne, S. Ozaki, and R. O. Stenerson. (1969) *High-Energy Neutrinos in the Utah Apparatus.* 11th International Conference on Cosmic Rays. A. Somogyi, ed. Budapest: Akadémiai Kiadó.

Hendrickson, J. B. (1990) Organic Synthesis in the Age of Computers. *Angew. Chem. Int. Ed. Engl.* **29:**1286–1295.

Hennessy, J. L., and D. A. Patterson. (1990) *Computer Architecture: A Quantitative Approach.* Morgan Kaufmann Publishers, Inc.: San Mateo.

Hiemenz, P. C. (1986) *Principles of Colloid and Surface Chemistry.* Second Edition. New York: Marcel Dekker.

Himpsel, F. J., J. A. Knapp, J. A. Van-Vechten, and D. E. Eastman. (1979) Quantum photoyield of diamond(111)—A stable negative-affinity emitter. *Phys. Rev. B.* **20:**624–627.

Ho, S. P., and W. F. DeGrado. (1987) Design of a 4-Helix Bundle Protein: Synthesis of Peptides Which Self-Associate into a Helical Protein. *J. Am. Chem. Soc.* **109:**6751–6758.

Hopfield, J. J., J. N. Onuchic, and D. N. Beratan. (1988) A Molecular Shift Register Based on Electron Transfer. *Science.* **241:**817–820.

Horn, R. G., and D. T. Smith. (1992) Contact Electrification and Adhesion Between Dissimilar Materials. *Science.* **256:**362–364.

Huang, D., M. Frenklach, and M. Maroncelli. (1988) Energetics of Acetylene-Addition Mechanism of Diamond Growth. *J. Phys. Chem.* **92:**6379–6381.

Hubbard, H. M. (1989) Photovoltaics Today and Tomorrow. *Science.* **244:**297–304.

Hudson, J. A. (1980) *The Excitation and Propagation of Elastic Waves.* Cambridge: Cambridge University Press.

Huheey, J. E. (1978) *Inorganic Chemistry.* New York: Harper and Row.

Huisgen, R. (1977) Tetracyanoethylene and Enol Ethers. A Model for 2 + 2 Cycloadditions via Zwitterionic Intermediates. *Acc. Chem. Res.* **10:**117–124.

Huse, W. D., L. Sastry, S. A. Iverson, A. S. Kang, M. Alting-Mees, D. R. Burton, S. J.

Benkovic, and R. A. Lerner. (1989) Generation of a Large Combinatorial Library of the Immunoglobulin Repertoire in Phage Lambda. *Science.* **246:**1275–1281.

Ichikawa, H., Y. Ebisawa, and A. Shigihara. (1985) Potential Energies of Rotation of Double Bond in Ethylene Molecule and Ion. *Bull. Chem. Soc. Jpn.* **58:**3619–3620.

Imanaka, T., M. Shibazaki, and M. Takagi. (1986) A new way of enhancing the thermostability of proteases. *Nature.* **324:**695–697.

Isaacs, N. S., and A. V. George. (1987) Chemical synthesis at high pressures. *Chem. Br.* **23:**47–51.

Israelachvili, J. N. (1992) *Intermolecular and Surface Forces.* Second Edition. New York: Academic Press.

Jaenicke, R. (1987) Folding and association of proteins. *Prog. Biophys. Molec. Biol.* **49:**117–217.

Janda, K. D., S. J. Benkovic, and R. A. Lerner. (1989) Catalytic Antibodies with Lipase Activity and R or S Substrate Selectivity. *Science.* **244:**437–440.

Janda, K. D., D. Schloeder, S. J. Benkovic, and R. A. Lerner. (1988) Induction of an Antibody That Catalyzes the Hydrolysis of an Amide Bond. *Science.* **241:**1188–1191.

Jenner, G. (1985) The Pressure Effect on Strained Transition States. *J. Chem Soc., Faraday Trans. 1.* **81:**2437–2460.

Johnson, M. D. (1983) Bimolecular Homolytic Displacement of Transition-Metal Complexes from Carbon. *Acc. Chem. Res.* **16:**343–349.

Johnston, R. L., and R. Hoffmann. (1989) Superdense Carbon, C_8: Supercubane or Analogue of γ-Si? *J. Am. Chem. Soc.* **111:**810–819.

Kaehler, T. (1990) Molecular Carpentry. *Foresight Update.* Issue 10, 6–7.

Kaehler, T. (1992) A Proposed Dictionary of Molecular Mounting Brackets. Draft paper.

Karfunkel, H. R., and T. Dressler. (1992) New Hypothetical Carbon Allotropes of Remarkable Stability Estimated by Modified Neglect of Diatomic Overlap Solid-State Self-Consistent Field Computations. *J. Am. Chem. Soc.* **114:**2285–2288.

Karplus, M., and R. N. Porter. (1970) *Atoms and Molecules.* Menlo Park, California: W. A. Benjamin, Inc.

Kaszynski, P., A. C. Friedli, and J. Michl. (1992) Toward a Molecular-Size "Tinkertoy" Construction Set. Preparation of Terminally Functionalized [*n*]Staffanes from [1.1.1]Propellane. *J. Am. Chem. Soc.* **114:**601–620.

Kauzman, W., and H. Eyring. (1940) *J. Am. Chem. Soc.* **62:**3113.

Kay, L. E., G. M. Clore, A. Bax, and A. M. Gronenborn (1990) Four-Dimensional Heteronuclear Triple-Resonance NMR Spectroscopy of Interleukin-1β in Solution. *Science.* **249:**411–414.

Kellis, J. T., Jr., K. Nyberg, D. Sali, and A. R. Fersht. (1988) Contribution of hydrophobic interactions to protein stability. *Nature.* **333:**784–786.

Kelly, A. (1973) *Strong Solids.* Oxford: Clarendon Press.

Kent, S. B. H. (1988) Chemical Synthesis of Peptides and Proteins. *Ann. Rev. Biochem.* **57:**957–989.

Kepner, G. R., and R. I. Macey. (1968) Membrane Enzyme Systems: Molecular Size Determination from Radiation Inactivation. *Biochim. Biophys. Acta.* **163:**188–203.

Kerr, J. A. (1973) Rate Processes in the Gas Phase. *Free Radicals.* J. K. Kochi, ed. New York: John Wiley & Sons.

Kerr, J. A. (1990) Strengths of Chemical Bonds. *CRC Handbook of Chemistry and Physics.* 71st Edition. D. R. Lide, ed. Boca Raton: CRC Press.

Keyes, R. W. (1985) What Makes a Good Computer Device? *Science.* **230:**138–144.

Keyes, R. W. (1989) Physics of digital devices. *Rev. Modern Phys.* **61:**279–287.

Kivelson, S. (1988) Intrinsic Conductivity of Conducting Polymers. *Synth. Met.* **22:**371–384.

Knight, T. F., Jr. (1991) Personal communication.

Knox, J. H. (1971) *Molecular Thermodynamics: An Introduction to Statistical Mechanics for Chemists.* New York: John Wiley & Sons.

Kogan, V. I., and V. M. Galitskiy. (1963) *Problems in Quantum Mechanics.* New York: Prentice-Hall.

Kohnke, F. H., J. P. Mathias, and J. F. Stoddart. (1989) Structure-Directed Synthesis of New Organic Materials. *Angew. Chem. Int. Ed. Engl. Adv. Mater.* **28:**1103–1110.

Krishnaswamy, M. R., M. G. K. Menon, V. S. Narasimham, H. Sesaki, S. Kino, S. Miyake, R. Craig, A. J. Parsons, and A. W. Wolfendale. (1969) *The Kolar Gold Field Neutrino Experiment.* 11th International Conference on Cosmic Rays. Budapest: Akadémiai Kiadó.

Kubiak, G. D., and K. W. Kolasinsky. (1989) Normally unoccupied states on C(111) (diamond) (2 × 1): Support for a relaxed π-bonded chain model. *Phys. Rev. B.* **39:**1381–1384.

Kuntz, P. J., E. M. Nemeth, J. C. Polanyi, S. D. Rosner, and C. E. Young. (1966) Energy Distribution Among Products of Exothermic Reactions. II. Repulsive, Mixed, and Attractive Energy Release. *J. Chem. Phys.* **44:**1168–1184.

Kuzmin, L. S., P. Delsing, T. Claeson, and K. K. Likharev. (1989) Single-Electron Charging Effects in One-Dimensional Arrays of Ultrasmall Tunnel Junctions. *Phys. Rev. Lett.* **62:**2539–2542.

Kyriacou, D. K. (1981) *Basics of Electroorganic Synthesis.* New York: John Wiley & Sons.

LaBarbera, M. (1990) Principles of Design of Fluid Transport Systems in Zoology. *Science.* **249:**992–1000.

Lakatos, I. (1978) *The methodology of scientific research programs.* Cambridge: Cambridge University Press.

Lakes, R. (1987) Foam Structures with Negative Poisson's Ratio. *Science.* **235:**1038–1040.

Landauer, R. (1961) Irreversibility and Heat Generation in the Computing Process. *IBM J. Res. Dev.* **3:**183–191.

Landauer, R. (1982) Uncertainty Principle and Minimal Energy Dissipation in the Computer. *Int. J. Theor. Phys.* **21:**283–297.

Landauer, R. (1987) Energy Requirements in Computation. *Appl. Phys. Lett.* **51:**2056–2058.

Landauer, R. (1988) Dissipation and noise immunity in computation and communication. *Nature.* **335:**779–784.

Landman, U., W. D. Luedtke, N. A. Burnham, and R. J. Colton. (1990) Atomistic Mechanisms and Dynamics of Adhesion, Nanoindentation, and Fracture. *Science.* **248:**454–461.

Lea, D. (1946) *Actions of Radiation on Living Cells.* London: Cambridge University Press.

Lee, T. J., A. Bunge, and H. F. Schaefer III. (1985) Toward the Laboratory Identification of Cyclopropenylidene. *J. Am. Chem. Soc.* **107:**137–142.

Lehn, J.-M. (1980) *Leçon Inaugural,* Collége de France, Paris.

Lehn, J.-M. (1988) Supramolecular Chemistry—Scope and Perspectives: Molecules, Supermolecules, and Molecular Devices. *Angew. Chem. Int. Ed. Engl.* **27:**89–112.

Leszczynski, J. F., and G. D. Rose. (1986) Loops in Globular Proteins: A Novel Category of Secondary Structure. *Science.* **234:**849–855.

Levin, R. L. (1985) Arynes. *Reactive Intermediates.* M. Jones, Jr. and R. A. Moss, eds. New York: John Wiley & Sons.

Levine, R. D., and R. D. Bernstein. (1987) *Molecular Reaction Dynamics and Chemical Reactivity.* Oxford: Oxford University Press.

Lide, D. R. (1990) *CRC Handbook of Chemistry and Physics.* Boca Raton: CRC Press.

Lii, J., and N. L. Allinger. (1989a) Molecular Mechanics. The MM3 Force Field for Hydrocarbons. 2. Vibration Frequencies and Thermodynamics. *J. Am. Chem. Soc.* **111:**8566–8575.

Lii, J., and N. L. Allinger. (1989b) Molecular Mechanics. The MM3 Force Field for Hydrocarbons. 3. The van der Waals' Potentials and Crystal Data for Aliphatic and Aromatic Hydrocarbons. *J. Am. Chem. Soc.* **111:**8576–8582.

Lii, J., and N. L. Allinger. (1991) The MM3 Force Field for Amides, Polypeptides and Proteins. *J. Comp. Chem.* **12:**186–199.

Likharev, K. K. (1982) Classical and Quantum Limitations on Energy Consumption in Computation. *Int. J. Theor. Phys.* **21:**311–325.

Lim, W. A., and R. T. Sauer. (1989) Alternative packing arrangements in the hydrophobic core of λ repressor. *Nature.* **339:**31–36.

Lindsey, J. S. (1991) Self-Assembly in Synthetic Routes to Molecular Devices. Biological Principles and Chemical Perspectives: A Review. *New J. Chem.* **15:** 153–180.

Lothe, J. (1962) Theory of Dislocation Mobility in Pure Slip. *J. Appl. Phys.* **33:** 2116–2125.

Maitland, G. C., M. Rigby, E. B. Smith, and W. A. Wakeham. (1981) *Intermolecular Forces: Their Origin and Determination.* Oxford: Clarendon Press.

Mao, H. K., Y. Wu, R. J. Hemley, L. C. Chen, J. F. Shu, and L. W. Finger. (1989) X-ray Diffraction to 302 Gigapascals: High-Pressure Crystal Structure of Cesium Iodide. *Science.* **246:**649–651.

Masters, C. (1981) *Homogeneous Transition-metal Catalysis.* London: Chapman and Hall.

Matthews, B. W., H. Nicholson, and W. J. Becktel. (1987) Enhanced protein thermostability from site-directed mutations that decrease the entropy of unfolding. *Proc. Natnl. Acad. Sci. U.S.A.* **84:**6663–6667.

McBride, J. M., B. E. Segmuller, M. D. Hollingsworth, D. E. Mills, and B. A. Weber. (1986) Mechanical Stress and Reactivity in Organic Solids. *Science.* **234:** 830–835.

McCabe, W. L., and J. C. Smith. (1976) *Unit Operations of Chemical Engineering.* Third Edition. McGraw-Hill Chemical Engineering Series. New York: McGraw-Hill.

McMillen, D. F., and D. M. Golden. (1982) Hydrocarbon Bond Dissociation Energies. *Ann. Rev. Phys. Chem.* **33:**493–532.

Mead, C., and L. Conway. (1980) *Introduction to VLSI Systems.* Reading, Massachusetts: Addison-Wesley.

Merkle, R. C. (1991) The program tube.c is available from the Xerox Palo Alto Research Center by anonymous FTP from parcftp, in the directory /pub/nano.

Merkle, R. C. (1992a) Computational nanotechnology. *Nanotechnology.* **2:**134–141

Merkle, R. C. (1992b) A Proof About Molecular Bearings. Draft paper.

Merkle, R. C. (1992c) Reversible Electronic Logic Using Switches. Draft paper.

Merkle, R. C. (1992d) Two Types of Mechanical Reversible Logic. Draft paper.

Merz, K. M., Jr., R. Hoffmann, and A. T. Balaban. (1987) 3,4-Connected Carbon Nets: Through-Space and Through-Bond Interactions in the Solid State. *J. Am. Chem. Soc.* **109:**6742–6751.

Miller, M. S., and K. E. Drexler. (1988a) Comparative Ecology: A Computational Perspective. *The Ecology of Computation.* B. A. Huberman, ed. Amsterdam: Elsevier Science Publishers, B. V.

Miller, M. S., and K. E. Drexler. (1988b) Markets and Computation: Agoric Open Systems. *The Ecology of Computation.* B. A. Huberman, ed. Amsterdam: Elsevier Science Publishers, B. V.

Mislow, K. (1989) Molecular Machinery in Organic Chemistry. *Chemtracts—Org. Chem.* **2:**151–174.

Mitchell, M. J., and J. A. McCammon. (1991) Free Energy Difference Calculations by Thermodynamic Integration: Difficulties in Obtaining a Precise Value. *J. Comp. Chem.* **12:**271–275.

Moss, R. A. (1989) Carbenic Reactivity Revisited. *Acc. Chem. Res.* **22:**15–21.

Moss, R. A., and M. J. Chang. (1981) Intermolecular Chemistry of a Dialkylcarbene: Adamantylidene. *Tetra. Lett.* **22:**3749–3752.

Musgrave, C. B., J. K. Perry, R. C. Merkle, and W. A. Goddard III. (1992) Theoretical studies of a hydrogen abstraction tool for nanotechnology. *Nanotechnology.* **2:** 187–195.

Mutter, M., E. Altmann, K.-H. Altmann, R. Hersperger, P. Koziej, K. Nebel, G. Tuchscherer, S. Vuilleumier, H.-U. Gremlich, and K. Müller. (1988) The Construction of New Proteins. Artificial Folding Units by Assembly of Amphiphilic Secondary Structures on a Template. *Helv. Chim. Acta.* **71:**835–847.

Nabarro, F. R. N. (1987) *Theory of Crystal Dislocations.* New York: Dover.

Nagle, J. K., J. S. Bernstein, R. C. Young, and T. J. Meyer. (1981) Charge-Transfer Excited States as Molecular Photodiodes. *Inorg. Chem.* **20:**1760–1764.

Nanis, L. (1984) Field Ion Microscopy. *Comprehensive Treatise of Electrochemistry.* Volume 8. R. E. White, J. O. Bockris, B. E. Conway, E. Yeager, eds. New York: Plenum.

Neidlein, R., V. Poignée, W. Kramer, and C. Glück. (1986) Synthesis and Spectroscopic Properties of Triafulvalene Derivatives. *Angew. Chem. Int. Ed. Engl.* **25:**731–732.

Nicolaou, K. C., N. A. Petasis, J. Uenishi, and R. E. Zipkin. (1982) The Endiandric Acid Cascade. Electrocyclizations in Organic Synthesis. 2. Stepwise, Stereocontrolled Total Synthesis of Endiandric Acids C–G. *J. Am. Chem. Soc.* **104:**5557–5558.

Noren, C. J., S. J. Anthony-Cahill, M. C. Griffith, and P. G. Schultz. (1989) A General Method for Site-Specific Incorporation of Unnatural Amino Acids into Proteins. *Science.* **244:**182–188.

Northrup, J. E., and M. L. Cohen. (1982) Reconstruction Mechanism and Surface-State Dispersion for Si(111)-(2 × 1). *Phys. Rev. Lett.* **49:**1349–1352.

Nugent, W. A., and J. M. Mayer. (1988) *Metal Ligand Multiple Bonds: The Chemistry of Transition Metal Complexes Containing Oxo, Nitrido, Imido, Alkylidene, or Alkylidyne Ligands.* New York: John Wiley & Sons.

Oosawa, F., and S. Asakura. (1975) *Thermodynamics of the Polymerization of Protein.* London: Academic Press.

Pabo, C. P., and E. G. Suchanek. (1986) Computer-Aided Model-Building Strategies for Protein Design. *Biochemistry.* **25:**5987–5991.

Pantoliano, M. W., R. C. Ladner, P. N. Bryan, M. L. Rollence, J. F. Wood, and T. L. Poulos. (1987) Protein Engineering of Subtilisin BPN': Enhanced Stabilization through the Introduction of Two Cysteines To Form a Disulfide Bond. *Biochemistry.* **26:**2077–2082.

Pantoliano, M. W., M. Whitlow, J. F. Wood, M. L. Rollence, B. C. Finzel, G. L. Gilliland, T. L. Poulos, and P. N. Bryan. (1988) The Engineering of Binding Affinity at Metal Ion Binding Sites for the Stabilization of Proteins: Subtilisin as a Test Case. *Biochemistry.* **27:**8311–8317.

Park Scientific Instruments. (1992) Product literature.

Parker, S. P. (1984) *McGraw-Hill Dictionary of Scientific and Technical Terms.* Third edition. New York: McGraw-Hill.

Párraga, G., S. J. Horvath, A. Eisen, W. E. Taylor, L. Hood, E. T. Young, and R. E. Klevit. (1988) Zinc-Dependent Structure of a Single-Finger Domain of Yeast ADR1. *Science.* **241:**1489–1492.

Patai, S. (1980) *The Chemistry of Ketenes, Allenes, and Related Compounds, Part 2.* New York: John Wiley & Sons.

Pate, B. B. (1986) The Diamond Surface: Atomic and Electronic Structure. *Surf. Sci.* **165:**83–142.

Perry, L. J., and R. Wetzel. (1984) Disulfide Bond Engineered into T4 Lysozyme: Stabilization of the Protein Toward Thermal Inactivation. *Science.* **226:**555–558.

Persson, B. N. J. (1987) The Atomic Force Microscope: Can it be used to study biological molecules? *Chem. Phys. Lett.* **141:**366–368.

Poland, D. C., and H. A. Scheraga. (1965) Statistical Mechanics of Noncovalent Bonds in Polyamino Acids. VIII. Covalent Loops in Proteins. *Biopolymers.* **3:**379–399.

Ponder, J. W., and F. M. Richards. (1987a) An Efficient Newton-like Method for Molecular Mechanics Energy

Minimization of Large Molecules. *J. Comput. Chem.* **8:**1016–1024.

Ponder, J. W., and F. M. Richards. (1987b) Tertiary Templates for Proteins. *J. Mol. Biol.* **193:**775–791.

Popper, K. R. (1963) *Conjectures and Refutations.* London: Routledge and Kegan Paul.

Porter, J., J. Dykert, and J. Rivier. (1987) Synthesis, resolution and characterization of ring substituted phenylalanines and tryptophanes. *Int. J. Peptide Protein Res.* **30:**13–21.

Prater, C. B., H. J. Butt, and P. K. Hansma. (1990) Atomic force microscopy. *Nature.* **345:**839–840.

Raabe, G., and J. Michl. (1989) Multiple Bonds to Silicon. *The Chemistry of Organic Silicon Compounds, Part 2.* S. Patai and Z. Rappoport, eds. New York: John Wiley & Sons.

Raiffa, H. (1968) *Decision Analysis: Introductory Lectures on Choices under Uncertainty.* Reading: Addison-Wesley.

Ranby, B., and J. F. Rabek. (1975) *Photodegradation, Photo-oxidation and Photostabilization of Polymers.* New York: John Wiley & Sons.

Rebek, J., Jr. (1987) Model Studies in Molecular Recognition. *Science.* **235:** 1478–1484.

Regan, L., and W. F. DeGrado. (1988) Characterization of a Helical Protein Designed from First Principles. *Science.* **241:**976–978.

Regueiro, M. N., P. Monceau, and J. Hodeau. (1992) Crushing C_{60} to diamond at room temperature. *Nature.* **355:**237–239.

Reiss, H., and C. S. Fuller. (1959) Diffusion Processes in Germanium and Silicon. *Semiconductors.* New York: Reinhold.

Rentzepis, P. M., and D. C. Douglass. (1981) Xenon as a solvent. *Nature.* **293:**165–166.

Richards, F. M. (1977) Areas, Volumes, Packing, and Protein Structure. *Annu. Rev. Biophys. Bioeng.* **6:**151–176.

Riddles, P. W., R. L. Blakeley, and B. Zerner. (1979) Ellman's Reagent: 5,5′-Dithiobis(2-nitrobenzoic Acid)—a Reexamination. *Anal. Biochem.* **94:**75–81.

Rigby, M., B. E. Smith, W. A. Wakeham, and G. C. Maitland. (1986) *The Forces Between Molecules.* Oxford: Clarendon Press.

Robinson, B. H., and N. C. Seeman. (1987) The design of a biochip: a self-assembling

molecular-scale memory device. *Protein Eng.* **1:**295–300.

Robinson, D. K., and K. Mosbach. (1989) Molecular Imprinting of a Transition State Analogue Leads to a Polymer Exhibiting Esterolytic Activity. *J. Chem. Soc., Chem. Commun.* Issue 14, 969–970.

Robinson, J. W. (1974) *Handbook of Spectroscopy.* Boca Raton: CRC Press.

Sakaki, H. (1980) Scattering Suppression and High-Mobility Effect of Size-Quantized Electrons in Ultrafine Semiconductor Wire Structures. *Japan. J. Appl. Phys.* **19:**L735–L738.

Salem, L., and C. Rowland. (1972) The Electronic Properties of Diradicals. *Angew. Chem. Int. Ed. Engl.* **11:**92–111.

Sasaki, T., S. Eguchi, M. Tanida, F. Nakata, and T. Esaki. (1983) Convenient Generation of Adamantylidenecarbene from (Bromomethylene)adamantane. An Efficient Method of Adamantylidenecyclopropanation. *J. Org. Chem.* **48:**1579–1586.

Sato, S. (1955) On a new method of drawing the potential energy surface. *J. Chem. Phys.* **23:**592.

Saunders, M., and R. M. Jarret. (1986) A New Method for Molecular Mechanics. *J. Comp. Chem.* **7:**578–588.

Saxena, S. C., and R. K. Joshi. (1989) *Thermal Accommodation and Adsorption Coefficients of Gases.* CINDAS Data Series on Material Properties. C. Y. Ho, ed. New York: Hemisphere.

Saxena, V. P., and D. B. Wetlaufer. (1970) Formation of Three-Dimensional Structure in Proteins. *Biochemistry.* **9:** 5015–5023.

Schaefer, H. F., III. (1986) Methylene: A Paradigm for Computational Quantum Chemistry. *Science.* **231:**1100–1107.

Schnabel, W. (1981) *Polymer Degradation.* New York: Hanser/MacMillan.

Schneiker, C. (1989) Nanotechnology with Feynman Machines: Scanning Tunneling Engineering and Artificial Life. *Artificial Life.* C. G. Langton, ed. Redwood City: Addison-Wesley.

Schoen, M., C. L. Rhykerd, Jr., D. J. Diestler, and J. H. Cushman. (1989) Shear Forces in Molecularly Thin Films. *Science.* **245:**1223–1225.

Scott, J. K., and G. P. Smith. (1990) Searching for Peptide Ligands with an Epitope Library. *Science.* **249:**386–390.

Scouten, W. H. (1987) A Survey of Enzyme Coupling Techniques. *Immobilization*

Techniques for Enzymes. K. Mosbach, ed. New York: Academic Press.

Seeman, N. C. (1982) Nucleic Acid Junctions and Lattices. *J. Theo. Biol.* **99:** 237–247.

Seeman, N. C. (1991) Construction of Three-dimensional Stick Figures from Branched DNA. *DNA Cell Bio.* **10:**475–486.

Shera, E. B., N. K. Seitzinger, L. M. Davis, R. A. Keller, and S. A. Soper. (1990) Detection of single fluorescent molecules. *Chem. Phys. Lett.* **174:**553–557.

Shevlin, P. B., and A. P. Wolf. (1970) The Formation of Carbon Atoms in the Decomposition of a Carbene. *Tet. Lett.* **46:**3987–3990.

Simon, H. A. (1981) *Sciences of the Artificial.* Second Edition. Cambridge, Massachusetts: MIT Press.

Singh, U. C., and P. A. Kollman. (1986) A Combined *Ab Initio* Quantum Mechanical and Molecular Mechanical Method for Carrying out Simulations on Complex Molecular Systems: Applications to the $CH_3Cl + Cl^-$ Exchange Reaction and Gas Phase Protonation of Polyethers. *J. Comp. Chem.* **7:**718–730.

Smith, J. L., W. A. Hendrickson, R. B. Honzatko, and S. Sheriff. (1986) Structural Heterogeneity in Protein Crystals. *Biochemistry.* **25:**5018–5027.

Soreff, J. (1991) Personal communications.

Soreff, J. (1992) Synthesis Paradigm. *Foresight Briefing.* Issue 4.

Stewart, J. J. P. (1990) MOPAC: A semiempirical molecular orbital program. *J. Comp.-Aided Mol. Design.* **4:**1–105.

Stillinger, F. H., and T. A. Weber. (1984) Packing Structures and Transitions in Liquids and Solids. *Science.* **225:**983–989.

Straatsma, T. P., and J. A. McCammon. (1991) Multiconfiguration thermodynamic integration. *J. Chem. Phys.* **95:**1175–1188.

Swaddle, T. W. (1986) Electron Transfer Reactions. *Inorganic High Pressure Chemistry.* Amsterdam: Elsevier.

Szajewski, R. P., and G. M. Whitesides. (1980) Rate Constants and Equilibrium Constants for Thiol-Disulfide Interchange Reactions Involving Oxidized Glutathione. *J. Am. Chem. Soc.* **102:** 2011–2026.

Tamor, M. A., and K. C. Hass. (1990) Hypothetical superhard carbon metal. **5:** 2273–2276.

Taniguchi, N. (1974) *On the Basic Concept of 'Nano-Technology'.* The International Conference on Production Engineering, Part II. Tokyo: Japan Society of Precision Engineering.

Tapley, B. D., and T. R. Poston. (1990) *Eschbach's Handbook of Engineering Fundamentals, Fourth Edition.* New York: John Wiley & Sons.

Teague, C. (1992) Editor, *Nanotechnology.* Conference proceedings issue. Vol. 2, No. 3–4. Institute of Physics.

ter Haar, G. R. (1988) Biological Effects of Ultrasound in Clinical Applications. *Ultrasound: its chemical, physical and biological effects.* K. S. Suslick, ed. New York: VCH Publishers, Inc.

Ternansky, R. J., D. W. Balogh, and L. A. Paquette. (1982) Dodecahedrane. *J. Am. Chem. Soc.* **104:**4503–4504.

Thyagarajan, B. S. (1968–1971) *Mechanisms of Molecular Migrations (Vol. 1–4).* New York: John Wiley & Sons.

Timoshenko, S. P., and J. N. Goodier. (1951) *Theory of Elasticity.* Third Edition. New York: McGraw-Hill.

Timoshenko, S., D. H. Young, and W. Weaver, Jr. (1974) *Vibration Problems in Engineering.* Fourth Edition. New York: John Wiley & Sons.

Timp, G., R. E. Behringer, and J. E. Cunningham. (1990) Suppression of impurity scattering in a one-dimensional wire. *Phys. Rev. B.* **42:**9259–9262.

Tjivikua, T., P. Ballester, and J. Rebek, Jr. (1990) A Self-Replicating System. *J. Am. Chem. Soc.* **112:**1249–1250.

Toffoli, T. (1981) Bicontinuous Extensions of Invertible Combinatorial Functions. *Math. Systems Theory.* **14:**13–23.

Toniolo, C. (1989) Structure of Conformationally Constrained Peptides: From Model Compounds to Bioactive Peptides. *Biopolymers.* **28:**247–257.

Trost, B. M. (1985) Sculpting Horizons in Organic Chemistry. *Science.* **227:**908–916.

Truhlar, D. G., and R. Steckler. (1987) Potential Energy Surfaces for Polyatomic Reaction Dynamics. *Chem. Rev.* **87:**217–236.

Tsuda, M., M. Nakajima, and S. Oikawa. (1986) Epitaxial Growth Mechanism of Diamond Crystal in CH_4–H_2 Plasma. *J. Am. Chem. Soc.* **108:**5780–5783.

Tuerk, C., and L. Gold. (1990) Systematic Evolution of Ligands by Exponential Enrichment: RNA Ligands to

Bacteriophage T4 DNA Polymerase. *Science.* **249:**505–510.

Utsumi, W., and T. Yagi. (1991) Light-Transparent Phase Formed by Room-Temperature Compression of Graphite. *Science.* **252:**1542–1544.

Uy, R., and F. Wold. (1977) Introduction of Artificial Crosslinks into Proteins. *Protein Crosslinking: Biochemical and Molecular Aspects.* M. Friedman, ed. New York: Plenum Press.

Vanderbilt, D., and S. G. Louie. (1985) Energy Minimization Calculations for Diamond (111) Surface Reconstructions. *The Structure of Surfaces.* M. A. Van Hove and S. Y. Tong, eds. Berlin: Springer-Verlag.

Verwoerd, W. S. (1981) A Study of the Dimer bond on the Reconstructed (100) Surfaces of Diamond and Silicon. *Surf. Sci.* **103:**404–415.

von Neumann, J., edited and completed by A. W. Burks. (1966) *Theory of Self-Reproducing Automata.* Urbana, Illinois: University of Illinois Press.

Watson, J. D., N. H. Hopkins, J. W. Roberts, J. A. Steitz, and A. M. Weiner. (1987) *The Molecular Biology of the Gene.* Fourth Edition. Menlo Park: Benjamin/Cummings.

Wayne, R. P. (1988) *Principles and Applications of Photochemistry.* Oxford: Oxford University Press.

Webb, T. H., and C. S. Wilcox. (1990) Improved Synthesis of Symmetrical and Unsymmetrical 5,11-Methanodibenxo[b,f][1,5]diaxocines. Readily Available Nanoscale Structural Units. *J. Org. Chem.* **55:**363–365.

Weiner, S. J., P. A. Kollman, D. A. Case, U. C. Singh, C. Ghio, G. Alagona, S. Profeta, Jr., and P. Weiner. (1984) A New Force Field for Molecular Mechanical Simulation of Nucleic Acids and Proteins. *J. Am. Chem. Soc.* **106:**765–784.

Weiner, S. J., P. A. Kollman, D. T. Nguyen, and D. A. Case. (1986) An All Atom Force Field for Simulations of Proteins and Nucleic Acids. *J. Comp. Chem.* **7:**230–252.

Weisenhorn, A. L., P. K. Hansma, T. R. Albrecht, and C. F. Quate. (1989) Forces in atomic force microscopy in air and water. *Appl. Phys. Lett.* **54:**2651–2653.

Weltner, W., Jr., and J. van Zee. (1989) Carbon Molecules, Ions, and Clusters. *Chem. Rev.* **89:**1713–1747.

Whitesides, G. M., J. E. Lilburn, and R. P. Szajewski. (1977) Rates of Thiol-Disulfide Interchange Reactions between Mono- and Dithiols and Ellman's Reagent. *J. Org. Chem.* **42:**332–338.

Whitesides, G. M., J. P. Mathias, and C. T. Seto. (1991) Molecular Self-Assembly and Nanochemistry: A Chemical Strategy for the Synthesis of Nanostructures. *Science.* **254:**1312–1319.

Wiberg, K. B., and F. H. Walker. (1982) [1.1.1]Propellane. *J. Am. Chem. Soc.* **104:** 5239–5240.

Wiesner, K., L. Poon, I. Jirkofsky, and M. Fishman. (1969) The total synthesis of optically active annotinine. *Can. J. Chem.* **47:**433–444.

Wildman, T. A. (1986) An *Ab Initio* Quantum Chemical Study of Hydrogen Abstraction from Methane by Methyl. *Chem. Phys. Lett.* **126:**325–329.

Wilks, H. M., K. W. Hart, R. Feeney, C. R. Dunn, H. Muirhead, W. N. Chia, D. A. Barstow, T. Atkinson, A. R. Clarke, and J. J. Holbrook. (1988) A Specific, Highly Active Malate Dehydrogenase by Redesign of a Lactate Dehydrogenase Framework. *Science.* **242:**1541–1544.

Williams, F. (1972) Early Processes in Radiation Chemistry and the Reactions of Intermediates. *The Radiation Chemistry of Macromolecules.* M. Dole, ed. New York: Academic Press.

Wisdom, J. O. (1963) The Refutability of "Irrefutable" Laws. *British J. Phil. Sci.* **13:** 303–306.

Xu, J. A., H. K. Mao, and P. M. Bell. (1986) High-Pressure Ruby and Diamond Fluorescence: Observations at 0.21 to 0.55 Terapascal. *Science.* **232:**1404–1406.

Zheng, C., C. F. Wong, J. A. McCammon, and P. G. Wolynes. (1988) Quantum simulation of ferrocytochrome c. *Nature.* **334:**726–728.

Index